SCOPE 19

The Global Biogeochemical Sulphur Cycle

SCOPE 19

The Global Biogeochemical Sulphur Cycle

Edited by
M. V. Ivanov
Institute of Biochemistry and Physiology of Microorganisms,
USSR Academy of Sciences, Pushchino-on-Oka, USSR

and

J. R. Freney
CSIRO Division of Plant Industry, Canberra, Australia

Associate Editors

I. E. Galbally
CSIRO Division of Atmospheric
Physics, Mordialloc, Australia

H. Rodhe
Department of Meteorology
University of Stockholm, Sweden

V. N. Kudeyarov
Institute of Agrochemistry and Soil
Science, USSR Academy of Sciences,
Pushchino-on-Oka, USSR

Yu. I. Sorokin
Institute of Oceanology
Gelendzhik, USSR

P. A. Trudinger
Baas Becking Geobiological
Laboratory, Canberra, Australia

T. A. Rafter
Wellington, New Zealand

Published on behalf of the
Scientific Committee on Problems of the Environment (SCOPE)
of the
International Council of Scientific Unions (ICSU)
by
JOHN WILEY & SONS
Chichester · New York · Brisbane · Toronto · Singapore

Library of Congress Cataloging in Publication Data:

Main entry under title:

The Global biogeochemical sulphur cycle.

  Based on the SCOPE/UNEP Workshop on the Global
Biogeochemical Sulfur Cycle, held at the Institute of
Biochemistry and Physiology of Microorganisms, USSR
Academy of Sciences, Pushchino, USSR, Oct. 15–19, 1979.
  'SCOPE 19.'
  Includes index.
  1. Sulphur cycle—Congresses.   I. Ivanov, M. V.
(Mikhail Vladimirovich)   II. Freney, J. R. (John
Raymond)   III. Scientific Committee on Problems of the
Environment.   IV. SCOPE/UNEP Workshop on the Global
Biogeochemical Sulfur Cycle (1979: USSR Academy of
Sciences)

QH344.G6   1983   574.5′222   82-8506
                                        AACR2

ISBN 0 471 10492 2

British Library Cataloguing in Publication Data:

The Global biogeochemical sulphur cycle. —
  (SCOPE; 19)
  1. Organosulphur compounds—Environmental
aspects—Congresses
  I. Ivanov, M. V.   II. Freney, J. R.
  III. Series
  363.7′38   QR97.0/

  ISBN 0 471 10492 2

Typeset by Preface Limited, Salisbury, Wiltshire
and printed by Page Bros. (Norwich) Limited.

International Council of Scientific Unions (ICSU)
Scientific Committee on Problems of the Environment (SCOPE)

SCOPE is one of a number of committees established by a non-governmental group of scientific organizations, The International Council of Scientific Unions (ICSU). The membership of ICSU includes representatives from 68 National Academies of Science, 18 International Unions, and 12 other bodies called Scientific Associates. To cover multi-disciplinary activities which include the interests of several unions, ICSU has established 10 Scientific Committees, of which SCOPE, founded in 1969, is one. Currently, representatives of 34 member countries and 15 Unions and Scientific Committees participate in the work of SCOPE, which directs particular attention to the needs of developing countries.

The mandate of SCOPE is to assemble, review, and assess the information available on man-made environmental changes and the effects of these changes on man; to assess and evaluate the methodologies of measurement of environmental parameters; to provide an intelligence service on current research; and by the recruitment of the best available scientific information and constructive thinking to establish itself as a corpus of informed advice for the benefit of centres of fundamental research and of organizations and agencies operationally engaged in studies of the environment.

SCOPE is governed by a General Assembly, which meets every three years. Between such meetings its activities are directed by the Executive Committee.

R. E. Munn
Editor-in-Chief
SCOPE Publications

Secretary-General: V. Plocq

Secretariat: 51 Bld de Montmorency
            75016 PARIS

Funds to meet SCOPE expenses are provided by contributions from SCOPE National Committees, an annual subvention from ICSU (and through ICSU, from UNESCO), an annual subvention from the French Ministère de l'Environnement et du Cadre de Vie, contracts with UN bodies, particularly UNEP, and grants from foundations and industrial enterprises.

Contents

ix

Foreword

SCOPE's Project on Biogeochemical Cycles, of which the present report is an important product, aims to build up a comprehensive basis of knowledge and understanding of the major biogeochemical cycles which together constitute the life-support system of our planet, determining as they do the composition of the atmosphere and the fertility of land and waters.

Widening and deepening our knowledge of these cycles has become especially urgent because man-made disturbances of the cycles are already quite substantial at a time when our demand for renewable resources for food, fuel, and fibre is increasing more and more rapidly. Our knowledge of budgets, fluxes, and transformation and transport mechanisms is still far from sufficient for predicting the limits to which the cycles can be safely exploited.

Interest in this complex problem was brought into focus by the USSR National Committee for SCOPE in a meeting which it organized in Moscow during November 1974. Much of the background thinking was published by Professor V. A. Kovda in *Biogeochemical Cycles in Nature and Their Human Disturbance* (1975), Nauka, Moscow (in Russian).

SCOPE's formal Project on Biogeochemical Cycles started in 1975 with an international interdisciplinary workshop on nitrogen, phosphorus, and sulphur cycles (the results of which were published in 1976 in SCOPE Report No. 7) and continued with another workshop and report on the carbon cycle (SCOPE Report No. 13). Further updating and synthesis is now taking place in co-ordinating units at Stockholm, on the nitrogen cycle, and at Hamburg, Stockholm, Woods Hole, and Brussels on the carbon cycle. Further specialized workshops were held on critical gaps in knowledge of these cycles and currently the project is focusing attention on the interactions of the major cycles. A report on this aspect is presented in SCOPE 21.

After the initial study in 1975 it soon became clear that a broader, more intensive effort would be required to complete our knowledge of the sulphur cycle, especially as the extensive work done in this area by our colleagues in the USSR had not yet been fully taken into account. Therefore, SCOPE gratefully accepted a proposal by the USSR Academy of Sciences to organize an international workshop at Pushchino so that a major and comprehensive report on the global sulphur cycle could be prepared; this workshop was held in October 1979. Following SCOPE's general strategy, it reviewed the draft contributions and appointed an editorial board to prepare the final text of the report to be published in Russian as well as in English.

SCOPE is greatly indebted to the authors of the various chapters of this report, not only for their hard work and high standards but also for their patience and generosity in the second phase of the work, i.e. of amending and harmonizing the chapters to produce a coherent report. In this regard the work of the Editorial Board has been invaluable; without their persistence in overcoming language and postal barriers the present result would not have been achieved.

The generous hospitality of the USSR Academy of Sciences and the financial support of UNEP in this study are gratefully acknowledged.

Finally, it is a great pleasure to announce that one of the recommendations of the Pushchino workshop has already been implemented. SCOPE, the USSR Academy, and UNEP have jointly set up a Coordinating Unit for the sulphur cycle, located in Pushchino and directed by Professor M. V. Ivanov. This Unit will continue to collect and collate information on the sulphur cycle and will be of general assistance to all workers and institutions active in this field. Thus, SCOPE is not only indebted to Professor Ivanov for his energetic leadership in preparing this report but also for his willingness to continue his contribution to the SCOPE Project on Biogeochemical Cycles.

J. W. M. LA RIVIÈRE
Chairman SCOPE Project on Biogeochemical Cycles

Preface

The proposal for a special SCOPE project on the global biogeochemical cycles of elements was initiated at the Paris SCOPE meeting in 1973. By the time of the Moscow meeting of this committee in 1974 it had become apparent that such a project had the support of many National SCOPE Committees and the Executive Committee decided to include a project on 'Biogeochemical Cycles' in the scientific programme of SCOPE (Kovda and La Rivière, 1976).

In December, 1975 the Swedish SCOPE Committee with financial support from the Royal Swedish Academy of Sciences, UNEP, and Shell sponsored the first international SCOPE workshop on biogeochemical cycles at Friiberghs Herrgård, Örsundsbro, Sweden. The proceedings of this workshop entitled *Nitrogen, Phosphorus and Sulphur—Global Cycles*, edited by B. H. Svensson and R. Söderlund was published as SCOPE Report 7, *Ecological Bulletin*, **22** in 1976. Soon after the publication of this report it became clear that a more detailed treatment of the sulphur cycle, dealing especially with the physical, chemical, and biological processes involved and providing more reliable estimates of the sulphur reservoirs and fluxes of anthropogenic and natural origin, was required.

In May 1977 at the Paris meeting of the SCOPE Executive Committee, Professor M. V. Ivanov, on behalf of the Soviet National SCOPE Committee, put forward a proposal to continue work on the synthesis of data on the global biogeochemical sulphur cycle. This initiative was seconded by the Executive Committee and later the same year by an open meeting of SCOPE in London. A working group of Soviet scientists consisting of D. I. Grigoryan, V. A. Grinenko, M. V. Ivanov, A. Yu Lein, A. A. Migdisov, A. L. Rabinovich, A. B. Ronov, A. G. Rozanov, A. G. Ryaboshapko, and I. I. Volkov was set up to plan the project. As they had been engaged in a study of the specific processes involved in the global biogeochemical sulphur cycle for many years, they agreed to the proposal to produce a monograph on this topic, but declined to prepare a review of reviews or to collect a series of articles written by individual experts. They decided instead to concentrate on an analysis of the primary analytical data they had collected for this cycle and to prepare a monograph with all chapters written to fit a common plan.

In April 1978 members of the Soviet working group met with representatives of the Swedish National SCOPE Committee in Stockholm to discuss the draft composition of the monograph and immediately afterwards work on the project commenced.

The approach was approved by the SCOPE Executive Committee at its meeting in Pushchino, USSR, in May 1978 and an International Advisory Committee consisting of J. R. Freney, P. Goldsmith, M. V. Ivanov, W. E. Krumbein, J. W. M. La Rivière, J. O. Nriagu, T. A. Rafter, and H. Rodhe were appointed to oversee the project.

In August 1979 a meeting of the International Advisory Committee was held in Canberra and a preliminary report on the project was presented by Professor Ivanov. This was followed in October 1979 by an international workshop in Pushchino, USSR, convened by SCOPE, UNEP, and the USSR Academy of Sciences. Scientists from 17 countries (see List of Participants) met to discuss the material which had been prepared. The present monograph has been written taking account of the comments, criticisms, and suggestions for improvement there offered. The workshop nominated an editorial board consisting of M. V. Ivanov, J. R. Freney, I. E. Galbally, V. N. Kudeyarov, T. A. Rafter, H. Rodhe, Yu I. Sorokin, and P. A. Trudinger to prepare the monograph for publication and invited Drs. J. R. Freney and C. H. Williams to prepare a chapter on 'The sulphur cycle in soil' to complete the overall cycle.

In conclusion, it should be pointed out that the organizational activity of SCOPE on global biogeochemical cycles has led to an intensification of work in this field in various countries, the production of a series of monographs, and increased discussion of these problems at international symposia.

On behalf of the authors and editorial committee we wish to thank all the scientists who participated in the discussions at the workshop, SCOPE, UNEP, and the USSR Academy of Sciences for financial assistance, the latter body for its hospitality during the workshop and editorial board meeting, and members of the Institute of Biochemistry and Physiology of Micro-organisms of the USSR Academy of Sciences who assisted with the organization of the workshop and the translation of the various forms of the monograph. We also wish to thank the other scientists who assisted in various ways with the improvement of the text. Special thanks are due to the project bibliographer, Miss D. I. Grigoryan, to the translator and project manager Mr V. D. Gorokhov, and to Professor T. Munn for reviewing the monograph and making numerous helpful comments which have improved the readability of this volume.

Kovda, V., and La Rivière, J. W. M. (1976) Preamble. In: Svensson, B. H., and Söderlund, R. (eds), Nitrogen, Phosphorus and Sulphur—Global Cycles, Scope Report 7, *Ecol. Bull.*, **22**, 9.

M. V. IVANOV and J. R. FRENEY

Authors and Participants in the SCOPE/UNEP Workshop on the Global Biogeochemical Sulphur Cycle held at the Institute of Biochemistry and Physiology of Micro-organisms USSR Academy of Sciences, Pushchino, USSR, 15–19 October 1979

Dr L. E. BÅGANDER
Department of Geology
University of Stockholm
Box 6801, S-11386, Stockholm
Sweden

Dr S. S. BELYAEV
Institute of Biochemistry and
  Physiology of Micro-organisms
USSR Academy of Sciences
Pushchino, Moscow Region
USSR

Professor M. E. BERLYAND
Central Geophysical Observatory
Leningrad, Ul. Karbysheva 7
USSR

Professor J. M. BREMNER
Department of Agronomy
Iowa State University of Science and
  Technology
Ames, Iowa 50011
USA

Professor T. D. BROCK
Department of Bacteriology
University of Wisconsin
1550 Linden Drive
Madison, Wisconsin 53706
USA

Dr E. N. CHEBOTAREV
Lacustrine Science Institute
USSR Academy of Sciences
197046 Leningrad, Petrovskaya
  Naberezhnaya 4
USSR

Professor E. T. DEGENS
Geologisch-Paläontologisches
  Institut
University of Hamburg
Bundesstrasse 55, 200, Hamburg 13
Federal Republic of Germany

Academician A. A.
  DORODNITSIN
USSR Academy of Science

*Contributors to the SCOPE report.

President of the Soviet National
  SCOPE Committee
Ul. Vavilova 40
117333, Moscow B-333
USSR

*Dr J. R. FRENEY
CSIRO Division of Plant Industry
PO Box 1600, Canberra City,
  ACT 2601
Australia

Dr J. P. FRIEND
Chemistry Department
Drexel University
Philadelphia, PA 19104
USA

Dr S. GODZIK
Institute of Environmental
  Engineering
Polish Academy of Science
Curie-Sklodowskiej 34, 41–800
  Zabrze
Poland

Dr V. D. GOROKHOV
Institute of Biochemistry and
  Physiology of Micro-organisms
USSR Academy of Sciences
Pushchino, Moscow Region
USSR

Dr D. I. GRIGORYAN
Institute of Biochemistry and
  Physiology of Micro-organisms
USSR Academy of Sciences
Pushchino, Moscow Region
USSR

*Professor V. A. GRINENKO
Institute of Geochemistry
USSR Academy of Sciences
117975, GSP-1, Moscow, V-334,
  Vorobjevskoe Shosse, 47[a]
USSR

Dr D. R. HITCHCOCK
Norton Lane
Farmington, Connecticut
USA

*Professor M. V. IVANOV
Deputy Director of the Institute of
  Biochemistry and Physiology of
  Micro-organisms
USSR Academy of Sciences
Pushchino, Moscow Region
USSR

Dr B. B. JØRGENSEN
Institute of Ecology and Genetics
Aarhus University
Ny Munkegade, DK-8000 Aarhus C
Denmark

Professor JAN JUDA
Institute of Environmental
  Engineering
Warsaw Technical University
00-61 Warsaw, Pl. Jednosci
  Robotniczej
Poland

Dr V. D. KORZH
Institute of Oceanology
USSR Academy of Sciences
Ul. Krasikova 23, 117218, Moscow
USSR

Dr R. KRAL
Landscape Ecology Institute CSAV
Prague
Czechoslovakia

Professor H. R. KROUSE
Department of Physics
The University of Calgary
Calgary
Canada

Professor V. N. KUDEYAROV
Institute of Agrochemistry and Soil
  Science

USSR Academy of Sciences
Pushchino, Moscow Region
USSR

Professor S. I. KUZNETSOV
Corresponding Member of the USSR
    Academy of Sciences
Institute of Microbiology
60 Let Octiabrya 7, korp. 2
117312 Moscow
USSR

Professor J. W. M. LA RIVIÈRE
International Institute for Hydraulic
    and Environmental Engineering
Oude Delft 95, POB 3015,
    2601 DA, Delft
The Netherlands

*Dr A. YU. LEIN
Institute of Biochemistry and
    Physiology of Microorganisms
USSR Academy of Sciences
Pushchino, Moscow Region
USSR

Professor A. P. LISITSIN
Corresponding Member of the USSR
    Academy of Sciences
Institute of Oceanology
Ul. Krasikova 23, 117218 Moscow
    B-218
USSR

Dr A. A. MATVEEV
Institute of Hydrochemistry
Rostov on Don, Pr. Stachki 192,
    korp. 3
USSR

*Dr A. A. MIGDISOV
Institute of Geochemistry
USSR Academy of Sciences
117975, GSP-1, Moscow, V-334,
    Vorobjevskoe Shosse, 47[a]
USSR

Dr NAKAI
Faculty of Science
Department of Earth Sciences
Nagoya University
Chikusa, Nagoya
Japan

Professor S. NEDIALKOV
Scientific Research and Coordination
    Center for Preservation and
    Restoration of the Environment
Bulgarian Academy of Sciences
Str. M. Gagarin N 2, 1113 Sofia
Bulgaria

Dr J. O. NRIAGU
Canada Center for Inland Waters
PO Box 5050, Burlington, Ontario
    L7R LA6
Canada

Professor E. A. OSTROUMOV
Institute of Oceanology
Ul. Krasikova 23, 117248 Moscow
    B-218
USSR

Dr B. OTTAR
Norwegian Institute for Air Research
Royal Norwegian Council for
    Scientific and Industrial Research
P.B. 130-N-2001, Lillestrom
Norway

Dr O. P. PETRENCHUK
Central Geophysical Observatory
Leningrad, Ul. Karbysheva 7
USSR

Dr J. PILOT
Bergakademie Freiberg
92 Freiberg (Sachs)
DDR

Dr V. A. POPOV
Institute of Applied Geophysics

Ul. Glebovskaya 20b, Moscow
USSR

*Dr A. P. RABINOVICH
Institute of Hydrochemistry
344090, Rostov on Don, Pr. Stachki,
   192, corp. 3
USSR

Dr T. A. RAFTER
16 Simla Crescent
Khandallah, Wellington
New Zealand

Dr M. REDLI
Hungarian Academy of Sciences
Roosevelt ter. 9, Budapest V
Hungary

Dr B. W. ROBINSON
Institute of Nuclear Sciences, DSIR
Lower Hutt
New Zealand

Professor H. RODHE
Arrhenius Laboratory
Department of Meteorology
University of Stockholm
S-10691 Stockholm
Sweden

Professor E. A. ROMANKEVICH
Institute of Oceanology
Ul. Krasikova 23, 117218 Moscow
USSR

*Professor A. B. RONOV
Corresponding Member of the USSR
   Academy of Sciences
Institute of Geochemistry
USSR Academy of Sciences
117975, GSP-1, Moscow, V-334,
   Vorobjevskoe Shosse, 47^a, USSR

*Dr A. G. ROZANOV
Institute of Oceanology
USSR Academy of Sciences

117218 Moscow V-218
Ul. Krasikova 23
USSR

*Dr A. G. RYABOSHAPKO
Institute of Applied Geophysics
USSR Academy of Sciences
Moscow, Ul. Glebovskaya 20^b
USSR

Professor K. K. SHOPAUSKAS
Institute of Physics
Lithuanian Academy of Sciences
Vilnius, Prospect Krasnoi Armii 231
USSR

Academician G. K. SKRYABIN
Director of the Institute of
   Biochemistry and Physiology of
   Micro-organisms
USSR Academy of Sciences
Pushchino, Moscow Region
USSR

Professor YU. I. SOROKIN
Institute of Oceanology
Blue Buhta, Krasnodar Region,
   Gelendzhik
USSR

Professor H. G. THODE
Department of Chemistry
McMaster University
West Hamilton, Ontario L8S 4K1
Canada

Dr R. W. TILLMAN
Department of Soil Science
Massey University
Palmerston North
New Zealand

Dr P. A. TRUDINGER
Baas Becking Geobiological
   Laboratory
PO Box 378
Canberra City, ACT 2601
Australia

*Dr I. I. VOLKOV
Institute of Oceanology
USSR Academy of Sciences
117218 Moscow V-218, Ul.
  Krasikova 23
USSR

*Dr C. H. WILLIAMS
CSIRO Division of Plant Industry
PO Box 1600 Canberra City,
  ACT 2601
Australia

Professor G. A. ZAVARZIN
Corresponding Member of the USSR
  Academy of Sciences

Vice-President of SCOPE;
Institute of Microbiology
60 Let Octiabrya, Prospect 7, korp.
  2, 117312 Moscow
USSR

Dr A. M. ZYAKUN
Institute of Biochemistry and
  Physiology of Micro-organisms
USSR Academy of Sciences
Pushchino, Moscow Region
USSR

The Global Biogeochemical Sulphur Cycle
Edited by M. V. Ivanov and J. R. Freney
© 1983 Scientific Committee on Problems of the Environment (SCOPE)

CHAPTER 1
Principal Reactions of the Global Biogeochemical Cycle of Sulphur

V. A. GRINENKO and M. V. IVANOV

1.1 INTRODUCTION

The element sulphur with atomic number 16 and atomic mass 32.06 is in group VI of Mendeleev's periodic system. Terrestrial sulphur is a mixture of four stable isotopes with mass numbers 32, 33, 34, and 36 occurring in the proportions of 95.02, 0.75, 4.21, and 0.02% respectively. It occurs in the earth's crust to the extent of 0.047% and is widely distributed in nature in free and combined forms; it can be found in a range of valence states from the highly reduced sulphide (-2) to the oxidized form in sulphate ($+6$). The most abundant forms of sulphur are sulphate, sulphide, polysulphide, and elemental sulphur.

1.2 LOW-TEMPERATURE CHEMICAL REACTIONS OF THE GLOBAL SULPHUR CYCLE

The global biogeochemical cycle is a complex network of chemical and biochemical reactions in which sulphur participates in various forms with different physicochemical properties and aggregation states.

Of the gaseous forms of sulphur involved in this cycle the most important are hydrogen sulphide, sulphur dioxide, and sulphur trioxide. These gases are formed by various natural processes and as a result of man's activity. In the atmosphere hydrogen sulphide is quickly oxidized to sulphur dioxide which can take part in numerous reactions including oxidation, hydration, and combination with ammonia to form sulphate aerosols. An intricate complex of homogeneous and heterogeneous reactions involving sulphur-containing gases occurs in the atmosphere; many of these reactions are discussed in Chapter 4.

1

All of the sulphurous gases are soluble in water and their aqueous solutions are endowed with pronounced acidic properties, as a consequence of which sulphur plays an important role in the cycling of other elements, particularly the metals.

Sulphur dioxide, in both gaseous and hydrated forms, is readily oxidized to sulphur trioxide in the presence of oxygen (equation 1),

$$2SO_2 + O_2 \rightleftharpoons 2SO_3 \qquad (1)$$

or reduced to elemental sulphur in the presence of reducing agents such as hydrogen sulphide (equation 2),

$$SO_2 + 2H_2S \rightleftharpoons 3S + 2H_2O \qquad (2)$$

Sulphur trioxide dissolves in water to form the chemically active sulphuric acid which, even when diluted, reacts vigorously with many metals and their oxides to produce sulphates. These reactions are of paramount importance in the processes of weathering and corrosion of metals, as well as in many technological processes. Most sulphates are fairly soluble in water and thus are widespread in nature.

Of the poorly soluble sulphates, the most important geochemically are the calcium sulphates, anhydrite ($CaSO_4$) and gypsum ($CaSO_4 \cdot 2H_2O$). Oxidized sulphur is mainly removed from the cycle in one of these forms; consequently, deposits of gypsum and anhydrite are the biggest reservoirs of natural sulphur, being formed principally in the first stages of sea-water evaporation during very hot weather.

Sulphides, which form in aqueous solutions of hydrogen sulphide, have quite different chemical properties and geochemical peculiarities. As can be seen from Fig. 1.1, the zone of stability of hydrogen sulphide and bisulphide ion for all pH values is situated in the field of low partial pressures of oxygen, i.e. under anaerobic conditions. Most metal sulphides, with the exception of the alkali and alkaline earth metals, are poorly soluble in water, and thus the geochemical behaviour of both sulphur-containing compounds and many metals changes at the aerobic–anaerobic interface. When anaerobic conditions change to aerobic, sulphides are oxidized to form soluble and migratory sulphates, but when the reverse occurs, poorly soluble sulphides are formed and metals are immobilized.

The mechanisms of these geochemical reactions are rather complex, but under low-temperature conditions they become even more complicated because of the involvement of micro-organisms. Examples of such processes are given in Chapter 5 where the formation and oxidation of iron disulphide (FeS_2) is discussed. It should be noted that iron disulphide, represented in nature mainly by the mineral pyrite, is one of the most abundant compounds of sulphur and is widely used for technological purposes. Together with elemental sulphur it is one of the raw materials used for the production of sulphuric acid.

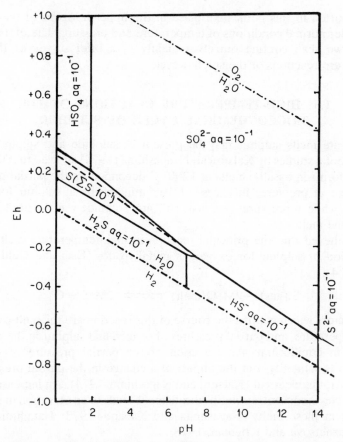

Fig. 1.1 The equilibrium distribution of sulphur species in water at 25 °C and 1 atm total pressure for activity of dissolved sulphur equal to 10^{-1}. The dashed lines indicate equal values of dissolved species within sulphur field (from Garrels and Christ, 1965)

Elemental sulphur (S^0) is the sole relatively stable natural form of sulphur with an intermediate state of oxidation. Under natural conditions it is formed mainly at the expense of hydrogen sulphide oxidation (equations 2 and 3).

$$2H_2S + O_2 \rightleftharpoons 2S + 2H_2O \qquad (3)$$

The zone of stability for sulphur (see Fig. 1.1) stretches slightly beyond the upper limit of hydrogen sulphide and bisulphide stability, which means that sulphur is unstable in the presence of oxygen. As with pyrite, micro-organisms appear to be of paramount importance for the low-temperature oxidation of elemental sulphur.

The major processes in the biogeochemical cycle of sulphur which have been reported in this book take place in the upper horizons of the earth's crust under normal conditions of temperature and pressure. Therefore in this chapter we shall confine ourselves mainly to a brief survey of the low-temperature reactions of the sulphur cycle.

1.3 HIGH-TEMPERATURE REACTIONS OF THE BIOGEOCHEMICAL CYCLE OF SULPHUR

In silicate melts sulphur may be present as sulphide and sulphate ions. Experimental studies by Katsura and Nagashima (1974) showed that the ratio of these forms in basaltic melts at 1200 °C depended on the partial pressure of oxygen: at pressures in excess of 10^{-8} atm the basic sulphur form was sulphate, while at pressures less than 10^{-8} atm sulphide was the predominant form in the melt.

It is believed that the principal reaction in high-temperature melts is the substitution of sulphur for oxygen in metal oxides (Esin and Geld, 1966; equation 4):

$$S_2(gas) + 2MeO(melt) \rightleftharpoons 2MeS + O_2 \qquad (4)$$

Major factors which control the course of this reaction are the melt composition, temperature, and partial pressures of oxygen and sulphur in the gaseous medium in equilibrium. An elevation of the partial pressure of sulphur increases its solubility, but the impact of a change in the partial pressure of oxygen is not as clear cut (Katsura and Nagashima, 1974). An increase in the FeO and Na_2O concentrations in silicate melts leads to an elevation in sulphur solubility in these melts (Nagashima and Katsura, 1973; Haughton *et al.*, 1974; Kuznetsova and Krigman, 1978).

Under conditions of rapid melt cooling, e.g. during submarine basalt eruptions, the erupted matter is rapidly crystallized and sulphur is isolated in the form of sulphide inclusions (Moore and Fabbi, 1971). However, during slow basalt cooling, in subaerial conditions, most sulphur compounds are lost in the form of sulphurous gases.

These gases form as a result of high-temperature interaction of water with the sulphides and ferrous silicates of melts (equations 5, 6, and 7):

$$FeS + H_2O \rightleftharpoons FeO + H_2S \qquad (5)$$

$$FeS + FeSiO_3 + H_2O \rightleftharpoons Fe_2SiO_4 + H_2S \qquad (6)$$

$$FeS + FeSiO_3 + 3H_2O \rightleftharpoons Fe_2SiO_4 + SO_2 + 3H_2 \qquad (7)$$

In the course of temperature and pressure variations sulphurous gases react with each other (equations 3, 8, and 9), as well as with the other magmatic gases:

$$SO_2 + 2CO \rightleftharpoons 2CO_2 + S \qquad (8)$$

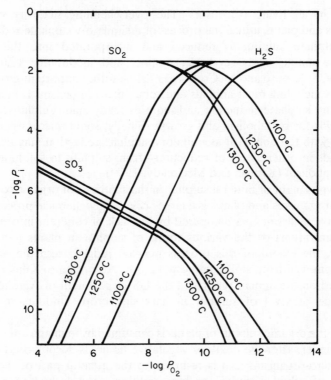

Fig. 1.2 Relationships between calculated partial pressures of SO_2, SO_3, and H_2S and calculated P_{O_2} at 2.1% $(SO_2)_i$ under various temperatures. $(SO_2)_i$ is the initial SO_2 concentration (from Katsura and Nagashima, 1974)

$$H_2S + 2H_2O \rightleftharpoons SO_2 + 3H_2 \qquad (9)$$

The calculated partial pressures of three basic components of magmatic gases (SO_2, SO_3, and H_2S) at different partial pressures of oxygen are presented in Fig. 1.2. Under real conditions the ratio of gases depends strongly on the composition of the magma (Matsuo, 1960), while the total content of sulphur compounds in magmatic gases decreases with decreasing temperature (Sokolov, 1966).

1.4 BIOLOGICAL PROCESSES FOR THE FORMATION AND DECOMPOSITION OF SULPHUR-CONTAINING ORGANIC COMPOUNDS

Sulphur is a constituent of a number of amino acids, and therefore may be referred to as one of the so-called biogenic elements whose metabolism is inherent in all living organisms. The overwhelming majority of micro-organisms and plants utilize the process of assimilatory sulphate reduction, in which sulphate sulphur is reduced and incorporated into the sulphur-containing amino acids—cysteine, cystine, and methionine (Zinder and Brock, 1978; Krouse and McCready, 1979). Another important group of live organisms including most animals and some micro-organisms is incapable of assimilatory sulphate reduction and requires ready-made sulphur-containing amino acids for metabolism and growth. Finally, some micro-organisms not endowed with the ability of assimilatory sulphate reduction may use sulphur compounds in other states of reduction—from sulphite to sulphide ions for their metabolism (Krouse and McCready, 1979).

The involvement of mineral sulphur in the synthesis of organic compounds by micro-organisms and plants is a large-scale biogeochemical process, whose quantitative assessment is hampered by the lack of coherent information on the concentrations of the various forms of sulphur in plants and animals. However, the extent of this process may be judged from the amount of sulphur removed from soil in harvested crops. According to Kilmer's (1979) data based on the annual reports of the US Department of Agriculture, the world-wide removal of sulphur in harvested crops amounts to 2.5 TgS annually.

Part of the organic sulphur of plants is consumed by man and animals, while the remainder, after the death of vegetative tissue, is decomposed by sapro-phytic micro-organisms and is returned to the mineral part of the cycle in the form of sulphate under aerobic conditions, and as hydrogen sulphide under anaerobic conditions.

During decomposition the sulphur-containing amino acids, cysteine and methionine, are released and then degraded by the action of certain enzymes. Cysteine, for example, reacts with cysteine desulphhydrase and ammonia and hydrogen sulphide are released. The reaction may be represented by equation (10) (Segal and Starkey 1969):

$$HS—CH_2—CH(NH_2)—COOH + H_2O \longrightarrow$$
$$\text{cysteine}$$

$$CH_3—CO—COOH + NH_3 + H_2S \quad (10)$$
$$\text{pyruvic acid}$$

In recent years, due to the discovery of organic sulphur compounds such as methyl mercaptan (CH_3SH) and dimethyl sulphide (CH_3SCH_3) in the atmosphere, scientists became interested in the biological reactions which lead to their production in the biosphere. These compounds are produced by many bacteria, yeasts, and moulds on media containing methionine and other sulphur-containing organic compounds (Krouse and McCready, 1979):

$$CH_3-S-CH_2-CH_2-CH(NH_2)-COOH + H_2O \longrightarrow$$

methionine

$$CH_3-CH_2-CO-COOH + NH_3 + CH_3SH \quad (11)$$

α-ketobutyric acid　　　　　　　　　　　　methyl
mercaptan

In addition, certain micro-organisms, in particular the fungi *Schizophyllum commune*, have been reported to produce dimethyl sulphide and methyl mercaptan on media containing glucose with sulphate as the sole source of sulphur (Young and Maw, 1958). A more detailed analysis of the literature on the microbiological formation of volatile organic sulphur compounds may be found in the reviews of Krouse and McCready (1979) and Zinder and Brock (1978).

1.5　OXIDATION OF SULPHUR COMPOUNDS BY MICRO-ORGANISMS

The oxidation of reduced sulphur is always accompanied by the release of energy, e.g. in the oxidation of thiosulphate (equation 12; Roy and Trudinger, 1970):

$$S_2O_3^{2-} + 2O_2 + H_2O \longrightarrow 2SO_4^{2-} + 2H^+ + 211 \text{ kcal} \quad (12)$$

Various groups of chemolithotrophic micro-organisms are capable of using this energy for synthesizing their organic constituents from carbon dioxide. Chemolithotrophic sulphur organisms have been classified into three important groups on the basis of their morphology: (1) Thiobacteriaceae, colourless coccoid, straight or curved rod-shaped bacteria; (2) Beggiatoaceae, colourless cells occurring in trichomes within which they are arranged in chains; and (3) Achromatoaceae, large spherical, ovoid or short cylindrical cells containing sulphur granules (Roy and Trudinger, 1970).

The *Beggiatoa* are filamentous organisms commonly found in marine and freshwater environments containing hydrogen sulphide. Micro-organisms of this group oxidize hydrogen sulphide to elemental sulphur which is often deposited inside the cells. When hydrogen sulphide in the media becomes unavailable, the accumulated sulphur is oxidized to sulphate and removed from the cells. The physiology of *Beggiatoa* has not been well studied and even the fact of obligatory chemolithotrophy has not been proven (Trudinger, 1979) especially as Strohl and Larkin (1978) have separated five groups of *Beggiatoa* from freshwater sediments, all of which grew heterotrophically.

In contrast to the filamentous sulphur-bacteria, numerous representatives of the *Thiobacillus* genus have been investigated in every detail (Ralph, 1979). Members of this group of bacteria which take an active part in the oxidation of natural compounds of sulphur, differ in their response to pH and temperature and in their ability to grow when organic compounds are present in the medium. The autotrophic bacteria *Thiobacillus thioparus, T. denitrificans*, and *T. neapolitanus*, together with the closely related *T. novellus* and *T. perometablis*, grow preferentially in alkaline and slightly acidic conditions, whereas *T. thiooxidans, T. denitrificans*, and *T. intermedius* prefer acidic and ultra-acidic conditions. Under these acidic conditions many reduced compounds of sulphur are unstable and thus elemental sulphur and metal sulphide are usually oxidized by these organisms.

Unlike other unicellular sulphur-bacteria, the acidophilic bacteria *T. ferrooxidans* and *Sulfolobus acidocaldarians* are also capable of oxidizing ferrous iron in acidic media. The peculiarity makes *T. ferrooxidans*, in particular, one of the most important geochemical agents participating in the aerobic decomposition of sulphide ores. This ability is used in the bacterial leaching of ores to recover metals, e.g. the extraction of iron from pyrite.

The biological and chemical reactions involved in the dissolution of pyrite may be represented by the following equations (Temple and Delchamps, 1953):

$$2FeS_2 + 2H_2O + 7O_2 \rightleftharpoons 2FeSO_4 + 2H_2SO_4 \tag{13}$$

$$4FeSO_4 + O_2 + 2H_2SO_4 \rightleftharpoons 2Fe_2(SO_4)_3 + 2H_2O \tag{14}$$

$$Fe_2(SO_4)_3 + FeS_2 \rightleftharpoons 3FeSO_4 + 2S \tag{15}$$

$$2S + 3O_2 + 2H_2O \rightleftharpoons 2H_2SO_4 \tag{16}$$

In this process the second and fourth reactions follow the biological pathway with the participation of *T. ferrooxidans* and *T. thiooxidans*, whereas the third reaction appears to be strictly chemical; metal sulphide is oxidized by ferric ion which in turn is reduced to ferrous ion. One of the most important reactions of the whole oxidation cycle in the natural dissolution of metal sulphides or the bacterial leaching of metals is the microbial oxidation of ferrous iron to ferric iron (equation 14).

Growing at the interface between the aerobic and anaerobic zones, the *Thiobacilli* and *Beggiatoa* play an important part in the oxidation of various reduced sulphur compounds to both elemental sulphur and sulphate (Trudinger, 1979).

1.6   OXIDATION OF REDUCED SULPHUR UNDER ANAEROBIC CONDITIONS

The importance of micro-organisms in oxidative reactions of the sulphur cycle is not, however, solely confined to aerobic conditions. Two groups of micro-organisms exist which are capable of oxidizing hydrogen sulphide, elemental sulphur, and other partially oxidized compounds of sulphur in the absence of free oxygen.

One group, represented by *Thiobacillus denitrificans*, are chemolithotrophic anaerobic bacteria which can oxidize reduced sulphur compounds in the absence of oxygen at the expense of nitrate which is simultaneously reduced to molecular nitrogen (equation 17):

$$5S_2O_3^{2-} + 8NO_3^- + H_2O \longrightarrow 10SO_4^{2-} + 4N_2 + 2H^+ \qquad (17)$$

The distribution and geochemical activities of *T. denitrificans* have not yet been studied in detail. Nevertheless it appears that these organisms are quite important for the oxidation of sulphur compounds in paddy soils (Baldensperger and Garcia 1975; Jacq and Roger, 1978) and in the upper horizons of marine and lacustrine reduced sediments where nitrate penetrates deeper than dissolved oxygen.

The second group of anaerobic bacteria which have the ability to oxidize reduced sulphur compounds are the photolithotrophs. Two families, Thiorhodaceae and Chlorobacteriaceae, have been recognized, with the best-known members being the purple sulphur bacteria, *Chromatium* and the green bacteria, *Chlorobium* (Roy and Trudinger, 1970). These organisms contain photosynthetic pigments analogous to those of green plants and fix carbon dioxide in the presence of light and a reduced sulphur compound which acts as an electron donor. Equation (18) represents the oxidation of hydrogen sulphide by the photolithotrophs:

$$2CO_2 + H_2S + 2H_2O \xrightarrow{\text{light}} 2CH_2O + SO_4^{2-} + 2H^+ \qquad (18)$$

Like all photosynthetic reactions these processes take place only in the presence of light which imposes certain restrictions on the distribution of the photosynthetic sulphur-bacteria. The zone of activity is limited to the upper part of the anaerobic zone of the biosphere into which light penetrates. This includes the uppermost horizons of shallow sediments of water bodies and the lower part of the photic zone of water columns containing hydrogen sulphide. In these ecosystems the photosynthetic bacteria grow in tremendous quantities and play an active part not only in the sulphur cycle but also in the carbon cycle.

More detailed information on the ecology and geochemical activity of these organisms can be found in the reviews of Pfennig (1975, 1977) and the monographs of Kondratyeva (1972) and Gorlenko *et al.* (1977).

1.7 REDUCTION OF SULPHATE BY MICRO-ORGANISMS

The low-temperature reduction of sulphate to hydrogen sulphide is conducted by an exclusive group of micro-organisms called desulphurizing or sulphate-reducing bacteria; the process is termed 'dissimilatory sulphate reduction'. This reduction may be regarded as a redox process in which sulphate is used as a terminal electron acceptor in the oxidation of organic compounds or hydrogen. Physiologically, this process provides the bacterial cell with the necessary energy for metabolism.

Typically, dissimilatory sulphate reduction is expressed formally in the following way (equation 19):

$$2CH_3—CHOH—COOH + H_2SO_4 \longrightarrow$$
$$2CH_3—COOH + 2CO_2—2H_2O + H_2S \quad (19))$$

Three genera of sulphate-reducing bacteria have been described on the basis of their morphology and physicobiological properties: the *Desulfovibrio* are heterotrophic, obligate motile anaerobes which are generally curved rods; *Desulfotomaculum* are heterotrophic, anaerobic rod-shaped organisms that form heat-resistant spores; and *Desulfomonas* are non-motile rods.

All known species of sulphate-reducing bacteria grow well on nutritive media which include sulphate and lactate or pyruvate. However, under natural conditions this group apparently utilizes a much wider spectrum of organic compounds. For example, *Desulfovibrio africanus* can grow on ethanol, and *D. vulgaris* metabolizes methanol, ethanol propanol, and butanol. (See note added in proof.)

Of great importance for understanding the final stages in the anaerobic decomposition or organic compounds is the work of Widdel and Pfennig (1977) who isolated a new species of sulphate-reducing bacteria, *Desulfotomaculum acetoxidans*, which is capable of oxidizing acetate to carbon dioxide. Before this work appeared, it was believed that acetate was metabolized exclusively by methane-producing bacteria, and numerous schemes were proposed for the anaerobic decomposition of organic matter which included a succession of sulphate-reducing and methane-producing bacteria.

Before closing this brief review of the major biogeochemical reactions of the global sulphur cycle, it should be stressed that many living organisms are of paramount importance for the cycling of sulphur at low temperatures and pressures. They not only speed up the oxidation of reduced sulphur compounds but also perform various redox reactions under anaerobic conditions.

1.8 FRACTIONATION OF THE STABLE ISOTOPES OF SULPHUR

Originally, the aim of isotope abundance measurements was to identify the natural isotopes and to gain an understanding of the atomic weights of the elements. More recently they have been used to interpret the long-term geochemical changes in nature, to determine the origins of mineral deposits, and to study the diagenesis of sulphur in modern environments (Chambers and Trudinger, 1978).

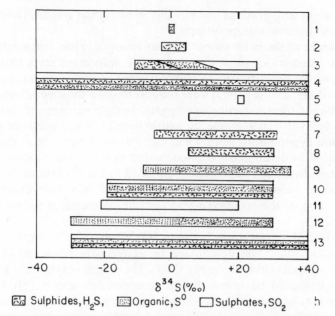

Fig. 1.3 Isotopic composition of sulphur in various natural substances. 1, Meteorites; 2, basic rocks; 3, volcanic rocks; 4, sedimentary rocks; 5, ocean-water; 6, evaporites; 7, zinc–lead deposits; 8, hydrogen sulphide from Devonian rocks, Alberta, Canada; 9, petroleum; 10, atmospheric compounds; 11, fresh water; 12, coal; 13, soil (from Krouse, 1970)

Most workers in these fields have studied the sulphur isotopes with mass numbers 32 and 34 because of their more favourable abundances; numerous studies have demonstrated marked variations in the relative proportions of these isotopes in sulphur compounds from natural sources (Fig. 1.3). The isotopic compositions are often expressed as $\delta^{34}S$ values which relate the isotopic composition of a sample to that of a standard according to equation (20):

$$\delta^{34}S(^{0}/_{00}) = \left[\frac{(^{34}S/^{32}S) \text{ sample}}{(^{34}S/^{32}S) \text{ standard}} - 1 \right] \times 1000 \qquad (20)$$

The standard for the analysis of terrestrial samples is usually troilite from the Canon Diablo meteorite which has a $\delta^{34}S$ value of $0^{0}/_{00}$. Positive values of $\delta^{34}S$ mean that the sample is enriched in the ^{34}S isotope while negative values denote an enrichment in ^{32}S compared to the meteorite standard.

In natural systems the fractionation of stable isotopes of sulphur occurs in the course of oriented chemical reactions under normal temperatures (termed the kinetic isotopic effect), and in isotope exchange reactions proceeding, as a rule, at raised temperatures and possibly in biological systems (referred to as the thermodynamic isotopic effect).

Knowledge of the behaviour of sulphur isotopes in low-temperature oxidation and reduction of sulphur compounds is important for a better understanding of the geochemistry of sulphur in the sedimentary process.

Harrison and Thode (1957) were the first to show the kinetic isotopic effect during the chemical reduction of sulphate. In their experiments sulphate was reduced to hydrogen sulphide by hydriodic acid in the presence of hydrochloric and hypophorous acids. Generally, no more than 2% of the sulphate was reduced to determine the fractionation factor of this reaction. In these experiments hydrogen sulphide was enriched in the ^{32}S isotope by 21–22^{0}/oo compared to sulphate, and the isotopic effect was independent of the concentration of the reducing agent and the temperature within the range 17–50 °C.

The effect of temperatures between 100 °C and 300 °C on the magnitude of the kinetic effect was determined by studying the reduction of sulphuric acid by hydrogen (Grinenko *et al.*, 1969). The results of this study are given in Fig. 1.4. It should be noted that at temperatures above 150 °C isotopic exchange may be induced between the reaction products which would lead to an isotope redistribution at the expense of the thermodynamic isotopic effect.

Grinenko and Thode (1970) demonstrated a kinetic isotopic effect during the partial reduction of sulphur dioxide by hydrogen sulphide (equation 21):

$$2H_2S + nSO_2 \longrightarrow 3S + 2H_2O + (n-1)SO_2 \qquad (21)$$

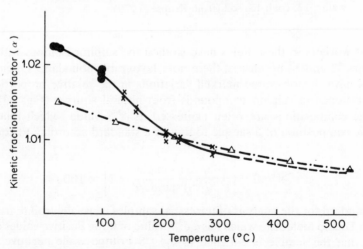

Fig. 1.4 Effect of temperature on the kinetic fractionation factor in the reduction of sulphate to sulphur dioxide and hydrogen sulphide. ●, Reduction to hydrogen sulphide; ×, reduction to sulphur dioxide; △, calculated curve (from Harrison and Thode, 1957 and Grinenko and Grinenko, 1974)

As expected, the remaining sulphur dioxide was enriched in the heavy sulphur isotope. The fractionation factor for the temperature interval from 25 °C to 280 °C calculated by the Rayleigh formula was 1.015 (Grinenko and Thode, 1970).

At ambient temperatures oxidation of reduced sulphur compounds is not generally accompanied by a substantial fractionation of sulphur isotopes. In experiments conducted by Vinogradov and Grinenko (1964) a stream of oxygen was passed into sulphur dispersed in water; sulphur was oxidized to sulphate which was slightly enriched in the light isotope while the heavy isotope accumulated in the residual sulphur. A significant fractionation of isotopes was observed only when more than 80% of the original sulphur was oxidized; the fractionation factor calculated from the experimental data did not exceed 1.0014. A slight fractionation of isotopes was observed during the reaction of water with elemental sulphur at 80–100 °C with the simultaneous production of hydrogen sulphide and oxidized sulphur compounds (Monster *et al.*, 1965). Compared to the original sulphur, the oxidized compounds were slightly enriched in ^{32}S and hydrogen sulphide was enriched with ^{34}S.

Therefore, the kinetic isotopic effects occurring in both oxidation and reduction reactions lead to products that are enriched in the light isotope and residues that are enriched with the heavy isotope of sulphur. During the chemical reduction of oxidized sulphur compounds, any hydrogen sulphide formed cannot be enriched in ^{32}S by more than 22‰ compared to the initial sulphur. However, the heavy isotope may accumulate in the residue and reach high values according to the Rayleigh formula.

$$N = N_0 \left(\frac{V_0}{V}\right)^{(\alpha-1)/\alpha}$$

where N_0 and N are the contents of ^{34}S in the original compound and residue, V_0 and V are the amounts of original compound and residue, and α is the fractionation factor.

An important conclusion that can be drawn from the review of work on the kinetic isotopic effect is that sulphate formed by oxidation of sulphide and elemental sulphur has essentially the same isotopic composition as the sulphur compounds undergoing oxidation.

Table 1.1 groups the theoretically calculated thermodynamic characteristics of various sulfur compounds; these data enable us to predict the distribution of sulphur isotopes in isotope exchange reactions at equilibrium. At equilibrium the heavy isotope of sulphur should accumulate to a large degree in oxidized compounds, because the fractionation factor is greater when there is a large difference between the oxidation states of sulphur (Table 1.1). The highest fractionation (82‰ at 27 °C) should be expected at equilibrium in a system containing sulphate and sulphide ions (Table 1.1).

When assessing thermodynamic isotopic effects from the data of Table 1.1,

Table 1.1   Isotopic properties of sulphur compounds (Sakai, 1968)

Compound	*Temperature (K)*					
	300	400	500	600	700	800
	1000 · ln f^a (^{o}/oo)					
SO_4^{2-} aq.	82.1	51.7	35.2	25.3	18.8	14.0
SO_3^{2-} aq.	71.2	43.1	28.6	20.4	15.2	11.7
SO_2 gas	43.0	27.6	19.1	13.9	10.6	8.2
H_2S	13.0	8.9	6.6	5.1	4.0	3.2
HS^-	9.2	6.7	5.1	3.7	3.0	2.4
$S_8 (1/8)^b$	15.8	8.9	5.7	4.0	2.9	2.2
$S_2 (1/2)$	10.1	6.0	4.0	2.8	2.1	1.6
$FeS_2 (1/2)$	18.4	10.9	7.0	4.8	3.6	2.7
ZnS	14.3	8.0	5.2	3.6	2.6	2.0
SiS^c	12.8	7.5	5.0	3.5	2.6	2.0
PbS	3.6	2.0	1.3	0.9	0.6	0.5
S^{2-} aq.	0	0	0	0	0	0

^{a}f is the reduced partition function ratio.
^{b}Values for 300 K are calculated for rhombic sulphur by Tudge and Thode (1950).
Values for other temperatures were calculated by assuming proportionality to
$1/T^2$.
^{c}Data of Hulston (1964).

it should be understood that these data allow the precise estimation of the degree of fractionation; they do not, however, enable us to determine whether a reaction will reach equilibrium. To obtain evidence for isotopic fractionation due to the thermodynamic effect, experimental studies on rates of isotopic exchange should be conducted.

The main problem in the interpretation of isotopic data on volcanic gases, fumarole emissions, and hydrothermal minerals is to decide whether equilibrium has been attained between oxidized and reduced compounds of sulphur at the various temperatures.

Thode *et al.* (1971) obtained evidence for a rather high rate of isotopic exchange between hydrogen sulphide and sulphur dioxide at temperatures above 500 °C, as did Grinenko and Thode (1970) for the temperature interval of 300–450 °C. The experimentally established dependence of the equilibrium constant for the distribution of sulphur isotopes between these compounds on temperature is shown in Fig. 1.5. These data show that sulphur dioxide is slightly more enriched in ^{34}S than hydrogen sulphide in the high-temperature gases, but at 300 °C the difference in isotopic composition amounts to 9°/oo. When a temperature drop occurs hydrogen sulphide and

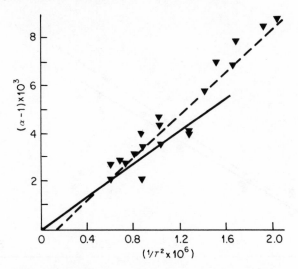

Fig. 1.5 Effect of temperature on the equilibrium constant for the isotope exchange reaction, $H_2{}^{34}S + {}^{32}SO_2 \rightleftharpoons H_2{}^{32}S + {}^{34}SO_2$. Continuous line—theoretical; broken line—experimental (α = equilibrium isotopic factor, T = absolute temperature) (from Thode *et al.*, 1971 and Grinenko and Thode, 1970)

sulphur dioxide react rapidly to form elemental sulphur. At 200 °C the rate of sulphur exchange between hydrogen sulphide and sulphur dioxide is quite low, and therefore the isotopic composition may change because of a kinetic isotopic effect. All of these effects result in a greater enrichment of the residual sulphur dioxide in the heavy sulphur isotope.

In acidic solutions rapid isotopic exchange has been demonstrated between hydrogen sulphide and sulphate at 200 °C. The dependence of the isotope fractionation factor for hydrogen sulphide and sulphate on temperature is illustrated in Fig. 1.6. These data suggest that in solution at high temperatures sulphate will be enriched in the ${}^{34}S$ isotope compared to the hydrogen sulphide present.

Some experimental data have been obtained showing that crystallization and recrystallization of sulphides above 250 °C proceed under conditions of isotopic equilibrium. This results in a definite distribution of sulphur isotopes between co-crystallizing sulphides, and such a distribution is temperature-dependent (Rye and Czamanske, 1969, Kajiwara and Krouse, 1971). The content of the ${}^{34}S$ isotope decreases in the series.

pyrite → (pyrotite-sphalerite) → chalcopyrite → galena

The greatest difference observed was for the pyrite–galena pair which amounted to only 4⁰/₀₀ at 250 °C. At ordinary temperatures, isotopic equili-

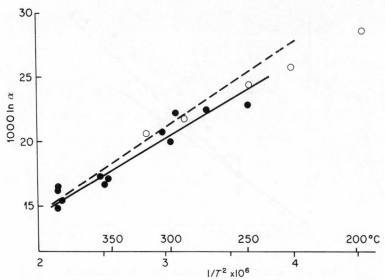

Fig. 1.6   Effect of temperature on equilibrium isotopic factor (α) for the exchange between hydrogen sulphide and sulphate. ●, from Igumnov *et al.* (1977); ○, from Robinson (1973); broken line, calculated curve (Sakai, 1968)

brium is not attained during the crystallization of sulphides and pyrite is only slightly enriched in the ^{32}S isotope (1–2°/oo) compared with the co-crystallizing sphalerite and chalcopyrite. Crystallization of gypsum from a saturated solution results in a 1.6°/oo increase in the ^{34}S content of the crystals compared to the solution (Thode and Monster, 1965).

In the geochemistry of isotopes special attention is paid to the processes of fractionation in biological systems, since these processes engender the greatest isotopic fractionation in nature.

The process of sulphate assimilation from solution by bacteria, algae, and plants is accompanied by a slight isotopic effect; the total sulphur of these organisms is enriched in ^{32}S by 1–3°/oo (Kaplan and Rittenberg, 1964; Mekhtieva and Pankina, 1968).

The partial decomposition of cysteine or other sulphur-containing amino acids in bacterial processes leads to a slight fractionation of sulphur isotopes; the first hydrogen sulphide released is enriched in ^{32}S by 5°/oo.

No detectable isotopic fractionation of sulphur occurs during the reduction of elemental sulphur to hydrogen sulphide by yeasts, nor during the oxidation of elemental sulphur by thiobacilli (Jones and Starkey, 1957; Kaplan and Rafter, 1958; Eremenko and Mekhtieva, 1961). However, oxidation of elemental sulphur and pyrite by mixed cultures produced sulphates which were enriched in ^{32}S by a maximum of 1.7‰ (Nakai and Jensen, 1964).

In the case of oxidation of sulphide by thiobacilli a slight enrichment in ^{32}S is observed for elemental sulphur and a greater one for sulphates (up to 18‰), while the heavy isotope accumulates in intermediary products (Kaplan and Rittenberg, 1964). Studies on photosynthetic oxidation of hydrogen sulphide by purple bacteria have produced conflicting results on isotopic fractionation. In one series of experiments the elemental sulphur produced showed a low enrichment in the heavy isotope, while in another series sulphur

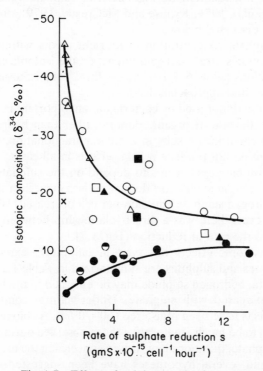

Fig. 1.7   Effect of sulphate reduction rate and electron acceptor on the isotopic composition of hydrogen sulphide formed during sulphate reduction by *Desulfovibrio desulfuricans*. O, lactate at 10–45 °C, 0.06 M SO_4^{2-}; $\triangle$, ethanol at 10–45 °C, 0.06 M SO_4^{2-}; ●, hydrogen at 10–45 °C, 0.06 M SO_4^{2-}; ×, lactate at 0 °C; ▲, lactate + 0.01 M SO_4^{2-}; □, lactate + 0.02 M SO_4^{2-}; ■, lactate + 0.12 M SO^{2-}; ◓, hydrogen + 0.02 M SO_4^{2-}; ◑, hydrogen + 0.18 M SO_4^{2-} (from Kaplan and Rittenberg, 1964)

was enriched in the light isotope by up to 10⁰/₀₀; in the latter experiment no
significant fractionation was observed in the sulphate (Kaplan and
Rittenberg, 1964). One may conclude from the discussion above that essentially no fractionation of sulphur isotopes occurs during the oxidation of
sulphur compounds.

The greatest fractionation of sulphur isotopes occurs during the microbial
reduction of sulphate. As early as 1951, Thode *et al.* reported that the bacterial reduction of sulphate produced hydrogen sulphide enriched in the light
isotope of sulphur. As a number of reviews have been published on the details
of sulphur isotope fractionation during bacterial sulphate reduction (e.g.
Grinenko and Grinenko, 1974; Krouse and McCready, 1979), only the basic
conclusions will be discussed below.

Variations in sulphate concentration in nutrient media within the range
1×10^{-3}–3×10^{-2} moles litre^{-1} have no impact on the isotopic effect. However, for concentrations below 6×10^{-4} moles litre^{-1}, the isotopic fractionation factor decreases and approaches 1.00.

The type of electron donor used by bacteria has an important impact on the
isotopic effect; viz. the use of organic donors rather than hydrogen under
otherwise identical conditions results in a greater fractionation. Changes in
pH and temperature do not produce regular effects in all cases.

In general the isotopic effect seems to depend on the sulphate reduction
rate: with the use of organic electron donors an isotopic effect increased as
the intensity of hydrogen sulphide emission per cell decreased. When hydrogen was used, however, there was a direct relationship between the degree
of fractionation and the rate of reduction (Fig. 1.7).

The maximum isotopic effects observed in laboratory experiments on
reduction of sulphates and sulphites are summarized in Table 1.2. As can be
seen from these data, hydrogen sulphide may be enriched in the light isotope
by up to 50⁰/₀₀ compared with sulphate. Under natural conditions even
greater isotopic effects have been registered (60–70⁰/₀₀). As biogenic isotopic
effects have proved to be much greater than those observed during the chemical reduction of sulphate, questions have arisen concerning the mechanisms of
the fractionation and several hypotheses have been suggested. At present
most investigators share the belief (Rees, 1973; Grinenko and Grinenko,
1974) that isotopic exchange occurs between the intermediary products of
reduction in the bacterial cell which produces the thermodynamic isotopic
effect under ordinary temperatures. The reality of such a mechanism is supported by recent experiments on the exchange of radioactive sulphur between
$H_2^{35}S$ and Na_2SO_4 in the presence of sulphate-reducing bacteria (Trudinger
and Chambers, 1973), and by the dependence of the isotopic composition of
oxygen in the residual sulphate on that of oxygen in the water in which the
reaction occurred.

The sulphur isotopic composition of natural samples can vary up to 160⁰/₀₀

19

Table 1.2 Maximum isotopic effects of various bacterial processes in laboratory experiments

Process	Organism	Initial reactant	Final product	Maximum isotopic effect	Reference[a]
Sulphate reduction	Desulfovibrio desulfuricans	SO_4^{2-}	H_2S	-46	1
	Desulfovibrio desulfuricans	SO_4^{2-}	H_2S	-35	2
	Desulfovibrio vulgaris	SO_4^{2-}	H_2S	-24	3
	Desulfotomaculum nigrificans	SO_4^{2-}	H_2S	-12	4
	Saccharomyces cerevisiae	SO_4^{2-}	H_2S	-25	5
	Desulfovibrio gigas	SO_4^{2-}	H_2S	0	6
Sulphate assimilation	Escherichia coli	SO_4^{2-}	Organic S	-2.8	1
	Saccharomyces cerevisiae	SO_4^{2-}	Organic S	-2.4	11
	Desulfovibrio desulfuricans	SO_4^{2-}	Organic S	-2.7	4
Sulphite reduction	Desulfovibrio desulfuricans	SO_3^{2-}	H_2S	-14	1
	Desulfovibrio desulfuricans	SO_3^{2-}	H_2S	-13	10
	Desulfovibrio vulgaris	SO_3^{2-}	H_2S	-33	3
	Saccharomyces cerevisiae	SO_3^{2-}	H_2S	-50	5
	Salmonella sp.	SO_3^{2-}	H_2S	-42	7
	Desulfotomaculum nigrificans	SO_3^{2-}	H_2S	-8	4
Cysteine decomposition	Proteus vulgaris	Cysteine	H_2S	-5.1	1
Chemosynthetic oxidation	Thiobacillus concretivorus	H_2S	S^0	-2.5	1
	Thiobacillus concretivorus	H_2S	SO_4^{2-}	-18	
	Thiobacillus concretivorus	H_2S	$S_xO_y^{2-}$	$+19$	
	Thiobacillus concretivorus	S^0	SO_4^{2-}	-1.4 to 0.4	1
	Thiobacillus denitrificans	S^0	SO_4^{2-}	-1.1 to $+0.4$	8
	Thiobacillus sp.	S^0	SO_4^{2-}	-2	9
Photosynthetic oxidation	Chromatium	H_2S	S^0	-10	1
	Chromatium	H_2S	SO_4^{2-}	0	
	Chromatium	H_2S	$S_xO_y^{2-}$	$+11$	

[a]1. Kaplan and Rittenberg (1964); 2. Chambers et al. (1975); 3. Kemp and Thode (1968); 4. McCready (1975); 5. McCready et al. (1974); 6. Smejkal et al. (1971); 7. Krouse and Sasaki (1968); 8. Mekhtieva (1964); 9. Nakai and Jensen (1964); 10. Harrison and Thode (1958); 11. McCready et al. (1974).

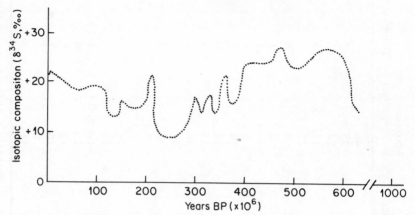

Fig. 1.8   Isotopic composition of sulphur in evaporites of different age (from Grinenko and Grinenko, 1974)

although most samples vary within 70⁰/₀₀ (Fig. 1.3). This range far exceeds the fractionation that can be related exclusively to chemical and physical processes. Undoubtedly, the major processes responsible for the fractionation are related to the biological activity during sulphur turnover in the sedimentary cycle.

The mechanisms of microbial isotopic fractionation, in which the light isotope is transferred to sulphide and the heavy isotope accumulates in sulphate, became operational some 2 billion years ago. This resulted in an accumulation of ^{34}S in sea-water and marine evaporites.

Variations in the sulphur isotopic composition of oceanic sulphate evaporites are shown in Fig. 1.8. It appears that the isotopic composition in evaporites of different age is irregular due to the differing rates of sulphate influx with river-water and sulphur exhalation from volcanoes and also because of the temporal variations in bacterial reduction and evaporite deposition.

NOTE ADDED IN PROOF

Since this report was proposed 5 additional genera of sulphate-reducing bacteria have been reported. Among these are organisms capable of completely oxidizing fatty acids from C_1 to C_{18} and some aromatic compounds. (Pfennig, N., Widdel, F. and Trüper, H. G. (1981) The dissimilatory sulphate-reducing bacteria. In: Starr, M. P., Stolp, H., Trüper, H. G., Balows, A., and Schlegel, H. G. (eds), *The Prokaryotes* Vol. 1. Springer-Verlag, Berlin. pp. 926–940.)

REFERENCES

Baldensperger, J. F., and Garcia, J-L. (1975) Reduction of oxidized inorganic nitrogen compounds by a new strain of *Thiobacillus denitrificans. Arch. Microbiol.*, **103**, 31–36.

Chambers, L. A., and Trudinger, P. A. (1978) Microbiological fractionation of stable isotopes: a review and critique. *Geomicrobiol. J.*, **1**, 249–293.

Chambers, L. A., Trudinger, P. A., Smith, J. W., and Burns, M. S. (1975) Sulfur isotope fractionation during sulfate reduction by dissimilatory sulfate-reducing bacteria. *Can. J. Microbiol.*, **21**, 1602–1607.

Eremenko, N. A., and Mekhtieva, V. L. (1961). Role of microorganisms in processes of fractionation of stable sulfur isotopes. *Geokhimiya*, No. 2, 174–180 [in Russian].

Esin, O. A., and Geld, P. V. (1966). *Physical Chemistry of Pyrometallurgical Processes.* Metallurgia, Moscow [in Russian].

Garrels, R. M., and Christ, C. L. (1965) *Solutions, Minerals and Equilibria.* Harper and Row, New York, 320 pp.

Gorlenko, V. M., Dubinina, G. A., and Kuznetsov, S. I. (1977) *Ecology of Aquatic Microorganisms.* Nauka, Moscow, 288 pp. [in Russian].

Grinenko, V. A., and Grinenko, L. N. (1974) *Geochemistry of Sulfur Isotopes.* Nauka, Moscow, 274 pp. [in Russian].

Grinenko, L. N., Grinenko, V. A., Zagryazhskaya, G. D., and Stolyarov, Yu, M. (1969) Isotopic composition of sulfide and sulfate sulfur in pyrite deposits of Levikha in connection with problems of their genesis. *Geologia rudnykh mestorozhd.*, No. 3, 26–39 [in Russian].

Grinenko, V. A., and Thode, H. G. (1970) Sulfur isotope effects in volcanic gas mixtures. *Can. J. Earth Sci.*, **7**, 1402–1409.

Harrison, A. G., and Thode, H. G. (1957) The kinetic isotope effect in the reduction of sulfate. *Trans. Faraday Soc.*, **53**, 1648–1651.

Harrison, A. G., and Thode, H. G. (1958) Mechanism of the bacterial reduction of sulfate from isotope fractionation studies. *Trans. Faraday Soc.*, **54**, 84–92.

Haughton, D. R., Roeder, P. L., and Skinner, B. J. (1974) Solubility of sulfur in mafic magmas. *Econ. Geol.*, **69**, 451–462.

Hulston, I. R. (1964) Variations in the ratios of the four stable sulphur isotopes in meteorites and their relations to chemical and nuclear effects. *Ph. D. Thesis*, McMaster University, Hamilton.

Igumnov, S. A., Grinenko, V. A., and Poner, N. B. (1977) Temperature dependence of $\delta^{34}S$ for H_2S and dissolved sulphates over the range 260–400 °C. *Geokhimiya*, No. 7, 1085–1087 [in Russian].

Jacq, V. A., and Roger, P. A. (1978) Evaluation of the probability of sulfate reduction in paddy soils by means of an in situ microbiological assay. *Cah. O.R.S.T.O.M. Sér Biol.*, **13**, 137–142 [in French].

Jones, G. E., and Starkey, R. L. (1957) Fractionation of stable isotopes of sulfur by microorganisms and their role in deposition of native sulfur. *Appl. Microbiol.*, **5**, 111–118.

Kajiwara, Y., and Krouse, H. R. (1971) Sulfur isotope partitioning in metallic sulfide systems. *Can. J. Earth Sci.*, **8**, 1397–1408.

Kaplan, I. R., and Rafter, T. A. (1958) Fractionation of stable isotopes of sulfur by thiobacilli. *Science* (Wash. DC), **127**, 517.

Kaplan, I. R., and Rittenberg, S. C. (1964) Microbiological fractionation of sulfur isotopes. *J. Gen. Microbiol.*, **34**, 195–212.

Katsura, T., and Nagashima, S. (1974) Solubility of sulfur in some magmas of the atmosphere. *Geochim. Cosmochim. Acta*, **38**, 517–531.

Kemp, A. L. W., and Thode, H. G. (1968) The mechanism of the bacterial reduction of sulphate and sulphite from isotope fractionation studies. *Geochim. Cosmochim. Acta*, **32**, 71–91.

Kilmer, V. J. (1979) Minerals and agriculture. In: Trudinger, P. A. and Swaine, D. J. (eds), *Biogeochemical Cycling of Mineral-Forming Elements*. Elsevier, Amsterdam, pp. 515–558.

Kondratyeva, E. N. (1972) *Photosynthesizing Bacteria and Bacterial Photosynthesis*. Izd. MGU, Moscow [in Russian].

Krouse, H. R. (1979) Sulfur isotope abundances in the environments surrounding the Teepee Greek gas processing plant, Peace River region, *Alberta Final Report*, No. 1. Alberta Environ. Res. Secretariat, 58 pp.

Krouse, H. R., and McCready, R. G. L. (1979) Biogeochemical cycling of sulfur. In: Trudinger, P. A. and Swaine, D. J. (eds), *Biogeochemical Cycling of Mineral-Forming Elements*. Elsevier, Amsterdam, pp. 401–431.

Krouse, H. R., and Sasaki, A. (1968) Sulfur and carbon isotopic fractionation by *Salmonella heidelberg* during anaerobic sulfite ion reduction in trypticase soy broth medium. *Can. J. Microbiol.*, **14**, 417–422.

Kuznetsova, S. Ya., and Krigman, L. D. (1978) Solubility of salts in silicate melts of natural magmas. *Geokhimiya*, No. 2, 238–247 [in Russian]

McCready, R. G. L. (1975) Sulfur isotope fractionation by *Desulfovibrio* and *Desulfotomaculum* species. *Geochim. Cosmochim. Acta*, **39**, 1395–1401.

McCready, R. G. L., Kaplan, I. R., and Din, G. A. (1974) Fractionation of sulfur isotopes by the yeast *Saccharomyces cerevisiae*. *Geochim. Cosmochim. Acta*, **38**, 1239–1253.

Matsuo, S. (1960) On the origin of volcanic gases. *J. Earth Sci. Nagoya Univ.*, **8**, 222–245.

Mekhtieva, V. L. (1964) Isotope fractionation of sulfur during bacterial oxidation with *Thiobacillus denitrificans*. *Geokhimya*, No. 1, 61–64 [in Russian].

Mekhtieva, V. L., and Pankina, R. G. (1968) Isotopic composition of sulfur in the aquatic flora and sulfates of water bodies. *Geokhimiya*, No. 6, 739–742 [in Russian].

Monster, J., Anders, E., and Thode, H. G. (1965) $^{34}S/^{32}S$ ratios for the different forms of sulphur in the Orgueil meteorite and their mode of formation. *Geochim. Cosmochim. Acta*, **29**, 773–779.

Moore, J. G., and Fabbi, B. F. (1971) An estimate of the juvenile sulfur content of basalt. *Contr. Miner. and Petrol*, **33**, 118–127.

Nagashima, S. A., and Katsura, T. (1973) The solubility of sulfur in Na_2O–SiO_2 melts under various oxygen partial pressures. *Bull. Chem. Soc. Japan*, **46**, 3099–3103.

Nakai, N., and Jensen, M. L. (1964) The kinetic isotope effect in the bacterial reduction and oxidation of sulfur. *Geochim. Cosmochim. Acta*, **28**, 1893–1912.

Pfennig, N. (1975) The phototrophic bacteria and their role in the sulfur cycle. *Plant Soil*, **43**, 1–16.

Pfennig, N. (1977) Phototrophic green and purple bacteria: a comparative, systematic survey. *Annu. Rev. Microbiol.*, **31**, 275–290.

Ralph, B. J. (1979) Oxidation reactions in the sulphur cycle. In: Trudinger, P. A. and Swaine, D. J. (eds), *Biogeochemical Cycling of Mineral-forming Elements*. Elsevier, Amsterdam, pp. 369–400.

Rees, C. E. (1973) A steady state model for sulphur isotope fractionation in bacterial reduction processes. *Geochim. Cosmochim. Acta*, **37**, 1141–1162.

Robinson, B. W. (1973) Sulphur isotope equilibrium during sulphur hydrolysis at high temperatures. *Earth Planet. Sci Lett.*, **18**, 443–450.

Roy, A. B., and Trudinger, P. A. (1970) *The Biochemistry of Inorganic Compounds of Sulphur*. Cambridge University Press, London, 400 pp.

Rye, R. O., and Czamanske, G. K. (1969) Experimental determination of sphalerite–galena sulfur isotope fractionation and application to the ores at Providencia, Mexico. *Am. Geol. Soc. Annual Meeting. Abs.*, 195–196.

Sakai, H. (1968) Isotopic properties of sulfur compounds in hydrothermal processes. *Geochem. J.*, **2**, 29–49.

Segal, W., and Starkey, R. L. (1969) Microbial decomposition of methionine and identity of the resulting sulfur compounds. *J. Bacteriol.*, **98**, 908–913.

Smejkal, V., Cook, F. D., and Krouse, H. R. (1971) Studies of sulfur and carbon isotope fractionation with microorganisms isolated from springs of western Canada. *Geochim. Cosmochim. Acta*, **35**, 787–800.

Sokolov, V. A. (1966) *Geochemistry of Gases of the Earth's Crust and the Atmosphere*. Nauka, Moscow [in Russian].

Strohl, W. R., and Larkin, J. M. (1978) Enumeration, isolation and characterization of *Beggiatoa* from freshwater sediments. *Appl. Environ. Microbiol.*, **36**, 755–770.

Temple, K. L., and Delchamps, E. W. (1953) Autotrophic bacteria and the formation of acid in bituminous coal mines. *Appl. Microbiol.*, **1**, 255–258.

Thode, H. G., and Monster, J. (1965) Sulfur-isotope geochemistry of petroleum, evaporites and ancient seas. In: Young, A., and Galley, J. E. (eds), *Fluids in Subsurface Environments*, Vol 4. Tulsa, Oklahoma, pp. 367–377.

Thode, H. G., Kleerekoper, H., and McElcheran, D. (1951) Isotope fractionation in the bacterial reduction of sulphate. *Research*, **4**, 581–582.

Thode, H. G., Cragg, C. B., Hulston, J. R., and Rees, C. E. (1971) Sulphur isotope exchange between sulphur dioxide and hydrogen sulphide. *Geochim. Cosmochim. Acta* **35**, 35–45.

Trudinger, P. A. (1979) The biological sulfur cycle. In: Trudinger, P. A. and Swaine, D. J. (eds), *Biogeochemical Cycling of Mineral-Forming Elements*. Elsevier, Amsterdam, pp. 293–313.

Trudinger, P. A., and Chambers, L. A. (1973) Reversibility of bacterial sulfate reduction and its relevance to isotope fractionation. *Geochim. Cosmochim. Acta*, **37**, 1775–1778.

Tudge, A. P., and Thode, H. G. (1950) Thermodynamic properties of isotopic compounds of sulphur. *Can. J. Res.*, **28**, 567–578.

Vinogradov, A. P., and Grinenko, V. A. (1964) Causes of an important dispersion of the sulfur isotope composition in sedimentary pyrites. In: *Chemistry of the Earth's Crust*, Vol. 2. Nauka, Moscow, pp. 581–588 [in Russian].

Widdel, F., and Pfennig, N. (1977) A new anaerobic sporing, acetate oxidising, sulfate-reducing bacterium *Desulfotomaculum* (emend.) *acetoxidans*. *Arch. Microbiol.*, **112**, 119–122.

Young, L., and Maw, G. A. (1958) *The Metabolism of Sulphur Compounds*. Methuen, London, 180 pp.

Zinder, S. H., and Brock, T. D. (1978) Microbial transformation of sulfur in the environment. In: Nriagu, J. O. (ed.), *Sulfur in the Environment* Pt 2. Wiley, Chichester, pp. 445–466.

The Global Biogeochemical Sulphur Cycle
Edited by M. V. Ivanov and J. R. Freney
© 1983 Scientific Committee on Problems of the Environment (SCOPE)

CHAPTER 2
The Sulphur Cycle in the Lithosphere

Part I RESERVOIRS

A. A. MIGDISOV, A. B. RONOV, and V. A. GRINENKO

2.1 INTRODUCTION

Estimation of the amount and isotopic composition of sulphur in sediments and the earth's crust is one of the most challenging problems to be solved before a satisfactory evaluation of the global sulphur balance can be made. The first problem involves the reliability of estimates of the distribution of various rocks in the sedimentary envelope and in the crystalline part of the crust lying on the earth's surface. A second, more complex problem is the determination of the structure of hypogene zones of the crust, and their petrographic and chemical composition (Ronov and Yaroshevsky, 1967, 1976).

A third problem is to obtain reliable estimates of the concentration of sulphur, and its isotopic composition, in the sedimentary envelope and the earth's crust. The geochemistry of sulphur is as yet imperfectly understood, even in the accessible regions of the planet. A reliable evaluation of the amount of sulphur in the lithosphere is made even more difficult by the irregular distribution of sulphur in the rocks forming various tectonic structures of the earth's crust and in rocks of different geological age (Ronov *et al.*, 1974; Grinenko *et al.*, 1973b; Granat *et al.*, 1976).

Granat *et al.* (1976) discuss the problems which affect the evaluation of the size of this sulphur reservoir. They are as follows: (1) analyses of sulphur in rocks are too few to represent the entire earth's crust; (2) because sulphur analyses are difficult to perform published figures show great variation; (3) factors playing a major part in one region may not be important in another.

The irregular distribution of sulphur in rocks (especially in sedimentary and metamorphic envelopes) and the scarcity of analyses are the main reasons for the large discrepancies between estimates of the concentration of sulphur in the earth's crust (Table 2.1).

The literature contains a surprisingly small number of systematic investigations of the forms and isotopic composition of sulphur in rocks of different regions. The sulphur isotopic composition of evaporites of various ages has been extensively investigated by Trofimov (1949), Thode and Monster (1964), and Holser and Kaplan (1966). This pioneering work was followed by numerous other studies (Eremenko and Pankina, 1972; Nielsen, 1973, etc.).

Numerous attempts have been made to obtain systematic data on the concentration of the different forms of sulphur in sediments of modern basins (Chapter 6), although even there the material presented will allow only a rough estimate of the global balance.

Recent work has been directed towards the evaluation of the concentration and isotopic composition of sulphur in metamorphic shields and magmatic rocks (including those in the ocean). However, the same reasons, i.e. the irregular distribution of sulphur in rocks and, in a number of cases, the limited distribution of the regions sampled (Dimroth and Kimberley, 1976; Cameron and Baumann, 1972, Cameron and Jonasson, 1972), cast doubt on the final conclusions.

Therefore, a systematic study of the distribution of sulphur and its isotopic composition, and the determination of their average values in rocks of the sedimentary envelope, metamorphic rocks of shields, basalts of oceanic bottom, etc. are still required to estimate the mass and isotopic sulphur balance of the outer envelopes of the earth. The immediate task is to obtain more reliable estimates of volumes and masses of rocks on the continents and

Table 2.1 Estimates of sulphur concentration in the earth's crust according to different authors

Author	Concentration (%S)
Vernadsky (1934)	0.150
Fersman (1977)	0.100
Goldschmidt (1954)	0.050
Rankama and Sahama (1952)	0.052
Vinogradov (1959)	0.037
Taylor (1964)	0.026
Wedepohl (1973)	0.031
Shaw (1969)	0.030
Ronov and Yaroshevsky (1967)	0.040
Ronov and Yaroshevsky (1976)	0.113
Holser and Kaplan (1966)	0.065
Granat et al. (1976)	0.033

(using the ever-growing information from the deep-sea drilling projects) in the oceans.

Attempts have been made to obtain systematic data on a number of large regions characterizing the basic structural types of the crust; i.e. platforms and geosynclines.

Similar attempts were made to acquire more reliable data on the upper part of the sedimentary envelope of modern seas and oceans and the basic rocks of oceanic ridges and beds. The Russian (East European) and Scythian platforms (Ronov and Migdisov, 1970; Ronov *et al.*, 1974), and Caucasian, Carpathian, and Ural geosynclines (Ronov *et al.*, 1965, 1974), were taken as 'model' regions in continents. Efforts were also undertaken to obtain data on the distribution of the various forms of sulphur and their isotopic composition in rocks of Ukrainian and Baltic Precambrian shields (Ronov and Migdisov, 1970; Ronov *et al.*, 1977; Grinenko and Grinenko, 1974).

Recent data allow us to take another step towards the more complete elucidation of the sulphur distribution in the crust. These data are based on an evaluation of volumes and masses of different layers of the crust within the most important tectonic zones of continents and oceans, and on the relationship between the different types of rocks composing them (Ronov and Yaroshevsky, 1976). Recent data evaluating volumes and masses of the major types of sedimentary formations and volcanogenic rocks of the sedimentary envelope of continents for each period of the Phanerozoic (Ronov *et al.*, 1976, Ronov, 1980) as well as unpublished data on the Upper Proterozoic horizon allow a new approach to the global analysis of elements; this has been demonstrated using carbon as an example (Ronov, 1976). Recent data on the area of magmatic and metamorphic groups of sedimentary rocks of different ages on the surface of modern continents are also very important (Blatt and Jones, 1975). Finally, data were obtained on the masses of volcanic rocks building up oceanic beds during each period of the Mesozoic (beginning from the Jurassic, J_3) and Cenozoic eras (Ronov *et al.*, 1979).

Information on the distribution of sulphur in various rock types has also been increasing. Several years ago (Ronov *et al.*, 1974), data were published on the average concentration of different forms of sulphur in the most important lithological types of sedimentary rocks and volcanites, representing all periods of the Mezo-Cainozoic part of the Caucasian geosyncline, Russian and Scythian platforms, and that on the Upper Proterozoic, Lower Palaeozoic and Middle–Upper Palaeozoic parts of the sedimentary envelope of the Russian platform (Ronov and Migdisov, 1970). New data were obtained on rocks of all systems and parts of the sedimentary cover of this platform, and through platform and geosynclinal paleobasins of different ages (Migdisov *et al.*, 1974; Zagryazhskaya *et al.*, 1973; Girin *et al.*, 1975; Grinenko *et al.*, 1974). The above publications on the sedimentary envelope have one undeniable point in their favour: they present average weighted sulphur contents in rocks of different lithological composition and age, taking into account rock thicknesses (Ronov *et al.*, 1972).

A need to account for an unequal distribution of sulphur in various structural zones of the earth's crust of different geological age required additional data on sedimentary formations and volcanogenic rocks from other regions. Very little data of this kind are available.

Even fewer data are available on the isotopic composition of different forms of sulphur in the sedimentary envelope of continents. Systematic information was obtained only for the East European and Caucasian regions (Ronov *et al.*, 1974; Grinenko *et al.*, 1973a, 1974). Information on the isotopic composition of sulphide sulphur in the sedimentary envelope of the German Democratic Republic is presented by Pilot *et al.* (1973). However, this work lacks data on the concentration of sulphide sulphur in the rocks under consideration. Information on other regions is fragmentary (Thode *et al.*, 1953; Vinogradov *et al.*, 1956). Later, we attempt to estimate the size of this reservoir and its isotopic composition using direct (factual) and indirect data. However, it should be noted that such evaluations are approximate and there is still a need to obtain more reliable data so that the entire sedimentary and other envelopes of the crust can be characterized.

From observations on the distribution of sulphur and its isotopic composition throughout the crustal envelopes we shall attempt to consider more comprehensively the sedimentary envelope which accumulates the so-called excess volatile elements (Rubey, 1951; Vinogradov, 1959; Ronov, 1980) in order to determine the interrelationship between its distribution in the envelope and its efflux from the hypogene source.

2.2   SULPHUR IN THE SEDIMENTARY ENVELOPE OF CONTINENTS

2.2.1   Sulphur Reservoirs in the Sedimentary Envelope of Platforms

A.   Sulphur distribution in the major types of sedimentary rocks of platforms

Sulphur redistribution and chemical modification occurred, at least during the Phanerozoic, in the dynamic system involving continental rocks, the ocean, and the sediments. The sedimentary processes of differentiation of matter occur most actively in tectonically quiet platforms zones of the crust. These processes are responsible for the heterogeneous distribution of sulphur, in its various forms, in rocks of the sedimentary envelope of the platform.

Up to the present, a systematic study has been made of the distribution of sulphur and its isotopic composition in the sedimentary envelope of two platform regions: (1) the East European (Russian) platform with the ancient

(Precambrian), folded basement and (2) the Scythian plate, characterizing the young platform zones (Ronov *et al.*, 1974; Ronov and Migdisov, 1970; Grinenko *et al.*, 1973b). Estimates of the average sulphur concentration and its isotopic composition in various genetic types of rocks and sediments of different geological ages, representing all stages of the history of the sedimentary envelope of these regions, are reasonably reliable. They are based on numerous analyses of composite and complex samples (Ronov *et al.*, 1972) —about 30,000 from the Russian platform and over 2600 from the Scythian plate taken from hundreds of holes evenly distributed over these regions. The distribution of sulphur in platform sediments is based on these data, but also involves information on other less studied regions.

The concentration of sulphur in its different forms varies in the different lithological types of platform rocks. Reduced sulphur is highest in clayey rocks, lower in siltstones and sandstones, and is usually minimal in carbonate rocks. Pyrite predominates as the reduced form of sulphur; organically bound forms account for no more than 10% of the value of the pyrite sulphur, and elemental sulphur is usually less than 5%. The monosulphide form is usually absent or is found only in trace quantities (0.001–0.003%). The sulphate sulphur concentration varies greatly; it may vary from a few per cent of the total sulphur in humid pelagic formations, to complete predominance in evaporite beds (Table 2.2).

For sulphate and reduced sulphur the conditions of sediment formation are the most important factors determining their distribution in the sedimentary envelope. Thus, the results of analyses of sulphur in sedimentary rocks formed under different climatic conditions (Table 2.2) clearly indicate the different levels of accumulation of sulphate and reduced sulphur. Sediments formed under arid conditions, represented by complexes with occurrences of evaporite and red-bed formations, are characterized by having only half of the reduced sulphur which accumulates in sediments of the humid zones. At the same time, the concentration of sulphate sulphur and the ratio between sulphate sulphur and reduced sulphur are substantially higher in the rocks formed in arid zones. The basic reasons for these differences are the lower concentration of organic matter in arid rocks, their greater oxidation, reflected by the Fe_2O_3/FeO ratio, and the high salinities of the basins (Table 2.2, Fig. 2.1). The same factors also affected the distribution of the different forms of sulphur in sediments of different facies complexes. Variations in amounts of pyrite sulphur generally correspond to changes in organic carbon concentrations and are inversely proportional to the Fe_2O_3/FeO ratio except for continental formations where, due to the low salinity of water and significant oxidation of rocks, accumulation of organic matter is not accompanied by an adequate enrichment of rocks with reduced sulphur. The high organic carbon concentrations and minimal values of Fe_2O_3/FeO ratios correspond to the maximum concentrations of pyrite and organic sulphur in

Table 2.2 Concentration and isotopic composition of various forms of sulphur in facies complexes of sedimentary rocks of the Russian platform

Climatic conditions	Facies complex	Volume of rocks (10^3 km³)	No. of assays	No. of samples	Clayey rocks[c]				Organic carbon (%)	$\dfrac{Fe_2O_3{}^a}{FeO}$
					Sulphate	Organic sulphur	Pyrite	Total sulphur		
Arid	Continental	418	46	1173	$\dfrac{0.34^b}{+10.3}$	0.015	$\dfrac{0.06}{-8.1}$	$\dfrac{0.41}{+7.0}$	0.20	4.18
	Saline lagoons	36	48	931	$\dfrac{0.41}{+20.0}$	0.010	$\dfrac{0.27}{-7.5}$	$\dfrac{0.69}{+8.8}$	0.25	1.69
Sea	Coastal	351	65	1226	$\dfrac{0.18(0.07)}{+1.4(+20)}$	$\dfrac{0.026}{+2.3}$	$\dfrac{0.41(0.51)}{-10.0}$	$\dfrac{0.62}{-6.2}$	0.83	0.84
	Pelagic	136	13	923	$\dfrac{0.05(0.03)}{+7.2(+20)}$	0.011	$\dfrac{0.25(0.27)}{-15.4}$	$\dfrac{0.31}{-11.8}$	0.24	1.26
	Average	941	172	4253	$\dfrac{0.24(0.20)}{+8.4(+12.6)}$	$\dfrac{0.018}{+2.3}$	$\dfrac{0.23(0.27)}{-10.4}$	$\dfrac{0.49}{-0.7}$	0.44	1.72

Humid	Continental	221	159	1343	$\dfrac{0.23}{-12.0}$	$\dfrac{0.030}{-2.9}$	$\dfrac{0.30(0.53)}{-7.1}$	$\dfrac{0.56}{-8.9}$	1.68
									0.94
	Sea								
	Coastal	593	257	3256	$\dfrac{0.19(0.01)}{-12.8(+20)}$	0.033	$\dfrac{0.61(0.79)}{-15.5}$	$\dfrac{0.83}{-14.9}$	0.96
									0.75
	Pelagic	230	71	1384	$\dfrac{0.11.}{-21.3}$	$\dfrac{0.044}{+3.7}$	$\dfrac{0.40(0.51)}{-21.6}$	$\dfrac{0.55}{-19.7}$	0.94
									0.85
	Average	1044	487	5983	$\dfrac{0.18(0.06)}{-13.8(+20)}$	$\dfrac{0.035}{0}$	$\dfrac{0.50(0.68)}{-15.5}$	$\dfrac{0.72}{-14.3}$	1.11
									0.81

Sandstones and siltstones[c]

Climatic conditions	Facies complexes	Volume of rocks (10^3 km^3)	No. of samples	No. of specimens	Sulphate	Pyrite	Total sulphur	Organic carbon (%)	$\dfrac{Fe_2O_3}{FeO}$
Arid	Continental	494	54	1104	0.37	0.13	0.50	0.12	2.65
	Saline lagoons	56	38	440	0.44	0.13	0.57	0.16	2.37
	Sea								
	Coastal	153	77	1591	0.13	0.23	0.36	0.20	1.09
	Pelagic	59	10	171	0.06	0.14	0.20	0.16	1.00
	Average	762	179	3306	0.30	0.15	0.45	0.14	1.97
Humid	Continental	400	102	1289	0.12	0.19	0.31	0.46	1.40
	Sea								
	Coastal	385	229	2367	0.12	0.33	0.45	0.33	0.87
	Pelagic	150	57	711	0.07	0.89	0.96	0.38	0.73
	Average	935	388	4367	0.11	0.36	0.47	0.39	1.03

Table 2.2 (continued)

Climatic conditions	Facies complexes	Volume of rocks (10³ km³)	No. of assays	No. of samples	Carbonate rocks[c] Sulphate	Pyrite	Total sulphur	Organic carbon (%)	Fe₂O₃/FeO
Arid	Continental	30	11	1120	0.08 / +12.9	0.06 / −7.3	0.14 / +4.3	0.15	1.21
	Saline lagoons	171	43	1908	3.80 / +19.1	0.10 / −11.5	3.90 / +18.3	0.14	0.56
	Sea Coastal	1158	83	2879	0.54 / +17.4	0.25 / −13.8	0.79 / +7.5	0.23	0.46
	Pelagic	450	79	4149	1.04 / +20.2	0.10 / −16.4	1.14 / +17.0	0.25	0.39
	Average	1809	216	10056	0.96 / +18.8	0.20 / −14.0	1.16 / +13.1	0.23	0.47
Humid	Continental				—	—	—	—	—
	Sea Coastal	335	52	432	0.09(0.04) / +1.8(+20)	0.37(0.42) / −16.0	0.46 / −12.5	0.37	0.30
	Pelagic	130	16	717	0.04(0.02) / +3.8(+20)	0.29(0.31) / −10.8	0.33 / −9.0	0.57	0.41
	Average	465	68	1149	0.08(0.03) / +2.1(+20)	0.35(0.40) / −14.8	0.43 / −11.7	0.43	0.33

[a] Fe_2O_3/FeO ratio with respect to pyrite iron.

[b] Numerator denotes sulphur concentration (%S w/w), denominator $\delta^{34}S$(‰). Values in parentheses are primary sulphide (prior to oxidation) and sulphate calculated from the isotopic composition.

[c] The sulphur isotopic composition was determined on a sample representative of the average volume-weighted composition of sedimentary rocks of the Russian platform.

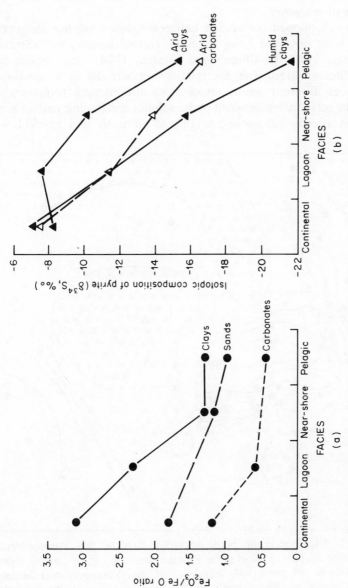

Fig. 2.1 Variations in Fe_2O_3/FeO ratios (a), and isotopic composition of pyrite (b) in sediments of arid and humid basins

rocks of near-shore marine sediments of both humid and arid zones. In carbonate sediments of arid zones, where the admixture of sulphates deposited due to salinization of basins is substantial, their concentrations are proportional to the accumulation of magnesium, chloride and other components of sea-water.

The above-mentioned correlation between reduced sulphur and organic matter has been observed many times by various investigators (Strakhov, 1960; Berner, 1970; Goldhaber and Kaplan, 1974, etc.), especially in modern sediments. Here also, the relationship between these components changes under different conditions of more ancient rock formation. The decline in the relationship between pyrite sulphur and organic carbon is most clear cut in continental sediments formed in fresh and brackish-water

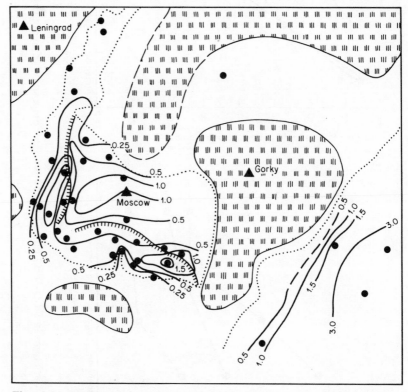

Fig. 2.2 Distribution of pyrite (%S) in clays of the Yasnopolyanski (Carboniferous) palaeobasin of the Russian platform. ⊞ , source region; ..., contour of preserved sediments of the Yasnopolyanski horizon; ⊔⊔⊔ line of maximum development of marine sediment; ●, drill-holes; 3.0——3.0 etc., isolines of pyrite concentration (%S)

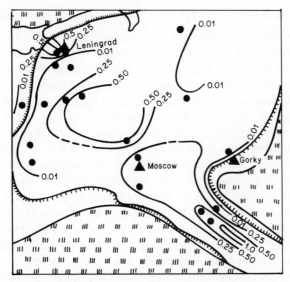

Fig. 2.3   Distribution of pyrite sulphur (%) in clays
of the Upper Proterozoic palaeobasin on the Russian
platform (designations are as given in Fig. 2.2)

palaeobasins. Continental sediments not only have the lowest S/C ratio but
their concentrations of reduced sulphur do not increase further at high levels
of organic carbon (Strakhov, 1960). A similar effect was observed in marine
sediments formed in highly reducing conditions. For example, the
relationships between pyrite sulphur and organic carbon in Ordovician schists
of the Russian platform and Cambrian black schists of Sweden (Bjorlike,
1974) are similar to that of the Black Sea sediments (Strakhov, 1960; Volkov,
1961). The deficiency of sulphate in water of peripheral zones of palaeobasins
is an important factor in continental conditions, and in both continental and
marine situations the lack of reactive iron may result in the lowering or even
prevention of pyrite formation. Recognition that this process has occurred
widely in the epicontinental basins of sedimentation during the geological
past has come from studies of the clayey sediments of the Carboniferous
(Yasnopolyanski horizon) and Upper Proterozoic (the Valdai series) periods
formed on the area of the Moscow syncline of the Russian platform (Figs. 2.2
and 2.3).

It should also be noted that rocks containing small quantitites of sulphur
and organic carbon are characterized by increased ratios of pyrite sulphur:
organic carbon. High values of this ratio are also characteristic of the clayey
formations of the Upper Proterozoic age.

B. Isotopic composition of sulphur in various genetic types of rocks

Pyritic sulphur distributed in various facies of clayey and carbonate rocks of the Russian platform tends to become lighter in the progression from continental to pelagic sediments regardless of their lithological composition or the climatic conditions of their formation (Table 2.2). The change in isotopic composition correlates well with the change in conditions of rock formation from oxidative to reducing which is reflected in the variations in the Fe_2O_3/FeO ratios (Fig. 2.1a,b). At the same time, this change corresponds to a gradual transition from sediments of freshwater basins to hyposaline and normal marine ones. The change in the isotopic composition of reduced sulphur with zones is shown on the isotopic palaeogeochemical map (Figs. 2.4 and 2.5) of the ancient basins of sedimentation (Migdisov *et al.*, 1974; unpublished data). This map and the plot (Fig. 2.6) of variations in $\delta^{34}S$ values with changes in facies conditions of sedimentation determined by independent (lithological, palaeogeographical, and palaeoecological) methods demonstrate an approximate correspondence of the isotopic composition of reduced sulphur of sedimentary rocks to the conditions of their formation.

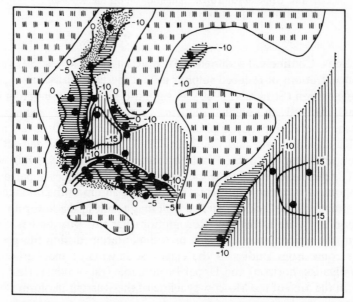

Fig. 2.4 Isotopic composition of pyrite sulphur in clays of the Yasnopolyanski palaeobasin.| ⬚ , Source region; . . . , contour of preserved sediments;   line of maximum development of marine sediments; ●, drill-holes; −10——−10 etc., isolines of $\delta^{34}S$ (⁰/₀₀) values;   , continental multicoloured sediments;   marine sediments;   continental and lagoon coal-bearing sediments

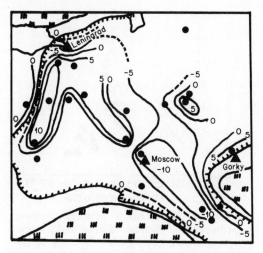

Fig. 2.5 Isotopic composition of pyrite sulphur ($\delta^{34}S$, $^0/_{00}$) in clays of the Upper Proterozoic palaeobasin. [⠿], Source region; ⠿⠿ , line of maximum development of marine sediments -5 —— -5 etc., isolines of $\delta^{34}S$ ($^0/_{00}$) values; ●, drill-holes

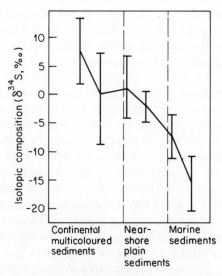

Fig. 2.6 Changes in isotopic composition of pyrite throughout the facies profile of the Carboniferous palaeobasin

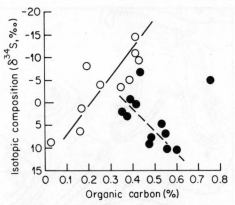

Fig. 2.7 Relationship between isotopic composition of pyrite sulphur and concentration of organic carbon in clays of the Upper Proterozoic palaeobasin. O, Marine sediments; ● sediments of brackish-water zones and continental sediments

Different relationships between $\delta^{34}S$ values of pyrite and organic carbon in rocks of different facies[*] also exist. Clays deposited under usual marine sediments are characterized by an increased content of ^{32}S pyrite as the quantity of carbon increases. However, in sediments formed under the conditions which existed in littoral plains and lacustrine environments, an increase in organic carbon has resulted in the reverse effect. Similar relationships are observed for continental and hyposaline versus marine clays in the other, more ancient (Upper Proterozoic), palaeobasin of the Russian platform (Fig. 2.7).

The isotopic zonation of reduced sulphur in the basin, and its dependence on the concentration of organic matter in sediments can be seen by comparing the above data with those obtained for surface sediments in modern basins of sedimentation. The possibility of direct estimation of such factors as salinity and sulphate content of water, and the redox potential of sediments in modern basins, allows for an understanding of the regularities of the isotopic composition in sediments of ancient basins. Surface sediments of two epicontinental basins were studied; the Baltic Sea basin and the Azov–Black Sea basin, including river, lake, marsh, and lagoon sediments within the range of the surrounding continental environment (Cherkovsky *et al.*, 1978).

The schematic map for the Azov–Black Sea basin (Fig. 2.8) and that for the Baltic Sea basin (Fig. 2.9) show the similarity in the distribution of $\delta^{34}S$ values of pyrite from sediments of modern and ancient sedimentary basins. Figures

---

[*]Facies—a term applied to the general appearance or nature of one part of a rock body as contrasted with other parts.

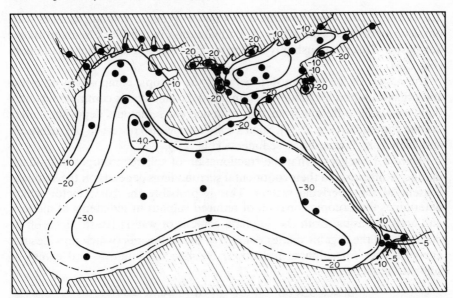

Fig. 2.8 Distribution of δ^{34}S ($^0/_{00}$) in pyrite from sediments of the Azov and Black Seas (from Migdisov *et al.*, 1974). .-.-., Hydrogen sulphide zone; -10——-10 etc., isolines of δ^{34}S ($^0/_{00}$) values; ●, sampling location; ▨ , land

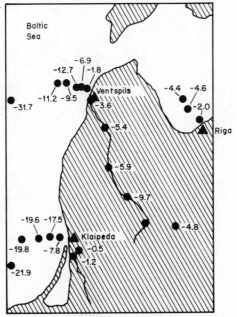

Fig. 2.9 Distribution of δ^{34}S ($^0/_{00}$) in pyrite from surface sediments of the eastern part of the Baltic Sea and neighbouring water-bodies

2.10A and 2.10B show how the $\delta^{34}S$ value of reduced sulphur changes with physicogeographical conditions of sedimentation in the Black Sea and Baltic Sea basins. This relationship is similar to that observed for ancient sediments (Fig. 2.6).

The nature of the relationship between the isotopic composition of pyrite and organic carbon for modern sediments deposited under continental conditions was quite different from that obtained for sediments deposited under marine conditions (Fig. 2.11). Such a contrast was also observed in Palaeozoic and even Late Precambrian sediments (Fig. 2.7).

Figure 2.12 suggests that the fractionation of sulphur isotopes in modern marine sediments and their continental surroundings depends on the sulphate content of the overlying waters. This is probably the basic factor which determines the isotopic zonation of reduced sulphur in sediments of modern and ancient basins. Both the sulphate content of waters (fresh, brackish or marine) and the organic matter content of the sediments (which provides the energy source for sulphate reduction) regulate the intensity and extent of sulphate reduction.

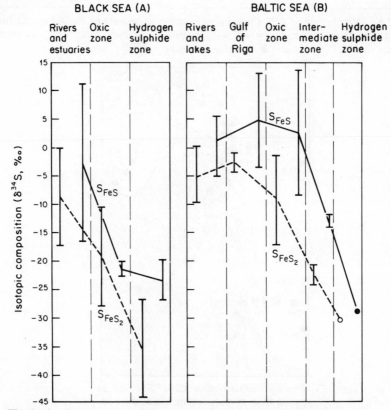

Fig. 2.10   Changes in isotopic composition of monosulphide and pyrite sulphur throughout facies profiles of Black and Baltic Sea sediments

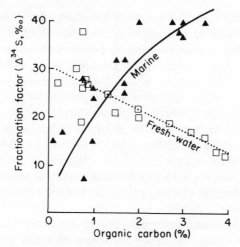

Fig. 2.11 Effect of organic carbon on the fractionation of sulphur isotopes in freshwater and brackish sediments of the Azov–Black Sea and Baltic Sea basins ($\Delta^{34}S = \delta^{34}S_{SO_4^{2-}} - \delta^{34}S_{FeS_2}$)

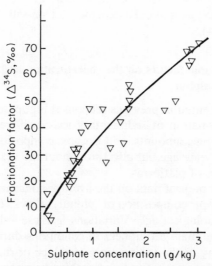

Fig. 2.12 Effect of sulphate concentration of overlying waters on the fractionation of isotopes in modern marine sediments (the Azov, Black Sea, and Baltic Sea basins, shelf and continental slope of the south-east Pacific Ocean) ($\Delta^{34}S = \delta^{34}S_{SO_4^{2-}} - \delta^{34}S_{FeS_2}$)

The nature of the bacterial sulphate reduction system (i.e. whether it is open or closed with respect to sulphate) and the extent of its reduction will determine the isotopic zonation, and its correspondence with facies (Vinogradov *et al.*, 1962; Kaplan *et al.*, 1963; Grinenko *et al.*, 1973a; Grinenko and Grinenko, 1974; Migdisov *et al.*, 1974). Evidently, under fresh and brackish-water conditions, sulphate and its reduction are at a minimum resulting in the maximum enrichment of pyrite with the heavy sulphur isotope. Increases in organic matter intensify sulphate reduction so that sulphate is further depleted and the reduction products become increasingly enriched in ^{34}S. The other extreme occurs in marine sediments where reduction occurs near the sediment surface with an unlimited supply of sulphate. An increase in the amount of organic matter in this case does not create a sulphate deficiency (Migdisov *et al.*, 1980).

The above factors which determine the extent of the fractionation of sulphur isotopes also dictate the facies conditions of sedimentation; i.e. a close relationship exists between facies conditions and isotopic composition of reduced sulphur in sedimentary rocks.

It is worth noting that the relationships considered above are characteristic of sediments of the entire sedimentary envelope of platforms, irrespective of when they were formed, at least during the last 10^9 years. The only exception is the increase in the value for the fractionation factor of sulphur isotopes in sea sediments ($\Delta^{34}S = \delta^{34}S_{SO_4^{2-}} - \delta^{34}S_{pyr}$) from 25–30‰ in the Upper Proterozoic to 30–35‰ in the Palaeozoic and 45–50‰ in modern seas Grinenko *et al.*, 1974).

C.   Influence of tectonic factors on the concentration and isotopic composition of sulphur

From the data presented in previous sections it is apparent that changes in the conditions of formation of sedimentary rocks will result in an uneven distribution of the forms, amounts, and isotopic composition of sulphur, not only in layers of the same age but also throughout the vertical section of the sedimentary envelope of platforms.

Tables 2.3 and 2.4 present data on the average concentrations of forms of sulphur, and the isotopic composition of sulphur, in rocks of the Russian platform according to geological age. Variations in these values are significant and should reflect the main differences in conditions during rock formation throughout the history of the sedimentary envelope of the given platform.

Investigation of the quantitative relationships between rocks and palaeogeographical conditions during their accumulation on the Russian platform, and on the interrelation between sedimentation, palaeogeographic development, and tectonic movements (Ronov *et al.*, 1969), showed that the location of areas of elevation and submersion changed with time from the Cambrian to

Table 2.3 Concentration and isotopic composition ($\delta^{34}S$, ‰) of pyrite and sulphate sulphur in clayey, carbonate, and sulphate rocks of different age from the Russian platform

	Rock							
	Clayey			Carbonate			Sulphate	
		Pyrite			Sulphate			Sulphate
Period	Volume (10³ km³)	(%S)	($\delta^{34}S$)	Volume (10³ km³)	(%S)	($\delta^{34}S$, ‰)	Volume (10³ km³)	($\delta^{34}S$, ‰)
Riphean and Vendean (PR₃)	498.2	0.300	+0.1	48.2	0.05	+19.0	—	—
Cambrian [ε]	97.2.	0.220	+1.8	—	—	—	—	—
Ordovician (O)	53.4	0.720	−16.4	34.9	0.02	—	—	—
Silurian (S)	124.6	0.430	−21.5	68.8	0.30	—	—	—
Lower Palaeozoic (Pz₁)	275.2	0.412	−15.4	103.7	0.21	+18.4	9.5	+17.6
Middle Devonian (D₂)	124.2	0.580	−12.2	78.1	1.05	+24.2	33.6	+23.3
Upper Devonian (D₃)	364.8	0.396	−3.8	524.0	1.03	+8.6	—	—
Lower Carboniferous (C₁)	117.3	1.330	−9.5	348.6	0.14	—	—	—
Middle Carboniferous (C₂)	98.2	0.260	−11.0	381.8	0.39	—	—	—
Upper Carboniferous (C₃)	—	—	—	233.6	1.93	+10.8	3.4	+10.3
Carboniferous (C)	205.5	0.871	−9.7	964.0	0.67	+10.6	3.4	+10.3
Lower Permian (P₁)	10.6	0.070	−6.5	199.0	5.64	+11.5	60.6	+11.3
Upper Permian (P₂)	164.3	0.235	−7.8	101.6	0.38	+7.4	26.0	+11.9
Permian (P)	174.9	0.225	−7.8	300.6	3.86	+11.4	86.6	+11.5
Middle and Upper Palaeozoic (Pz₂₋₃)	869.4	0.500	−7.9	1866.7	1.30	+14.3	133.1	+14.8
Triassic (T)	49.7	0.013	−14.9	—	—	—	5.0	+22.1
Lower Jurassic (J₁)	19.0	0.290	−11.9	—	—	—	—	—
Middle Jurassic (J₂)	46.1	0.336	−12.9	0.5	0.09	—	0.5	+15.1
Upper Jurassic (J₃)	43.1	0.579	−12.8	26.6	0.08	—	0.5	+15.1
Jurassic (J)	108.2	0.425	−19.3	27.1	0.08	—	—	—
Lower Cretaceous (K₁)	55.0	0.381	−26.7	180.9	0.04	+4.9	—	—
Upper Cretaceous (K₂)	35.5	0.345	−16.7	208.8	0.05	+4.9	5.5	+21.5
Mesozoic (Mz)	284.4	0.321	−11.6	9.4	0.10	—	—	—
Palaeozoic (Pz)	27.4	0.496	−8.2	12.3	0.08	—	—	—
Neogene (N)	35.2	0.287	−10.1	21.7	0.09	+13.5	0.2	+20.7
Tertiary system (Tr)	62.6	0.397	−10.1	—	—	+13.5	0.2	+20.7
Quarternary system (Q)	13.3	0.024	−10.1	—	—	—	—	—
Cainozoic (Kz)	75.9	0.317	−10.1	21.7	0.09	+13.5	0.2	+20.7

Table 2.4 Distribution of sulphur in sedimentary rocks of the Russian platform as a function of geological age

Stratigraphic interval	Ratio of clay : sand : carbonate : sulphate (% by volume)	Volume (10³ km³)	No. of assays	No. of samples in assays	Pyrite (%S)	Sulphate (%S)	Total sulphur (%S)	Organic carbon (%C)	Sulphate S / Pyrite S
Riphean and Vendean (R + V)	43.7 : 52.1 : 4.2 : 0	1141.3	67	2571	0.23	0.08	0.31	0.19	0.26
Cambrian [Є]	52.7 : 47.3 : 0 : 0	184.6	40	947	0.17	0.07	0.24	0.18	0.41
Ordovician (O)	53.2 : 12.0 : 34.8 : 0	100.3	31	874	0.52	0.14	0.66	2.41	0.27
Silurian (S)	61.4 : 4.7 : 33.9 : 0	202.9	12	915	0.44	0.16	0.60	0.20	0.36
Lower Palaeozoic (Pz₁)	56.4 : 22.3 : 21.3 : 0	487.8	83	2736	0.36	0.13	0.49	0.65	0.36
Middle Devonian (D₂)	35.5 : 39.4 : 22.3 : 2.8	349.7	214	3240	0.34	0.93	1.27	0.30	2.74
Frasnian stage (D₃fr)	35.2 : 26.4 : 36.1 : 2.3	762.2	246	6012	0.30	0.55	0.85	0.40	1.83
Famennian stage (D₃fm)	24.8 : 7.1 : 64.0 : 4.1	388.8	59	2631	0.17	2.72	2.89	0.25	16.00
Upper Devonian (D₃)	31.7 : 19.9 : 45.5 : 2.9	1151.0	305	8643	0.26	1.28	1.54	0.35	4.92
Devonian (D)	32.6 : 24.4 : 40.1 : 2.9	1500.7	519	11883	0.28	1.20	1.48	0.34	4.29
Lower Carboniferous (C₁)	21.7 : 13.8 : 64.5 : 0	540.6	52	1207	0.50	0.17	0.67	0.61	0.34
Middle Carboniferous (C₂)	16.9 : 10.1 : 73.0 : 0	522.8	62	1591	0.22	0.31	0.53	0.18	1.41
Upper Carboniferous (C₃)	0 : 0 : 98.6 : 1.4	237.0	15	526	0.18	2.23	2.41	0.07	12.39
Carboniferous (C)	15.8 : 9.8 : 74.1 : 0.3	1300.4	129	3324	0.33	0.61	0.94	0.34	1.85
Lower Permian (P₁)	3.9 : 0 : 73.7 : 22.4	270.2	18	866	traces	9.00	9.00	0.06	8.00
Upper Permian (P₂)	35.8 : 36.3 : 22.2 : 5.7	458.2	56	1458	0.19	1.52	1.71	0.93	
Permian (P)	24.0 : 22.8 : 41.3 : 11.9	728.4	73	2324	0.12	4.29	4.41	0.17	35.75
Middle and Upper Palaeozoic (Pz₂₋₃)	24.6 : 18.7 : 52.9 : 3.8	3529.5	721	17531	0.27	1.62	1.89	0.30	6.00
Palaeozoic (Pz)	28.5 : 19.2 : 49.0 : 3.3	4017.3	804	20267	0.28	1.43	1.71	0.34	5.11

Triassic (T)	44.1:51.5:0:4.4	112.7	16	208	0.06	0.99	1.05	0.09	16.50
Lower Jurassic (J_1)	67.9:32.1:0:0	28.0	14	78	0.34	0.20	0.54	0.97	0.59
Middle Jurassic (J_2)	61.5:37.8:0.7:0	74.9	70	572	0.37	0.08	0.45	0.87	0.22
Upper Jurassic (J_3)	42.2:31.2:26.1:0.5	102.1	111	812	0.48	0.22	0.70	0.91	0.46
Jurassic (J)	52.8:33.8:13.2:0.2	205.0	195	1462	0.42	0.16	0.58	0.90	0.38
Lower Cretaceous (K_1)	56.1:43.9:0:0	98.1	68	658	0.37	0.16	0.53	1.06	0.43
Upper Cretaceous (K_2)	14.3:13.1:72.6:0	249.1	92	815	0.13	0.05	0.18	0.26	0.38
Cretaceous (K)	26.1:21.8:52.1:0	347.2	160	1473	0.20	0.08	0.28	0.48	0.40
Mesozoic (Mz)	37.4:30.5:31.3:0.8	664.9	371	3143	0.24	0.26	0.50	0.65	1.08
Palaeocene (f_1)	15.8:78.7:5.5:0	18.3	19	166	0.28	0.08	0.36	0.50	0.29
Lower and Middle Eocene (f_2^{1-2})	22.6:73.8:3.6:0	30.2	100	614	0.40	0.16	0.56	1.35	0.40
Upper Eocene (f_2^3)	40.4:30.3:39.6:0	24.3	74	467	0.34	0.15	0.49	0.55	0.44
Eocene (f_2)	30.5:54.3:15.2:0	54.5	174	1081	0.37	0.15	0.52	0.99	0.41
Oligocene (f_3)	27.7:72.0:0.3:0	28.6	116	951	0.21	0.23	0.44	0.38	1.10
Palaeogene (f)	27.0:63.7:9.3:0	101.4	309	2198	0.31	0.17	0.48	0.73	0.55
Miocene (N_1)	26.7:37.9:34.8:0.6	32.3	81	416	0.19	0.20	0.39	0.31	1.05
Pliocene (N_2)	60.9:36.6:2.5:0	43.7	18	122	0.16	0.12	0.28	0.49	0.75
Neogene (N)	46.3:37.2:16.2:0.3	75.9	99	538	0.17	0.16	0.33	0.42	0.94
Tertiary system (Tr)	35.4:52.3:12.2:0	177.3	408	2736	0.26	0.16	0.42	0.59	0.62
Quaternary system (Q)	66.5:33.5:0:0	20.0	35	219	0.03	0.04	0.07	0.26	1.35
Cainozoic (Kz)	38.5:50.4:11.0:0.1	197.3	443	2955	0.23	0.15	0.38	0.56	0.65
Mesozoic and Cainozoic (Mz + Kz)	37.6:35.1:26.6:0.7	862.2	814	6098	0.24	0.25	0.49	0.55	1.04
					-19.5^a	$+7.9^a$	-5.5		
Phanerozoic (Ph)	30.1:22.0:45.1:28	48795	1618	26365	0.27	1.21	1.48	0.38	4.48
Total or average (R – Q)	32.7:27.7:37.3:2.3	6020.8	1685	28936	0.26	1.01	1.27	0.34	3.88

^{a}Numerator denotes sulphur concentration (%S); denominator $\delta^{34}S$ ($^0/_{00}$).

the Quaternary. There is a definite relation between these fluctuations corresponding to the recurrence of tectonic cycles, the interchange of transgressions and regressions of seas, which determined the relative proportions of land and sea masses. As the analysis of data on the area and amount of rocks formed under different palaeogeographic conditions showed, their interchange is periodic, and conforms to the same general scheme describing the tectonic history of the platform. It is worthy of note that the interchange of land and sea masses on the Russian platform coincides with their interchange over the entire continental part of the earth (Ronov, 1968). Thus the amount of sulphur in rocks of the Russian platform and its isotopic composition (which are determined by the conditions of sediment formation), are not a distinctive feature of this region but reflect the changing conditions on the earth's surface during its development.

There are certain anomalies relating to the concentration and isotopic composition of the different forms of sulphur. Strakhov (1963) presented evidence for the migration of arid zones during the history of the earth, as the result of which, a substantial part of the Russian platform was located at times in either the arid (Hercynian cycle) or the humid (Alpine cycle) zones. This influence of palaeogeographic conditions showed up initially in the amount of organic matter concentrated in sediments of the basins of salt or freshwater lagoons. As a result, significant amounts of sea-water sulphate or reduced sulphur were deposited under such conditions (Table 2.2). It was because of this fact that earlier estimations of the sulphur balance in the sedimentary envelope (Grinenko *et al.*, 1973b; Ronov and Yaroshevsky, 1976) overestimated the role of sulphate sulphur in carbonate rocks of platforms, since the carbonates of the Russian platform are mainly Hercynian.

Sulphur is irregularly distributed across the stratigraphic section of the Russian platform: for example, 8-fold or 140-fold differences, respectively, may occur in the average concentrations of reduced and sulphate sulphur in different age complexes of clayey rocks. This probably reflects the influence of periodic changes in conditions of sediment formation mentioned above (Table 2.3).

Let us now consider the change in reduced and oxidized forms of sulphur. The change in the concentration of pyrite sulphur (Fig. 2.13) is clearly cyclic, and its maximum concentrations occur in the middle stages of geotectonic cycles. In accordance with the successive changes in palaeogeographic conditions during transgressions and regressions (Ronov *et al.*, 1969), the concentrations of the different forms of reduced sulphur correspond to periods of maximum development of littoral plains, freshened lagoons, and shallow parts of the shelf where organic matter usually accumulates. On the other hand, the epochs of aridity, salting of basins and formation of evaporites, are characterized by low concentrations of pyrite sulphur and organic carbon. Changes in pyrite sulphur and organic carbon across the stratigraphic section

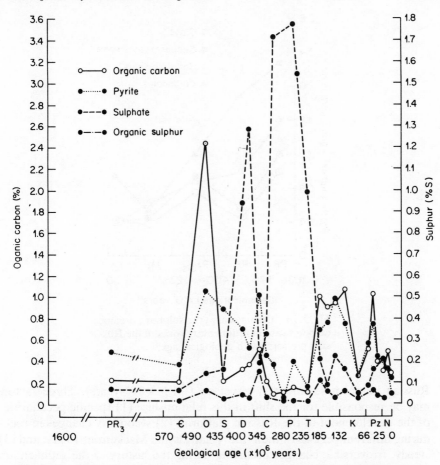

Fig. 2.13   Change in sulphur and organic carbon concentrations in sedimentary rocks of the Russian platform with geological age (designations are given in Table 2.4)

are broadly synchronous, but some variations in this relationship do occur. Holland (1973) studied the relationship between these variations and the oxygen content of the atmosphere during the Phanerozoic period. He analysed the data on the Russian platform (Ronov and Migdisov, 1970) to determine the relationship between sulphur and organic matter at various stages of the Phanerozoic history of the sedimentary envelope. He noted that all of the rocks of the Palaeozoic age are characterized by higher pyrite sulphur : organic carbon ratios than those of Mesozoic and Cainozoic age. Figure 2.14 shows that a change in this relationship with time occurs in all types of rocks during the whole period of development of the

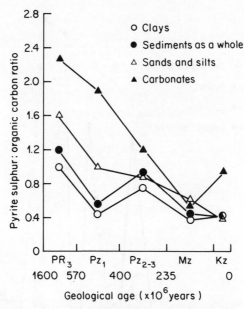

Fig. 2.14   Changes in pyrite sulphur : organic
carbon ratios in sedimentary rocks of the Russian platform with geological age

Russian platform sedimentary envelope ($\sim 1.5 \times 10^9$ years). Three factors
may be responsible for the shift in this relationship: (1) specific peculiarities
of the development of the Russian platform; (2) secondary changes in rocks
during different periods of their 'life' (Garrels and Mackenzie, 1971); and (3)
steady, irreversible changes occurring during the history of the sedimentary
envelope. Comparisons with data on other platforms suggest that it is possible
to eliminate the first of these. The similar trend in this relationship in different
regions is indicative of the common character of this phenomenon in the
entire sedimentary envelope. It may be assumed that differences in organic
matter concentration and composition in ancient and younger sedimentary
terrains (Ronov, 1958; Ronov and Migdisov, 1970; Galimov *et al.*, 1975;
Ronov, 1976) are mainly responsible for the changes in the pyrite sulphur :
organic carbon ratios.

The rate of sedimentation is one of the main manifestations of tectonic
activity in the region of sedimentation. Strakhov (1960) and Berner (1964)
consider that low rates of sedimentation favour the complete reduction of
sulphate and thus the accumulation of reduced sulphur. However, under the
conditions of sedimentation on platforms the facies environment exerts a
more powerful influence on the concentration of pyrite. As a result, several
complexes of rocks can be identified, each of which is characterised by a

certain pyrite-sedimentation rate relationship: (1) rocks containing the largest amounts of pyrite sulphur and corresponding to horizons of coal bearing facies, inflammable schists, and bituminous clays; (2) rocks of a normal marine genesis; and (3) predominantly continental and lagoon strongly oxidized formations. The same dependence is observed for the distribution of organic carbon. A similar relationship between pyrite concentration and sedimentation rate has also been found for clayey formations of other platform regions; this is especially true of the Oslo region (Bjorlike, 1974).

The isotopic composition of pyrite sulphur changes greatly across the stratigraphic section of the platform envelope; the changes being assumed to correspond with changes in sedimentation conditions (Table 2.3, Fig. 2.15). In sediments of different ages the $\delta^{34}S$ values for pyrite varied from $+1.8\%$ to -26.7%. The clear-cut recurrence in the change of these values can be seen

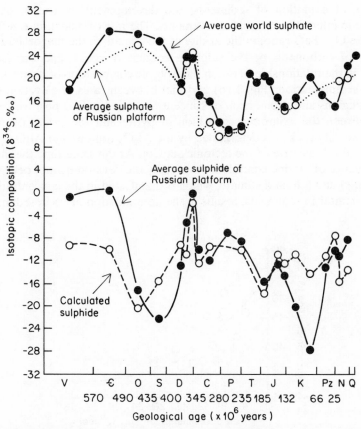

Fig. 2.15 Effect of geological age on the isotopic composition of evaporite sulphur, and of sulphide and sulphate sulphur in rocks of the Russian platform

from this figure, and to a first approximation, it corresponds to tectonic cycles (Migdisov *et al.*, 1972). Pyrite sulphur from sediments of the middle stages of geotectonic cycles is characterized by a concentration of the light sulphur isotope, while the rocks of the initial and later stages of these cycles accumulate the heavier isotope. Maxima in the concentration of the light isotope of pyrite sulphur occur at different times to the maxima in concentrations of reduced sulphur and organic carbon in rocks (Fig. 2.15).

Some interesting conclusions could be drawn from a comparison of the data on the isotopic composition of pyrite sulphur and those on the intensity of tectonic movements, the distribution of various types of rocks and the palaeogeographic conditions of their concentration at different stages in the history of the Russian platform sedimentary envelope (Ronov *et al.*, 1969), and the chemical composition of its sediments. All these data suggest that the concentration of ^{34}S in the reduced sulphur of sedimentary rocks depends on the rate of oxidation of sediments, the development of aridity, and the increase in intensity of tectonic movements. The rate of oxidation is reflected in the Fe_2O_3 : FeO ratio, in the aridity of the climate in the magnesium oxide content of carbonates or the sulphate content of rocks, and the tectonic activity in the relationship between the isotopic composition of pyrite and the rate of sedimentation (Fig. 2.16) and that between rates of elevation of the source regions and submersion of concentration zones. The positive correlation between the isotopic composition of pyrite and the extent of mature sandy and silty rocks (as determined by the Al_2O_3 content and Al_2O_3 : Na_2O ratio) is also indicative of the tectonic activity. At the same time the increasing influence of marine conditions (especially the development of deep-water sediments) and a humid climate, which is manifested in the concentration of organic matter in sediments, results in the accumulation of ^{32}S in sediments.

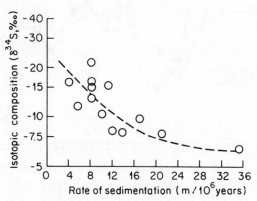

Fig. 2.16   Effect of sedimentation rate on the Russian platform on the isotopic composition of pyrite sulphur

From the above-mentioned relationships an empirical equation can be derived which enables us to estimate the isotopic composition of reduced sulphur from tectonic and palaeogeographic characteristics plus data on the forms of iron and sulphur in rocks:

$$\delta^{34}S_{pyr} = 2.406\delta^{34}S_{SO_4^{2-}} + 1.418\frac{Fe_2O_3}{FeO} - 0.52\frac{V_{subm}}{V_{elev}} - 1.714C_{org}(\%)$$

where V_{subm} = rate of submersion of regions of sediment accumulation, and V_{elev} = rate of elevation of source regions.

Figure 2.15 shows the comparison between calculated and observed values. The main discrepancy between these sets of data lies in the shift of the maximum concentration of ^{32}S in the Lower Palaeozoic and the low accumulation indicated for Cretaceous sediments.

The distribution of sulphate over the Russian platform exhibits a maximum coinciding with the Hercynian period of its development when arid conditions were predominant (Strakhov, 1963). Sulphate sulphur is extremely high (disregarding sulphate rocks) in arid carbonate terrains especially in the Lower Permian (Table 2.3). High concentrations of sulphate are also observed in carbonates of the Middle and Upper Devonian, and the Upper Carboniferous. The relationship between these sulphate concentrations in carbonate sediments and periods of evaporite formation is obvious and is confirmed by the strong correlation between the distribution of sulphate and the distribution of gypsum and anhydrite complexes over the Russian platform. Sulphate typically occurs in all types of rocks formed during periods of dolomite formation. The isotopic compositions of sulphates are different in various types of sediments and at various stratigraphic levels of the platform envelope. Thus, the isotopic composition of sulphate sulphur in carbonates deposited during arid climatic periods is the same as that in evaporites of the same age (Table 2.3). Part of the sulphate in carbonates of humid zones came from the oxidation of reduced sulphur. This is also characteristic of clayey layers where the isotopic composition of sulphate sulphur is similar to that in evaporites formed under extremely dry conditions. Evaporites from this platform are similar to those found in other regions of the world (Thode and Monster, 1964; Holser and Kaplan, 1966; Nielsen and Ricke, 1964; Eremenko and Pankina, 1972; Nielsen 1973, etc.).

It is interesting to compare fluctuations in the isotopic composition of sulphates and changes in the isotopic composition of reduced sulphur of the platform with those of the fractionation factor ($\Delta^{34}S = \delta^{34}S_{SO_4^{2-}} - \delta^{34}S_{pyr}$).

Changes in $\delta^{34}S_{SO_4}$ and $\delta^{34}S_{pyr}$ are not parallel contrary to Nielsen (1973) and Schidlowski *et al.* (1977), therefore, fluctuations in $\Delta^{34}S$ values should occur. Moreover, the curve relating the isotopic composition of sulphides of the Russian platform to geological age is, to a first approximation, the mirror image of the curve for evaporite sulphur. Evidently, the tectonically stable stages of the Caledonian, Hercynian, and Alpine cycles are characterized by a maximum fractionation of the sulphur isotopes. Taking into account that the sulphate curve is similar to the well-known curve of $\delta^{34}S$ variation in the world evaporites (e.g. Holser and Kaplan 1966; Nielsen and Ricke, 1964) such a correlation may take place when changes in the isotopic composition of reduced sulphur reflect the regional specificity of the Russian platform.

D. The distribution of sulphur in sedimentary rocks of platforms. Influence of global movements on changes in sulphur concentration and its isotopic composition

Movements of a general global character which affect the distribution of sulphur in rocks of the Russian platform need special consideration. In an earlier section we discussed the common character of the changes in palaeogeographic conditions on the Russian platform and continents. This common character and the dynamic nature of transgressions and regressions on the Russian platform and continents are illustrated by Fig. 2.17. Further evidence is provided by the parallel changes with time of pyrite and organic

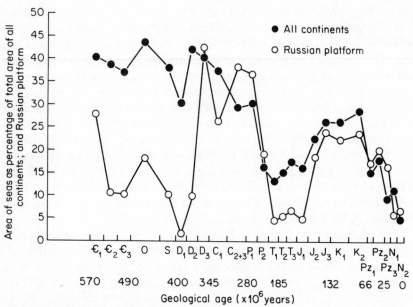

Fig. 2.17   Change in areas of seas on the Russian platform and continents with time (from Ronov, 1976; Ronov *et al.*, 1979)

carbon (Fig. 2.18) incorporated into the envelope of the Russian platform and the changes in organic carbon in the sedimentary envelope as a whole (Ronov, 1976). This is also confirmed by comparing data obtained for other platform regions of the world. Each region may have specific features of its development, but development as a whole should comply with the general picture characteristic of continental platforms. Such systematic data on the distribution of the different forms of sulphur are lacking, except for the information on the Russian platform considered above, and the Scythian plate (Ronov *et al.*, 1974). Unfortunately, the sedimentary envelope of the Scythian plate comprises the Mezosoic and Cainozoic stages of development only. Results of a systematic study in the Oslo region (Bjorlike, 1974) characterize only a small part of the sedimentary envelope, the Lower Palaeozoic. A summary of the available data on the distribution of sulphur and organic carbon in the sediment sections of a number of platforms is presented in Tables 2.5 and 2.6. Not all the analyses used in this table are equally reliable. For instance, the data on the Siberian platform and West Siberian plate consist partially of direct analyses of sulphur and estimations of pyrite iron.

Even the more rigorous data, obtained by reliable analytical methods, for the West European platform (Ricke, 1960) and Turanian plate suffer from

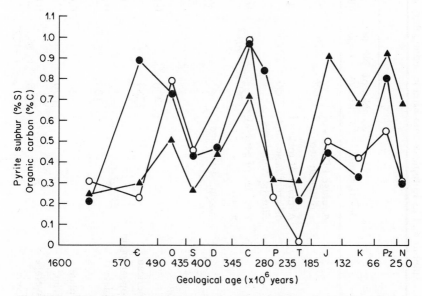

Fig. 2.18 Changes in concentrations of pyrite sulphur and organic carbon in sedimentary rocks with time. ▲———▲, organic carbon in continental rocks (Ronov, 1976); ○———○, pyrite from Russian platform, ●———● pyrite of other platforms

Table 2.5 Sulphur and its isotopic composition in sedimentary and effusive rocks of ancient and young platforms

Platform	Description	No. of assays	No. of samples	Clayey rocks					
				Sulphate	Monosulphide	Organic sulphur	Elemental sulphur	Pyrite	Total sulphur
Russian	Average of rocks of different age	589	9766	0.237		0.032[a] / −2.5	0.012	0.405 / −8.4	0.692
Russian	Average of rocks of different facies and climatic complexes	659	10236	0.210 / −0.4[b]		0.027 / −4.3		0.370 / −14.0	0.607 / −8.9
Russian	Average of general assays	603	9017	0.225 / −1.6	0.001			0.365 / −10.5	0.591 / −7.1
Russian	Accepted average			0.224 / −1.0	0.001	0.030 / −3.4	0.012	0.380 / −11.0	0.647 / −7.2
Oslo region		95							0.585
Siberian		420							0.590
North American		1289							0.148
West European plate				0.073				0.356	0.429
Scythian plate		106	2617	0.144 / −6.6	0.003	0.035 / −6.9	0.010	0.393 / −15.3	0.585 / −12.6
Turanian plate		528		0.020				0.760	0.780
West Siberian		1079							0.470
Total or average		765	12853 +3422[d]	0.095 / −3.8	0.002	0.027 / −5.2	0.009	0.396 / −13.2	0.529 / −11.5

Table 2.5 (continued)

Platform	Description	Sandstones and siltstones					Carbonate rocks				
		No. of assays	No. of samples	Sulphate	Pyrite	Total[c] sulphur	No. of assays	No. of samples	Sulphate	Pyrite	Total[c] sulphur
Russian	Average of rocks of different age	686	7922	0.140	0.200	0.340	368	10926	1.100 / +14.2	0.170	1.270
Russian	Average of rocks of different facies and climatic complexes	567	7673	0.200	0.230	0.430	284	11205	0.780 / +18.44	0.194 / −14.2	0.974 / +11.9
Russian	Average of general assays	616	7488	0.140 / +5.9	0.171 / −6.0	0.311	368	10078	0.790	0.116	0.906
Russian	Accepted average			0.160 / +5.9	0.200 / −6.0	0.391 / −1.1			0.890 / +16.3	0.160 / −14.2	1.072 / +11.1
Oslo region		1				0.022	103				0.368
Siberian		46		0.020	0.480	0.500	803		0.040	0.220	0.260
North American		152				0.096	1359		0.040	0.090	0.130
West European plate							8		0.021	0.032	0.053
Scythian plate		128	3450	0.051	0.109	0.150					
West Siberian		841				0.160					
Total average		814	10372	0.052 / +5.9	0.200 / −6.8	0.252 / −4.8	435	11804	0.214 / +16.3	0.126 / −14.2	0.240 / +5.0

Table 2.5 (continued)

Platform	Description	Gypsum and anhydrites		Salts		Siliceous rocks		Effusives	
		No. of assays	Total sulphur	No. of assays	Total sulphur	No. of assays	Total sulphur	No. of assays	Total sulphur
Russian	Average of rocks of different age	14 from 322 samples	20.320 / +15.1	7	0.390	128	0.200	596	0.060
Russian	Accepted average		20.820 / +15.1		0.390		0.200		0.060
Siberian				1	0.220			52	0.050
North American		47	17.180	3	0.310				
Total or average		61	19.000 / +19.5[e]	11	0.307	128	0.200	648	0.055

[a] Numerator denotes sulphur concentration (%S); denominator, $\delta^{34}S$ ($^0/oo$).
[b] Low $\delta^{34}S$ values in clay rocks, sandstones and siltstones compared with evaporites indicate admixture of secondary sulphur.
[c] Total sulphur includes organic and monosulphide sulphur; 0.026%S in sandstones, 0.021%S and 0.001%S in carbonates, respectively.
[d] Number of separate analyses
[e] From Table 2.17

Table 2.6 Pyrite, total sulphur and organic carbon in rocks of the sedimentary envelope of platforms

Age	Russian platform		North American platform		Siberian platform		Oslo region		Scythian platform		West-Siberian plate		Average		
	Pyrite (%S)	Organic carbon (%C)	Total sulphur (%S)	Organic carbon (%C)	Pyrite (%S)	Organic carbon (%C)	Total sulphur (%S)	Organic carbon (%C)	Pyrite (%S)	Organic carbon (%C)	Pyrite (%S)	Organic carbon (%C)	Pyrite (%S)	Organic carbon (%C)	Pyrite S / Organic C
Upper Proterozoic	0.23	0.19	—		0.08	0.24	1.457	5.0	—	—	—	—	0.15	0.22	0.68
Cambrian	0.17	0.18	—		0.47	0.30	0.496	0.5	—	—	—	—	0.70	1.54	0.45
Ordovician	0.52	2.41	0.07	0.7	0.55	0.26	0.284	0.2	—	—	—	—	0.41	0.97	0.42
Silurian	0.44	0.20	0.02	0.35	0.44	0.20	0.431	0.5	—	—	—	—	0.30	0.24	1.25
Lower Palaeozoic	0.36	0.65	0.05	0.63	0.49	0.28	—	—	—	—	—	—	0.43	0.51	0.84
Devonian	0.28	0.34	0.12	0.6	tr	0.14	—	—	—	—	—	—	0.13	0.36	0.36
Carboniferous	0.33	0.34	0.10	0.8	0.30	0.85	—	—	—	—	—	—	0.24	0.66	0.36
Permian	0.12	0.17	0.11	0.3	0.85	0.88	—	—	—	—	—	—	0.36	0.45	0.80
Middle and Upper Palaeozoic	0.27	0.30	0.11	0.6	0.54	0.79	—	—	—	—	—	—	0.31	0.56	0.55
Triassic	0.06	0.09	0.17	—	0.29	0.47	—	—	—	—	—	—	0.17	0.28	0.61
Jurassic	0.42	0.90	0.17	0.4	—	—	—	—	0.258	0.54	0.55	1.87	0.35	0.93	0.38
Cretaceous	0.20	0.48	0.15	0.9	—	—	—	—	0.223	0.60	0.27	0.76	0.21	0.68	0.31
Mesozoic	0.24	0.55	0.16	0.76	—	—	—	—	0.236	0.58	0.33	1.00	0.24	0.63	0.38
Palaeogene	0.31	0.73	—		—	—	—	—	0.388	0.93	0.35	0.70	0.35	0.79	0.44
Neogene	0.26	0.59	0.37		—	—	—	—	0.380	0.54	—	—	0.34	0.56	0.61
Cainozoic	0.23	0.56	0.37	1.7	—	—	—	—	0.385	0.78	0.35	0.70	0.33	0.94	0.35
Average	0.26	0.34	0.16	0.43	0.37	0.44	0.431	0.5	0.298	0.67	0.33	0.87	0.31	0.54	0.57

the drawbacks that they cover a narrow age interval or a small region (Table 2.5).

However, the data presented in Fig. 2.18 and Table 2.5 on changes in pyrite sulphur in clayey rocks of different platforms correlate well. Other data on concentrations of pyrite and organic matter in Cambrian layers of the Oslo region (1.58% S, 6% C) and the Siberian platform (0.80% S), and the high concentration of pyrite in Triassic clays of the Siberian platform (1.6% S) (average from seven analyses for one region) diverge from the general relationship, suggesting that some data in Fig. 2.18 and Table 2.5 do not adequately represent the platform as a whole. When these aberrant values are taken into account the *average* sulphur concentration for all platforms changes in the same way as the pyrite sulphur and organic carbon concentrations in the Russian platform, and with changes in the concentration and amount of organic carbon in the sedimentary envelope of continents. Therefore, data on the distribution of sulphur in rocks of the Russian platform are roughly characteristic of platforms as a whole.

It is more difficult to check the reliability of changes in the isotopic composition of reduced sulphur in the sedimentary envelope of continental platforms, since few investigations of the distribution of the sulphur isotopes in sedimentary rocks have been made. The literature contains dispersed and sometimes systematic data (Thode *et al.*, 1953; Vinogradov *et al.*, 1956; Pilot *et al.*, 1973, etc). on the isotopic composition of inclusions and concretions of pyrite from sedimentary layers of different ages. They differ from sulphides formed at the earliest stages of diagenesis of sediments by having increased δ^{34}S values. Such values can be as high as those found at later stages due to processes of diagenetic and epigenetic modification of sediments.

These changes in isotopic composition occurring during the lithification of sediments begin at δ^{34}S values corresponding to those of dispersed sulphide. This is why in sulphides with a wide range of δ^{34}S values one observes the 'inheritance' of the isotopic composition of sulphur distributed in the rock (Fig. 2.19), and the dependence of the isotopic composition of concretions on the rate of the sediment formation and its facies (Fig. 2.20; Zagryazhskaya *et al.*, 1973). In this connection, a definite correlation is observed between changes in the isotopic composition of dispersed sulphide and concretions of sedimentary rocks of the same region (Vinogradov *et al.*, 1956).

Analogous data on other regions of the world are cited by many investigators. Of particular importance is the work of Thode *et al.* (1953) which presented carefully selected results of isotopic analyses of sulphur in sedimentary sulphides of various ages. Their conclusion regarding the accumulation of the light sulphur isotope was supported by the studies of rocks of the Russian platform (Grinenko *et al.*, 1973b; Table 2.3). The analyses of Thode *et al.* (1935) characterize sulphides from formations of different age in North

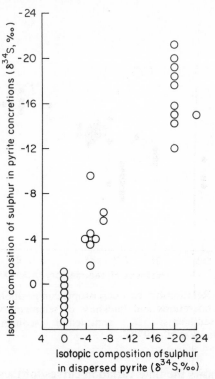

Fig. 2.19 Relationship between the isotopic composition of sulphur in concretions and that in dispersed pyrite of rocks (from Zagryazhskaya *et al.*, 1973)

America and some other regions represented not only by concretions, but also by sulphides dispersed in rocks. These data reflect the isotopic composition of a major part of sulphide sulphur in sedimentary rocks.

Analyses of sulphides from sedimentary rocks cited in numerous publications point to a wide range of $\delta^{34}S$ values. Many of these formations show significant accumulations of ^{34}S, testifying to their formation at relatively late stages of diagenetic and epigenetic processes. This is especially characteristic of Ordovician, Silurian, and Triassic sulphide formations in the German Democratic Republic (GDR) (Pilot *et al.*, 1973), as well as the Permian concretions of the Russians platform (Vinogradov *et al.*, 1956).

We attempted to generalize the available data on various important regions (Europe, North America, Asia) by calculating the average sulphur isotopic composition of sulphide formations for each geologic system. The average value was then calculated from the values for those regions studied (Table

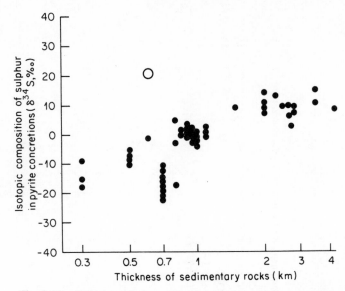

Fig. 2.20   Relationship between istopic composition of sulphur
in pyrite concretions and thickness of sedimentary rocks. ●,
Marine sediments; ○, continental sediments (from
Zagryazhskaya *et al.*, 1973)

2.7). Figure 2.21 shows the time variation curves for these data compared to
the average δ^{34}S value of dispersed pyrite sulphur in the Russian and Scythian
platforms and Caucasian geosyncline.

The curves show that δ^{34}S variations in sulphides as a function of the age of
the rock for all regions are very similar. The only divergence from the cyclic
variations is the accumulation of ^{34}S in the European sulphide formations of
the Lower palaeozoic (mainly due to the data for the GDR territory). After
the Devonian a better correlation of δ^{34}S minima and maxima is observed.
The closest similarity is shown for the Russian platform and North American
region. Their average δ^{34} values are practically identical. Since many of
Thode's North American samples were dispersed sulphides, such a result
appears to be very significant. Despite the limited number of samples from
the Urals, Caucasus, Kazakhstan, Sayany, and Japan, the changes in δ^{34}S
values of Asian sulphides are similar to those of other continents.

This suggests that the variations in δ^{34}S values of sulphate and sulphide in
sedimentary rocks, and the changes with time of the fractionation factor for
the sulphur isotopes are natural and are determined by the peculiarities of the
tectonic development of continents at various stages of their Phanerozoic
history. It is not surprising, therefore, that the amounts and concentrations of
the different forms of sulphur and their isotopic compositions in the Russian
platform rocks are associated and related to the dynamics of the tectonic

Table 2.7 Average isotopic composition of sulphide sulphur inclusions and concretions in sedimentary rocks of different geological age from various regions

Age	Europe	Asia	North America	Average
Upper Proterozoic	$\dfrac{+2.7(1)^a}{2}$		$\dfrac{+4.4(2)}{83}$	$\dfrac{+3.8(3)}{84}$
Cambrian	$\dfrac{-7.5(1)}{2}$	$\dfrac{-8.5(2)}{2}$	$\dfrac{+3.6(2)}{2}$	$\dfrac{-3.5(5)}{6}$
Ordovician	$\dfrac{-3.4(2)}{20}$	—	$\dfrac{-2.2(4)}{43}$	$\dfrac{-2.6(6)}{63}$
Silurian	$\dfrac{+15.5(1)}{29}$	$\dfrac{-21.7(1)}{2}$	$\dfrac{-12.3(2)}{3}$	$\dfrac{-7.7(4)}{34}$
Devonian	$\dfrac{+2.7(3)}{29}$	$\dfrac{+5.1(1)}{5}$	$\dfrac{-5.4(2)}{6}$	$\dfrac{+0.4(6)}{40}$
Carboniferous	$\dfrac{-8.0(2)}{9}$	$\dfrac{-3.6(2)}{11}$	$\dfrac{-9.7(2)}{13}$	$\dfrac{-7.1(6)}{21}$
Permian	$\dfrac{-3.3(4)}{36}$	$\dfrac{-1.7(1)}{9}$	$\dfrac{-6.4(2)}{2}$	$\dfrac{-4.0(7)}{47}$
Triassic	$\dfrac{+18.0(1)}{8}$	—	$\dfrac{-5.8(2)}{13}$	$\dfrac{+2.1(3)}{21}$
Jurassic	$\dfrac{-6.4(4)}{14}$	$\dfrac{-5.1(4)}{97}$	—	$\dfrac{-5.8(8)}{111}$
Cretaceous	$\dfrac{-32.0(3)}{6}$	$\dfrac{-10.7(3)}{11}$	$\dfrac{-20.7(1)}{1}$	$\dfrac{-21.3(7)}{18}$
Tertiary	$\dfrac{-5.5(2)}{11}$	$\dfrac{-9.9(2)}{25}$	$\dfrac{+15.0(1)}{20}$	$\dfrac{-3.2(5)}{56}$
References	1, 16, 17, 19, 21	2, 3, 11, 12, 13, 14, 15, 21, 22, 23	4, 5, 6, 7, 8, 9, 10, 18, 19, 20	

[a]Numerator denotes $\delta^{34}S$ value $(^0/oo)$. Value in parentheses denotes number of regions studied. Denominator denotes number of analyses.

References: 1, Anger *et al.* (1966); 2, Bogdanov *et al.* (1971); 3, Bogush *et al.* (1972); 4, Botoman and Faure (1976); 5, Boyle *et al.* (1976); 6, Buddington *et al.* (1969); 7, Burnie *et al.* (1972); 8, Chrismas *et al.* (1969); 9, Cole (1975); 10, Dechow (1960); 11, Dolzhenko (1976); 12, Girin *et al.* (1975); 13, Grinenko *et al.* (1969); 14, Palamarchuk *et al.* (1972); 15, Palamarchuk *et al.* (1976); 16, Pilot (1970); 17, Pilot *et al.* (1973); 18, Sasaki and Krouse (1966); 19, Thode *et al.* (1953); 20, Tupper (1960); 21, Vinogradov *et al.* (1956); 22, Yamamoto *et al.* (1968); 23, Zagryazhshaya *et al.* (1973).

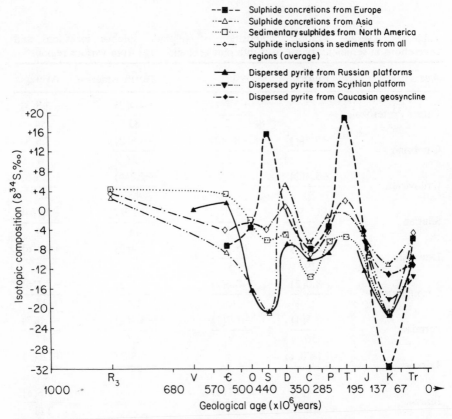

Fig. 2.21   Comparison of average $\delta^{34}S$ values of sulphide inclusions and concretions from sedimentary rocks of different age from North America, Europe, and Asia with data on isotope composition of dispersed pyrite from the Russian and Scythian platforms and Caucasian geosyncline

regime of continental platforms as a whole and the entire sedimentary envelope (Ronov, 1976). For instance, the enlargement of areas of sediment accumulation on platforms of all continents is reflected in the increased amount of ^{32}S in pyrite from the Russian platform (Fig. 2.22).

It is interesting to consider the relationship between the isotopic composition of reduced sulphur and the amount of siliceous rocks in the sedimentary envelope (Fig. 2.23). Siliceous formations are mostly biogenic and are indicative of definite conditions of sedimentation which, by and large, are favourable for the fractionation of sulphur isotopes. Many investigators assume that volcanic processes are the source of silica for these formations. There is a relationship between the amount of sulphur and the intensity of global volcanic processes, but it is concealed by the influence of facies conditions and

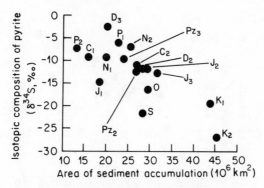

Fig. 2.22 Relationship between isotopic composition of pyrite sulphur from the Russian platform and area of sediment accumulation on continental platforms

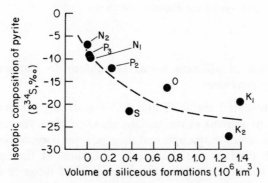

Fig. 2.23 Relationship between isotopic composition of pyrite sulphur and volume of siliceous formations on continents

apparently results from the strong relationship between pyrite formation and the amount of organic matter. It has been mentioned above that the distribution of organic matter in the terrains of the Russian platform of different ages correlates with the changes in amount of organic carbon in the sedimentary envelope. The dependence of these changes on the endogenous influx of carbon has been considered recently (Ronov, 1976, 1980).

The close agreement between changes in the fractionation factor for sulphate and sulphide isotopes in sediments and that of carbonate oxygen (Dontsova *et al.*, 1972; Migdisov *et al.*, 1973) should also be noted (Fig. 2.24).

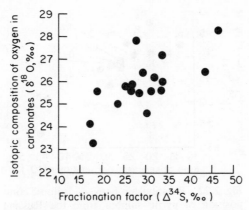

Fig. 2.24 Relationship between fractiona-
tion factor ($\Delta^{34}S = \delta^{34}S_{SO_4^{2-}} - \delta^{34}S_{FeS_2}$) for sulphur isotopes in sediments and the isotopic
composition of oxygen in synchronous carbonates of the Russian platform

E. Concentration of different forms of sulphur in platform zones of sedimentary envelopes

Data were presented above (Table 2.5) concerning the concentration of different forms of sulphur in the various rocks of platform zones. Based on these results, the average amounts and their isotopic compositions were calculated. Although discrepancies between estimations of the sulphur content were large, on average the estimates approximated those of rocks from regions which have been studied systematically.

The model of the chemical structure of the crust made by Ronov and Yaroshevsky (1976) served as the source of data on volumes and masses and their relationship between rocks. Some corrections were made in accordance with new data (Ronov, 1980) on the formations of the late Precambrian sedimentary terrains (see Table 2.13). The relationships between the main type of rocks composing evaporite terrains also change in accordance with new data on the relative distribution and volumes of rocks in salt-bearing basins (Zharkov, 1974, 1978; Ronov, 1980).

The use of these corrected data enables us to estimate the amount of sulphur buried in platform zones of the sedimentary envelope which amounts to 26.3×10^8 TgS. The average $\delta^{34}S$ value of the total sulphur of platforms does not differ significantly from the meteorite standard and equals $-1.5^0/oo$ The sulphur concentration of platform sedimentary rocks (0.487%) corresponds closely to the estimates of Holser and Kaplan (1966) for the sedimentary envelope as a whole (0.49%). The concentrations of pyrite and sulphate in platform rocks are similar (0.28 and 0.21% S, respectively; see Table 2.13). By subtracting from the total sulphate sulphur the amount that exists as secondary sulphate, it is found that the concentration of pyrite sulphur (0.321%) is greater than that of primary sulphate (0.166%). However, even in this case, although the concentrations of oxidized sulphur are lower than our previous estimates (Grinenko *et al.*, 1973b; Ronov and Yaroshevsky, 1976), the high sulphate concentrations in platform terrains correspond to a quiet tectonic regime on platforms where the bulk of evaporites accumulated.

The dynamics of formation of the platform sedimentary sulphur reservoir have also been studied. Table 2.6 presents, along with the data on the Russian and Scythian platforms, information on the sulphur content of sedimentary envelopes of a number of platforms zones. The average content of pyrite, in all regions where the entire sedimentary envelopes were comprehensively studied, varied only slightly and approximated 0.3% S, except for data on the North American platform where methods for sulphur determination were not indicated (Hill *et al.*, 1967; White, 1959).

As can be seen from Table 2.6, the average concentration of pyrite in sediments of different ages of almost all the other platforms correspond to those of the Russian platform. The Cambrian sediments in two of three regions under consideration were exceptions. In the black shales of the Oslo region, and the bituminous clays and carbonates of the Siberian platform, reducing conditions existed and an intensive accumulation of sulphides occurred. Other platform regions with Cambrian sediments were not found and thus the average estimates for sediments of this age may be overstated. The same situation applies to the Permian sediments where solitary analyses of continental sediments of the Siberian platform had a great effect on the average value for the concentration of reduced sulphur.

Based on the data obtained, we were able to calculate the amount of reduced sulphur in the sedimentary Phanerozoic rocks of platforms at various stages of the development of their envelope (Table 2.8). The rate of pyrite accumulation in platform sediments was calculated for various periods of the Phanerozoic eon, and Fig. 2.25 shows the changes in these values with time. It also shows the changes in the amount of organic carbon buried in the sedimentary envelope of Phanerozoic age (Ronov, 1976). It can be seen that the pattern of change for pyrite in the sedimentary envelope of platforms is similar to that for organic carbon in continental sediments and that the maximum accumulation of reduced sulphur in platform sediments occurred in the Ordovician, Carboniferous, and Cretaceous periods, i.e. in the middle stages of the geotectonic cycles. The most potent sulphur flows are characteristic of the Alpine cycle.

Table 2.8 Amounts and rates of accumulation of pyrite sulphur and organic carbon in sedimentary envelopes of platforms (using data of Table 2.6 and rock masses from Ronov, 1976)[a]

Age	Duration of period (10^6 years)	Sedimentary rocks	Amount (10^9 Tg)		Accumulation rate (Tg/year)	
			Pyrite sulphur	Organic carbon	Pyrite sulphur	Organic carbon
Cambrian	70	47.5	0.158(0.081)	0.731(0.086)	2.26(1.16)	10.44(1.23)
Ordovician	55	22.5	0.092(0.117)	0.218(0.542)	1.68(2.12)	3.97(9.85)
Silurian	35	13.9	0.042(0.062)	0.033(0.028)	1.19(1.77)	0.96(0.80)
Lower Palaeozoic	160	83.9	0.292(0.260)	0.982(0.656)	1.82(1.62)	6.13(4.10)
Devonian	55	43.2	0.056(0.121)	0.156(0.147)	1.02(2.2)	2.83(2.67)
Carboniferous	65	33.2	0.080(0.110)	0.219(0.113)	1.23(1.69)	3.37(1.78)
Permian	45	32.4	0.117(0.039)	0.146(0.056)	2.60(0.87)	3.24(1.24)
Middle–Upper Palaeozoic	165	108.8	0.175(0.270)	0.431(0.316)	1.58(1.64)	3.26(1.92)
Triassic	50	43.1	0.073(0.026)	0.121(0.039)	1.47(0.52)	2.41(0.78)
Jurassic	53	58.7	0.205(0.247)	0.546(0.528)	3.88(4.66)	6.76(9.96)
Cretaceous	66	97.7	0.206(0.196)	0.666(0.470)	3.12(2.96)	9.79(7.12)
Mesozoic	169	199.7	0.484(0.469)	1.333(1.037)	2.86(2.78)	6.65(6.14)
Palaeogene	41	36.4	0.12 (0.113)	0.288(0.266)	3.11(2.76)	7.37(6.49)
Neogene	23	19	0.065(0.049)	0.106(0.112)	2.81(2.13)	4.63(4.86)
Cainozoic	64	554	0.192(0.162)	0.349(0.378)	3.00(2.53)	6.38(5.91)
Phanerozoic	558	447.8	1.221(1.161)	3.036(2.387)	2.1 (2.08)	5.44(4.28)

[a]Estimates based on average contents in different platforms; values in parentheses based on results for the Russian platform.

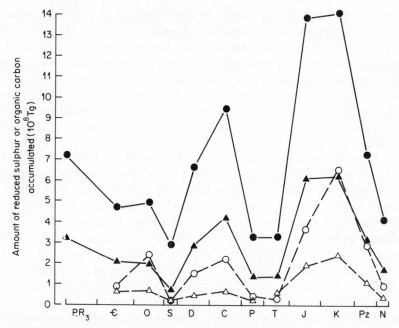

Fig. 2.25 Changes in the amounts of reduced sulphur and organic carbon accumlated in the sedimentary envelope of platforms and continents as a whole in the Phanerozoic. Continents: reduced sulphur (▲———▲), organic carbon (●———●); Platforms: reduced sulphur (△———△), organic carbon (○———○)

2.2.2 Sulphur in Sedimentary Rocks of Geosynclinal Zones

A. Difference between geosynclinal and platform zones in the concentration and isotopic composition of different forms of sulphur

Specificity of the structure and development of geosynclines, with characteristic large amplitudes of vertical movement, high rates of sedimentation, and intensive volcanic activity, affects the distribution of sulphur in the sedimentary rocks of these regions.

It was mentioned above that even in platforms which are tectonically quiet regions of the crust, changes in the rate of sedimentation affected the sulphur distribution and its isotopic character. The maximum influence of this factor, must have occurred during sedimentation in geosynclinal zones.

Ronov *et al*. (1974) studied the influence of tectonic and volcanic activity on sulphur distribution and isotopic composition in sedimentary rocks of the Alpine (J_1–N) Caucasian geosyncline. They considered a number of tectonic zones, beginning from the Russian platform, where the tectonic regime was already stable in the Riphean period, to the Scythian plate, established in the

Jurassic period, and through zones of the Caucasian geosyncline, including the north (outer) and south (inner) miogeosynclines of the Great Caucasus, the Trans-Caucasian middle massif, and the eugeosyncline of the Minor Caucasus. The rate of sedimentation was at a minimum on the Russian platform, increased gradually until it reached a maximum in the South miogeosyncline of the Great Caucasus, then decreased dramatically in the Trans-Caucasian middle massif where it approximated the platform sedimentation rates and then increased again in the eugeosyncline of the Minor Caucasus.

The data presented in Tables 2.9 and 2.10 show that usually more than half (60–70%) of the total sulphur of Meso-Cainozoic geosynclinal sedimentary rocks is incorporated in dispersed pyrite. The amount of sulphate sulphur is substantially less, and it is only in arid sediments (and not in all cases) that this quantity exceeds the amount of sulphur as pyrite. A significant part of the sulphate sulphur of rocks, especially clayey ones, is secondary sulphate formed by the oxidation of pyrite. Almost all of the sulphate sulphur in clays of humid sediments is secondary. The concentration of organically bound sulphur rarely reaches 0.1%, elemental sulphur is always less than 0.010–0.020%, and monosulphide sulphur is negligible.

In the geotectonic profile beginning from the platform and moving to the depth of the geosyncline, the concentration of sulphur in various forms decreases and the isotopic composition becomes heavier (Fig. 2.26). Sedimentation rates increase in the same direction. Changes in sedimentation rates are usually reflected in the distribution of sulphur in sedimentary rocks. When a maximum occurs in the sedimentation rate at a given stage of the development of the geosyncline, the pyrite sulphur concentration is at a minimum and the $\delta^{34}S$ value is at a maximum (e.g. Lower Cretaceous). On the other hand the slowly accumulating marine sediments of stable zones inside the geosyncline (e.g. Trans-Caucasian middle massif in the Lower Cretaceous) have a high reduced sulphur concentration but a low ^{34}S content. This indicates that the sedimentation rate is an important factor regulating the sulphur concentration in sedimentary terrains. The higher the rates of sedimentation and submersion of sediments, the more difficult is sulphate diffusion from the overlying water to the sediment; also the earlier the sulphate reduction system becomes closed, the faster the initial sulphate is exhausted. The influence of this factor may limit the manifestation of the facies and climatic zonation which is reflected in the distribution of sulphur and its isotopic composition in platform sediments. Nevertheless, the facies and climatic factors also affect sulphur distribution within the geosyncline.

In a number of cases it is possible to follow these changes in sulphur content and isotopic composition, even at high rates of sedimentation. The Middle Jurassic period (Aalen) of Dagestan (Zagryazhskaya *et al.*, 1973) provides an example of such facies zonation in the distribution of $\delta^{34}S$ values in pyrite.

Table 2.9 Changes in concentration and isotopic composition of sulphur in sedimentary and volcanogenic sedimentary terrains of the Mesozoic and Cainozoic epochs through the geotectonic profile: Russian platform—eugeosyncline of the Minor Caucasus

Age	Tectonic Zone	% of total volume of sediments in platforms or Caucasian geosyncline	Number of samples in assays	Pyrite (%S)	Pyrite $\delta^{34}S$ (‰)	Sulphate[b] (%S)	Sulphate[b] $\delta^{34}S$ (‰)	Organic sulphur $\delta^{34}S$ (‰)	Pyrite plus sulphate (%S)	Pyrite plus sulphate $\delta^{34}S$ (‰)
J_1–N_2	Russian platform	68.7	4032	0.263	−21.2	0.099	−12.4	−8.0	0.362	−12.8
J_1–N_2	Scythian platform	31.3	5945	0.256	−15.1	0.097	−8.8	−6.9	0.353	−13.4
J_1–N_2	Average for Russian and Scythian platforms[a]	100.0	9977	0.261	−19.3	0.098	−11.3	−7.7	0.359	−17.1
J_1–N_2	North miogeosyncline and fore trough of the Great Caucasus	44.9	12333	0.222	−13.6	0.181	−14.9	−4.3	0.402	−14.2
J_1–Pg_2	South miogeosyncline of the Great Caucasus	19.0	4932	0.078	−6.6	0.046	−2.4	−0.1	0.124	−5.0
J_1–Pg_2 Pg_3–N_2	Riono-Kurnisky inter-montane trough	2.9	1918	0.069	−19.3	0.070	−8.5	−4.3	0.139	−13.9
J_1–N_2	Eugeosyncline and inter-montane trough of the Minor Caucasus	15.4	1805	0.182	−20.6	0.137	−14.7	—	0.319	−18.1
J_1–N_2	Average for the Caucasian geosyncline[a]	17.8	4482	0.101	−5.3	0.081	+7.0	−1.2	0.182	+0.2
		100.0	25470	0.162	−13.3	0.127	−11.4	−2.5	0.289	−12.5

[a] Average values were calculated to be proportional to the volumes of sediments of corresponding zones.
[b] Low $\delta^{34}S$ value of sulphate compared to evaporites indicates a significant part of secondary sulphate.

Table 2.10   Average concentration and isotopic composition of sulphur in major rock types of Alpine and Hercynian geosynclines of the Caucasus

Rock	Alpine geosyncline of Caucasus					Hercynian geosyncline of Caucasus				
	Number of samples in assays	Pyrite		Sulphate		Number of samples in assays	Pyrite		Sulphate	
		(%S)	$\delta^{34}S$ ($^o/oo$)	(%S)	$\delta^{34}S^a$ ($^o/oo$)		(%S)	$\delta^{34}S$ ($^o/oo$)	(%S)	$\delta^{34}S^a$ ($^o/oo$)
Silty–sandy	7528	0.123	−4.6	0.041	−2.9	277	0.101	−3.5	0.033	−1.8
Clayish	9664	0.251	−11.4	0.123	−11.2	184	0.095	−3.1	0.015	−0.5
Carbonate	5492	0.047	−7.9	0.044	+5.4	93	0.038	−5.8	0.038	+11.0
Sulphate	246	0.557	—	23.800	+9.8	—	—	—	—	—
Volcanic	1455	0.057	+2.3	0.010	+5.9	—	—	—	—	—

[a] Low $\delta^{34}S$ value indicates significant part of secondary sulphate.

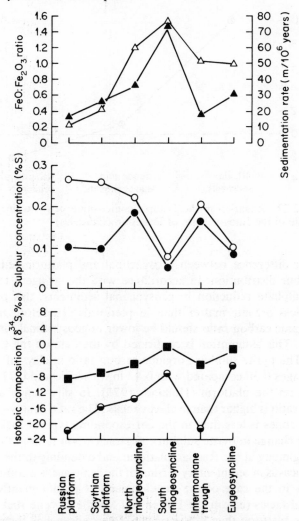

Fig. 2.26   Sedimentation rates (△), FeO/Fe₂O₃ ratios
(▲), sulphur concentrations in pyrite (○) and sulphate
(●), and isotopic composition of organic sulphur (■)
and pyrite (●) in sedimentary and volcanic rocks of the
tectonic profile of the Russian platform–eugeosyncline
of Minor Caucasus

However, a comparison of the changes in the facies profile of Dagestan with
those in the platforms suggests that the higher δ^{34}S values in geosyncline
palaeobasins resulted from higher rates of sedimentation (Fig. 2.27). This
confirms the idea (Ronov *et al*., 1974) that the rate of sedimentation restricts
the appearance of low δ^{34}S values which are determined by sedimentation
conditions.

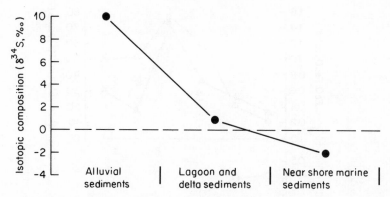

Fig. 2.27   Zonation in the isotopic composition of pyrite in the facies profile of the Jurassic basin of Dagestan (Caucasus)

Another difference between geosynclinal and platform sediments lies in their sulphur distribution. In accordance with the stringent tectonic limitations of sulphate reduction in geosynclinal sediments, this process should consume less organic matter than in platforms. Therefore the pyrite sulphur : organic carbon ratio should be lower in geosynclinal than in platform sediments. This assumption is confirmed by the data on the Caucasian geosyncline. The pyrite sulphur : organic carbon ratio in clays of this deformed zone averages 0.30 compared with 0.42 in Mesozoic and 0.77 in Palaeozoic sediments on the platform (Holland, 1973). In silty–sandy and carbonate rocks this ratio is higher than in clayey rocks. The ratio in silty–sandy rocks of deformed zones is less than in the corresponding type of rocks on the platform. The change in pyrite sulphur : organic carbon ratio in the geosynclinal profile beginning at the Russian platform and extending to the eugeosyncline of the Caucasus in sedimentary terrain of the same age is a good illustration of this. Only in the case of the development of predominantly coal-bearing continental facies (e.g. in the Lower and Middle Eocene rocks) is this ratio less on the platform than in the synchronous geosyncline terrains.

Confirmation is provided by the relationships between pyrite or total sulphur and organic carbon in Palaeozoic rocks of the Caucasian geosyncline (Ronov *et al.*, 1974). In these geosyncline sediments the pyrite sulphur : organic carbon ratio varies from 0.6 in silty–sandy rocks to 0.2 in shales, giving an average about 0.3 as in the case of rocks of the Alpine geosyncline.

B.   Total sulphur in the geosynclinal part of the sedimentary envelope

In this section we attempt to evaluate the total amount of sulphur in geosynclinal zones of the sedimentary envelope. Published data are not systematic; some reports give the concentration of the different forms of sulphur, while others give only the total sulphur concentration of rocks. These data were assembled after checking, using values weighted for rock thicknesses and volumes for the Palaeozoic and Meso-Cainozoic geosynclines of the Caucasus (Table 2.11). In addition to this material, published data and results from unfinished studies on the sedimentary terrains of the Carpathian and Ural geosynclines were included. The values for the distribution of the different forms of sulphur in geosynclinal rocks correspond to the average ratio between the most comprehensively studied young Caucasian geosyncline and an older one, the Palaeozoic deformed zone of the same region. These values are in agreement with the most reliable published data and supplementary material from a large geosyncline region (West European geosyncline, the Urals) covering a sufficiently long time interval Pz, R–T, Mz–Kz, etc.

The data obtained indicate the basic differences in the sulphur concentration and its isotopic composition between the geosynclinal sedimentary envelope and platform sediments (see Table 2.13) the former has lower sulphate (mainly secondary), and pyrite concentrations, an accumulation of the ^{32}S isotope in sulphate and of ^{34}S in pyrite. A comparison of the values obtained for geosynclinal terrains with those of platforms, shows that the concentration of sulphate in geosynclinal terrains was lower than that in platforms (0.063% versus 0.209%). More striking differences between concentrations of sulphate (0.018–0.166%) are shown in geosynclines and platforms by excluding the secondary sulphate formed by the oxidation of sulphide.

Even though the volume of geosynclinal rocks is 2.5 times greater than that of platform rocks, the amount of sulphate sulphur present is less than in platform terrains (8.4 versus 11.3 $\times$ 10^8 Tg, and for primary sulphate 2.4 and 9.0 $\times$ 10^8 Tg, respectively).

Such a distribution of sulphate sulphur in structural zones of the sedimentary envelope results from the concentration of the basic reserves of evaporites in platform terrains (the amount of evaporite sulphur in them is five times greater than that in geosynclines), and the greater dispersion of sea-water sulphate in platform sediments (especially in arid carbonates).

The amount of pyrite in geosynclines only slightly exceeds that in sedimentary rocks of platforms, which suggests that the basic process of sulphide formation occurs on platforms.

The dynamics of sulphur accumulation in the sedimentary envelope of geosynclinal zones is also of interest. Unfortunately, no data are available which allow us to evaluate the rates of sulphur accumulation in sediments for each period of the Phanerozoic. Based on the few data available (Table 2.11), it is possible to describe the dynamics of this process only for long periods of the history of sedimentary envelopes. According to these estimates both sulphate and pyrite increase during the Phanerozoic. The rate of

Table 2.11 Concentration (%S) and isotopic composition ($\delta^{34}S$, ‰) of different forms of sulphur in sedimentary and effusive rocks of different age in geosynclinal regions

Geosynclines, age of sediments	Clayey rocks				Sandstones and siltstones			Carbonate rocks			Gypsum and anhydrites	Siliceous rocks	Effusive rocks		
	Sulphate	Organic	Pyrite	Total	Sulphate	Pyrite	Total	Sulphate	Pyrite	Total	Total	Total	Sulphate	Pyrite	Total
Caucasus (J_1–N_2)	$\dfrac{0.212^a}{-11.2}$	$\dfrac{0.036}{-2.5}$	$\dfrac{0.248}{-11.4}$	$\dfrac{0.496}{-10.7}$	$\dfrac{0.092}{-2.9}$	$\dfrac{0.159}{-4.6}$	$\dfrac{0.251}{-4.0}$	$\dfrac{0.028}{+5.4}$	$\dfrac{0.088}{-7.9}$	$\dfrac{0.116}{-4.7}$	$\dfrac{23.8}{+9.8}$	0.249	$\dfrac{0.010}{+5.9}$	$\dfrac{0.057}{+2.3}$	$\dfrac{0.067}{+2.8}$
Carpathians (K–f)	0.125		0.452	0.577	0.178						16.72 $\dfrac{13.80}{+22.3}$				
Crimea (K–N_1)	0.100		0.077	0.177				0.056		0.180					
Fergana (J–K)				0.079											
South Mangyshlak (K_Z)				0.254											
Himalayas (f–N)	0.056				0.074			0.074							
South Sakhalin (J–K_1)				0.578									0.020		
California (P_{2-3})	$\dfrac{0.015}{-0.5}$	0	$\dfrac{0.095}{-3.1}$	$\dfrac{0.110}{-2.7}$	$\dfrac{0.033}{-1.8}$	$\dfrac{0.101}{-3.5}$	$\dfrac{0.134}{-3.1}$	$\dfrac{0.038}{+11.0}$	$\dfrac{0.038}{-5.8}$	$\dfrac{0.076}{+2.6}$					
Caucasus (D–P)															$\dfrac{0.079}{+3.0}$

Region															
Urals (R–S)	0.071		0.103	0.174	0.055	0.099	0.154	0.017	0.064	0.081	19.57				0.058
Central Kazakhstan (PR$_3$–Є$_2$)														0.084	
Appalachians (PR$_3$–O)				0.099											0.030
Altai–Sayan region (PR$_3$–Є)						0.082				0.051				0.086	
Donbass (C–P$_1$)			0.280								18.55				
Verkhoyan region (C–P)				0.040			0.023							0.317	
West European geosyncline (Pz)		0.055	0.173	0.288	0.004	0.137	0.141	0.033	0.051	0.084					
Himalayas, Nepal (Є–S)										0.078					
Japan (M$_Z$ – K$_Z$)		0.061	0.056	0.117											
England (C)			0.839				0.476			0.305					
Caledonides, Sweden (O)				0.111											
Czechoslovakia, Barandien (Є–D)							0.080								
Average for geosynclines	0.074 / −5.8	0.026 / −2.5	0.177 / −7.2	0.277 / −6.4	0.060 / −2.4	0.110 / −4.0	0.170 / −3.4	0.056 / +8.2	0.065 / −6.8	0.121 / +0.1	18.49 / +19.5	0.184	0.008 / +5.9	0.043 / +2.3	0.051 / +2.9

[a]Numerators and denominators denote concentrations and isotopic composition respectively.

Table 2.12 Concentration, amount and rate of accumulation of sulphur in the geosynclinal part of the sedimentary envelope in different stages of the geological history

Age complex	Mass of sedimentary rocks of geosynclines (10^9 Tg)	Duration of complex (10^6 years)	Concentration in rocks Pyrite (%S)	Amount in rocks Pyrite (10^8 TgS)	Rate of accumulation Pyrite (TgS year^{-1})
Lower Palaeozoic	266.0	160	0.091	2.421	1.513
Middle and Upper Palaeozoic	418.6	165	0.193	8.079	4.89
Mesozoic and Cainozoic	465.8	233	0.213	9.921	4.25
Geosynclinal part of sedimentary envelope (Phanerozoic)	1150.4	558	0.177	20.44	3.66

Table 2.13 Distribution of sulphur and its isotopic composition in the sedimentary envelope of the earth[a]

	Distribution of rocks by volume (%)	Sulphate			Pyrite[b]			Total sulphur			Organic carbon (%)
Type of rock		(%S)	Mass (10⁸ TgS)	δ³⁴S (⁰/₀₀)	(%S)	Mass (10⁸ TgS)	δ³⁴S (⁰/₀₀)	(%S)	Mass (10⁸ TgS)	δ³⁴S (⁰/₀₀)	
Continents											
Platforms ($V = 220 \times 10^6$ km³, $M = 0.54 \times 10^{12}$ Tg)											
Sandstones and siltstones	22.3	0.052	0.63 (0.30)[c]	+ 5.9	0.200	2.41 (2.74)	− 6.8	0.252	3.04	− 4.2	0.37
Clay and clay schists	46.3	0.095	2.37 (0.68)	− 3.8	0.434	10.85 (12.54)	−13.2	0.529	13.22	−11.5	0.92
Carbonate	24.3	0.214	2.81 (2.54)	+16.3	0.126	1.65 (1.92)	−14.2	0.340	4.46	+ 5.0	0.33
Gypsum and anhydrites	0.5	19	5.13	+19.5	—	—	—	19	5.13	+19.5	None
Salts	2.0	0.307	0.33	+19.5	—	—	—	0.307	0.33	+19.5	None
Siliceous	0.2	0.070	0.01		0.130	0.01		0.200	0.02		
Effusives	4.4	0.010	0.02	− 1.2	0.045	0.11	− 1.2	0.055	0.13	− 1.2	None
Sum or average for platforms	100.0	0.209	11.30 (8.99)	+12.6	0.278	15.03 (17.34)	−12.2	0.487	26.33	− 1.5	0.59

Table 2.13 (continued)

Type of rock	Distribution of rocks by volume (%)	Sulphate Mass (10^8 TgS)	Sulphate (%S)	Sulphate δ^{34}S (⁰/₀₀)	Pyrite[b] Mass (10^8 TgS)	Pyrite[b] (%S)	Pyrite[b] δ^{34}S (⁰/₀₀)	Total sulphur Mass (10^8 TgS)	Total sulphur (%S)	Total sulphur δ^{34}S (⁰/₀₀)	Organic carbon (%)
Continents											
Geosynclines ($V = 540 \times 10^6$ km^3, $M = 1.35 \times 10^{12}$ Tg)											
Sandstones and siltstones	19.2	1.56 (0.11)	0.060	− 2.4	2.85 (4.30)	0.110	− 4.0	4.41	0.170	− 3.4	0.38
Clay and clay schists	40.9	4.09 (0.11)	0.074	− 5.8	11.20 (15.18)	0.203	− 6.6	15.29	0.277	+ 6.4	0.96
Carbonate	19.2	1.45 (0.83)	0.056	+ 8.2	1.69 (2.31)	0.065	− 6.8	3.14	0.121	+ 0.1	0.18
Gypsum and anhydrites	0.04	1.00	18.49	+19.5	—	—	—	1.0	18.49	+19.5	0.06
Salts	0.16	0.07	0.307	+19.5	—	—	—	0.07	0.307	+19.5	0.38
Siliceous	1.1	0.07	0.052		0.20	0.132		0.27	0.184		0.38
Effusives	19.4	0.21	0.008	+ 5.9	1.13	0.043	+ 2.3	1.34	0.051	+ 2.9	None
Sum or average for geosynclines	100.0	8.45 (2.40)	0.063	+ 0.8	17.07 (23.12)	0.126	− 5.6	25.52	0.189	− 3.5	0.51

	Volume (10^6 km^3)	Mass (10^{12} Tg)	Sulphate			Pyrite			Total Sulphate			Organic carbon (%)
			(%S)	Mass (10^8 TgS)	δ^{34}S ($^o/oo$)	(%S)	Mass (10^8 TgS)	δ^{34}S ($^o/oo$)	(%S)	Mass (10^8 TgS)	δ^{34}S ($^o/oo$)	
Continents as a whole	760	1.89	0.105	19.75 (11.39)	+ 7.6	0.169	32.10 (40.46)	− 8.7	0.274	51.85	− 2.5	0.53
Subcontinents, (shelves, and slope)	160	0.40	0.105	4.20 (2.40)	+ 7.6	0.169	6.76 (8.56)	− 8.7	0.274	10.96	− 2.5	0.53
Continents and subcontinents	920	2.29	0.105	23.95 (13.79)	+ 7.6	0.169	38.86 (49.02)	− 8.7	0.274	62.81	− 2.5	0.53 0.46^d
Ocean sediments, seismic layer I^e	120	0.19	0.126	2.39	+19.4	0.038	0.72	−43.6	0.165	3.14	+ 4.8	0.10^d
Sedimentary envelope of earth	1040	2.48	0.106	26.34 (16.18)	+ 8.7	0.159	39.58 (49.74)	− 9.3	0.265	65.95	− 2.2	0.53
Ocean water	1370	1.40	0.090	12.60	+20.0	—	—	—	0.090	12.60	+20.0	0
Sedimentary envelope and ocean	2410	3.88	0.100	38.94 (28.78)	+12.4	0.102	39.58 (49.74)	− 9.3	0.202	78.55	+ 1.4	0.34

^{a}Data on volume and mass of rocks from Ronov and Yaroshevsky (1976) and Ronov (1980); density of sediment taken as 2.5 gm cm^{-3}.
^{b}Pyrite includes monosulphide, organic and elemental sulphur.
^{c}Values in parentheses are corrected for primary sulphide oxidized to sulphate.
^{d}Data from Trotsyuk (1979).
^{e}Values for pyrite sulphur in seismic layer I agree with those calculated from organic carbon (Trotsyuk, 1979) and S/C ratio of Goldhaber and Kaplan (1974).

sulphur accumulation in the platform and geosynclinal sediments tends to increase in the same way. It is clearly seen from a comparison of Tables 2.8 and 2.12 that these rates are more than 1.5 times higher in geosynclines. This is evidently due to higher rates of sediment accumulation. As in the case of platforms (Table 2.8), the total mass of reduced sulphur in geosynclines (Table 2.12) (obtained when studying the dynamics of the sulphur pool formation in this structure) proved to be somewhat higher compared to the ultimate estimates (Table 2.13).

2.2.3 Sulphur in Sedimentary Rocks of Continents

A. Amounts and rates of accumulation of reduced sulphur

The total amount of pyrite sulphur in continental sedimentary formations, obtained from the summation of pyrite in platforms and geosynclinal parts of the sedimentary envelope, amounts to 32.1×10^8 TgS (Table 2.13). The summation of data presented in Tables 2.8 and 2.12 yields a value of 31.1×10^8 TgS. This does not include estimates for Upper Proterozoic rocks which, based on indirect data (ratio with organic carbon), are less than $2–3 \times 10^8$ TgS.

This value for pyrite sulphur was also confirmed by other calculations: the total amount of pyrite sulphur in Phanerozoic sedimentary terrains was calculated from estimates of the concentration of sulphur in its different forms in various lithological formations (taking into account platforms and geosynclines) and the total amount of each formation (Table 2.14) (Ronov, 1976, 1980). Although this calculation did not involve the effect of age on sulphur distribution in sediments and the Upper Proterozoic part of the sedimentary envelope, the value obtained, 31.6×10^8 TgS, agrees well with those given above.

Hosler and Kaplan (1966) and Grinenko *et al.* (1973b) estimated the amount of pyrite sulphur to be 27×10^8 TgS, while Ronov and Yaroshevsky (1976) assessed the value to be slightly higher at 34×10^8 TgS. Calculation of the sulphur flux to sediments during the Phanerozoic eon is difficult, because of the absence of data on geosynclinal sediments where changes in the pyrite flux to sediments are available only for long time intervals. However, based on the close agreement between pyrite sulphur and organic carbon fluxes (Fig. 2.25 shows them in platform sediments) it is possible to estimate these fluxes using the ratio between these elements.

Differences in pyrite sulphur : organic carbon ratios with time (in sediments of the Russian platform) and in space (between geosyncline and platform formations) have already been mentioned. It is difficult to estimate the changes with time in this ratio in all rocks of the sedimentary envelope. We have restricted ourselves to an approximation approach. Taking into account

Table 2.14   Assessment of the reservoir of sulphur in the sedimentary envelope of continents from various lithological formations[a]

Type of formation	Volume (10^6 km^3)	Concentration (%S)		Mass (10^8 TgS)	
		Sulphate	Pyrite	Sulphate	Pyrite
Continental					
Disintegrated	97.2	0.21	0.12	5.0	2.8
Glacial	4.3	0.01	0.02	0.0	0.0
Coal-bearing	17.0	0.15	0.80	0.6	3.3
Carbonate	144.0	0.11	0.13	4.0	4.7
Marine clay–sandy	261.0	0.08	0.33	5.1	20.8
Salt-bearing (salts)	5.1	0.30	—	0.0	—
Salt-bearing (gypsum and anhydrite)	1.3	19.0	—	5.5	—
Volcanogenic	104.3	0.01	0.04	0.3	1.1
Siliceous	4.4	0.07	0.13	0.0	0.0
All types of formations	639.3	0.128	0.198	20.5	31.6

[a]After Ronov (1980).

the fact that this ratio in Palaeozoic platform rocks is higher than that in younger ones (0.77 in Pz and 0.45 in Mz and Cz, Holland, 1973) and the volume distribution of these rock age groups on platforms, an average ratio of pyrite sulphur to organic carbon in platform sediments of 0.6 was obtained. This agrees well with empirical data presented in Table 2.6. For geosynclinal sediments a ratio of 0.3 was accepted. Based on data for the distribution of geosynclinal and platform rocks in the sedimentary envelope of continents this ratio is 0.40 a value that was used to estimate the distribution of pyrite sulphur with time (Table 2.15). Reliability of the data obtained was checked by comparison with empirical data on sulphur masses in large age units obtained from analyses of the different forms of sulphur (Tables 2.8, 2.12). There was reasonable agreement between these values (Table 2.15). Only in the Middle and Upper Palaeozoic sedimentary terrains were the results somewhat higher. We pointed out above some overestimates of the mass of sulphur from data on the stratigraphically most complete sections which did not allow for all the empirical material. Moreover, fluctuations in the pyrite sulphur : organic carbon ratio with time are not excluded. However, the volume of the total pyrite sulphur reservoir obtained by this indirect method is surprisingly close to estimates made by direct methods. This is also confirmed by a comparison of the fluctuations in reduced sulphur and organic carbon with time with those of platform sediments comprising about half of the total pyrite sulphur reservoir (Fig. 2.25).

Reduced sulphur in the sedimentary envelope is heavier (δ^{34}S = $-8.7‰$)

Table 2.15 Distribution of reduced sulphur in the sedimentary envelope of continents[a]

| Stratigraphic interval | Duration (10^6 years) | Mass | | | | Rate of pyrite accumulation (TgS year^{-1}) | Average pyrite concentration (%S) |
| | | Rocks (10^8 Tg) | Organic carbon (10^8 Tg) | Pyrite (10^8 TgS) | | | |
				From organic carbon	From tables 2.8, 2.12		
Cambrian	70	1 672	3.7	1.47	—	2.1	0.09
Ordovician	55	1 105	4.8	1.92	—	3.5	0.17
Silurian	35	717	1.8	0.72	—	2.1	0.10
Lower Palaeozoic	160	3 499	10.3	4.11	4.56	2.6	0.12
Devonian	55	2 308	7.3	2.92	—	5.3	0.13
Carboniferous	65	1 663	9.3	3.72	—	5.7	0.22
Permian	45	1 303	3.2	1.28	—	2.8	0.10
Middle and Upper Palaeozoic	165	5 274	19.8	7.92	9.83	4.8	0.15
Triassic	50	1 420	3.2	1.28	—	2.6	0.09
Jurassic	53	1 699	13.6	5.44	—	10.3	0.32
Cretaceous	66	2 488	13.7	5.48	—	8.3	0.22
Mesozoic	169	5 607	30.5	12.20	—	7.2	0.22
Palaeogene	41	910	7.2	2.88	—	7.0	0.32
Neogene	23	691	4.1	1.64	—	7.1	0.24
Cainozoic	64	1 601	11.3	4.52	—	7.1	0.28
Mesozoic and Cainozoic	223	7 209	41.8	16.72	16.68	7.5	0.23
Phanerozoic	558	15 981	71.9	28.75	31.1	5.2	0.18

[a]Volumes and masses of sediments and organic carbon from Ronov (1976, 1980).

than estimated by Holser and Kaplan (1966) ($\delta^{34}S = -12^0/_{00}$) and the $\delta^{34}S$ value agrees with that obtained by Grinenko *et al.*, 1973b). It is impossible to determine the isotopic composition of sulphur buried in sediments of continents at different stages of geological history, but it is evident that the character of such changes, with some exceptions, follow those determined for the Russian platform. Evidently, the $\delta^{34}S$ values should occupy an intermediate position between those for the Russian platform and average pyrite inclusions from platform and geosynclinal regions of various continents (Fig. 2.21).

We have made approximate estimates of $\delta^{34}S$ values of pyrite in sediments of various ages on the basis of data condensed in Tables 2.4, 2.7, 2.9, 2.10. These estimates are given in Table 2.16.

Estimates of the reduced sulphur reservoir considered in this chapter are obtained from the results of sulphide sulphur analysis in sedimentary rocks, and they do not include the sulphide sulphur that has been oxidized and is actually in the form of dispersed sulphates. As noted above, such 'secondary' sulphates are an important part of sulphate sulphur in rocks (especially in terrigenic ones). The reservoir of primary pyrite (sulphide) sulphur should be increased, while the reservoir of primary sulphates should be decreased by the magnitude of this form of sulphate. One may assess this amount with the help of data on the isotopic composition of sulphur in pyrite and primary sulphate. Tentative estimates for the rocks of the sedimentary envelope are summarized in Table 2.13. The mass of oxidized sulphides is large enough to increase the reservoir of the primary reduced sulphur by one-third, while the reservoir of sulphate sulphur is reduced by almost one-half and in this case characterizes the disseminated primary sulphate of sedimentary rocks (analogous to evaporite).

Table 2.16 Average $\delta^{34}S$ values ($^0/_{00}$) in sedimentary sulphides from different periods and regions (from analyses of dispersed pyrite, sulphide inclusions and concretions in sedimentary rocks)

Stratigraphic interval	Europe	Asia	North America	Average
Upper Proterozoic	+ 0.1	+ 2.7	+ 4.4	+ 2.4
Cambrian	− 2.8	− 8.5	+ 3.6	− 2.6
Ordovician	− 9.9	—	+ 2.2	− 6.0
Silurian	− 3.0	−21.7	−12.3	−12.3
Devonian	− 2.0	+ 5.1	− 5.4	− 0.8
Carboniferous	− 8.8	− 3.6	− 9.7	− 7.4
Permian	− 5.6	− 1.7	− 6.4	− 4.6
Triassic	(+18.0)	—	− 5.8	− 5.8
Jurassic	− 8.1	− 4.0	—	− 6.0
Cretaceous	−23.8	−11.6	−20.7	−18.7
Tertiary	− 9.4	− 9.7	(+15.0)	− 9.6

B. Sulphate

From the data given in Table 2.13 it can be seen that the amount of sulphate sulphur in sedimentary rocks is about two-thirds of the amount of pyrite sulphur. Sulphates of evaporite terrains account for a little more than one-third of the total, and the remainder (13.2 from 19.8×10^8 TgS) occurs in the dispersed state in sedimentary rocks. It has already been mentioned that the previous estimation (Grinenko *et al.*., 1973b) greatly overstated the size of this reservoir (39.6×10^8 TgS). Holser and Kaplan (1966), using analyses of carbonate rocks from the Russian platform, estimated the amount of sulphate in evaporites to be 51×10^8 TgS. This is approximately eight times larger than our estimate of 6.5×10^8 TgS calculated from the volume of evaporites and sulphate concentrations in Palaeozoic basins (Zharkov, 1974, 1978), and the less detailed information from Meso-Cainozoic rocks (A. B. Ronov, V. E. Khain, and A. N. Balukhovsky, unpublished data).

These data are presented in Table 2.17 along with the average $\delta^{34}S$ values for evaporites of each age and the weighted average for sulphate in the sedimentary envelope ($\delta^{34}S = +19.5^0/oo$). This estimate is higher than the statistical average proposed by Holser and Kaplan (1966) ($\delta^{34}S = +17^0/oo$) though it does not reach values postulated by Holland (1973) ($\delta^{34}S = +21^0/oo$). It can be seen from Table 2.17 that different amounts of gypsum and anhydrite were deposited in evaporite basins during each geological period: the maximum amount was recorded in the Cambrian and the minimum in the Silurian.

At present we cannot evaluate the changes which occur in the fluxes of dispersed sulphate in sedimentary rocks with time. Part of this reservoir represents dispersed evaporite which should be added to the amount of evaporite estimated above, and according to estimates of Table 2.13 its mass is 4.8×10^8 TgS. The other part of the sulphate sulphur is oxidized pyrite (8.4×10^8 TgS) and should be added to the pyrite reservoir.

Calculations based on the average $\delta^{34}S$ values for sulphate and pyrite suggest that 31% of the total reservoir of sulphate sulphur was evaporite, 45% came from the oxidation of pyrite, while 'marine' sulphate accounted for only 24% of the total. Unfortunately, these are only average values, and the ratio of these forms of sulphate sulphur changes with time and the structural zone of the crust.

Part of the oxidized pyrite sulphur appears to change with geological time. Evidence for this comes from the convergence of the $\delta^{34}S$ values of sulphate to those of pyrite in the progression from Palaeozoic to Mesozoic and Cainozoic rocks (Grinenko *et al.*, 1973b). At the same time, sulphates dispersed in rocks should change with time in proportion to changes in evaporite masses, as seen from the example of Palaeozoic terrains of the Russian platform. This justifies the use of data on masses and accumulation rates as characteristic of all sulphate sulphur.

Table 2.17 Volume, mass, and isotopic composition of sulphate sulphur in sedimentary terrains of continents

Stratigraphic interval	Volume (10^6 km³) Gypsum and anhydrite	Total evaporite	Mass of sulphate rock (10^8 Tg)	Rate of accumulation of evaporite sulphur (TgS year⁻¹)	δ^{34}S of sulphate (°/oo)	Source of volume data
Upper Proterozoic	0.060	0.360	1.38	0.06	+16.3	Ronov (1980)
Cambrian	0.239	1.525	5.50	1.50	+29.3	Zharkov (1978)
Ordovician	0.020	0.045	0.46	0.16	+28.5	Zharkov (1978)
Silurian	0.006	0.032	0.01	0.003	+26.2	Zharkov (1978)
Devonian	0.151	0.311	3.47	1.18	+20.1	Zharkov (1978)
Carboniferous	0.068	0.240	1.56	0.46	+16.7	Zharkov (1978)
Permian	0.149	1.430	3.43	1.44	+10.8	Zharkov (1978)
Triassic	0.142	0.550	3.27	1.24	+18.8	Ronov (1980)
Jurassic	0.168	0.750	3.86	1.38	+16.6	Ronov (1980)
Cretaceous	0.129	0.570	2.97	0.86	+16.6	Ronov (1980)
Palaeogene	0.064	0.250	1.47	0.68	+18.1	Ronov (1980)
Neogene	0.129	0.410	2.97	2.45	+22.4	Ronov (1980)
Total or average for period (PR_3–N)	1.325	6.473	30.35	—	+19.5	

C. Sulphur balance and fluxes to the sedimentary envelope

The rates of accumulation of sulphate and pyrite which were estimated in previous sections, are presented in Fig. 2.28. It can be seen from this figure that the maximum rates of pyrite and evaporite sulphur accumulation, in sulphate rocks, are not coincident in time. This results from the different climatic conditions required for their accumulation and from the shift in maximum evaporite formation towards the initial and final tectonic stages (Ronov *et al*., 1969). In addition to these episodic changes the rates of accumulation of pyrite and sulphate sulphur tend to increase with the age of rocks (Grinenko *et al*., 1973b). In addition, the light isotope of sulphur (^{32}S) tends to increase in pyrite, irrespective of whether the dispersed pyrite of rocks from the Russian platform, or the average data on pyrite inclusions are considered. They were assumed to be associated with the development of life and the increase in free oxygen in the atmosphere (Grinenko *et al*., 1973b).

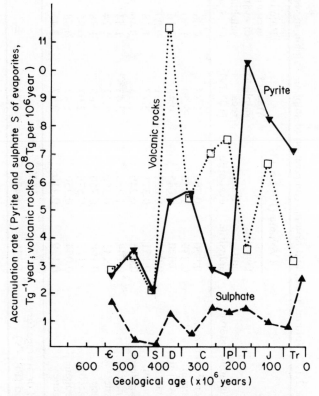

Fig. 2.28   Change in sedimentation rates of sulphide and sulphate sulphur in evaporites and volcanic rocks of the sedimentary cover of continents during the Phanerozoic

However, an increase in sulphur flux to the sedimentary envelope requires a renewal of the sulphur sources. Ronov (1976) assumed that for carbon the source was renewed at the expense of CO_2 influx during volcanic activity. The relationship between carbonate formation and volcanic activity is sufficiently clear, but that between organic carbon and the rate of pyrite formation appears to have been obscured by climatic factors. A relationship between sulphur and volcanic intensity has not been found, possibly due to the buffering activity of the ocean; nor was it observed when comparing the rate of increase in evaporite sulphur and the rate of accumulation of pyrite sulphur. The greatest differences were found for the Mesozoic–Cainozoic period. However, the sharp increase in the intensity of volcanic activity (Ronov *et al.*, 1979) in oceans, which is significantly greater than on continents, might explain the initiation of sulphur accumulation in the sedimentary envelope.

Figure 2.29 presents data on the isotopic composition of evaporite sulphur (Holser and Kaplan, 1966; Nielsen, 1973; Claypool *et al.*, 1980) and previously adopted estimates of the isotopic composition of sedimentary sulphides (Table 2.16). Provided the direction of variations of $\delta^{34}S_{pyr}$ in continental sedimentary rocks in time is confirmed by subsequent research, then there arises the necessity of taking into account some non-homogeneity of the sulphur isotope fractionation between sulphide and sulphate reservoirs in the course of geological time in models devoted to the sulphur cycle (Holser and Kaplan, 1966; Nielsen, 1973; Garrels, 1975; Schidlowski *et al.*, 1977; Rees, 1970; and others). It should be noted that the important role of the sedimentation rate of evaporites in influencing the isotopic composition of sulphur in the oceans and the necessity of accounting for it in modelling, as pointed out by Rees (1970), is well confirmed by the correlation between rates of accumulation of sediments and the isotopic composition of sedimentary sulphides (Fig. 2.29).

Corrections introduced during the evaluation of the amount of evaporite sulphur result in a substantial change to the total sulphur balance in the sedimentary envelope, compared to earlier calculations (Holser and Kaplan, 1966; Grinenko *et al.*, 1973b; Ronov and Yaroshevsky, 1976). The average isotopic composition of sulphur in the sedimentary envelope parallels these changes. There was only a slight concentration of the light isotope (Table 2.13) compared to the meteorite standard ($\delta^{34}S = -2.5‰$).

The distribution of sulphur and its isotopic composition in the sedimentary envelope of subcontinents (continental slope and shelf) are similar to that in continental rocks. This was confirmed by the data on organic carbon in rocks from the continental slope obtained from cores of the Deep Sea Drilling Project (Trotsyuk, 1979). The concentration of pyrite sulphur calculated from its relationship with organic carbon, 0.17% S, was in good agreement with the accepted value of 0.169%. The total amount of sulphur in the sedimentary envelope of continents and subcontinents amounts to 5.19×10^9 TgS.

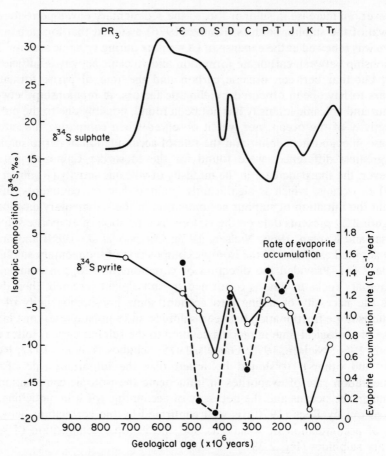

Fig. 2.29 Variation in isotopic composition of the sedimentary sulphide and sulphate sulphur of evaporites (Claypool *et al.*, 1980) and rate of evaporite accumulation in the Phanerozoic

2.2.4 Sulphur in the Sedimentary Layer of the Oceanic Part of the Earth's Crust (Seismic Layer I)

This layer consists of unconsolidated sediments, of average depth 0.4 km (Khain *et al.*, 1971). The rocks of this layer have been fairly well studied although the data on sulphur distribution in cores from the Deep Sea Drilling Project are rather scarce. Less representative are data on the isotopic composition of sulphur in these sediments (Krouse *et al.*, 1977; Sweeney and Kaplan, 1980; Migdisov *et al.*, 1979).

There are more data available on the surface horizon of oceanic basins which include analyses of the different forms of sulphur and their isotopic composition (Lein *et al.*, 1976; Volkov *et al.*, 1976; Migdisov *et al.*, 1978). To evaluate the amount of sulphur in sediments of the ocean, use was made of the data on the surface distribution of the different types of sediments (Table 2.13). The different rates of accumulation and changes in reduced sulphur concentration down to the depth of layer I were accounted for by calculating sulphide sulphur from the organic carbon concentrations.

The average concentrations of organic carbon in these sediments were obtained from the Deep Sea Drilling Project (Trotsyuk, 1979), and the coefficient suggested by Goldhaber and Kaplan (1974) was used as the average pyrite sulphur : organic carbon ratio. It might be well to point out that for the subcontinental region, the two available values for the inner part of the ocean were close to each other (0.038% and 0.036% of reduced sulphur). The amount of sulphur in sediments of layer I was calculated from 229 analyses of the different forms of sulphur and 75 analyses of their isotopic composition. This reservoir is very small, and contains 0.31×10^9 TgS (Table 2.19). Sulphate sulphur predominates in this layer and its isotopic composition resembles that of sulphur of marine water. Reduced sulphur in this layer amounts to 0.07×10^9 TgS and it has a very light isotopic composition ($\delta^{34}S = -43.6\%$). Such a value is characteristic of pelagic sediments of the ocean that have low rates of sulphate reduction. Overall, the sulphur of layer I sediments is only slightly enriched with ^{34}S; its $\delta^{34}S$ value is $+4.8\%$.

2.2.5 Entire Sedimentary Envelope of the Earth

The sedimentary envelope of the earth contains about 6.6×10^9 TgS; its isotopic composition shows a slight accumulation of ^{32}S ($\delta^{34}S = -2.2\%$, Table 2.13). The sulphur reservoir in sediments together with the ocean amounts to 7.85×10^9 TgS and its isotopic composition in the surface envelopes is close to that of the meteoritic standard ($\delta^{34}S = +1.4\%$, Table 2.13). The value obtained is close to that of the mean isotopic composition of volcanic sulphur (Borisov, 1970). From this we may conclude that there is only a small fractionation of the sulphur isotopes as sulphur moves from a deep reservoir to the outer envelope of the earth.

Intensive processes of sulphur fractionation occurring in the surface layer of the globe are balanced and point to deep sulphur as the source supplying sulphur to the cycle by volcanism and weathering.

2.3 SULPHUR IN THE 'GRANITE' ENVELOPE OF CONTINENTS

The 'granite' layer occupies over one-third of the total mass of the continental crust (Table 2.18). The composition of the metamorphic envelope of continents was determined from rocks composing the Precambrian shields.

Table 2.18 Volume, mass and average thickness of the envelopes (after Ronov, 1980)

Geosphere	Envelope or strata	Volume (10^8 km³)	Average thickness (km)	Mean density (gm cm⁻³)	Mass (10^{12} Tg)	% of the total volume Earth	% of the total volume Earth's crust	% of the total mass Earth	% of the total mass Earth's crust
Earth's crust	Sedimentary envelope	11.1	2.2	2.5	2.7	0.10	10.9	0.05	9.5
	Granitic strata	30.4	14.0	2.7	8.2	0.28	29.8	0.14	28.8
	Basaltic strata	60.4	21.7	2.9	17.5	0.56	59.2	0.29	61.4
Mantle		8 990	2880	4.5	4016	82.9	—	67.1	—
Earth's core		1 750	3471	11.0	1936	16.2	—	32.4	—
Entire earth		10 842	6371	5.52	5980	100.0	—	100.0	—

Estimation of sulphur distribution in these rocks is based on original data from (1) various metamorphic rocks of the Baltic and Ukrainian shields (Grinenko *et al.*, 1973b; Ronov *et al.*, 1977, unpublished data), (2) sulphur determinations of the Canadian shield (Shaw *et al.*, 1967; Fahrig and Eade, 1968), and (3) numerous analyses described in the literature. Altogether, about 7000 samples of the 'granite' envelope were considered. The isotopic composition of its sulphur was based on results of 648 assays.

Meta-sedimentary rocks represent a special problem (Dimroth and Kimberley, 1976; Cameron and Baumann, 1972). From analyses of 460 Archaean shales of the Canadian shield, Cameron and Baumann (1972) determined the average sulphur concentration to be about 1%. Analogous data are presented in other studies which explained the reasons for the high sulphur content in Archaean meta-sedimentary formation. It should be noted that samples for these assays were taken near deposits (Dimroth and Kimberley, 1976), many of which are characterized by the development of sulphide facies. Cameron and Garrels (1980) in their recent study outlined the possibility of an important accumulation of sulphur in the Archaean shales due to volcanism and hydrothermal activity. Assays of samples taken regionally and covering large areas (Shaw *et al.*, 1967; Fahrig and Eade, 1968) did not indicate a substantial enrichment with sulphide. Analyses of 1094 samples of Precambrian shales gave a mean value of 0.132% for the sulphur concentration in this type of rock. Estimation of the influence of metamorphic processes on the distribution of sulphur and its isotopic composition within this envelope suggested a substantial loss of sulphur with increasing intensity of metamorphism. The isotopic composition of sulphur in these rocks approached that in hypogene rocks (Ronov *et al.*, 1977). These and other data reported in the literature (Grinenko and Grinenko, 1974, 1975; Wedepohl, 1973; etc.) suggest a substantial sulphur efflux during metamorphism to outer envelopes of the earth's crust.

Overall the sulphur concentration in the metamorphic envelope of continents (0.094% S) is only one-third of that in the sedimentary envelope (Table 2.19). As shown in the previous section, the amount of sulphate sulphur in sedimentary rocks is about two-thirds of the amount of pyrite sulphur. In the granite envelope, however, there is a fivefold difference between the amounts of the two forms. The difference between their isotopic compositions, however, is insignificant ($\delta^{34}S_{pyr} = +1.9\%$; $\delta^{34}S_{SO_4^{2-}} = +2.8\%$). Our estimates for the concentration of sulphur and its isotopic composition in the 'granite' envelope differ from those obtained by Holser and Kaplan (1966) who (after Poldervaart, 1955) seem to have overestimated the mass of granite rocks and the concentration of the heavy isotope of sulphur ($\delta^{34}S = +10\%$). They also obtained a heavier isotopic composition for sulphur in the crust. We determined the amount of sulphur in the metamorphic envelope of continents to be 7.79×10^9 TgS.

Table 2.19 Concentration, isotopic composition and amount of sulphur in the earth's crust and hydrosphere (after Ronov, 1980, and sections 2.2–2.6)

Type of earth's crust	Envelopes or strata	Volume (10^6 km^3)	Mass (10^{12} Tg)	Sulphate			Pyrite			Total		
				(%S)	δ^{34}S (⁰/oo)	Mass (10^9 TgS)	(%S)	δ^{34}S (⁰/oo)	Mass (10^9 TgS)	(%S)	δ^{34}S (⁰/oo)	Mass (10^9 TgS)
Continental and subcontinental	Sedimentary	920	2.29	0.105[a]	+ 7.6	1.98	0.169	− 8.7	3.21	0.274	− 2.5	5.19
	'Granitic'	3 040	8.24	0.016	+ 2.8	1.35	0.078	+ 1.9	6.44	0.094	+ 2.0	7.79
	'Basaltic'	4 100	11.88	0.017	+ 1.3	2.09	0.057	+ 1.3	6.72	0.074	+ 1.3	8.81
	Total crust of continents and subcontinents	8 060	22.41	0.026	+ 4.3	5.42	0.076	− 0.5	16.37	0.102	+ 0.7	21.79
Oceanic	Sedimentary (layer I)	120	0.19	0.126	+19.4	0.24	0.038	−43.6	0.07	0.165	+ 4.8	0.31
	Basalts (layer II)	360	1.00	0.015	+ 3.3	0.15	0.050	+ 1.4	0.50	0.065		0.65
	Ultrabasic rocks (layer III)	1 690	4.90	0.010	+ 0.4	0.49	0.023	− 0.5	1.13	0.033	− 0.2	1.62
	Total oceanic crust	2 170	6.09	0.013	+ 6.1	0.88	0.031	− 1.9	1.70	0.044	+ 0.8	2.58
	Total earth's crust	10 230	28.50	0.022	+ 4.3	6.30	0.063	− 0.5	18.07	0.086	− 0.7	24.37
	Water of oceans	1 370	1.40	0.090	+20.0	1.26	—	—	—	0.090	+20.0	1.26
	Earth's crust and oceans	11 600	29.90	0.025	+ 6.9	7.56	0.063	− 0.5	18.07	0.085	+ 1.6	25.63

[a]Part of the sulphate sulphur was derived from oxidation of sulphide

2.4 SULPHUR IN THE 'BASALT' LAYER OF THE CONTINENTAL CRUST

The determination of the structure and composition of this part of the crust is very difficult. Until recently, the nature of the rocks forming the 'basalt' layer was not known. Ronov and Yaroshevsky (1976) from a study on the rates of transference of seismic waves in the basalt layer suggested that it was composed largely of basic rocks, and that the ratio between 'granite' and basic rocks in this layer was 1 : 2. They proposed that the composition of Archaean basic rocks be used for this layer since the oldest rocks correspond to the deepest layers in the crust. The data that formed the basis of this calculation are the same as those used for characterizing the 'granite' layer, also consisting predominantly of Precambrian rocks. The values obtained suggest that the 'basalt' envelope has a lower sulphur concentration, 0.074% S, than the 'granite' envelope, 0.094% S, and that the isotopic composition of the total sulphur is similar to that of the meteoritic standard. The total amount of sulphur in this envelope was calculated to be 8.8×10^9 Tg S, which is very close to that estimated for the 'granite' envelope (Table 2.19).

Altogether, the crust of continents and subcontinents contains 21.8×10^9 TgS, 75% of which is in the reduced state with a $\delta^{34}S$ value of $-0.5‰$. The remainder, the oxidized fraction, has an average $\delta^{34}S$ value of $+4.3‰$.

The main characteristic feature of sulphur in continental envelopes is the increase in sulphur concentration with progression from deep to surface zones. The content of oxidized sulphur increases in the same way. With this progression there is a clear-cut fractionation of the sulphur isotopes between sulphide and sulphate: in the deeper zones they have an identical isotopic composition, but in the progression to the surface there is an ever-increasing accumulation of ^{34}S in the oxidized sulphur and an enrichment of ^{32}S in sulphides of the sedimentary envelope. These changes in the continental crust are analogous to those occurring in sulphide and sulphate in the progression from Archaean rocks to Phanerozoic formation (Grinenko *et al.*, 1973b).

2.5 SULPHUR IN ROCKS OF THE OCEANIC CRUST

2.5.1 Seismic Layer II

The sulphur isotopic composition and sulphur distribution in sediments that constitute seismic layer I, have already been considered in the section of the sedimentary envelope. Seismic layer II has a predominance of basalt rocks; sedimentary rocks playing only a subordinate part. Ronov and Yaroshevsky (1976) proposed a ratio of 2 : 1 for the proportion of basalt to sedimentary rocks in this layer. Evaluation of the forms and amounts of sulphur in basalts of oceans is based on analyses of the forms of sulphur (Grinenko *et al.*, 1975, 1978; Migdisov *et al.*, unpublished data), the total sulphur content (Moore and Fabbi, 1971; Naldrett *et al.*, 1977, 1978) and its isotopic composition (Kanehira *et al.*, 1973; Grinenko *et al.*, 1975; Migdisov *et al.*, unpublished data). Altogether 455 samples and 47 isotopic determinations were used to evaluate the distribution and isotopic composition of sulphur in seismic layer II. The amount of sulphur in this layer amounts to 0.65×10^9 TgS. Reduced sulphur, which predominates, has a δ^{34}S value of $+1.4^0/oo$, which is close to that of the meteoritic sulphur; the remaining oxidized fraction, amounting to 0.15×10^9 TgS, is slightly enriched in ^{34}S, having a δ^{34}S value of $+3.3^0/oo$.

2.5.2   Seismic Layer III

Seismic layer III, consisting of ultrabasic rocks and basalts, occupies 1690×10^6 km^3 (Ronov and Yaroshevsky, 1976). Analyses of the different forms of sulphur and their isotopic composition in ultrabasic rocks were made on rocks taken from the Pacific and Indian oceans (Grinenko *et al.*, 1975; see Table 2.19).

2.5.3   Sulphur in the Oceanic Crust as a Whole

Altogether the oceanic crust contains, according to our estimates, 2.58×10^9 TgS, which is one order less than that of continents, 21.79×10^9 TgS. The average isotopic compositions of sulphur in continental and oceanic crust are identical, but there is an increase in the concentration of both forms of sulphur in the continental crust.

2.6   SULPHUR IN ROCKS OF THE ENTIRE EARTH'S CRUST

The total amount of sulphur in the crust amounts to 24.37×10^9 TgS, with three-quarters being in reduced form, while the sulphur in the crust and hydrosphere together amounts to 25.63×10^9 TgS. The isotopic composition of the reduced sulphur does not differ appreciably from the meteorite standard. The isotopic composition of total sulphur also approximates this value (δ^{34}S $= +1.6^0/oo$). Our estimate of the total amount of sulphur is slightly larger than that of Holser and Kaplan (1966). However, there is a large variation in the values obtained for the average isotopic composition of sulphur. The latest calculations do not confirm the large-scale isotopic fractionation of sulphur during exhalations on the earth's surface proposed by Ault and Kulp (1959) and Grinenko *et al.* (1973b). The statistical analysis of the δ^{34}S data on volcanogenic sulphur (Borisov, 1970) suggests a definite though comparatively slight fractionation of isotopes during volcanogenic processes ($\sim 2\%_{00}$).

Part II CYCLING

A. Yu. Lein

2.7 INTRODUCTION

Before considering the cycling of sulphur in the lithosphere, the cyclic nature of rock formation and decomposition in the earth's crust is described. This cycle includes the succession of three principal stages. The first stage—formation of rocks—is followed by a second one during which rocks undergo metamorphic changes due to tectonic shifts, such as uplifting of the oceanic crust, the sinking of rocks in geosynclinal regions, etc. In the third terminal stage rocks are destroyed by weathering and erosion.

During the last decade models of the rock cycle have been developed by various authors. The most accepted is that of Garrels and Mackenzie (1971, 1972), which was later corrected and expanded by Garrels and co-workers (Garrels and Perry, 1974; Garrels *et al.*, 1975). The model of Garrels and Mackenzie (1971) is based on the principal stages in the formation and decomposition of earth crust rocks. The same approach was used by Kempe (1979) for examining the carbon cycle in rocks of the earth's crust.

In assessing the global biogeochemical sulphur cycle, the same three stages of formation, metamorphism, and decomposition of sulphur-containing rocks are considered. However, the processes of sedimentation and diagenesis of the sulphur components of sedimentary rocks are given sufficient attention in the chapters devoted to the sulphur cycle in continental reservoirs (Chapter 5) and oceans (Chapter 6). For the present chapter the size of the sulphur pool in the lithosphere and the major fluxes between three main reservoirs—lithosphere, hydrosphere, and atmosphere—are assessed.

In this chapter tectonic processes that may lead to the redistribution of sulphur in rocks of continental and oceanic sectors of the earth's crust are discussed briefly. Detailed consideration is given to sulphur emission resulting from volcanic activity on continents and island arcs, since volcanic sulphur in addition to biogenic sulphur, provides the natural sulphur flux to the atmosphere; it is against this background that the emission of anthropogenic sulphur should be assessed.

Estimates of sulphur fluxes from crustal rocks to the other reservoirs (hydrosphere, atmosphere, soil, etc.) are made on the basis of recent data on weathering and erosion, to complete the survey of the sulphur cycle in the lithosphere.

2.8 DISTRIBUTION OF SULPHUR IN THE EARTH'S CRUST

Estimation of the sulphur pool in the earth's crust is considered in detail in sections 2.2–2.6. The present section briefly reviews information on sulphur contents in various zones and rocks of the earth's crust necessary for the quantitative assessment of sulphur fluxes in the lithosphere.

Attempts have been made by many geochemists at various times to assess the distribution of sulphur in the earth's crust (Table 2.1). The results of one of the most comprehensive studies of the distribution of sulphur in rocks and reservoirs of the earth's crust conducted by Holser and Kaplan (1966) are condensed in Table 2.20. Data obtained by Ronov and co-workers on the

Table 2.20 Sulphur distribution in various reservoirs of the earth's crust (after Holser and Kaplan, 1966)

Rock type or reservoir	Mass (10^9 Tg)	Mean sulphur concentration (%)	Sulphur content (10^6 Tg)
Deep oceanic rocks			
Sediments	300	0.025	75
Mafic rocks (to Moho)	4 400	0.063	2 300
Sedimentary rocks			
Sandstone	280	0.090	250
Shale	750	0.27	2 000
Limestone	290	0.13	380
Evaporite	30	17	5 100
Volcanic	120	0.04	50
Connate water	140	0.019	27
Total evaporites and connate water	170		5 100
All other sedimentary rocks	1 440	0.19	2 700
All sediments	1 600	0.49	7 800
Fresh water	0.3	0.0011	0.003
Ice	35	0.00003	0.006
Atmosphere	5.3		(3.6×10^{-6})
Seas	1 420	0.090	1 280
Total in sedimentary rocks and seas	3 020		9 100
Continental igneous and metamorphic rocks (to Moho)			
Granitic	10 500	0.021	2 200
Mafic	8 700	9.053	4 600
Total	21 000	0.032	6 800
Total in the crust	28 420	0.065	18 200

distribution of sulphur in the sedimentary mantle of the earth are summarized in Table 2.13.

The studies cited above show that sulphur occurs in two major forms in sedimentary rocks: pyrites (>90% of the reduced sulphur) and sulphates. The remaining forms of sulphur are (1) organic sulphur (0.1–10% of the total reduced sulphur), (2) elemental sulphur (0.01–5% of the reduced sulphur), (3) base metal sulphides, and (4) metastable iron sulphides (from trace amounts to 0.003%). The maximum concentration of reduced sulphur is found in argillaceous rocks of platform land areas (0.434%), while the minimum concentration is found in effusive rocks (0.045%, Table 2.13).

Sulphate sulphur varies within quite broad limits: from fractions of a per cent in the humic pelagic formations to solid sulphate layers in evaporites.

Values for the concentration and amount of sulphur in rocks of the earth's crust are given in Table 2.19. These data show that most of the sulphur is contained in rocks of continents and subcontinents. The total pool of sulphur in continental rocks amounts to 21.8×10^9 TgS, compared with the total in rocks of the earth's crust—24.4×10^9 TgS, and in the earth's crust including sulphates in oceanic water—25.6×10^9 TgS. These corrected estimates of sulphur distribution and amounts in the earth's crust exceed the values calculated by Holser and Kaplan (1966), though they are of the same order of magnitude.

2.9 SULPHUR FLUX FROM THE MANTLE

The amount of endogenic matter supplied to the earth's crust by the eruptions of basalt volcanoes, occurring over the ocean floor, was assessed by Lisitsin (1978) as 3×10^3 Tg year^{-1}. The value of the annual flux of matter from the mantle presented by Kempe (1979), 5×10^3 Tg year^{-1}, is close to the previous one, and was calculated from the total mass of the earth's crust, 24×10^{12} Tg (Garrels and Mackenzie, 1971) and the age of crustal rocks taken as 4.5×10^9 years. Using these rates of basalt generation and a sulphur concentration for basalts of layers II and III of the oceanic crust of 0.04% (Table 2.19), one obtains a sulphur flux from the mantle of 1.2–2.1 TgS year^{-1}. We also attempted to assess the amount of volatile sulphur released in the basalt–water interphase during basalt eruptions. If we assume that 3% of the total weight of erupted material in both submarine and land eruptions is volatile (Strakhov, 1976), then we conclude that the amount of volatile material is 90–150 Tg year^{-1}. Adopting a SO_2 concentration in volatiles of 10% (see section 2.10), we estimate that the amount of sulphur emitted in the basalt–water interphase is 4.5–7.5 TgS year^{-1}. In addition to these two hypogene sulphur fluxes due to basaltic eruptions, sulphur may also enter the sediments of the ocean floor through other pathways of submarine volcanism. The description of sulphur 'termite-hills' (see Chapter 6), the discovery of recent fumarole formations near the Esmeralda submarine volcano (Gorshkov *et al.*, 1980; Gavrilenko, *et al.*, 1980), and the findings of heavy-isotopic sulphur in the East Pacific submarine ridge (Migdisov *et al.*, 1979), may be cited as direct evidence of sulphur influx from the core to the spreading areas of the ocean floor. However, quantitative estimation of the sulphur flux from submarine volcanism is not yet possible.

2.10   SULPHUR FLUX DUE TO VOLCANISM IN CONTINENTS AND ISLAND ARCS

We did not estimate the sulphur flux between the oceanic and continental crust, since it is generally assumed that the masses of basaltic matter annually entering the oceanic crust in areas of mid-oceanic ridges due to volcanic processes are compensated by the burial of a similar amount of matter in geosyncline regions. Furthermore, the uplifting of the continental crust is compensated by matter sinking in the course of continental denudation.

The sulphur flux to the atmosphere due to andesitic and acidic volcanism of continents and island arcs is of particular importance in the global redistribution of the matter of the earth's crust. Two processes are mainly responsible for the natural sulphur supply to the atmosphere—biogenic and volcanic emissions. An estimate of the natural emission of sulphur to the atmosphere is important in assessing the magnitude of the anthropogenic flux in the global sulphur cycle.

The ocean spreading and subduction theory suggests that the material erupted by volcanoes of island arcs was formed at the expense of depressions in areas of subduction and melting of oceanic basalts and bedded sedimentary rocks. The rocks formed correspond to andesites in chemical composition. Some volcanic products return to the ocean in the form of pyroclastic material, while the volatile compounds are emitted to the atmosphere. The total flux of sedimentary material from the ocean in zones of subduction is assessed at 4.28×10^3 Tg year^{-1}, with an annual subduction rate of 2.3 cm and a density of dry residue of 2.14 gm cm^{-3} (Kempe, 1979).

If we accept 0.2–0.5% S to be the sulphur concentration in sediments of the shelf and continental slope (see Chapter 6), which are subject to the processes of subduction, then we can assess the sulphur flux from the ocean to zones of active volcanism to range from 8.6 to 21.4 TgS year^{-1}. This is a minimum estimate, since we do not take into account sulphur contained in basalts and ultrabasic rocks, which also melt and supply sulphur for andesitic volcanism.

The mass of andesite ash annually returned to oceanic sediments is assessed as 2.5×10^3 Tg year^{-1} (Lisitsin, 1978). The sulphur content in effusives (andesites) is 0.055% (see Table 2.13), and the sulphur flow with the pyroclastic matter is 1.37 TgS year^{-1}. Consequently, the minimum flux of volcanogenic sulphur to the atmosphere in the form of volatile components is from 7.2 to 20 TgS year^{-1}.

Below, we consider another method for estimating volcanic sulphur emission to the atmosphere based on the chemical analysis of gases, ashes, and rocks, their volumes, amounts of volatiles in rocks, and amounts of sulphur in volatiles.

There is no general agreement on the relative contribution of sulphur-containing gases of volcanic origin to the composition of atmospheric gases. Goldberg (1973) holds that the amount of volcanic sulphur emitted to the atmosphere each year is comparable with the sulphur emission from other natural sources. Other authors (e.g. Mészáros and Várhelyi, 1975) state that the contribution of volcanic sulphur to the atmosphere is much smaller than the biogenic flux. Such a discrepancy emphasizes just how little we know about the size of the volcanic sulphur flux to the atmosphere. This is primarily related to analytical problems concerning the composition of volatile compounds in areas of active volcanism. Another difficulty is caused by the high variability in the amount of sulphur compounds, which depends (even within the limits of the same volcano) on the composition and properties of the magma, the temperature of the eruption, and a number of other parameters. In this context, it is appropriate to cite a remark by Shepherd who devoted his whole life to the development of reliable methods of analysis for volcanic gases, and died convinced that no single analysis of volcanic gas 'worthy of being published' had been performed (Macdonald, 1975). Difficulties of analysis have been largely overcome by the introduction of chromatographic and spectrometric methods. There has also been a significant improvement in gas sampling techniques.

Sulphur is released by volcanoes in gaseous form, mainly as sulphur dioxide although smaller amounts of hydrogen sulphide are also discharged. Negligible amounts of sulphur are emitted in elemental form, as sulphur trioxide, or other oxidized forms.

2.10.1 Assessment of the Volcanic Sulphur Flux

Two independent methods were used to assess sulphur emissions to the atmosphere and to the surface of the globe: (1) estimation of sulphur emission during active volcanic eruptions, and (2) assessment of the sulphur flux from fumarole gases.

Most estimates of global sulphur emission are based on the computations of Sapper (1929), who reported that the mean amount of matter erupted annually by 450 volcanoes during the past four centuries was of the order of 0.98 km^3 of lava, ash, and other pyroclastic material.

Table 2.21 shows how various authors obtained different estimates of the sulphur flux to the atmosphere by using different volumes or weights of igneous rocks, and different concentrations of volatile components and sulphur gases in volatiles. It should be noted that very different values for the

Table 2.21 Sulphur emitted with volcanic material during eruption

Reference	Volume (km³ year⁻¹)	Density (gm cm⁻³)	Weight (10⁻³ Tg)	Volatile components			Annual emmission of SO₂ (TgS year⁻¹)
				Rock		SO₂ content (% w/w of volatile components)	
				(%)	Amount (Tg)		
1	0.98	3	3000	0.5	15	10	1.5
2	0.80	2.8	2200	5.0	110	^{a}	4.0
3	0.98	3	3000	2.5	7.5	10	7.5
4	—	—	—	—	—	—	9.0
5	—	—	—	—	—	—	34.0

[a]Molar composition of gas: SO_2—1, H_2O—95, CO_2—4.
References: 1, Kellogg et al. (1972); 2, Friend (1973); 3, Cadle (1975); 4, Granat et al. (1976); 5, Barteles (1972).

Table 2.22 Annual emission of sulphur dioxide due to fumarolic activity ($TgS\ year^{-1}$)

SO_2	S	Reference
7.0	3.5	Stoiber and Jepsen (1973)
30.0	15.0	Okita and Shimozuru (1975)
27.0	13.5	Naughton *et al.* (1975/76)

concentration of volatile components—from 0.5% to 5% of the magma weight were used (Table 2.21). The reason for the preference of a particular value is not always given, and this problem should be discussed further.

The value for the SO_2 content equal to 10% of the total weight of volatiles used by Kellogg *et al.*, (1972) and Cadle (1975), seems to be an overestimate. This value was based on analyses of Shepherd (1938) and Macdonald (1975) for gases of Hawaiian volcanoes with typical basaltic volcanism. In andesitic volcanoes, the water vapour content of the volatiles is markedly higher (70–95%) while the gas content, including SO_2, is lower.

Another group of researchers (Table 2.22) estimated the total flux of volcanic sulphur against the background of sulphur gas emission during periods of fumarole activity only. It should be noted that the flux values of Stoiber and Jepsen (1973) seem to be underestimates, because these authors used the data for only 100 volcanoes, whereas the total number of active volcanoes in the world may be 578 (Vlodavets, 1966) or even 850 (Lisitsin, 1978).

In addition, the estimate of the global sulphur cycle from data obtained for the Kilauea volcano (Naughton *et al.*, 1975/76) is probably too large, since the highest SO_2 content in volatile compounds is found in the basaltic volcanoes of the Hawaiian Islands (see above); these data then should not be used for calculation of the flux from all volcanoes.

Our assessment of the emission of gaseous sulphur compounds from volcanic sources to the atmosphere is based on the critical revision of all available material obtained during the studies of both fumarole activity of volcanoes and sulphur emission in various types of eruptions.

2.10.2 Sulphur Emission During Fumarole Activity of Volcanoes

Most of the time the so-called active volcanoes manifest fumarole activity. In the literature one can find thousands of analyses of fumarole gases and condensates from laboratories all over the world. White and Woring (1965) state that attempting to collate the results of all these analyses in the same units was 'a hard job'.

The most systematic material on the analyses of fumarole gases and condensates can be found in publications of Soviet and Japanese volcanologists.

Table 2.23   Sulphur dioxide release from volcanoes with known fumarole emissions

Volcano	Fumarole emission	SO_2 released (TgS year^{-1})	Reference
Ebeko	50–60 m^3 sec^{-1}	0.02	Markhinin (1967)
Kudryavyi	—	0.01	Vlasov (1971)
Larderello	26 Tg year^{-1}	0.092	Rittmann (1960)
Tolbachik (1976)	10 m^3 sec^{-1}	0.006	Menyailov *et al.* (1980)

The results of analyses demonstrate the variability in composition within any one fumarole or volcano in different periods of activity. The use of such data for the assessment of the global sulphur flux proved to be impossible because of the lack of information on gaseous emission and the total amount of fumarole activity in volcanic areas. For example, rough estimates for emissions during fumarole activity are known for the Kurile volcanoes of Ebeko and Kudryavyi (Table 2.23). Using the data of Rittmann (1960) one can

Table 2.24   Sulphur dioxide emission to the atmosphere from the fumarole fields of some volcanoes

Volcano	Year of analysis	Rate of SO_2 emission		Reference
		(Tg × 10^{-4}) day^{-1}	Tg year^{-1}	
Osima	1971	3.45	0.124	Okita and Shimozuru (1975)
Asama	1972	1.42	0.051	Okita and Shimozuru (1975)
Asama	1974	7.87	0.787	Okita and Shimozuru (1975)
Santjagito	1973	4.20	0.151	Stoiber and Jepsen (1973)
Fuego	1973	0.04	0.014	Stoiber and Jepsen (1973)
Pacaya	1973	2.60	0.094	Stoiber and Jepsen (1973)
St. Cristobal	1973	3.60	0.130	Stoiber and Jepsen (1973)
Telika	1973	0.02	0.007	Stoiber and Jepsen (1973)
Molotambo	1973	0.05	0.018	Stoiber and Jepsen (1973)
Masaya	1973	1.80	0.065	Stoiber and Jepsen (1973)
Karymsky	1975	1.85	0.066	Berlyand (1975)
Average		2.53		

calculate the sulphur flow in the steam jets of the Larderello volcano in Toscana (Table 2.23).

Of greater interest are the results in recent years of remote measurements of sulphur concentrations in fumaroles for a number of volcanoes. The data compiled in Table 2.24 enable us to assess the mean annual SO_2 emission from the results obtained during the investigation of 10 different volcanoes.

For a daily average emission of 2.53×10^{-4} $TgSO_2$ (Table 2.24), we calculated the annual sulphur dioxide flow from fumarole terrains of an active volcano as 0.092 $TgSO_2$ and the total emission from 578 active volcanoes (Vlodavets, 1966) as 53.2 $TgSO_2$ $year^{-1}$ or 26.6 TgS $year^{-1}$.

2.10.3 Emission of Sulphur Gases During Volcanic Eruptions

A number of scientists (Koyama *et al.*, 1965; Stoiber and Jepsen, 1973) believe that following volcanic eruptions most sulphur compounds are leached from the erupted rocks by rain-water, and thus are not released to the atmosphere but enter river runoff. However, analyses of aqueous extracts of fresh ashes after the eruption of some Kurile volcanoes showed that they contained only one-third of the sulphur and the remainder had been lost to the atmosphere (Markhinin, 1967).

Table 2.25 contains a list of the 17 most important volcanic eruptions of the twentieth century. The total volume of erupted material during the past 75 years exceeds 56 km^3, or 0.73 km^3 $year^{-1}$. If one takes into account that a considerable amount of volcanic matter is discharged during small-scale eruptions, then the value of 1 km^3 of matter per year used by many authors seems to be fairly reasonable (Table 2.21).

To assess the flux of volatile components we assume (as did Strakhov, 1976) a value of 3% of the total weight of erupted matter. This value was obtained by Markhinin (1967) when calculating the energy necessary for the discharge for volcanic ash. However, to assess the sulphur flux one should know the content of sulphur compounds. Judging from the 26 analyses of gases from basaltic lavas of Hawaiian volcanoes, the SO_2 content is about 10% (Miyake, 1965). These data may be used for 15% of all eruptions of Hawaiian or effusive-type volcanoes, characterized by a quiet eruption of fluid basaltic lava.

Very few data exist on the emission of sulphur gases during the explosive eruptions of volcanoes. According to Abramovsky *et al.*, (1977), some 1.6 Tg of sulphur dioxide were emitted to the atmosphere over nine months of 1975 during the eruption of Tolbachik volcano. At the same time about 100 Tg of volatile compounds and some 3600 Tg of pyroclastic material were discharged. On the basis of these data we assume that during explosive andesitic volcanism, characteristic of 85% of all active volcanoes, the SO_2 concentration in the volatile components is of the order of 2%.

Table 2.25 Large (catastrophic) volcanic eruptions during the twentieth century from Abdurakhmanov *et al.*, 1976; Basharina, 1966; Markhinin, 1967; and others)

Year of eruption	Volcano	Volume of igneous rock (km^3)
1902	Santa Maria	5.4
1902/3	Sufrier Vincent	1.0
1905	Motobanu	1.0
1907	Ksudach	3.0
1912	Katmai	28.0
1914	Minai-Ivoshima	1.2
1914	Sakuradzima	2.2
1930	Sarychev	0.5
1931	Severgin	1.5
1944/45	Klyuchevskaya Sopka	0.8
1947/48	Gecla	1.7
1950/51	Ambrin	0.8
1954	Miyamuragir	0.5
1956	Bezymyanny	3.0
1961	Bezymyanny	1.8
1964	Shiveluch	1.3
1968	Sierro-Negro	0.7
1975	Tolbachik	1.8
	Total	56.2

Thus, for the mean annual supply of erupted volcanic matter of 1 km^3 and a weight of 3×10^3 Tg, the total flux of volatile components to the atmosphere amounts to 90 Tg (3% of the total weight of erupted matter). Of this volatile material 15%, or 13.5 Tg, contain 10% SO_2 and the remaining 76.5 Tg released in andesitic eruptions contain about 2% SO_2. From this it is concluded that every year about 2.9 Tg sulphur gases (SO_2), or 1.45 Tg of sulphur, are emitted to the atmosphere. Adding this to the sulphur flux from fumaroles (see above), we estimated that the annual flux of volcanic sulphur compounds from the lithosphere to the atmosphere is about 28 Tg of sulphur.

2.11 SULPHUR FLUXES IN THE NATURAL WEATHERING OF CONTINENTS

It is impossible to construct a model of the rock cycle without taking into account the volumes and masses of weathered rocks, and the pathways and rates of the transport of matter. This is particularly important for the construction of a model of such cyclic constituents of the earth's crust as sulphur, carbon, and nitrogen. There is also an urgent need for separating and assessing the anthropogenic flux of substances on their way from the lithosphere, through the atmosphere, soil, and hydrosphere back to the lithosphere.

Several attempts have been made to describe the sulphur cycle in nature (Eriksson, 1960; Junge, 1963; Holser and Kaplan, 1966; Robinson and Robbins, 1970; Kellogg *et al.*, 1972; Friend, 1973; Granat *et al.*, 1976; Krouse and McCready, 1979; etc.). These reviews were mainly concerned with the atmospheric sulphur cycle. Only Friend (1973) and Granat *et al.* (1976) estimated the flux of sulphur due to weathering. The total sulphur flux due to river runoff was estimated by Friend as 136 TgS year^{-1}. He included the sulphur leached from the pedosphere (89 TgS year^{-1}), fluxes of volcanogenic sulphur (5 TgS year^{-1}) and the sulphur flux due to weathering of rocks (42 TgS year^{-1}). All estimates were obtained indirectly. Results obtained by Friend form the basis of the sulphur cycle in the hydrosphere considered by Krouse and McCready (1979). The model of Granat *et al.* (1976) is based on the sulphur balance in the pedosphere during the pre-industrial period (Fig. 2.30). Assuming the mass of weathered rocks to be 10^4 Tg year^{-1}, and the sulphur content 0.33%, a value of 33 TgS year^{-1} for the weathering of rocks in the pedosphere during the pre-industrial period was obtained. Almost the same value was obtained for the anthropogenic flux (Granat *et al.*, 1976).

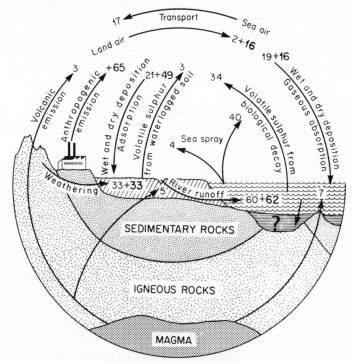

Fig. 2.30 The global sulphur cycle; the transfers shown are in TgS year^{-1}. Small type denotes the transfers estimated to have prevailed before man had a significant influence on the sulphur cycle. The bold figures give estimates of man's additions in various activities (from Granat *et al.*, 1976)

We think that both values are greatly underestimated since Granat *et al.* (1976) estimated only the sulphur flux to the ocean due to river runoff. The sulphur fluxes to continental reservoirs due to weathering of rocks was not taken into account and the role of other processes of weathering was underestimated. In Chapters 5 and 6, it will be shown that twice as much sulphur (135.5 Tg) is accumulated in sediments of the oceans and the mediterranean seas than was estimated by Granat *et al.* (1976).

We propose a new variant for the calculation of the sulphur flux during weathering of continental rocks based on improved values for masses of weathered matter and on a large number of new estimates of the sulphur content in various types of rocks, regions, and in the earth's crust as a whole (Tables 2.13 and 2.19).

2.11.1   Weathering of Mineral Matter from the Earth's Crust

The first attempts to assess the annual transfer of mineral matter from land to ocean on a world scale were made during the last century (Geicki, 1862; Reade, 1889; Penk, 1894). The principal method for estimating the mass of weathered rock is to assess the particulate and dissolved mineral matter in river runoff (Solisbary, 1909; Clarke, 1924; Lopatin, 1950; Alekin and Brazhnikova, 1964; Corbel, 1964; etc.) Bondarev (1974) proposed an equation to balance land weathering based on the well-known equation of Lopatin (1952):

$$\Delta M = - (R_T + G_T + S_T + V_T + Gl + A_T - W - Vl - N - B - F - C + T)$$

where:

ΔM = quantity of weathered land mass; R_T = flux of colloidal matter in runoff and G_T = flux of suspended and solid matter in runoff (total, 14.1×10^3 Tg year^{-1}); S_T = flux of dissolved matter in runoff $(1.6-1.7 \times 10^3$ Tg year^{-1}); V_T = flux of matter transported by wind $(2.0-4.0 \times 10^3$ Tg year^{-1}); Gl = flux of glacial exaration in areas of development of the contemporary glacial cover $(2.2-2.3 \times 10^3$ Tg year^{-1}); A_T = flux of matter removed by marine abrasion $(0.7-1.0 \times 10^3$ Tg year^{-1}); T = flux of technogenic weathering (combustion of mineral fuels) $(2.6 \times 10^3$ Tg year^{-1}); W = flux of gaseous compounds to the atmosphere during weathering $(0.1-0.6 \times 10^3$ Tg year^{-1}); Vl = flux of volcanic products $(1.8 \times 10^3$ Tg year^{-1}); N = flux of salts transported by wind from the ocean to the land; B = flux of biogenic matter; F = flux of mineral water from deep underground water; C = flux of matter to the land surface from the cosmos. The total loss of mineral matter from the land surface is $23.2-25.7 \times$ Tg year^{-1} and the total gain is $2.9-4.4 \times 10^3$ Tg year^{-1}.

Thus, according to Bondarev (1974), the land surface loses $20.3-21.3 \times 10^3$ Tg of its matter annually.

A substantially higher value for land weathering (48.95×10^3 Tg year^{-1}) was obtained by Gorshkov (1980). Unlike earlier authors, he introduced, as a credit item, the flux of land matter not only to oceans but also to inland seas. About 18.2×10^3 Tg of matter is accumulated in inland reservoirs annually, of which 13.4×10^3 Tg accumulates in man-made basins. The values for some other items were also changed. We took Gorshkov's (1980) estimate for the fluxes of mineral matter in land weathering (Table 2.26) as a basis for estimating the sulphur fluxes. The only change we introduced concerns the flux of dissolved matter in river runoff (estimated by Gorshkov as 3.4×10^3 Tg year^{-1}). This flux includes dissolved organic matter in colloidal forms, minerals, ions of salts transferred to the land from the ocean, the ion flux of mineralized deep waters transferred along tectonic faults from depths below sea-level, and the flux of carbonate ions of atmospheric origin. In total, this flux amounts to 1.7×10^3 Tg year^{-1} and should not be used for estimation of the sulphur flux in river runoff. The other part of the flux of dissolved matter in river runoff (1.7×10^3 Tg year^{-1}) represents the value of true chemical weathering or the ionic flux (Bondarev, 1974). From the data given in Table 2.26 it is evident that an anthropogenic component is present, to a variable extent, in all fluxes except for the flux of the morainic matter. The processes listed in Table 2.26 embrace the transfer not only of rock masses but also of mineral masses of the soil envelope of the earth (pedosphere). The impact of soil may be neglected only when considering glacial exaration, moraine abrasion, and direct underground flux from the lithosphere to the ocean. For all other cases it may arbitrarily be assumed that 10% of the total mass of transferred mineral matter (ΔM), is extracted from the soil layer, since arable land and perennial plants occupy a surface of 15×10^6 km^2; i.e. 10–11% of the continents not covered by glaciers (Vorobyov *et al.*, 1972). Some values given in Table 2.26 are comparable with estimates of Garrels and Mackenzie (1971), while others are considerably higher than those cited in the works of that period. For instance, from the balance of weathered material calculated by Judson (1968), about 9.6×10^3 Tg of matter is transferred annually by rivers, winds, and ice. Our estimate of these three items is at least twice that amount (23.6×10^3 Tg year^{-1}, Table 2.26).

Estimates of fluxes of matter carried by river runoff obtained by Degens *et al.* (1976) and Gorshkov (1980) agree closely, even though the authors used different approaches.

The rate of removal of land matter in the natural processes shown in Table 2.26 amounts to 51.2×10^3 Tg year^{-1}. The loss of land matter is partially compensated by an influx of matter from outside (see Table 2.27), but the major mass of this material enters the soil rather than the lithosphere.

Table 2.26   Fluxes of mineral matter in land weathering processes

| Process | Flux (10^3 Tg year^{-1}) | |
	After Garrels and Mackenzie (1971)	After Gorshkov (1980) and present author
River runoff		
Suspended and solid matter	12.5	17.5
Dissolved matter		1.7[b]
Underground runoff		
Dissolved matter		1.0
Exaration of moraine material	1.85	2.4
Marine abrasion	0.25	0.7
Wind erosion	0.054	2.0
Total flux of matter to ocean		25.3
Transfer of dissolved matter to continental water bodies		18.2
Emission of gaseous compounds to the atmosphere		7.7
Total removal of matter in natural processes		51.2

[a] After the data of Degens *et al.* (1976) and Bolin *et al.* (1979) for the Black Sea depression.
[b] After Bondarev (1974).

Table 2.27   Influx of matter to the earth's crust (after Gorshkov, 1980)

Sources or processes	Flux (Tg year^{-1})
Cosmic matter	1
Cyclic salts	580
Flux of salts into minerals at the surface of the earth	1 862
Accumulation of matter during peat-bog formation	100 (600)[a]
Flux of soil humus	1 500
	4 043 (4543)

[a] From Bolin *et al.* (1979).

2.11.2   Flux of Sulphur Due to Natural Weathering of Continents

Having obtained the masses of matter transferred from land to other reservoirs (hydrosphere, atmosphere, and pedosphere) in the process of weathering (Table 2.26), we are now able to assess the amount of sulphur transported in the separate components of this flux. We used the data of Blatt and Jones (1975) on the various types of rocks occurring at the surface of continents undergoing weathering and leaching (Table 2.28), with the sulphur content in each type of rock (Tables 2.13 and 2.19). Using this approach we assessed the rate of sulphur removed from the lithosphere by weathering as 112.43 TgS year^{-1} (Table 2.28).

To calculate the sulphur removed in each of the weathering processes, we used the data of Table 2.26. The ratios of the fluxes of weathered materials to the total removal of matter from the lithosphere have been determined, and the amount of sulphur in each of these fluxes has been calculated on the basis of the total sulphur efflux from the lithosphere during weathering (Table 2.29).

We assumed that all the sulphur is unstable in hypergene conditions and is leached from particulate products of weathering during transfer. In other words, we think that the particulate matter of river runoff undergoes mechanical and chemical weathering simultaneously. Hence, the sulphur flux from the lithosphere to the ocean due to river runoff may be estimated from the total flux of the transferred particulate and ionic matter, and amounts to 37.94 TgS year^{-1} (Table 2.29).

In two cases—when assessing the sulphur fluxes to the ocean in underground water and to the ocean via the atmosphere—we used a different approach. Matter transported by underground runoff was assessed from the small amount of factual material representative only of the Baltic Sea basin (Gorshkov, 1980). Therefore, we used the value of 9.2 TgS year^{-1} cited in Table 2.29 for this sulphur flux. Sulphur released during the decay of organic matter in lithosphere rocks (flux to the hydrosphere via the atmosphere) was

Table 2.28 Sulphur content and weathering of the various rocks in the continental surface

| Rock type | Crystalline | | | | |
	Effusive	Intrusive	Metamorphic and Precambrian	Sedimentary	Total
Rock (%)[a]	8	9	17	66	100
Total sulphur (%)[b]	0.055	0.074	0.094	0.274	
Rock weathered (10^3 Tg year^{-1})	4.33	4.87	9.2	35.71	54.1
Sulphur removed due to weathering (TgS year^{-1})	2.38	3.60	8.65	97.8	112.43

[a] After Blatt and Jones (1975).
[b] From Tables 2.13 and 2.19.

Table 2.29 Mass of transported material and sulphur fluxes from the weathering of the earth's crust due to natural causes

	Mass of transported matter (10^3 Tg year^{-1}) (see Table 2.26)	Sulphur flux (TgS year^{-1})	
Process		From land	From the lithosphere (90% of runoff from land)
Particulate runoff[a]	17.5	42.15	37.94
Ionic runoff[a]	1.7		
Underground runoff[a]	1.0	9.2	9.2
Exaration	2.4	5.27	5.27
Abrasion	0.7	1.54	1.54
Wind erosion[a]	2.0	4.39	3.95
Total runoff to ocean	25.3	62.55	57.90
Runoff to continental water bodies[a]	18.2	39.95	35.95
Emission of components to hydrosphere via the atmosphere	7.7	9.9	8.91
Total runoff of matter and total sulphur flux due to natural processes	51.2	112.4	102.8

[a]Weathering processes involving lithospheric rocks and soil. Total runoff of matter due to weathering and erosion is 5.65×10^3 Tg year^{-1}, or 10% of the total mass of weathered land matter.

calculated as a difference between the total runoff of sulphur and the total sulphur removed with other fluxes (Table 2.29).

A comparison of the sulphur fluxes obtained above with those in continental water bodies (Chapter 5) is of interest. Sulphur runoff to continental reservoirs calculated from the weathering processes budget is 35.95 TgS year^{-1}. The magnitude of an analogous flux calculated in Chapter 5 (see Table 5.25) on the basis of the sulphur content of sediments in lakes and water reservoirs and runoff to continental water-bodies was 35 TgS year^{-1}, which agrees well with the above value.

Sulphur runoff due to the leaching of rocks and soils was assessed at 87.3 TgS year^{-1} (Table 5.25). In total, the sulphur flux with river runoff to the ocean and inland seas during land weathering amounts to 82.1 TgS year^{-1}. If we take into account the sulphur fluxes during exaration, abrasion and wind erosion, this value will be 93.3 TgS year^{-1} (Table 2.29). In other words, two absolutely different approaches to the estimation of the annual natural sulphur fluxes from continents give rather close results.

2.11.3 Sulphur Flux to the Surface of Continents

A. Sulphur flux from outer space

Every year about 1.0 Tg of matter from outer space is deposited on earth (Garrels and Mackenzie, 1971); 86% of this is chrondritic material (0.46 Tg year^{-1}). If the sulphur content of cosmic matter is 0.002% (Meyer, 1977), then the sulphur flux from outer space is below 10^{-3} TgS year^{-1}.

B. Sulphur flux with cyclic salts

This flux was assessed in Chapter 4 as 6 TgS year^{-1}. This sulphur and that released during the decay of organic matter, includes highly soluble sulphur compounds which are quickly leached from continents to oceans and continental reservoirs. Consequently, one may consider them as transit fluxes and not credit items for the sulphur budget of the lithosphere.

C. Sulphur flux to the lithosphere during peat-bog formation

The amount of peat deposited each year is estimated by Bolin *et al.* (1979) to be 600 Tg year^{-1}. The sulphur content in plants is about 0.1%, hence the flux to the lithosphere in the course of peat-bog formation could amount to 0.6 TgS year^{-1}. Sulphur fluxes to the lithosphere due to sediment accumulation in marine and continental water bodies are discussed in Chapters 5 and 6.

2.12 FLUX OF SULPHUR FROM CONTINENTS BY ANTHROPOGENIC WEATHERING

An outstanding Soviet geochemist, Vernadsky (1967), stated that 'undoubtedly, with the advent of man on the surface of our globe there appeared a new tremendous geologic force'.

2.12.1 Anthropogenic Weathering of the Continental Lithosphere

A. Mechanical weathering

About 20×10^6 Tg of matter are shifted annually as a result of man's industrial activity (Ryabchikov, 1970, 1972; Kirichenko, 1963; and others); this includes mining of minerals and construction materials, construction of canals, dams, dikes, roads, conduits, slope terracing, earthwork for foundations, levelling of construction sites, ploughing, accumulation and burial of various wastes, etc. (Bondarev, 1974). Except for agricultural ploughing, man's activities disturb the lithospheric cover of continents and increase the potential of the shifted rock masses to weathering and leaching. It has been calculated that the natural tectonic forces shift only $5-10 \times 10^6$ Tg of matter per year, i.e. half as much as man does (Bondarev, 1974).

Ores and bedrocks mined annually (20×10^3 Tg) should be given special consideration, as they contain more sulphur than rocks used for construction (Sidorenko, 1978). If we assume that the mean sulphur concentration of continental rocks is 0.1% (Table 2.19), then 20 TgS year^{-1} is transported in the material shifted by man.

B. Fuel weathering

By fuel weathering we mean the mining and combustion of solid mineral fuels: coal, peat, and oil-shales. In 1967 the world output of coal and lignite was 2740 Tg year^{-1} (Goldberg, 1973). A figure close to this, 2800 Tg year^{-1}, is mentioned in surveys for 1978–79 (Sidorenko, 1978; Moiseev *et al.*, 1979; Shvedov, 1976; and others). During the past century the world output of coal and lignite has increased 10-fold, i.e. from 250 to 2800 Tg year^{-1} (Moiseev *et al.*, 1979). It is likely that a decrease in oil consumption, due to the exhaustion of oil resources and political instability in major oil-producing regions, will be compensated for by an increase in consumption of coal during the next two decades. The consumption of coal could amount to 4.36×10^3 Tg year^{-1} by the year 2000 (Shvedov, 1976).

At present, fuel consumption is 2.6×10^3 Tg year^{-1} (Table 2.30), which is 5% of the global weathering of continental rocks (Table 2.26).

Table 2.30   Weathering of land due to fuel production (after Bondarev, 1974)

Type of fuel	World output (Tg year^{-1})	Mean ash content (%)	Weathering due to fuel production (Tg year^{-1})
Coal (1972)	2141	15	1820
Lignite (1970)	765	25	574
Peat (1967)	198	12.5	173
Oil-shale (1967)	100	55	45
Total			2612

2.12.2   Magnitude of Anthropogenic Weathering of the Continental Lithosphere

In the weathering of land in other than glacial areas, 47% is contributed by man (Bondarev, 1974; Gorshkov, 1980). The anthropogenic flux includes transfers of suspended, solid, and dissolved mineral matter, and aeolian transport to continental water bodies and oceans.

Douglas (1967) found a relationship between the rate of weathering and monthly and yearly average precipitation. Based on this dependence he believes that the magnitude of weathering under natural conditions is of the order of several m^3 per km^2 per year. In some regions, the anthropogenic factor may increase this value 50 times. Degens *et al.* (1976), having studied the lithology and age of modern thin-layer sediments of the Black Sea and the average rate of weathering of rocks in the Black Sea catchment basin (0.063 mm year^{-1}), also came to the conclusion that 50% of the material transferred by rivers is due to the anthropogenic influence. If this holds true, the contribution of anthropogenic weathering will amount to 25.6×10^3 Tg year^{-1}; including fuel weathering it totals 28.2×10^3 TG year^{-1}.

Summing up, we may conclude that 53.8×10^3 Tg of matter is lost from the lithosphere each year, of which 25.6×10^3 Tg year^{-1} is due to natural weathering processes and more than 28.2×10^3 Tg year^{-1} is lost as a result of anthropogenic processes. It should be noted that in the latter case we do not take into account the loss of matter during processing and use of minerals.

2.12.3 Contribution of the Anthropogenic Flux to the Weathering of the Lithosphere

A. Mineral sources of sulphur

Information on the types of sulphur-containing raw materials and sulphur output from these sources is condensed in Table 2.31. The sulphur output in the USA was 9.5 TgS in 1977 (Meyer, 1977). This constituted about 18% of the world sulphur production, which amounted to 52 TgS year^{-1} (Nriagu,

Table 2.31   Annual output of sulphur-containing raw materials (Meyer, 1977; Nriagu, 1978)

	Output (TgS year^{-1})	
Type	USA	World
Native sulphur	7.0	17.0
Processing of previously extracted sulphur	0.6	?
Pyrite ores	0.3	12.0
Gypsum and other types of sulphur-containing raw materials, including sulphur gases recovered during coal and gas combustion	1.6	8.0
By-products from new sources	—	15.0
Total	9.5	52.0

Table 2.32   Global uses of sulphur in industry and agriculture
(after Meyer, 1977)

Field of use	(%)	Sulphur used (TgS year^{-1})
Fertilizers	49	25.5
Pesticides	2	0.5
Chemical industry	19	9.9
Direct application to soil	70	35.9
Others	30	15.6
Total	100	52

1978). The use of sulphur in industry and agriculture is shown in Table 2.32 which shows that more than 70% of the sulphur is directly applied to soil (Gittinger, 1975; Meyer, 1977). Thus, there is the potential for much of this sulphur to be lost in river runoff.

The total flux of sulphur in river runoff from the lithosphere to oceans and continental water bodies is 131.16 Tg year^{-1} (Table 2.33, of which 57.6% is of anthropogenic origin. The mineral matter in present river runoff contains 0.33% S, whereas before man's interference it contained 0.105%. In the near future, all countries are expected to increase sulphur consumption and production of sulphuric acid (Fig. 2.31), particularly the socialist and developing

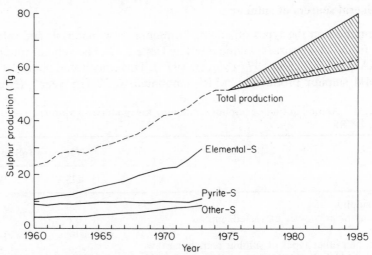

Fig. 2.31   Production of sulphur from various sources since 1960. The projected worldwide production of sulphur in the coming decade is expected to lie within the hatched area; the historical trend is shown as the broken line (after Nriagu, 1978)

Table 2.33 Annual sulphur flux from the lithosphere in weathering processes (from Tables 2.29, 2.30, 2.31, 2.32)

		Sulphur flux (TgS year^{-1})			
		Natural weathering	Anthropogenic weathering (50% of flux due to natural weathering)	Other anthropogenic processes	Total flux from the lithosphere
Natural and anthropogenic flux to ocean	River runoff (particulate + ionic)	37.94	18.97	52.0	89.94
	Underground runoff	9.2	—	—	9.2
	Exaration	5.27	—	—	5.27
	Abrasion	1.54	—	—	1.54
	Wind erosion	3.95	1.98	—	3.95
Total runoff to ocean		57.90	20.95	52.0	109.9
Runoff to continental water bodies		35.95	17.98	—	35.95
Release of components to the hydrosphere via atmosphere		8.91	4.45	—	8.91
Flux to soil		10.0	—	—	10.0
Anthropogenic	Mining and processing of ferrous and non-ferrous metals	—	—	10.0	10.0
	Mineral waste from cargo ships	—	—	0.7	0.7
	Combustion of solid fuels	—	—	42.0	42.0
Total removal of sulphur		112.8	43.4	104.7	217.5

countries. The USSR Ministry of Chemical Industry reported that 32 Tg of mineral fertilizers had been produced during the 35 years prior to 1973 and 38 Tg in the five years from 1973 to 1978. Kostandov (1978) estimated that the annual production of sulphuric acid would increase from 9 Tg in 1975 to 30 Tg in 1980.

The consumption of sulphur in industry and agriculture in the USA was 9.1 TgS year^{-1} in 1968, and by the year 2000 it is expected to reach 23–37 TgS year^{-1} (Meyer, 1977).

The growth in world production and consumption of sulphur for the next decade was assessed by Nriagu (1978) as 80 TgS year^{-1} in 1985, an increase of 28 TgS; consequently the total flux of sulphur from the lithosphere in river runoff will increase to 159.2 TgS year^{-1}. This will result in greater sulphate pollution of rivers and an increase in the sulphur concentration of river runoff to 0.4%.

2.12.4   Emission of Sulphur During Fuel Combustion and Metallurgical Process

A.   Sulphur emission during combustion of solid fuel

Sulphur is contained in coal and lignite in the form of organic compounds, pyrite, and sulphate. Generally, the sulphate sulphur does not change during combustion and remains in the ash. Earlier, organic sulphur was thought to be the predominant form of sulphur in coal and lignites. In recent years, it has been established that about 50% of the sulphur in coal is present as highly dispersed pyrite impregnating carbonized plant residues.

Table 2.34   Annual consumption of fuels and release of sulphur during combustion

| Type of fuel | Fuel consumption (10^3 Tg year^{-1}) | | Sulphur | |
	Goldberg (1973)	Sidorenko (1978)	Concentration (%) (Yurovsky, 1960; Meyer, 1977; Pankina, 1978)	Amount emitted (TgS year^{-1})
Coal, lignite (peat) schists	2.7	2.8	1.5(1–14)	42
Oil	1.63	} >4	0.1–3	} 18[a]
Natural gas	0.66		0.1–40	
Total				70

[a]Difference between total anthropogenic sulphur emission to the atmosphere (70 TgS year^{-1}, Chapter 4) and that emitted during combustion of fuel (42 TgS year^{-1}) and smelting (10 TgS year^{-1}).

The total content of sulphur in coal may vary from fractions of a per cent to 5%; in some coal basins (Ruhr, Donbass) coals may contain up to 8–14% of sulphur (Yurovsky, 1960). During combustion, organic and pyrite sulphur, which make up 95% of the total sulphur in coal (Kiyoura *et al.*, 1970; Kellogg *et al.* 1972; and others) is oxidized to sulphur dioxide (96%), and sulphur trioxide (4%), and emitted to the atmosphere. The sulphur pool size in coal is estimated to be 20×10^3 Tg for a coal reservoir of 1300×10^3 Tg (Meyer, 1977); i.e. the mean sulphur content in coal is assumed to be 1.5%. For an output of solid fuel equal to 2.8×10^3 Tg year^{-1} (Table 2.34), the sulphur flux resulting from its combustion will be 42 TgS year^{-1} (Tables 2.33 and 2.34). The foreseeable consumption of coal for the year 2000 is 4.36×10^3 Tg (Moiseev *et al.*, 1979), and consequently the amount of sulphur released will increase to 87 TgS year^{-1}.

B.   Emission of sulphur to the atmosphere from metallurgical processes

Sulphide sulphur is present in all ores of non-ferrous metals. In some pyrite ores the sulphur content amounts to 45%. During the smelting of copper, zinc, nickel, lead, and other metals, sulphide sulphur is oxidized to sulphur dioxide and released to the atmosphere. For four of the most industrialized countries—the USA, Great Britain, West Germany, and the USSR—sulphur emission by ferrous and non-ferrous industries is assessed as 8.6 TgS year^{-1} (see Table 4.10). This metallurgical sulphur flux amounts to 10–20% of the total sulphur flux to the atmosphere from each country. For Canada, the level of emission from non-ferrous metal smelting is especially high and amounts to 65% of the total sulphur emission to the atmosphere. Hence, we may assume that a minimum figure of about 10 TgS year^{-1} is transferred from the lithosphere to the atmosphere as a result of mining and smelting of ferrous and non-ferrous metals. In future, this flux will increase as industry increases unless efficient scrubbers are installed at smelters.

2.13   THE ANTHROPOGENIC AND TOTAL SULPHUR FLUX FROM WEATHERING PROCESSES

The anthropogenic contribution to weathering is 43.4 TgS year^{-1} (Table 2.33). The purely anthropogenic flux of sulphur due to weathering, industrial sewage (for example, cleaning of ship's holds), leaching of fertilizers, combustion of fuel, and metal smelting is 104.7 TgS year^{-1} (Table 2.33). The total anthropogenic sulphur flux due to the weathering processes in continental rocks is, therefore, 148.1 TgS year^{-1} i.e. 68.1% of the total sulphur removal with land matter, 217.5 TgS year^{-1} (Table 2.33). This does not include the 0.6 TgS year^{-1} for peat-bog formation.

2.14   SULPHUR FLUX FROM THE LITHOSPHERE DURING THE COMBUSTION OF LIQUID AND GASEOUS FUELS

This flux is assessed from data on the anthropogenic sulphur emission to the atmosphere (Chapter 4) and from the amounts of sulphur released during combustion of solid fuel and metal smelting. It accounts for 18 TgS year^{-1} (Table 2.34).

2.15   BALANCE OF SULPHUR IN THE LITHOSPHERE

The natural sulphur flux in the lithosphere is composed of three basic

Table 2.35   Sulphur budget for the lithosphere (TgS year^{-1})

Input		Output	
NATURAL SULPHUR FLUXES			
Sediment accumulation		*Volcanic emission*	
In ocean	100	*on continents and island arcs*	
In lakes and		Ash flow to ocean	1.37
reservoirs	16.5	Volatiles to atmosphere	28.0
In continental areas			
without runoff to		*Weathering of continental rocks*	
ocean	19	River runoff to ocean	37.94
	135.5	Underground runoff to ocean	9.2
		River runoff to continental	
Absorption from		water bodies	35.95
atmosphere by rocks	4.5	Release from organic	
Peat formation, from soil	0.6	compounds to atmosphere	
From outer space	0.00001	and ocean	8.91
From mantle	1.2–2.1	For soil formation	10.0
	142	Exaration	5.27
		Abrasion	1.54
		Wind erosion	3.95
			142.13
ANTHROPOGENIC SULPHUR FLUXES			
		Fuel weathering	
		Combustion of solid fuels	42
		Combustion of liquid and	
		gaseous fuels	18
		Disposal of solid wastes	0.7
		Extraction during mining and	
		processing of metals	10
		Output of sulphur containing	
		raw materials	52
			122.7

processes: (1) deposition and diagenesis; (2) land weathering; (3) volcanism. It is assumed here that the sulphur accumulation in the lithosphere is balanced by the sulphur runoff due to weathering and volcanism. The magnitudes of fluxes to and from the rocks of the lithosphere are given in a condensed form in Table 2.35. The total sulphur influx from natural processes is 142 TgS year^{-1} and the total natural efflux from continental rocks is 142.13 TgS year^{-1}. Owing to man's industrial activity, about 122.7 TgS year^{-1} are removed from the continental lithosphere through the atmosphere and river runoff. This flux is expected to reach 160–200 TgS year^{-1} by the year 2000, because of the growth of sulphuric acid production and use, and the increase in fertilizer production and industrial development.

REFERENCES

Abdurakhmanov, A. I., Firstov, P. P., and Shirokov, V. A. (1976) Possible relationship of volcanic eruptions and eleven year cycling of the solar activity. *Byull. Vulkanol. Stn.*, No. 52, 3–10 [in Russian].

Abramovsky, B. P., Alimov, V. N., Ionov, V. A., Nazarov, I. M., Patrakeev, S. I., Fedotov, S. A., and Chirkov, V. P. (1977) Emission of gases and aerosols to the atmosphere during the eruption of the volcano Tolbachik (Kamchatka). *Dokl. Akad. Nauk SSSR*, **237**, No. 6, 1479–1482 [in Russian].

Alekin, O. A., and Brazhnikova, L. V. (1964) *Runoff of Dissolved Compounds from the Territory of the USSR*. Nauka, Moscow, 144 pp. [in Russian].

Anger, G., Nielsen, H., Puchelt, H., and Ricke, W. (1966) Sulphur isotopes in the Rammelsberg ore deposit (Germany). *Econ. Geol.*, **61**, 511–536.

Ault, W. V., and Kulp, I. L. (1959) Isotopic geochemistry of sulphur. *Geochim. Cosmochim. Acta*, **16**, 201–235.

Barteles, O. G. (1972) An estimate of volcanic contributions to the atmosphere and volcanic gases and sublimates as the source of the radioisotopes ^{10}Be, ^{35}S, ^{32}P and ^{22}Na. *Health Phys.*, **22**, 387–392.

Basharina, L. A. (1966) Exhalation of basalt and andesite lavas of volcanoes of the Kamchatka. In: *Recent Volcanism*, vol. 1. Nauka, Moscow, pp. 139–146 [in Russian].

Berner, R. A. (1964) An idealized model of dissolved sulfate distribution in recent sediments. *Geochim. Cosmochim. Acta*, **28**, 1497–1503.

Berner, R. A. (1970) Sedimentary pyrite formation. *Am. J. Sci.*, **268**, 1–23.

Beryland, M. E. (1975) *Today's Problems of Atmospheric Diffusion and Contamination*. Hydrometeoizdat, Leningrad, 448 pp. [in Russian].

Bjorlike, K. (1974) Depositional history and chemical composition of lower paleozoic epicontinental sediments from the Oslo Region. *Nor. Geol. Unders.*, No. 305, 81 pp.

Blatt, H., and Jones, R. L. (1975) Proportions of exposed igneous metamorphic and sedimentary rocks. *Geol. Soc. Am. Bull.*, **86**, 1085–1088.

Bogdanov, Yu. V., Golubchina, M. N., Prilutsky, R. E., Toksubaev, A. I., and Feoktistov, V. I. (1971) On some peculiarities of the isotopic composition of sulfur in iron sulfides from the Paleozoic sediments of Djezkazgan. *Geokhimiya*, No. 11, 1376–1378 [in Russian].

Bogush, I. A., Rabinovich, A L., and Veselovsky, N. V. (1972) Genetic peculiarities of sulfur isotope composition in the Callovian of the North Caucasus. *Dokl. Akad. Nauk SSSR*, **205**, No. 2, 414–417 [in Russian].

Bolin, B., Degens, E. T., Kempe, S., and Ketner, P. (1979) *The Global Carbon Cycle*. SCOPE Report 13. John Wiley and Sons, Chichester, 491 pp.

Bondarev, L. G. (1974) *Perpetual Motion. Man and Preliminary Transfer of Matter*. Mysl, Moscow, 158 pp. [in Russian].

Borisov, O. G. (1970) Source of native volcanogenic sulfur. *Geokhimiya*, No. 3, 332–343.

Botoman, G., and Faure, G. (1976) Sulfur isotope composition of some sulfide and sulfate minerals in Ohio. *Ohio J. Sci.*, **76**, 66–71.

Boyle, R. W., Wanless, R. K., and Stevens, R. D. (1976) Sulfur isotope investigation of the barite, manganese and lead–zinc–copper–silver deposits of the Walton–Cheverie Area, Nova Scotia, Canada. *Econ. Geol.*, **71**, 749–762.

Buddington, A F., Jensen, M. L., and Mauger, R. L. (1969) Sulfur isotopes and origin of Northwest Adirondack sulfide deposits. *Geol. Soc. Am. Mem.*, **115**, 423–451.

Burnie, S. W., Schwartz, H. P., and Crocket, J. H. (1972) A sulfur isotopic study of the White Pine Mine, Michigan. *Econ. Geol.*, **67**, 895–914.

Cadle, R. D. (1975) Volcanic emissions of halides and sulphur compounds to the troposphere and stratosphere. *J. Geophys. Res.*, **80**, 1650–1652.

Cameron, E. M., and Baumann, A. (1972) Carbonate sedimentation during the Archean. *Chem. Geol.*, **10**, 17–30.

Cameron, E. M., and Garrels, R. M. (1980) Geochemical composition of some Precambrian shales from the Canadian Shield. *Chem. Geol.*, **28**, 181–197.

Cameron, E. M., and Jonasson, I. R. (1972) Mercury in the Precambrian shales of the Canadian shield. *Geochim. Cosmochim. Acta*, **36**, 985–1005.

Cherkovsky, S. L., Migdisov, A. A., and Grinenko, V. A. (1978) Isotopic composition of sulfur in sediments of inland seas and their water sinks. In: *All-Union Symposium on Stable Isotopes in Geochemistry. 7th Synthesis of reports*. Moscow, p. 300 [in Russian].

Chrismas, L., Baadsgaard, H., Folinsbee, R. E., Fritz, P., Krouse, H. R., and Sasaki, A. (1969) Rb/Sr, S and O isotopic analyses indicating source and date of contact metasomatic copper deposits, Craigmont, British Columbia, Canada. *Econ. Geol.*, **64**, 479–488.

Clarke, F. W. (1924) *The Data of Geochemistry*. Government Printing Office, Washington, 841 pp.

Claypool, G., Holser, W. T., Kaplan, I. R., Sakai, H., and Zak, I. (1980) The age curves of sulfur and oxygen isotopes in marine sulfate and mutual interpretation. *Chem. Geol.*, **28**, 199–260.

Cole, R. D. (1975) Sedimentology and sulfur isotope geochemistry of Green River Formation (Eocene), Unita Basin, Utah, Piceance Creek Basin, Colorado. *Diss. Abstr. Int. B.*, **36**, 1618–1619.

Corbel, J. (1964) L'érosion terrestre, étude quantitative. *Ann. Geogr.*, No. 398, 385–412.

Dechow, W. E. (1960) Geology of sulfur isotopes and the origin of the Health Steele ore deposits, Newcastle, N.B., Canada. *Econ. Geol.*, **55**, 533–556.

Degens, E. T., Paluska, A., and Eriksson, E. (1976) Rates of soil erosion. In: Svensson, B. H., and Söderlund, R. (eds.), *Nitrogen, Phosphorus and Sulphur—Global Cycles*. SCOPE Report 7, *Ecol. Bull. (Stockholm)*, **22**, 185–191.

Dimroth, E., and Kimberley, M. M. (1976) Precambrian atmospheric oxygen: evidence in the sedimentary distributions of carbon, sulfur, uranium and iron. *Can. J. Earth Sci.*, **13**, 1161–1185.

Dolzhenko, V. N. (1976) Isotopic composition of sulfide sulfur in ores and bedding rocks of a scarn deposit in Kirgizia. In: *Regional Geochemistry of the Tyan-Shan Ridge*. Ilim., Frunze, pp. 126–133 [in Russian].

Dontsova, E. I., Migdisov, A. A., and Ronov, A. B. (1972) Problem of causes of variation of oxygen isotopic composition in carbonate thicknesses of the sedimentary envelope. *Geokhimiya*, No. 11, 1317–1324 [in Russian].

Douglas, I. (1967) Man, vegetation and the sediment yields of rivers. *Nature (Lond.)*, **215**, 925–928.

Eremenko, N. A., and Pankina, R. G. (1972) On evolution of salt composition of the ocean on the basis of $^{32}S/^{34}S$ ratio of sulfate sulfur. In: *Geokhimia. Collect. Art.* Nauka, Moscow, pp. 36–41 [in Russian].

Eriksson, E. (1960) The yearly circulation of chloride and sulfur in nature; meteorological, geochemical and pedological implications, Pt. 2. *Tellus*, **12**, 63–109.

Fahrig, W. F., and Eade, K. E. (1968) Chemical evolution of the Canadian shield. *Can. J. Earth Sci.*, **5**, 1247–1252.

Fersman, A. E. (1977) *Essays on Mineralogy and Geochemistry*. Nauka, Moscow, 192 pp. [in Russian].

Friend, J. P. (1973) The global sulfur cycle. In: Rasool, S. I. (ed.), *Chemistry of the Lower Atmosphere*. Plenum Press, New York, pp. 177–201.

Galimov, E. M., Migdisov, A. A., and Ronov, A. B. (1975) Variations of isotopic composition of carbonate and organic carbon in sedimentary rocks in the course of the Earth's history. *Geokhimiya*, No. 2, 323–342 [in Russian].

Garrels, R. M. (1975) *Turnover of Carbon, Oxygen and Sulfur in the Course of Geological Periods*. Nauka, Moscow, 47 pp. [in Russian].

Garrels, R. M., and Mackenzie, F. T. (1971) *Evolution of Sedimentary Rocks*. Norton and Company, New York, 397 pp.

Garrels, R. M., and Mackenzie, F. M. (1972) A quantitative model for the sedimentary rock cycle. *Marine Chem.*, **I**, 27–41.

Garrels, R. M., and Perry, E. A. (1974) Cycling of carbon, sulfur and oxygen through geologic time. In: Goldberg, E. D. (ed.), *The Sea*, Vol. 5. Wiley, New York, pp. 303–336.

Garrels, R. M., Mackenzie, F. T., and Hunt, C. (1975) *Chemical Cycles and the Global Environment: Assessing Human Influences*. Kaufmann, Los Altos, Calif., 208 pp.

Gavrilenko, G. M., Gorshkov, A. P., and Skripko, K. A. (1980) Activation of the gasohydrothermal activity of the underground volcano Esmeralda in January 1978 and its influence on the chemical composition of sea water. *Vulkanol. Seismol.*, No. 2, 19–29 [in Russian].

Geicki, A. (1862) Cited in Bondarev, L. G. (1964) *Perpetual Motion. Man and Preliminary Transfer of Matter*. Mysl, Moscow, 158 pp.

Girin, Yu. P., Grinenko, V. A., Zagryazhskaya, G. D., and Savina, L. I. (1975). Pyrite sulfur isotope composition as index of the formation conditions of analcimes in the Gelatsky basin. *Geokhimiya*, No. 8, 1242–1249 [in Russian].

Gittinger, L. B. (1975) Sulfur. In: Lefond, S. J. (ed.), *Industrial Minerals and Rocks*. Am. Inst. Min. Metall. Pet. Eng., New York, pp. 1103–1125.

Goldberg, E. (1973) Role of man in the principal sedimentary cycle. In: *International Geochemical Congress 1st Proceedings. Sedimentary Processes*, vol. 4, No. 2. VINITI, Moscow, pp. 225–235 [in Russian].

Goldhaber, M. B., and Kaplan, I. R. (1974) The sulfur cycle. In: Goldberg, E. D. (ed.), *The Sea*, Vol. 5. Wiley, New York, pp. 569–655.

Goldschmidt, V. M. (1954) *Geochemistry*. The Clarendon Press, Oxford, 730 pp.

Gorshkov, A. P., Abramov, V. A., and Sapozhnikov, E. A. (1980) Geological structure of the underground volcano Esmeralda. *Vulkanol. Seismol.*, No. 4, 65–78 [in Russian].

Gorshkov, S. P. (1980) The cycle of products of land denudation. In: Ryabchikov, A. M. (ed.), *Cycle of Matter in Nature and its Variation under the Impact of Man's Economic Activity*. Izd. MGU, Moscow, pp. 34–55 [in Russian].

Granat, L., Rodhe, H., and Hallberg, R. O. (1976) The global sulphur cycle. In: Svensson, B. H., and Söderlund, R. (eds), *Nitrogen, Phosphorus and Sulphur—Global Cycles*. SCOPE Report 7. *Ecol. Bull (Stockholm)*, **22**, 89–134.

Grinenko, L. N., and Grinenko, V. A. (1975) Sulfur sources in the pre-Cambrian sulfide ores and rocks. In: *Problems of Sedimentary Geology of the Pre-Cambrian*, Vol. 2, No. 4. Nedra, Moscow, pp. 124–131 [in Russian].

Grinenko, L. N., Ivlev, A. I., and Popsilenko, V. I. (1972) Isotopic composition of sulfide sulfur and its importance for assessment of nickel content in metamorphic rocks of the granulitic complex of the Kola Peninsula in the region of Salnaya tundra. In: *Scientific Bases of Geochemical Methods for Prospecting Minerals and Assessment of Ore Contents in Magmatic and Metamorphic Complexes of the Pre-Cambrian Apatites*, pp. 55–61 [in Russian].

Grinenko, L. N., Grinenko, V. A., Zagryazhskaya, G. D., and Stolyarov, Yu. M. (1969) Isotopic composition of sulfide and sulfate sulfur in pyrite deposits of Levikha in connection with problems of their genesis. *Geol. Rudn. Mestorozhd.*, No. 3, 26–39 [in Russian].

Grinenko, V. A., and Grinenko, L. N. (1974) *Geochemistry of Sulfur Isotopes*. Nauka, Moscow, 274 pp. [in Russian].

Grinenko, V. A., Cherkovsky, S. L., and Migdisov, A. A. (1974) Regularities of variation of the sulfur isotopic composition in sediments of recent and ancient basins. In: *All-Union Symposium on Geochemistry of Stable Isotopes. 5th Synthesis of Reports*, vol. I. Moscow, pp. 1–2 [in Russian].

Grinenko, V. A., Migdisov, A. A., and Barskaya, N. V. (1973a) Sulfur isotopes in the sedimentary cover of the Russian Platform. *Dokl. Akad. Nauk SSSR*, **210**, No. 2, 445–448 [in Russian].

Grinenko, V. A., Migdisov, A. A., and Ronov, A. B. (1973b) Geochemistry of sulfur in the sedimentary envelope of the Earth's crust. In: *International Geochemical Congress 1st Proceedings. Sedimentary Processes*, vol. 4, pt. 1. VINITI, Moscow, pp. 141–157 [in Russian].

Grinenko, V. A., Dmitriev, L. V., Migdisov, A. A., and Sharaskin, A. Ya. (1975) Content and isotope composition of sulfur in magmatic and metamorphic rocks of the middle-oceanic mountain ridges. *Geokhimiya*, No. 2, 199–205 [in Russian].

Grinenko, V. A., Migdisov, A. A., Girin, Yu. P., Bogdanov, Yu. A., and Starostin, V. I. (1978) On sulfur sources in basic rocks and sediments of active zones of the south-eastern Pacific. In: *All-Union Symposium on Stable Isotopes in Geochemistry. 7th Synthesis of Reports*. Moscow, 299 pp. [in Russian].

Hill, T. P., Werner, M. A., and Horton, M. J. (1967) Chemical composition of sedimentary rocks in Colorado, Kansas, Montana, Nebraska, North Dakota, South Dakota and Wyoming. *U.S. Geol. Surv. Prof. Pap.*, No. 561, 241 pp.

Holland, H. D. (1973) Systematics of the isotopic composition of sulfur in the oceans during the Phanerozoic and its implications for atmospheric oxygen. *Geochim. Cosmochim. Acta.*, **37**, 2605–2616.

Holser, W. T., and Kaplan, I. R. (1966) Isotope geochemistry of sedimentary sulfates. *Chem. Geol.*, **1**, 93–135.

Judson, S. (1968) Erosion of the land. *Am. Sci.*, **56**, 356–374.

Junge, C. E. (1963) Sulfur in the atmosphere. *J. Geophys. Res.*, **68**, 3975–3976.

Kanehira, K., Yui, S., Sakai, H., and Sasaki, A. (1973) Sulphide globules and sulphur isotope ratios in the abyssal tholeiite from the mid-Atlantic Ridge near $30°$ N latitude. *Geochem. J.*, **7**, 89–96.

Kaplan, I. R., Emery, K. O., and Rittenberg, S. C. (1963) The distribution and isotopic abundance of sulfur in recent marine sediments off southern California. *Geochim. Cosmochim. Acta*, **27**, 297–331.

Kellogg, W. W., Cadle, R. D., Allen, E. R., Lazrus, A. L., and Martell, E. A. (1972) The sulfur cycle. *Science* (Wash. D.C.), **175**, 587–596.

Kempe, S. (1979) Carbon in the rock cycle. In: Bolin, B., Degens, E. T., Kempe, S., and Ketner, P. (eds), *The Global Carbon Cycle*. SCOPE Report 13. John Wiley and Sons, Chichester, pp. 343–375.

Khain, V. E., Levin, L. E., and Tuliani, L. I. (1971) Volume of sedimentary thickness and supposed reserves of hydrocarbons in the system of depressions of the World Ocean. *Dokl. Akad. Nauk SSSR*, **201**, No. 5, 1201–1202 [in Russian].

Kirichenko, I. P. (1963) Geotechnology as new science in the geologic cycling. In: *Interaction Between Sciences During Studies of the Earth*. Isv. Akad. Nauk SSSR, Moscow, pp. 321–324 [in Russian].

Kiyoura, R., Kuronuma, H., Uwanishi, G., Kojima, H., Iguchi, S., Kuramoto, T., and Munidasa, M. (1970) Some opinions on sulphur dioxide as an atmospheric pollutant. *Bull. Tokyo Inst. Technol.*, No. 98, 117–120.

Kostandov, L. A. (1978) Chemical industry during the years of the Soviet Power. *Vestn. Akad. Nauk SSSR*, No. 5, 41–53 [in Russian].

Koyama, T., Nakai, N., and Kamata, E. (1965) Possible discharge rate of hydrogen sulfide from polluted coastal belts in Japan. *J. Earth. Sci. Nagoya Univ.*, **13**, 1–11.

Krouse, H. R., and McCready, R. G. L. (1979) Biochemical cycling of sulfur. In: Trudinger, P. A., and Swaine, D. J. (eds), *Biogeochemical Cycling of Mineral-forming Elements*. Elsevier, Amsterdam, pp. 401–431.

Krouse, H. R., Brown, H. M., and Farquharson, R. B. (1977) Sulfur isotope compositions of sulfides and sulfates, DSDP Leg. 37. *Can. J. Earth Sci.*, **14**, 787–793.

Lein, A. Yu., Kudryavtseva, A. I., Matrosov, A. G., and Zyakun, A. M. (1976) Isotope composition of sulfur compounds in sediments of the Pacific Ocean. In: *Biogeochemistry of Diagenesis of Oceanic Sediments*. Nauka, Moscow, pp. 179–185 [in Russian].

Lisitsin, A. P. (1978) *Processes of Oceanic Sedimentation. Lithology and Geochemistry*. Nauka, Moscow, 392 pp. [in Russian].

Lopatin, G. V. (1950) Erosion and runoff of alluviums. *Priroda* (Mosc.), No. 7, 19–28 [in Russian].

Lopatin, G. V. (1952) *River Alluviums (Formation and Transfer)*. Geografizdat, Moscow, 366 pp. [in Russian].

Macdonald, G. (1975) *Volcanoes*. Mir, Moscow, 431 pp. [in Russian].

Markhinin, E. K. (1967) *Role of Volcanism in Formation of the Earth's Crust*. Nauka, Moscow, 255 pp. [in Russian].

Menyailov, I. A., Nikitina, L. P., and Shapar, V. N. (1980) *Geochemical Peculiarities of Exhalation of the Big Tolbachinsky Fissure Eruption*. Nauka, Moscow, 235 pp. [in Russian].

Mészáros, E., and Várhelyi, G. (1975) On the concentration, size distribution and residence time of sulfate particles in the lower troposphere. *Idojaras*, **79**, No. 5, 267–273.

Meyer, B. (1977). *Sulfur, Energy and Environment*. Elsevier, Amsterdam, 448 pp.

Migdisov, A. A., Cherkovsky, S. L., and Grinenko, V. A. (1974) Dependence of sulfur isotope composition in humid sediments on conditions of their formation. *Geokhimiya*, No. 10, 1482–1502 [in Russian].

Migdisov, A. A., Girin, Yu. P., and Grinenko, V. A. (1978) Sulfur in sediments of the south-east Pacific. In: *All-Union Symposium on Stable Isotopes in Geochemistry. Synthesis of Reports*. Moscow, pp. 298–299 [in Russian].

Migdisov, A. A., Grinenko, V. A., and Barskaya, N. V. (1972) Variation of isotopic composition of reduced sulfur forms in the course of geologic history. In: *All-Union Symposium on Geochemistry of Stable Isotopes. 4th Synthesis of Reports.* Moscow, pp. 13–15 [in Russian].

Migdisov, A. A., Dontsova, E. I., Kuznetsova, L. D., and Ronov, A. B. (1973) Probable causes of evolution of oxygen isotopic composition in outer envelopes of the Earth. In: *International Geochemical Congress 1st Proceedings. Sedimentary Processes*, Vol. 4, No. I. VINITI, Moscow, pp. 173–184 [in Russian].

Migdisov, A. A., Bogdanov, Yu. A., Lisitsin, A. P., Gurvich, E. G., Lebedev, A. I., Lukashin, V. N., Gordeev, V. V., Girin, Yu. P., and Sokolova, E. G. (1979) Geochemistry of metalliferous sediments. In: *Metalliferous Sediments of the South-East Pacific.* Nauka, Moscow, pp. 122–201. [in Russian].

Migdisov, A. A., Girin, Yu. P., Galimov, E. M., Grinenko, V. A., Barskaya, N. V., Krivitsky, V. A., Sobornov, O. P., and Cherkovsky, S. L. (1980) Major and minor elements and sulfur isotopes of the Mesozoic and Cenozoic sediments at Sites 415 and 416, Leg 50, Deep Sea Drilling Project. In: *Init. Rep. of the Deep Sea Drilling Project*, vol. 50. Washington, pp. 675–689.

Miyake, Y. (1965) *Elements of Geochemistry.* Maruzen, Tokyo.

Moiseev, N. N., Krapivin, V. F., Svirezhev, Yu. M., and Tarko, A. M. (1979) Progress in the construction of a model of dynamic processes in the biosphere. *Vestn. Akad. Nauk SSSR*, No. 10, 88–104 [in Russian].

Moore, J. G., and Fabbi, B. F. (1971) An estimate of the juvenile sulfur content of basalt. *Contrib. Mineral and Petrol.*, **33**, 118–127.

Naldrett, A. J., Goodwin, A. M., and Fisher, T. L. (1977). Sulfur in Leg 37 basalts. In: *Init. Rep. of Deep Sea Drilling Project*, vol. 37. Washington, pp. 561–562.

Naldrett, A. J., Goodwin, A. M., Fisher, T. L., and Ridler, R. H. (1978) The sulfur content of Archean volcanic rocks and a comparison with ocean floor basalts. *Can. J. Earth Sci.*, **15**, 715–728.

Naughton, J. J., Finalyson, J. B., and Lewis, V. A. (1975/76) Some results from recent chemical studies at Kilauea Volcano, Hawaii. *Bull. Volcanol.* **39**, No. 1, 64–69. Special Issue (Hawaii Institute of Geophysics Contribution No. 584).

Nielsen, H. (1973) Model estimates of sulfur isotope balance in ancient oceans. In: *International Geochemical Congress. 1st Proceedings. Sedimentary Processes*, vol. 4, No. 1. VINITI, Moscow, pp. 127–140 [in Russian].

Nielsen, H., and Ricke, W. (1964) Schwefel-Isotopenverhältnisse von Evaporiten aus Deutschland; Ein Beitrag zur Kenntnis von $\delta^{34}S$ im Meerswasser-Sulfat. *Geochim. Cosmochim. Acta*, **28**, 577–591 [in German].

Nriagu, J. O. (1978) Production and uses of sulfur. In: Nriagu, J. O. (ed.), *Sulfur in the Environment*, Pt. 1. Wiley, New York, pp. 1–21.

Okita, T., and Shimozuru, D. (1975) Remote sensing measurements of mass flow of sulfur dioxide gas from volcanoes. *Bull. Volcanol. Soc. Jpn*, **19**, 151–157.

Palamarchuk, S. F., Shor, G. M., Golubchina, M. N., Prilutsky, R. E., and Toksubaev A. I. (1976) Variation of sulfur isotope composition in sulfides from paleogenic deposits in north Kazakhstan. *Tr. Vses. Nauchno-Issled. Geol. Inst.*, **250**, 75–78 [in Russian].

Palamarchuk, S. F., Shor, G. M., Symslov, A. A., Golubchina, M. N., and Prilutsky, R. E. (1972) On sulfur isotope composition in sulfides from Chegansky deposits of the Karatau. *Geokhimiya*, No. 11, 1402–1405 [in Russian].

Pankina, R. G. (1978) *Geochemistry of Sulfur Isotopes of Petroleum and Organic Matter.* Nedra, Moscow, 247 pp. [in Russian].

Penk, A. (1894), cited in Bondarev, L. G. (1974). *Perpetual Motion. Man and Preliminary Transfer of Matter.* Mysl, Moscow, 158 pp. [in Russian].

Pilot, J. (1970) *Isotopengeochemie, Situation, Konzeptionen, Entwicklung, Möglich-*

keiten. VEB Deutscher Verlag für Grundstoffindustrie, Leipzig, 170 pp. [in German].

Pilot, I., Rösler, H., and Müller, E. (1973) Geochemical evolution of sea water and marine sediments in the Phanerozoic. In: *International Geochemical Congress. 1st Proceedings. Sedimentary Processes*, vol. 4, No. 1. VINITI, Moscow, pp. 158–171 [in Russian].

Poldervaart, A. (1955) Chemistry of the earth's crust. *U.S. Geol. Surv. Prof. Pap.*, **62**, 114–119.

Rankama, K., and Sahama, Th. G. (1952) *Geochemistry.* Univ. of Chicago Press, Chicago, 911 pp.

Reade, M. (1889), cited in Bondarev, L. G. (1974). *Perpetual Motion. Man and Preliminary Transfer of Matter.* Mysl, Moscow, 158 pp. [in Russian].

Rees, C. E. (1970) The sulphur isotope balance of the ocean: an improved model. *Earth Planet. Sci. Lett.*, **7**, 366–370.

Ricke, W. (1960) Ein Beitrag zur Geochemie des Schwefels. *Geochim. Cosmochim. Acta*, **21**, 35–80.

Rittmann, A. (1960) *Vulkane und ihre Tätigkeit.* Ferdinand Enke Verlag, Stuttgart [in German].

Robinson, E., and Robbins, R. C. (1970) Gaseous sulfur pollutants from urban and natural sources. *J. Air Pollut. Control. Assoc.*, **20**, 233–235.

Ronov, A. B. (1958) Organic carbon in sedimentary rocks (in connection with their oil-bearing capacity). *Geokhimiya*, No. 5, 409–423 [in Russian].

Ronov, A. B. (1968) Probable changes in the composition of sea water during the course of geological time. *Sedimentology*, **10**, 25–43.

Ronov, A. B. (1976) Volcanism, carbonate accumulation, life (regularities of global geochemistry of carbon). *Geokhimiya*, No. 8, 1252–1277 [in Russian].

Ronov, A. B. (1980) *Sedimentary Envelope of the Earth (Quantitative Regularities of Structure, Composition and Evolution).* Nauka, Moscow, 79 pp. [in Russian].

Ronov, A. B., and Migdisov, A. A. (1970). Evolution of the chemical composition of rocks of shields and sedimentary envelope of the Russian and North American Platforms. *Geokhimiya*, No. 4, 403–438 [in Russian].

Ronov, A. B., and Migdisov, A. A. (1971) Geochemical history of the crystalline basement and the sedimentary cover of the Russian and North American Platforms. *Sedimentology*, **16**, 137–185.

Ronov, A. B., and Yaroshevsky, A. A. (1967) Chemical composition of the Earth's crust. *Geokhimiya*, No. 11, 1285–1309 [in Russian].

Ronov, A. B., and Yaroshevsky, A. A. (1976) New model of chemical composition of the earth's crust. *Geokhimiya*, No. 12, 1763–1795 [in Russian].

Ronov, A. B., Khain, V. E., and Balukhovsky, A. N. (1979) Comparative estimate of the volcanism intensity on continents and in oceans. *Izv. Akad. Nauk SSSR, Ser. Geol.*, No. 5, 5–12 [in Russian].

Ronov, A. B., Migdisov, A. A., and Barskaya, N. V. (1969) Tectonic cycles and regularities in the development of sedimentary rocks and paleogeographic environments of sedimentation of the Russian Platform. *Sedimentology*, **13**, 179–212.

Ronov, A. B., Migdisov, A. A., and Khain, V. E. (1972) On reliability of quantitative methods of investigation in lithology and geochemistry. *Litol. Polezn. Iskop.*, No. 1, 3–26 [in Russian].

Ronov, A. B., Migdisov, A. A., and Lobach-Zhuchenko, S. (1977) Problems of evolution of the chemical composition of sedimentary rocks and regional metamorphism. *Geokhimiya*, No. 2, 163–168 [in Russian].

Ronov, A. B., Girin, Yu. P., Kazakov, G. A., and Ilyukhin, M. N. (1965) Comparative geochemistry of geosynclinal and platform thicknesses. *Geokhimiya*, No. 8, 961–979 [in Russian].

Ronov, A. B., Khain, V. E., Balukhovsky, A. N., Seslavinsky, K. B. (1976) Variation of distribution, volumes and accumulation rate of sedimentary and volcanogenic deposits in the Phanerozoic (within the limits of recent continents). *Izv. Akad. Nauk SSSR, Ser. Geol.*, No. 12, 5–12 [in Russian].

Ronov, A. B., Grinenko, V. A., Girin, Yu. P., Savina, L. I., Kazakov, G. A., and Grinenko, L. N. (1974) Influence of tectonic regime on concentration and isotope composition of sulfur in sedimentary rocks. *Geokhimiya*, No. 12, 1772–1798.

Rubey, W. W. (1951) Geologic history of sea water. *Geol. Soc. Am. Bull.*, **62**, 1111–1147.

Ryabchikov, A. M. (1970) Anthropogenic factor of changes in the geosphere. *Vestn. Mosk. Univ. Ser. V Geogr.*, No. 2, 90–96 [in Russian].

Ryabchikov, A. M. (1972) *Structure and Dynamics of the Geosphere, Its Natural Development and Changes Provoked by Man.* Mysl., Moscow, 223 pp. [in Russian].

Sapper, K. (1929) *Vulkankunde.* Stuttgart, 424 pp. [in German].

Sasaki, A., and Krouse, H. R. (1966) $^{32}S/^{34}S$ variations in springs of the Canadian Rockies. *Trans. Am. Geophys. Union*, **47**, 495.

Schidlowski, M., Junge, C. E., and Pietrek, H. (1977) Sulfur isotope variations in marine sulfate evaporites and the Phanerozoic oxygen budget. *J. Geophys. Res.*, **82**, 2557–2565.

Shaw, D. M. (1969) *Geochemistry of Microelements of Crystalline Rocks.* Nedra, Leningrad, 82 pp. [in Russian].

Shaw, D. M., Reilly, G. A., Muysson, J. R., Pattenden, G. E., and Campbell, F. E. (1967) An estimate of the chemical composition of the Canadian Precambrian shield. *Can. J. Earth Sci.*, **4**, 829–853.

Shepherd, E. S. (1938) The gases in rocks and some related problems. *Am. J. Sci.* **35A**, 311–351.

Shvedov, V. P. (1976) Progress of energetics and pollution of the biosphere. In: Styro, B. (ed.), *Physical Aspects of Atmospheric Pollution.* Mokslas, Vilnius. pp. 5–9 [in Russian].

Sidorenko, A. V. (1978) Protection of the environment and rational use of natural resources in the USSR. *Vestn. Akad. Nauk SSSR*, No. 5, 100–107 [in Russian].

Solisbary, R. D. (1909) Cited in Bondarev, L. G. (1974). *Perpetual Motion. Man and Preliminary Transfer of Matter.* Mysl, Moscow, 158 pp. [in Russian].

Stoiber, R. E., and Jepsen, A. (1973) Sulfur dioxide contributions to the atmosphere by volcanoes. *Science* (Wash. D.C.), **182**, 577–578.

Strakhov, N. M. (1960) *Principles of the Lithogenesis Theory*, Izv. Akad. Nauk SSSR, Moscow, vol. 1, 212 pp; vol. 2, 572 pp.

Strakhov, N. M. (1963) *Types of Lithogenesis and Their Evolution in the History of the Earth.* Gosgeoltekhizdat, Moscow, 535 pp. [in Russian].

Strakhov, N. M. (1976) *Problems of Geochemistry of Contemporary Oceanic Lithogenesis.* Nauka, Moscow, 299 pp. [in Russian].

Sweeney, R. E., and Kaplan, I. R. (1980) Diagenetic sulfate reduction in marine sediments. *Mar. Chem.*, **9**, 165–174.

Taylor, S. R. (1964) Abundance of chemical elements in the continental crust: a new table. *Geochim. Cosmochim. Acta*, **28**, 1273–1285.

Thode, G., and Monster, D. (1964) Distribution of sulfur isotopes in evaporites and ancient oceans. In: *Chemistry of the Earth's Crust*, vol. 2. Nauka, Moscow. pp. 589–600. [in Russian].

Thode, H. G., Macnamara, J., and Fleming, W. H. (1953) Sulfur isotopes in nature and geological and biological time scales. *Geochim. Cosmochim. Acta*, **3**, 235–243.

Trofimov, A. V. (1949) Isotopic composition of sulfur in meteorites and terrestrial objects. *Dokl. Akad Nauk SSSR*, **66**, 181–184 [in Russian].

Trotsyuk, V. Ya. (1979) Geochemical preconditions for gas and oil formation in the Mesozoic–Cainozoic sedimentary thickness of the World Ocean. *Izv. Akad. Nauk SSSR, Ser. Geol.*, No. 5, 132–142 [in Russian]

Tupper, W. M. (1960) Sulfur isotopes and the origin of the sulfide deposits of the Bathurst–Newcastle area of Northern New Brunswick. *Econ. Geol.*, **55**, 1676–1707.

Vernadsky, V. T. (1934) *Essays on Geochemistry*. ONTY, Moscow, Leningrad, 380 pp. [in Russian].

Vernadsky, V. T. (1967) *Biosphere*. Mysl., Moscow [in Russian].

Vinogradov, A. P. (1959) *Chemical Evolution of the Earth*. Izv. Akad. Nauk SSSR, Moscow, 44 pp. [in Russian].

Vinogradov, A. P., Grinenko, V. A., and Ustinov, V. I. (1962) Isotope composition of sulfur compounds in the Black Sea. *Geokhimiya*, No. 10, 851–873 [in Russian].

Vinogradov, A. P., Chupakhin, M. S., Grinenko, V. A., and Trofimov, A. V. (1956) Isotopic composition of sulfur and age of pyrites of sedimentary genesis. *Geokhimiya*, No. 1. 96–105 [in Russian].

Vlasov, G. M. (1971) *Volcanic Sulfur Deposits and Some Problems of Hydrothermal Ore Formation*. Nauka, Moscow, 360 pp. [in Russian].

Vlodavets, O. I. (1966) Volcanic activity on the Earth in historical time. In: *Recent Volcanism*. vol. 1. Nauka, Moscow. pp. 7–17 [in Russian].

Volkov, I. I. (1961) Regularities of formation and transformation of sulfur compounds in sediments of the Black Sea. In: *Recent Marine and Oceanic Sediments*. Izv. Akad. Nauk SSSR, Moscow, pp. 577–596 [in Russian].

Volkov, I. I., Rozanov, A. G., Zhabina, N. N., and Fomina, L. S. (1976) Sulfur compounds in sediments of the Californian Gulf and the adjoining part of the Pacific. In: *Biogeochemistry of Diagenesis of the Oceanic Sediments*. Nauka, Moscow, pp. 136–170. [in Russian]

Vorobyov, S. A., Nazarenko, V. I., and Shubin, V. F. (1972) Agriculture. In: *Big. Sov. Enc.*, 3rd ed, vol. 9. Moscow, pp. 464–469 [in Russian].

Wedepohl, K. (1973) Accumulation of volatile elements in sediments; degasation during metamorphism. In: *International Geochemical Congress 1st Proceedings, Sedimentary Processes*, vol. 4, No. 1. VINITI, Moscow, pp. 90–99. [in Russian].

White, W. E., and Woring, G. A. (1965) Volcanic emanations. In: *Geochemistry of Recent Post-Volcanic Processes*. Mir, Moscow, 9–48 [in Russian].

White, W. A. (1959) *Chemical and Spectrochemical Analysis of Illinois Clay Materials*. Ill. State Geol. Surv. Circ., No. 282, 55 pp.

Yamamoto, M., Ogushi, N., and Sakai, H. (1968) Distribution of sulfur isotopes, selenium and cobalt in the Yanahara ore deposits, Okayamaken, Japan. *Geochem. J.*, **2**, 137–156.

Yurovsky, A. Z. (1960) *Sulfur in Coals*. Izv. Akad. Nauk SSSR, 295 pp. [in Russian].

Zagryazhskaya, G. D., Girin, Yu. P., Grinenko, V. A., and Savina, L. I. (1973) Variation of sulfur isotope composition in the course of formation of sedimentary bed of the Lower and Middle Jurassic of the Caucasian geosyncline. *Geokhimiya*, No. 10, 1447–1459 [in Russian].

Zharkov, M. A. (1974) *Paleozoic Salt-bearing Formations of the World*. Nedra. Moscow, 392 pp. [in Russian].

Zharkov, M. A. (1978) *History of the Paleozoic Salt Accumulation*. Nauka, Novosibirsk, 272 pp. [in Russian].

The Global Biogeochemical Sulphur Cycle
Edited by M. V. Ivanov and J. R. Freney
© 1983 Scientific Committee on Problems of the Environment (SCOPE)

CHAPTER 3
The Sulphur Cycle in Soil

J. R. FRENEY and C. H. WILLIAMS

3.1 INTRODUCTION

The sulphur in soils was derived mainly from the weathering of plutonic rocks. Sulphides in primary minerals were released and oxidized to sulphate during the weathering process (Whitehead, 1964). Some of the sulphate found its way directly to the oceans in rivers and much of the remainder was incorporated into organic forms by plants or other living organisms or accumulated in the soil as relatively insoluble sulphate salts in semi-arid or arid climates. Some was reduced back to sulphides in anaerobic environments. Soil sulphur now exists in a great variety of forms and oxidation states, takes part in a variety of chemical and biological reactions, and interacts with the lithosphere, hydrosphere, and atmosphere either naturally or as a result of man's interference.

This chapter presents information on the natural cycle of sulphur in the soil, in particular, its reservoirs, the pathways of gains and losses, fluxes into and out of the system, and the influence of man on this cycle.

3.2 THE NATURAL CYCLE OF SULPHUR IN SOIL

3.2.1 Forms of Soil Sulphur

In well-drained, well-aerated soils most of the sulphur in the surface horizons is in organic forms; any inorganic sulphur present is almost entirely in the form of sulphates (e.g. Donald and Williams, 1954; Fedorov, 1954; Walker and Adams, 1958; Williams and Steinbergs, 1958; Williams, 1975b). Organic sulphur generally decreases with depth in the profile except in soils with organic matter accumulation in the B-horizon and in soils which have formed under conditions where erosion–deposition cycles have led to buried profiles. In contrast sulphate–sulphur commonly increases with depth in the profile (e.g. Williams, 1974) and the occurrence of gypsum in subsoil horizons of many soils is a recognized pedological feature.

Sulphides and other sulphur compounds of lower oxidation state than sulphate, including elemental sulphur, can be formed under anaerobic conditions such as occur in tidal swamps and waterlogged soils (Starkey, 1966; Bloomfield, 1969; Begheijn *et al.*, 1978).

A. Organic sulphur

Sulphur is an essential component of soil organic matter in which it is closely associated with carbon and nitrogen. Although appreciable variation occurs in the C : N : S ratio between individual soils, the mean ratios for groups of related soils are often similar (Whitehead, 1964; Williams, 1967). The similarity in C : N : S ratio for surface and subsurface soils seems to be world-wide (see Table 3.1) although, clearly, significant differences occur; these have been attributed to differences in parent material, to differences in soil-forming factors (such as climate, vegetation, leaching intensity, drainage, and temperature), and to differences in cultivation and management practice (Walker and Adams, 1958, 1959; Williams and Steinbergs, 1958; Williams *et al.*, 1960; Lowe, 1965; Bettany *et al.*, 1973; Neptune *et al.*, 1975; Biederbeck, 1978; see Tables 3.2, 3.3, 3.4 and 3.5).

Table 3.1 Relationships between carbon, nitrogen, and sulphur in soils from different regions of the world

Region	C : N : S	Reference
Africa		
Tropical Africa	<100 : 10 : 1.00	Dabin (1972)
North America		
Minnesota	122 : 10 : 1.48	Evans and Rost (1945)
Iowa	109 : 10 : 1.54	Tabatabai and Bremner (1972a,b)
Oregon	145 : 10 : 1.01	Harward *et al.* (1962)
South America		
Brazil	194 : 10 : 1.6	Neptune *et al.* (1975)
Asia		
India	80 : 10 : 0.96	Bhardwaj and Pathak (1969)
Australasia		
Eastern Australia	150 : 10 : 1.26	Williams and Steinbergs (1958)
Western Australia	118 : 10 : 1.20	Barrow (1969a)
New Zealand	120 : 10 : 1.30	Walker and Adams (1958)
Europe		
Scotland	140 : 10 : 1.40	Williams *et al.* (1960)

Table 3.2 Effect of weathering on the C:N:S ratios of soils derived from similar parent materials (graywacke or loess derived from greywacke) in New Zealand (Walker and Adams, 1959)

	C : N : S ratio
Weakly weathered soils	172 : 10 : 1.1
Moderately weathered soils	180 : 10 : 1.2
Strongly weathered soils	206 : 10 : 1.3

Table 3.3 Carbon, nitrogen, and sulphur ratios of Scottish soils derived from different parent materials (Williams *et al.*, 1960)

Parent material	C : N : S ratio
Granite	169 : 10 : 1.45
Slate	148 : 10 : 1.42
Old Red Sandstone	130 : 10 : 1.37
Basic Igneous	140 : 10 : 1.37
Calcareous	113 : 10 : 1.27
Mean	140 : 10 : 1.38

Table 3.4 Change in carbon : nitrogen : sulphur ratios in a chronosequence of New Zealand soils due to superphosphate application and pasture growth (Walker *et al.*, 1959)

Years under pasture	C : N : S ratio
0 (virgin scrub)	300 : 10 : 1.2
1.5	210 : 10 : 1.9
3	210 : 10 : 1.8
5	160 : 10 : 1.6
8	130 : 10 : 1.4
15	140 : 10 : 1.4
25	110 : 10 : 1.2

In general there seems to be less variation in the N : S ratio, both between soil groups and as a result of cultural treatment, than in the C : S ratio. Most agricultural soils have N : S ratios in the range 6.6 to 10 : 1 and C : S ratios in the range 60 to 150 : 1, and it seems reasonable to assume a mean world-wide ratio of C : N : S 130 : 10 : 1.3 for them. Wider ratios, however, seem likely in native grass- and woodland soils which may have ratios of the order of

Table 3.5   Variation in carbon:nitrogen:sulphur ratios with soil type in soils from North Queensland, Canada, and the USSR

Region	Soil type	C:N:S ratio	References
North Queensland	Solodic	133:10:1.3	Probert (1977)
	Neutral red		
	duplex	148:10:1.3	
	Red earth	200:10:1.1	
	Yellow earth	221:10:1.2	
	Krasnozem	143:10:1.7	
	Euchrozem	171:10:1.2	
Canada	Brown chernozem	93:10:1.64	Bettany *et al.* (1973)
	Dark brown		
	chernozem	94:10:1.54	
	Black chernozem	113:10:1.41	
	Grey black		
	transitional	125:10:1.32	
	Grey wooded	119:10:1.56	
	Brown chernozem	120:10:1.41	Lowe (1965, 1969)
	Black chernozem	125:10:1.30	
	Grey wooded	217:10:0.80	
	Gleysol	156:10:2.00	
USSR			
Bashkir ASSR,			
Pre-Urals	Chernozems	115:10:1.15	Balyanin and Evstigneeva (1974)
Central Volga,	Greyzems	109:10:1.90	Val'nikov (1970)
Tatar ASSR	(Orthic and Mollic)	106:10:2.18	
		108:10:1.77	
		103:10:1.88	
		104:10:1.75	
Central Volga,	Chernozems	112:10:1.59	Val'nikov *et al.* (1971)
Chuvash ASSR		113:10:1.44	
		115:10:1.34	
Uzbekistan	Xerosols, irrigated		Gulimov and Mukhanova (1976)
	(Haplic)	71:10:2.97	
		81:10:2.96	
		54:10:3.10	
	(Takyric)	73:10:11	
		67:10:10.5	
		76:10:10.9	

200:10:1. Peats and organic soils may have ratios intermediate between these values; approximately 160:10:1.22 (Williams and Steinbergs, 1958).

Current knowledge of the chemical nature of the organic sulphur is inadequate. Part of the organic sulphur occurs in amino-acid form. Cystine and methionine have been isolated in trace amounts from extracts of soils (Putnam and Schmidt, 1959; Paul and Schmidt, 1961) and their related compounds, cysteic acid, methionine sulphoxide, and methionine sulphone, have been identified in soil hydrolysates (Sowden, 1955, 1956, 1958; Stevenson, 1956). In two podzolic soils, for example, amino-acid sulphur has been found to account for 21–30% of the total organic sulphur (Freney *et al.*, 1972). Approximately 60% of the amino-acid sulphur was cystine sulphur.

Apart from the amino-acid sulphur, present knowledge only permits a grouping of compounds on a very broad basis which is believed to relate to the chemical bonding of the sulphur in them. This grouping is based on the reactivity of the soil organic sulphur with certain reducing agents.

These are as follows:

(1) Organic sulphur which is reduced to H_2S by hydriodic acid. This sulphur is not bonded directly to carbon and is believed to be mainly ester sulphates (Freney, 1961; Houghton and Rose, 1976).

(2) Organic sulphur which is not reduced by hydriodic acid. This sulphur is believed to be bonded directly to carbon (Freney *et al.*, 1970).

(3) Organic sulphur which is reduced to inorganic sulphide by Raney nickel. This makes up a substantial proportion of the carbon-bonded fraction and may consist mainly of sulphur in the form of amino acids (Freney *et al.*, 1975).

In general the hydriodic acid-reducible sulphur accounts for 30–70% of the organic sulphur (Williams and Steinbergs, 1959; Freney, 1961; Lowe and De Long, 1963; Cooper, 1972; Tabatabai and Bremner 1972a; Bettany *et al.*, 1973; Neptune *et al.*, 1975). In some soils the percentage of hydriodic acid reducible sulphur in the total organic sulphur remains constant with depth in the profile (Williams, 1975b), but in others the percentage increases with depth (Tabatabai and Bremner, 1972a; Williams, 1975b). Whether this increase reflects changes in the chemical composition of the organic matter or merely reflects errors in the chemical method of determination is not certain. Much of the hydriodic acid reducible sulphur is associated with the high-molecular-weight (humic acid) fraction of the organic matter (Freney *et al.*, 1969).

The Raney nickel reducible fraction forms part of the carbon-bonded fraction and probably consists mainly of the amino-acid sulphur in the soil organic matter (Freney *et al.*, 1975). This fraction has been found to account for up to 60% of the total organic sulphur (Lowe and De Long, 1963; Lowe, 1965, 1969; Freney *et al.*, 1970; Tabatabai and Bremner, 1972a; Neptune *et al.*, 1975).

The hydriodic acid reducible fraction appears to be readily hydrolysed to inorganic sulphate by acid or alkali, and consequently has been considered by some workers to represent the most labile fraction of the soil organic sulphur (see for example Biederbeck, 1978), while the fraction not reduced by either hydriodic acid or Raney nickel (which may account for up to 40% of the total organic sulphur) because of its resistance to degradation by drastic chemical treatments (Lowe, 1964, 1965) has been considered to be of little value as a source of mineralizable sulphur (Lowe, 1965). However, these chemical fractions all appear to be too broad to allow such generalizations as all three seem to be involved in cycling processes.

B. Inorganic sulphur

The main inorganic forms of sulphur are, sulphate in aerobic soils and sulphide in flooded or anaerobic soils.

Sulphate

Sulphate occurs in soils as water-soluble salts (mainly sodium, magnesium and calcium sulphates), sulphate ions adsorbed on soil colloids, or as insoluble compounds.

Water-soluble sulphate

The amounts of water-soluble sulphate vary greatly both within the soil profile itself and between soil types, depending upon the influence of factors such as leaching intensity, drainage, and the input of soluble sulphate in rain-water, irrigation water and fertilizers. Generally the surface horizons of well-drained soils contain only small amounts of soluble sulphates although under arid or poorly drained conditions high levels may accumulate. The amounts of soluble sulphate in the surface horizons of crop and pasture soils usually fluctuate seasonally (Simon-Sylvestre, 1965; Williams, 1968) as a result of the interactions of plant uptake, removal by leaching, and the impact of environmental factors on the rate of mineralization of organic sulphur. Fluctuation in amounts of soluble sulphate in subsoil horizons may also occur as a result of leaching.

Soluble sulphate often increases with depth in the profile (Williams, 1974; Probert, 1974, 1977) and in some soils, especially soils from semi-arid regions, may reach levels as high as 10%S in the deeper horizons (see Stace *et al.*, 1968). The occurrence of free gypsum in the subsoil horizons of many soils including black earths, red–brown earths, desert loams and grey, brown, and red clays is a well-known pedological feature.

Subsoil sulphate is readily available to deep-rooted species and is important in reducing the incidence of sulphur deficiency in cereal crops and deep-rooted pasture species (Lichtenwalner *et al.*, 1923; Lipsett and Williams, 1971; Gillman, 1973; Walker and Gregg, 1975).

Adsorbed Sulphate

The capacity of soils to adsorb sulphate varies greatly. In many soils, including those with pH greater than 6 and many light-textured soils, adsorption capacity is negligible or non-existent but in others, especially acid soils with moderate clay content, sulphate adsorption is very important. Adsorption capacity frequently increases with depth in the profile (see for example Table 3.6) and plays an important role in retaining sulphate against leaching and in determining the distribution of sulphate in the profile.

Sulphate adsorption is fully reversible and is concentration dependent (Kamprath *et al.*, 1956; Chao *et al.*, 1962b; Chang and Thomas, 1963). It depends greatly on the amounts and nature of the clay minerals (Ensminger, 1954; Kamprath *et al.*, 1956; Berg and Thomas, 1959; Neller, 1959) and is greatest in soils containing large amounts of aluminium and iron hydroxides, especially the former (Ensminger, 1954; Berg and Thomas, 1959; Chao *et al.*, 1962a,c, 1964; Aylmore *et al.*, 1967; Haque and Walmsley, 1973, 1974a,b). It is generally negligible at soil pH above 6 and increases with decrease in pH below this (Kamprath *et al.*, 1956; Williams and Steinbergs, 1962; Chao *et al.*, 1964) and is also influenced by the nature of the exchangeable cations (Chao *et al.*, 1963).

The amounts of adsorbed sulphate present are determined to a large degree by the sulphate adsorption capacity of the soil (Williams, 1974), especially in the subsoil horizons (see Table 3.6) and are influenced by the annual rainfall, the amounts of sulphate received in rain-water, and the amounts added in fertilizers (Barrow, 1969b; Hasan *et al.*, 1970; Toxopeus, 1970).

Adsorbed sulphate is readily available to plants and in many soils provides a major source of supply of plant available sulphur (Freney and Spencer, 1960; Fox *et al.*, 1964b; Williams and Steinbergs, 1964; Barrow, 1967a; Hasan *et al.*, 1970). The retention of sulphate by adsorption also enhances both the current and residual availability of fertilizer sulphate, especially in the case of deep-rooted species which can exploit subsoil sulphate.

Adsorbed sulphate is readily displaced by phosphate (Ensminger, 1954) so that phosphatic fertilizers tend to displace it from surface soil.

Insoluble sulphates

Of the insoluble sulphates the occurrence of sulphate as an impurity in calcium carbonate is the most important. Naturally occurring calcium carbonate normally contains sulphate as a co-precipitated or co-crystallized impurity which ranges in amount from 25 to 3000 ppmS (Williams and Steinbergs, 1962). Calcareous materials separated from soils have been found to contain between 200 and 2200 ppmS. Calcareous soils thus invariably contain insoluble sulphate in this form which in highly calcareous soils, such as calcareous

Table 3.6 The distribution of sulphur and the sulphate adsorption capacity in a krasnozem from Berry, NSW (Williams, 1974)

Soil	Depth (cm)	pH	Clay (%)	Nitrogen (%N)	Total sulphur	CaCl$_2$-soluble sulphate	Phosphate-soluble sulphate	Organic sulphur	Relative sulphate adsorption capacity
						(μgS g^{-1})			
Berry	0–4	5.3	43	0.53	752	2	25	726	88
	4–8	5.1	43	0.38	600	<1	25	576	150
Krasnozem	8–15	5.0	45	0.31	502	<1	27	475	231
Basalt	15–30	5.0	45	0.26	432	<1	28	404	254
1900 mm	30–45	5.0	46	0.17	358	<1	66	292	316
	45–60	4.8	47	0.10	431	1	177	254	459
	60–75	4.8	50	0.04	473	4	262	211	530
	75–90	4.7	52	0.03	642	9	440	202	698
	90–105	4.7	58	0.03	786	10	612	174	904
	105–120	4.7		0.01	941	11	804	137	1100

Table 3.7 The distribution of sulphur in soil profiles (Williams, 1974)

Soil	Depth (cm)	pH	Nitrogen (%N)	Total sulphur	Water-soluble sulphate	CaCO$_3$- sulphate	Total organic sulphur	Calcium carbonate (%)
					(μgS g^{-1})			
Griffith	0–10	8.7	0.084	136	2	11	123	3.2
Brown	10–20	8.9	0.052	116	1	31	84	10.0
Solonized soil	20–30	8.9	0.037	116	1	43	72	14.8
Aeolian deposits	30–45	8.9	0.022	112	4	48	60	15.1
390 mm	45–60	8.9	0.018	110	5	56	49	17.7
	60–75	9.0	0.014	118	6	55	57	18.1
Olary, SA	0–2	8.9	0.062	120	12	3	105	<0.1
Solonetz	2–15	9.2	0.016	69	6	0	63	0
180 mm	15–30	9.0	0.027	111	5	3	103	0
	30–45	9.5	0.021	125	8	16	101	0.3
	45–60	9.1	0.021	274	98	103	73	6.9
	60–90	9.0	0.017	616	229	270	117	21.6
	90–120	9.0	0.015	709	291	259	159	23.1
	120–150	9.0	0.015	1353	958	327	68	23.1
	150–180	9.0	0.009	894	448	450	0	19.5

sands, may contribute as much as 1200 ppmS to the total. This often accounts for over 95% of the total sulphur in surface horizons (Williams and Steinbergs, 1962). In subsoil horizons this form frequently accounts for 40–50% of the total sulphur (see for example Table 3.7).

The occurrence of insoluble barium sulphate in some soils, is quite likely, but this has not been generally established. Beattie and Haldane (1958) have identified barytes in small concentrations separated from the deeper subsoil horizons of a red earth in southern New South Wales, but in most soils the total barium content suggests that barium sulphate, if present, would be too small in amount to permit positive identification.

It is also likely that sulphate could occur in soils as basic aluminium and iron sulphates. Basic aluminium sulphates have been prepared in the laboratory by precipitation of aluminium hydroxide in the presence of sulphate ions (Bassett and Goodwin, 1949; Hsu and Bates, 1964; Singh, 1967; Adams and Hajak, 1978). When such precipitations are carried out under appropriate conditions in the presence of clay minerals basaluminite $[Al_4(OH)_{10}SO_4 \cdot 5H_2O]$ or alunite $[KAl_3)OH)_6(SO_4)_2]$ are formed (Singh, 1967; Singh and Brydon, 1967, 1969; Kodama and Singh, 1972) indicating the possibility that such compounds could be formed in soil. On the basis of such studies Adams and Rawajfih (1977) have proposed that retention of sulphate by acid soils could be a consequence of the precipitation and dissolution of aluminium hydroxy sulphates and their iron analogues rather than by adsorption. Wolt and Adams (1979) have shown that basaluminite was readily available to plants when added to a soil of pH 6.5, but that alunite was unavailable. Basaluminite has been isolated from materials occurring in weathered sandstone (Hollingworth and Bannister, 1952). Jarosite $[KFe_3(OH)_6(SO_4)_2]$ and coquinbite $[Fe_2(SO_4)_3 \cdot 5H_2O]$ have been identified in separates from drained tidal marsh ('cat-clay') soils (Clark *et al.*, 1961; Fleming and Alexander, 1961; Schwertmann, 1961; Spek, 1950) although their formation and occurrence in normal well-aerated soils has not been demonstrated. It seems likely that basic aluminium (and possibly iron) sulphates could occur in normal acid soils.

Compounds of Lower Oxidation State

In well-drained, well-aerated soils most of the inorganic sulphur occurs as sulphate and the amounts occurring as compounds of lower oxidation states are negligible (Freney, 1961; Neptune *et al.*, 1975). Under anaerobic conditions—as in poorly drained or waterlogged soils and especially in tidal swamps—considerable amounts of sulphides may be formed (Wiklander and Hallgren, 1949; Harmsen *et al.* 1954; Green, 1957; Fleming and Alexander, 1961; Whitehead, 1964; Brümmer *et al.*, 1971). Sulphite, thiosulphate and even elemental sulphur may also occur (Smittenberg *et al.*, 1951), but usually sulphides make up the greater proportion of the reduced compounds (Brümmer *et al.*, 1971). In poorly drained soils, reduced forms of sulphur are largely confined to the subsoil depths below the water-table.

Sulphides formed under waterlogged conditions in paddy soils can sometimes cause toxicity problems in rice crops (Green, 1957; Vámos, 1964). Tidal marsh conditions favour the accumulation of sulphur, much of it as sulphides and polysulphides of iron (Harmsen *et al.*, 1954), often resulting in total soil sulphur contents in excess of 1%S (Hart, 1959; Fleming and Alexander, 1961; Brümmer *et al.*, 1971). Values as high as 5.5% total sulphur have been reported in some South Carolina soils (Fleming and Alexander, 1961).

Draining such soils and establishing aerobic conditions generally results in rapid oxidation of the sulphides to sulphate with the formation of sulphuric acid and severe acidification of the soil leading to pH values below 2.0 in some cases (Wiklander and Hallgren, 1949; Hart, 1959; Fleming and Alexander, 1961; Moormann, 1963; Walker, 1972; Spek, 1950). Application of lime is thus often necessary in the reclamation of these soils, but in some cases the lime requirements are so high as to make reclamation by drainage uneconomical so that in some cases reclamation attempts must exclude drainage (Moormann, 1963). In some soils some of the sulphide may persist in the soil for many years after the initial drainage and reclamation (Harmsen *et al.*, 1954). Basic iron sulphates, such as jarosites, are often formed in the drained, oxidized subsoils (see above).

Tidal marsh, or acid–sulphate soils, are especially common in tropical regions, but they also occur on a less extensive scale in temperature regions. Particularly large areas of them occur in South-East Asia (Vietnam, Thailand, Pakistan, Indonesian Sumatra and Borneo), tropical Africa and tropical South America (Moormann, 1963).

3.2.2 Geographical Distribution of Sulphur in the Soils of the World

The total sulphur content of soils varies widely and it is difficult to make any simple generalizations. In the surface horizons of mineral soils the sulphur content generally ranges from less than 20 to 2000 ppm; in peats and swamp soils concentrations of up to 5000 ppm or more are found. In saline soils values in excess of 6000 ppmS have been recorded, and in tidal marsh soils values as high as 3000–35 000 ppmS have been reported. Most agricultural soils have sulphur contents ranging from 50 to 1000 ppm in the surface 15 cm (see Table 3.8).

In subsoils wide variation again occurs, depending especially on drainage and rainfall, and values ranging from less than 10 ppm in the subsoils of siliceous sands to as high as 10 000 ppm in the deeper horizons of soils from arid regions, have been reported.

Table 3.8   Total sulphur concentrations in the surface layers of soil from different regions of the world

Country or region	Soil	Range (μgS g^{-1})	References
Africa			
East Africa	Forest	230–750	Hesse (1957)
Tropical Africa		20–300	Dabin (1972)
Ivory Coast	Tropical ferruginous	30–120	Dabin (1972)
Malawi		35–139	Laurence et al. (1976)
Nigeria		18–180	Watson (1964), Oke (1967), Bromfield (1972)
Senegal	Saline	6400[a]	Dabin (1972)
Tanzania		120–439	Singh et al. (1979)
Zambia, Rhodesia		60–100	Grant et al. (1964)
America			
(a) North			
Canada			
Alberta	Chernozems and podzols	80–700	Lowe (1965)
British Columbia	Grassland	286–928	Chae and Lowe (1980)
	Forest	162–2328	Chae and Lowe (1980)
	Organic	1122–30430	Chae and Lowe (1980)
	Agricultural	214–438	Chae and Lowe (1980)
Quebec	Mineral	236–1385	Lowe and DeLong (1963)
	Organic	1620–6450	Lowe and DeLong (1963)
Saskatchewan	Agricultural	88–760	Bettany et al. 1973)
Eastern Canada	Agricultural	80–2070	Mackenzie et al. (1967)
USA			
Carolinas	Tidal marsh	3000–35000	Fleming and Alexander (1961)
Hawaii	Volcanic ash	180–2200	Hasan et al. (1970)
Iowa	Agricultural	56–618	Tabatabai and Bremner (1972a,b), Widdowson and Hanway (1974)
Kansas	Agricultural	150–910	Swanson and Latshaw (1922)
Minnesota	Agricultural	32–940	Evans and Rost (1945), Rehm and Caldwell (1968)
Oklahoma	Agricultural	40–1040	Harper (1959)
Oregon	Agricultural	82–784	Harward et al. (1962)

Region	Country	Soil type	Range	Reference
(c) South	Nicaragua	Volcanic	496–1324	Burbano and Blasco (1975)
	West Indies		110–510	Haque and Walmsley (1973, 1974b)
	Brazil		27–398	McClung et al. (1959), Neptune et al. (1975)
	Chile	Volcanic ash	391–1104	Schalscha et al. (1972)
Asia	India			
	Bihar		127–2045	Ahmed and Jha (1969)
	Gujarat		42–113	Reddy and Mehta (1970)
	Punjab		130–279	Kanwar and Takkar (1964)
	Rajasthan		91–3250	Ruhal and Paliwal (1978), Shukla and Gheyi (1971)
	Uttar Pradesh		93–189	Bhardwaj and Pathak (1969), Bhan and Tripathi (1973)
	Varanasi		154–207	Tiwari and Ram (1973)
	North-east India	Rice	110–272	Virmani and Kanwar (1971)
			112–275	Venkateswarlu et al. (1969)
	Indonesia	Saline	200–1490	Ismunadji and Zulkarnaini (1978)
	Iraq	Grassland	380–6450	Delver (1962)
	Japan		100–3180	Tsuji (1975)
	Malaysia	Acid sulphate	1300–4700	Chow and Ng (1969)
Australasia	Australia	Mineral	24–1100	Freney and Williams (1980)
		Basalt[b]	106–1100	
		Acid igneous rocks[b]	24–417	
		Sedimentary or metamorphic rocks[b]	32–340	
		Transported materials[c]	40–430	
		Calcareous sands	560–1030	
		Organic	390–1860	
	New Zealand	Agricultural	240–1360	Walker and Adams (1958), Metson and Blackmore (1978)
Europe	Bulgaria	Grey soils, chernozems	149–372	Nikolov (1964)
	England		112–3392	Williams (1975a), Massoumi and Cornfield (1964), Whitehead (1964)

Table 3.8 (continued)

Country or region	Soil	Range (μgS g^{-1})	References
Finland	Mineral	130–1140	Korkman (1973)
	Organic	530–1810	Korkman (1973)
France		56–512	Simon–Sylvestre (1969b)
Holland	Tidal marsh	100–3600	Harmsen et al. (1954)
Ireland		132–670	Gallagher (1969)
Italy		70–4210	Olivero (1960), Galoppini (1964), Pallotta et al. (1964)
Poland	Black earths	335[a]	Sklodovski (1969)
	Podzols	152[a]	
	Chernozems	283[a]	
Romania	Brown, podzols, chernozems	160–440	Davidesco and Palovski (1965)
Scotland		220–1790	Little (1957), Williams et al. (1960)
Sweden		132–1675	Johansson (1959)
West Germany		20–800	Buchner (1958)
Europe			
USSR			
Central Volga	Podzoluvisols[d]	145–305	Valnikov and Mishin (1974)
Tatar ASSR	Greyzems	143–707	
	Chernozems	431–1059	
Lower Volga	Chernozems	550–657	Ogoleva and Vershinina (1976)
Volgograd region	Kastanozems	256–362	
	Solonetzes	385[a]	
Central Chernozemic zone	Chernozems	756–1088	Aderikhin and Tikhova (1969)
	Greyzems	612–648	
Uzbekistan	Zerosols	220–1030	Gulimov and Mukhanova (1976)
Ukraine		320–465	Krupskii and Mamontova (1974)
		306–562	Krupskii et al. (1971)
Zei-Burya		200–8000	Shkonde (1957)
Chuvash	Chernozems	510–620	Valnikov et al. (1971)

[a]Mean value.   [b]Parent material.   [c]Alluvium, parna etc.   [d]FAO/UNESCO World Soil Map (Kovda, 1973).

Table 3.9   The total amounts of sulphur present in the surface metre of some Australian soils

Soil group	Total sulphur (tonnes ha^{-1})
Podzolic soils	0.1–1.5
Solodic soils	0.2–1.5
Solonized brown soils	1.4–27
Red–brown earths	1–6
Krasnozems	3–10
Black earths	3–45
Alpine humus	10
Calcareous sand	35
Grey, brown, and red clays	2–25
Calcareous red earths	1.3–550
Desert loams	30–450

As indicated previously, most of the sulphur in the surface layers of well-drained agricultural soils is organic in form while that in the subsoil horizons is predominantly inorganic sulphate.

The total amount of sulphur in the surface metre of Australian soils has been estimated to range between 0.1 and over 500 tonnes S ha^{-1} depending on soil type (Table 3.9; Freney and Williams, 1980) and at least similar ranges could be expected for soils from other regions.

3.2.3   Transformations of Sulphur Compounds in Soil

Since the first weathering of the primary rocks following the cooling of the earth's crust the sulphides of primary minerals have been converted, under well-drained conditions, to sulphate by the processes of soil formation. This sulphate has been converted to many and varied compounds by micro-organisms, plants, and animals. As sulphur exists in various states of oxidation, e.g. sulphate, +6; sulphite, +4; thiosulphate, +6 and −2 or +4 and 0; elemental sulphur, 0; disulphide, −1; and sulphide, −2 (Starkey, 1966), a great diversity of transformations is possible.

These transformations are frequently represented by a cycle which depicts an oversimplified view, because in reality there is not a cycle but an interwoven network of interrelated reactions.

It is generally believed that most of the transformations of sulphur in soil are carried out by micro-organisms, although strictly chemical reactions are also possible. The microbial transformation of sulphur in soil can be grouped into four main categories: (1) immobilization, or assimilation of sulphur into organic compounds by plants or micro-organisms; (2) mineralization, i.e. decomposition of organic sulphur compounds; (3) oxidation of sulphur and inorganic sulphur compounds; (4) reduction of sulphate and incompletely oxidized inorganic compounds of sulphur. The types of transformations are affected by the oxidation state of the sulphur and the environmental conditions, in particular the availability of oxygen (Starkey, 1966).

A. Soil-forming processes

In the lithosphere both sulphide- and sulphate-containing rocks are widespread (Roy and Trudinger, 1970) although in magma and igneous rocks sulphur is believed to be present primarily in the form of sulphides (ZoBell, 1963). The most abundant metallic sulphides in igneous rocks appear to be pyrite (FeS_2), pyrrhotite (Fe_5S_6–$Fe_{16}S_{17}$), chalcopyrite ($CuFeS_2$), pentlandite ($FeNiS_3$), and bornite (Cu_5FeS_4), but sphalerite (ZnS), galena (PbS), molybdenite (MoS_2), argentite (Ag_2S), cinnabar (HgS), chalcocite (Cu_2S), and other metallic sulphides are also found in limited amounts (ZoBell, 1963).

In the absence of water and oxygen most of these metallic sulphides are stable, but in their presence the minerals can be rapidly oxidized to sulphate, producing at the same time hydrogen ions and releasing metal ions to solution (Ivanov, 1968; Ralph, 1980). This has been found to occur when sulphides are exposed naturally or as a result of man's intervention. Some elemental sulphur may also be produced during this weathering process (ZoBell, 1963; Trudinger, 1975). Chemical and electrochemical reactions appear to be involved in these degradative processes, and while some reactions are speeded up by the presence of certain organisms (Ivanov, 1968; Ralph, 1980) there seems to be some doubt about the active role of micro-organisms in the weathering of reduced sulphides (Trudinger, 1975). Although bacteria can attack sulphide minerals, the relative importance of bacterial and chemical reactions is difficult to assess.

The weathering of pyrite is believed to be one of the main natural sources of sulphuric acid. This acid can then react with other mineral sulphides to release hydrogen sulphide which may be dissolved in water, may escape to the atmosphere, or may be oxidized to sulphur and sulphuric acid.

If the dilute sulphuric acid contacts limestones calcium sulphate will be formed. This is one way in which metallic sulphates are formed in the zone of weathering. Other sulphates such as barite ($BaSO_4$), celestite ($SrSO_4$), angelsite ($PbSO_4$), astrakhanite ($Na_2SO_4 \cdot MgSO_4 \cdot 4H_2O$), kainite ($KCl \cdot MgSO_4 \cdot 3H_2O$), kieserite ($MgSO_4 \cdot H_2O$), glauberite ($Na_2SO_4 \cdot CaSO_4$) and mirabilite ($Na_2SO_4 \cdot 10H_2O$) can be formed through the processes of recombination, base exchange, evaporation, hydration, etc. (ZoBell, 1963).

Little research has been carried out on the effect of the soil-forming factors—climate, organisms, topography, parent material, and age—on the amounts and distribution of sulphur in soil, although they probably would affect sulphur in much the same way as they affect nitrogen (Jenny, 1941).

The nature of the parent material seems to affect the type and amounts of inorganic sulphur in soils (Williams, 1974; Neptune *et al.*, 1975; Scott, 1976; Probert, 1980), but does not appear to be related to the accumulation of organic sulphur (Williams, 1974).

Barrow *et al.* (1969) found relationships between some soil-forming factors and the capacity of soils to adsorb sulphate. Soils formed from basic rocks had a greater capacity to adsorb sulphate than those formed from acidic rocks. This was probably due to the higher contents of iron oxides in the soils formed from the more basic parent materials.

Topography has been shown to have an effect on soil sulphur in certain situations—for example, Barrow and Spencer (1959) found that lateral movement of applied sulphate occurred on soils with a permeable surface layer over a less permeable subsoil on moderate slopes.

Bettany *et al.* (1973), in a study of cultivated soils from grassland and forested areas developed under semi-arid and subhumid conditions, observed the effects of climate and vegetation on the distribution of sulphur. They found that the concentration of total sulphur in the soils along a south-west to north-east transect of the agricultural region of Saskatchewan increased from the arid Chernozemic Brown zone to the Chernozemic Black zone and then decreased to the leached Grey Wooded zone. The C : N : S ratios, however, increased from 58 : 6.4 : 1 to 129 : 10.6 : 1 with the progression from the arid Chernozemic Brown soils to the leached Grey Wooded soils. These differences appear to be the result of different degrees of humification.

Scott and Anderson (1976) found no consistent effects of drainage conditions on the organic sulphur concentration or composition of Scottish soils, although the freely drained soils derived from basic igneous drift contained more organic sulphur than the poorly drained samples.

B. Microbial transformations

The four main microbial transformations that operate in soil (immobilization, mineralization, oxidation, and reduction) are mainly concerned with the internal cycling of sulphur from one pool to another, but they can also affect outputs from the pedosphere by converting immobile forms of sulphur into mobile forms and vice versa. For example, complex organic molecules can be converted to inorganic sulphate (mineralization) and leached to the hydrosphere, or gaseous sulphur compounds can be formed and lost to the atmosphere. Alternatively, inorganic sulphur can be converted into complex organic molecules (immobilization) and thus protected from leaching.

Immobilization and Mineralization

In the process of immobilization, micro-organisms absorb inorganic sulphate and convert it into organic forms for the synthesis of microbial tissue.

The opposing processes of immobilization and mineralization occur together in soil where organic sulphur mineralized is used for the synthesis of new cell material and only that sulphur not needed for synthesis is released as inorganic sulphur. If an abundant supply of carbon is available for energy then all of the inorganic sulphate in soil will be converted to organic forms, but if little carbon is available then inorganic sulphate will be released from the organic matter.

Both plants and micro-organisms are responsible for converting inorganic sulphate to organic sulphur compounds. Plants absorb inorganic sulphate and convert it to organic sulphur compounds which are then returned to the soil in plant residues (Williams and Donald, 1957). Micro-organisms working alone also incorporate significant amounts of sulphate into organic forms. Freney et al. (1971) found that when inorganic sulphate was added to soil there was steady incorporation of sulphur into organic matter over a period of 24 weeks and up to 50% of the 10 μgS g^{-1} soil added was converted into organic forms. Addition of glucose increased the incorporation to 82%.

By considering the carbon : sulphur ratio of the added material, attempts have been made to determine whether inorganic sulphate will be immobilized or released when organic residues are added to soil. Stotzky and Norman (1961) concluded from studies using glucose as a carbon source, that a C : S ratio of 900 or lower would provide adequate sulphur for maximum microbial activity. Therefore, sulphur should not be limiting for the decomposition of residues under natural conditions, because most plant residues have a C : S ratio less than 900.

However, other workers have found that immobilization of sulphur will occur during the decomposition of plant residues with C : S ratios well below 900 : 1 (Barrow, 1960; Massoumi and Cornfield, 1965; Stewart and Whitfield, 1965; Stewart et al., 1966a). For example, Barrow (1960) found that if the C : S ratio was below 200, sulphate usually accumulated in the soil; when the ratio was above 400, sulphate was incorporated into organic matter; and when the ratio was between 200 and 400, sulphate could either be incorporated into or released from the organic matter.

This suggested that the organic sulphur was not readily available for the requirements of the micro-organisms, or that the organisms involved had a higher sulphur requirement for the decomposition of plant residues than for glucose. Stewart et al. (1966b), working with pure compounds, showed that micro-organisms required more sulphur for the maximum decomposition of cellulose (1 part S : 300 parts C) than for glucose (1 part S : 900 parts C) and thus the latter explanation is more likely.

It is believed that micro-organisms are mainly responsible for the conversion of organic sulphur compounds to inorganic sulphate in soil (Alexander, 1961). Any variable which affects the growth of micro-organisms should therefore affect the mineralization of sulphur. In agreement with this supposition it has been shown that temperature, moisture, pH (Chaudhry and Cornfield, 1967a,b; Williams, 1967), and availability of food supply (Barrow, 1960) all affect the mineralization reaction (Blair, 1971; Biederbeck, 1978).

Mineralization of sulphur was markedly suppressed at 10 °C, but increased with increasing temperature from 20 °C to 40 °C and then decreased again with further temperature increase (Williams, 1967).

Sulphur mineralization was considerably retarded at moisture contents appreciably below or above field capacity (Williams, 1967), and the optimum for mineralization seemed to be at 60% of the maximum water-holding capacity (Chaudhry and Cornfield, 1967a).

The amount of sulphur mineralized from untreated soils does not appear to be directly related to soil type, to carbon, nitrogen, or sulphur contents, to C : S, N : S or C : N ratios, to soil pH, or to mineralizable N (Williams, 1967; Simon-Sylvestre, 1969b; Haque and Walmsley, 1972, 1974b; Jones *et al.*, 1972; Tabatabai and Bremner, 1972b). The pattern of release of sulphur from soil organic matter also does not appear to be related to any particular soil property, but is probably due to the chemical nature of the decomposing fraction of the soil organic matter.

The mineralization of sulphur in the presence of plants is often greater than in their absence (Freney and Spencer, 1960; Barrow, 1967a; Cowling and Jones, 1970; Nicolson, 1970; Jones *et al.*, 1972) and this may be due to the greater proliferation of organisms under plants.

Cycles of wetting and drying are known to accelerate the breakdown of soil organic matter (Birch, 1960a,b) and Barrow (1961) suggests that this is a major factor involved in the increased availability of sulphur under field conditions. A cumulative effect of wetting and drying on sulphur mineralization was found by Williams (1967) in laboratory studies, but cycles of wetting and drying of planted soils in the greenhouse did not stimulate mineralization of sulphur (Freney *et al.*, 1975).

Seasonal changes in sulphate sulphur in bare soils, cultivated fields, and soils under pasture have been studied in a number of locations (Barrow, 1966, 1969a; Simon-Sylvestre, 1965, 1967a,b, 1969a, 1972; Williams, 1968). In soil under subterranean clover pasture, sulphate accumulated in the surface soil during summer immediately after senescence of the pasture. High concentrations were maintained throughout the summer–autumn period and these decreased to low values in the winter and spring. The higher values in summer probably resulted from mineralization of soil organic sulphur under favourable moisture and temperature conditions and lack of plant uptake. The low values in winter and spring were probably due to leaching, plant uptake, and low rates of mineralization at low soil temperatures (Williams, 1968). Barrow (1966), however, suggests that the high levels of mineral sulphur in summer are due to the release of sulphur from organic matter by desiccation (as had been shown by Freney, 1958; Barrow, 1961; Williams,

1967; Tabatabai and Bremner, 1972b). Williams (1968), however, believed that only small amounts of sulphate were released from under the subterranean clover pasture by this mechanism. The sulphate that accumulates in summer, whatever the mechanism responsible for its accumulation, is vulnerable to leaching or erosion by sudden summer storms.

Little information is available on the total amount of sulphur mineralized from organic matter each year and which may therefore be lost from the soil reservoir by leaching, or absorbed by plants and removed in plant or animal products. However, it is generally believed that 1–3% of the organic sulphur is converted to inorganic sulphate annually.

Oxidation of Sulphur and Inorganic Sulphur Compounds

Sulphur-oxidizing organisms

The best-known, and usually considered to be the most important group of sulphur-oxidizing microorganisms are the autotrophic bacteria belonging to the genus *Thiobacillus*. This group can use elemental sulphur and incompletely oxidized inorganic sulphur compounds as specific electron donors for the assimilation of carbon dioxide. Carbon dioxide can supply all of their carbon requirements.

There are a number of species of *Thiobacillus*. Most, including some of the better-known species, *T. thiooxidans, T. thioparus, T. denitrificans*, and *T. ferrooxidans* are obligate autotrophs, but a few are facultative autotrophs, being able to use either oxidizable sulphur or organic substances as their energy source. Although the latter group are less well known, Vitolins and Swaby (1969) found that in 288 Australian soils one species of facultative autotroph (*T. intermedius*) was more numerous than all other species of *Thiobacilli* together. Most thiobacilli are strictly aerobic, but *T. denitrificans* is a facultative aerobe, being able to use nitrate under anaerobic conditions as an electron acceptor in place of oxygen in the oxidation process (Starkey, 1966; Weir, 1975).

The *Thiobacillus* group of bacteria has been more closely studied than the heterotrophic bacteria, fungi, and actinomycetes, whose importance in the oxidation of sulphur in soil has often been overlooked. Vitolins and Swaby (1969) found that heterotrophic yeasts and several genera of heterotrophic and facultative autotrophic bacteria were far more numerous than the strict autotrophs, and could play an important role in many soils. Strict autotrophs were generally low in numbers, absent from over one-third of the soils and poor competitors when oxidizable sulphur was added to soils. They were, however, more efficient than the heterotrophs when conditions suited them (Starkey, 1966).

A host of organisms can take part in sulphur oxidation, some of which carry the process only a step or two, but the overall reaction results in the production of sulphuric acid (Tisdale and Nelson, 1966).

The mechanism of oxidation of sulphur by autotrophic bacteria is obscure despite much investigation. Several pathways have been suggested and the one given below was proposed by Vishniac and Santer (1957).

$$S^0 \rightarrow S_2O_3^{2-} \rightarrow S_4O_6^{2-} \rightarrow S_3O_6^{2-} \rightarrow SO_3^{2-} \rightarrow SO_4^{2-} \tag{1}$$

Factors affecting the oxidation of elemental sulphur in soil

The oxidation of elemental sulphur in soil is affected by (a) the population of micro-organisms, (b) soil moisture and aeration, (c) soil temperature, (d) soil pH, and (e) particle size.

Microbial population   Soils differ markedly in their ability to oxidize sulphur and this is commonly ascribed to differences in microbial populations. The rate of oxidation of elemental sulphur in soil can often be increased markedly by addition of a suitable inoculum. Inoculation either with *Thiobacillus thiooxidans* (Li and Caldwell, 1966) or heterotrophic organisms (Vitolins and Swaby, 1969) produced a marked improvement in the initial rate of oxidation in pot and incubation studies.

Soil moisture and aeration   The most rapid oxidation of elemental sulphur takes place at a moisture level near to field capacity, i.e. in the soil moisture range that is optimum for the growth of agricultural plants. The rate of oxidation is markedly depressed at very low or very high soil moisture levels (Moser and Olson, 1952).

Soil aeration is intimately associated with soil moisture and thus at high moisture levels oxygen will be limiting for sulphur oxidation (Tisdale and Nelson, 1966).

Soil temperature   Microbial oxidation of sulphur will occur at temperatures as low as 4 °C, but the process is slow below 10 °C. Increasing the temperature up to 40 °C increases the rate of sulphur oxidation, but between 55 °C and 60 °C the organisms responsible for this oxidation are inactivated (Tisdale and Nelson, 1966). Optimum temperatures are not the same for all of the micro-organisms involved in the oxidation of sulphur, but temperatures between 25 °C and 40 °C are close to the optimum for most of the organisms (Weir, 1975).

pH   *Thiobacillus thioparus* and *T. denitrificans* grow best at reactions close to neutrality, whereas *T. thiooxidans* and *T. ferrooxidans* develop best in the pH ranges 2.0–3.0 and 2.2–4.7, respectively (Starkey, 1966). Addition of lime to soil can increase sulphur oxidation (Fox *et al.*, 1964a), have no effect (Shedd, 1928), or inhibit oxidation (Hart, 1959). These conflicting results may be due to an effect on pH or to an indirect effect through a reduction in solubility of other elements (Weir, 1975).

Particle size   Because elemental sulphur is insoluble in water it appears that its rate of oxidation would be a function of the surface area exposed to micro-organisms. Thus the smaller the particle size, the greater the surface area and the greater the rate of oxidation. This has been frequently demonstrated in incubation studies, pot and field trials (for references see Weir, 1975).

Oxidation of sulphide

The oxidation of iron sulphides in soil is more complex than the oxidation of elemental sulphur. The oxidation appears to consist of chemical as well as biological processes. The oxidations of both ferrous sulphide (FeS) and pyrite (FeS_2) appear to be basically similar.

The following sequence has been proposed for the oxidation of pyrite (Temple and Delchamps, 1953). The first step in the oxidation is non-biological and ferrous sulphate is formed (equation 2)

$$2FeS_2 + 2H_2O + 7O_2 \rightarrow 2FeSO_4 + 2H_2SO_4 \tag{2}$$

This reaction is then followed by the bacterial oxidation of ferrous sulphate, a reaction which is normally carried out by *T. ferrooxidans* (equation 3)

$$4FeSO_4 + O_2 + 2H_2SO_4 \rightarrow 2Fe_2(SO_4)_3 + 2H_2O \tag{3}$$

Subsequently, ferric sulphate is reduced and pyrite is oxidized by what appears to be a strictly chemical reaction (equation 4)

$$Fe_2(SO_4)_3 + FeS_2 \rightarrow 3FeSO_4 + 2S \tag{4}$$

Elemental sulphur so produced may then be oxidized by *T. thiooxidans* (equation 5) and the acidity generated favours the continuation of the process

$$2S + 3O_2 + 2H_2O \rightarrow 2H_2SO_4 \tag{5}$$

The main difference between the oxidation of iron sulphide and pyrite is the very much greater resistance of the latter to attack by atmospheric oxygen (Weir, 1975).

Reduction of Sulfate

Sulphate is unstable in anaerobic environments where conditions are favourable for bacterial development. It serves as a hydrogen acceptor for sulphate-reducing bacteria and is reduced to sulphide. This dissimilatory sulphate reduction is the distinctive property of a small number of bacteria that are strict anaerobes. Two principal genera are recognized: *Desulfovibrio* and *Desulfotomaculum* (Roy and Trudinger, 1970).[*]

These bacteria use a variety of fermentation products and molecular hydrogen to reduce sulphate, they tolerate high concentrations of salt and hydrogen sulphide, and function best in the pH range 5.5–9.0. Nitrate or low temperature retards the reaction (Starkey, 1966; Ponnamperuma, 1972).

Whereas oxidation of sulphur results in an increase in H-ion concentration, reduction of sulphate leads to an increase in pH (equation 6)

$$Na_2SO_4 + Fe(OH)_3 + 9H \text{ (organic matter)} \rightarrow FeS + 2NaOH + 5H_2O \quad (6)$$

The sodium hydroxide produced reacts with carbon dioxide from the decomposition of organic matter to produce carbonate and bicarbonate.

Most of the sulphide that accumulates as iron sulphide in waterlogged soils, bogs, ditches, and marine sediments is produced by sulphate-reducing bacteria. Hydrogen sulphide formed in the soil reacts with iron compounds to form FeS; under certain conditions the disulphide, usually pyrite, is formed more slowly (Bloomfield, 1969).

Partly reduced sulphur, such as that in sulphite, polythionate, thiosulphate, and elemental sulphur, is reduced more readily than sulphate. Many bacteria and other micro-organisms, in addition to the sulphate-reducing bacteria, produce sulphide from these substances (Starkey, 1966).

The pathway of reduction of sulphate to sulphide by *Desulfovibrio* has been discussed recently by Kelly (1972) and Roy and Trudinger (1970) and is described by scheme 7

$$SO_4^{2-} \xrightarrow{\text{ATP}} APS \xrightarrow{\text{2e}} SO_3^{2-} \xrightarrow{\text{6e}} S^{2-} \quad (7)$$

where APS = adenylylsulphate.

C. Nature of the gaseous sulphur compounds produced in soil and emitted to the atmosphere

Little information is available on the nature of the volatile sulphur compounds produced in soil or on the rates of emission of these volatiles to the atmosphere under natural conditions. Much of the information available comes from laboratory studies with pure organic sulphur compounds and isolated strains of micro-organisms.

[*]See footnote Chapter 1, page 23.

Hydrogen Sulphide

As hydrogen sulphide is often produced under laboratory conditions (e.g. Kondo, 1923; Almy and James, 1926; Tarr, 1933a,b; Tamiya, 1951a,b; Kadota and Ishida, 1972), and is formed in waterlogged soils as a result of protein decomposition or sulphate reduction (Vámos, 1964), many workers have assumed that hydrogen sulphide will be emitted to the atmosphere when organic matter is decomposed, or when sulphate is reduced under water-logged conditions. Because no data on the rate of hydrogen sulphide emission to the atmosphere from land or sea surfaces were available, this emission was used to balance the flows of sulphur in models of the global sulphur cycle (Eriksson, 1963; Junge, 1963; Robinson and Robbins, 1968; Kellogg et al., 1972; Friend, 1973; Cadle, 1975).

However, apart from a recent paper by Jørgensen et al. (1978) on hydrogen sulphide emission from coastal environments there does not appear to be any evidence for significant emission of hydrogen sulphide from soil in a natural environment (Bremner and Steele, 1978). No trace of this gas could be detected in the atmosphere above soil when inorganic and organic sulphur compounds, plant materials, animal manures, and sewage sludges were incubated in aerobic or waterlogged soils (Lewis and Papavizas, 1970; Banwart and Bremner, 1975a,b, 1976a,b). Even when sulphate and sulphide were added to a sandy soil, with a low capacity for sorption of hydrogen sulphide, under waterlogged conditions, no emission of hydrogen sulphide was detected during subsequent incubation (Bremner, 1977). Harter and McLean (1965) were also unable to detect hydrogen sulphide emission from a water-logged soil which produced large quantities of sulphide (>2000 μg g^{-1} soil).

On the other hand, some publications indicate that hydrogen sulphide has been emitted during incubation studies with soil under aerobic or anaerobic conditions (Bloomfield, 1969; Swaby and Fedel, 1973; Sachdev and Chhabra, 1974; Siman and Jansson 1976a). However, these publications show either that the emission of hydrogen sulphide was extremely small, or that the results are of doubtful validity because of the analytical technique used to detect hydrogen sulphide or because the incubation conditions were atypical of natural environments (Bremner and Steele, 1978).

In general it appears that soils have a high capacity to sorb hydrogen sulphide (Smith et al., 1973) and to hold hydrogen sulphide formed in situ from reduction of sulphate or decomposition of organic residues; this may be due to reaction of hydrogen sulphide with cations to form metallic sulphides. However, there may be certain situations where soils are high in sulphate and available organic matter but deficient in the cations required for the precipitation of metallic sulphides when the soils become waterlogged. Hydrogen sulphide may be emitted under those conditions or when organic residues are allowed to decompose on the surfaces of soils (Bremner and Steele, 1978).

Under certain conditions it appears that even living plants can emit hydrogen sulphide directly to the atmosphere. Wilson *et al.* (1978) found that hydrogen sulphide was emitted from squash, pumpkin, cantaloupe, corn, soybean, and cotton. The emission occurred after the roots were treated with sulphate, was higher if the roots were mechanically injured, was completely light-dependent and increased with light intensity. The maximum rates of emission observed were ~8 nmol $min^{-1} g^{-1}$ fresh weight. Spálený (1977) found that Norway spruce (*Picea abres* (L) Karsten) seedlings treated with 5 g potassium sulphate per day for 7 days exhaled 2.22 μg of hydrogen sulphide per hour per kg dry weight of needles when illuminated. No hydrogen sulphide was emitted when the seedlings were kept in darkness. Other plants such as tomato, beans, and grapevines have been reported to exhale hydrogen sulphide after treatment with sulphur dioxide (de Cormis, 1968a,b).

In summary then, there seems to be very little evidence in the literature to support the allocation by Eriksson (1963), Junge (1963), Robinson and Robbins (1968), Kellogg *et al.* (1972) and Friend (1973) of high flux rates for the emission of hydrogen sulphide and land surfaces.

Volatile Organic Compounds of Sulphur

Micro-organisms isolated from a variety of habitats and cultured under artificial conditions produce a variety of volatile organic sulphur compounds, including carbon disulphide, carbonyl sulphide, and a number of homologous sulphides and disulphides (Kadota and Ishida, 1972; Bremner and Steele, 1978).

A much smaller number of compounds has so far been isolated from soils or during the decomposition of organic compounds, plant materials, animal manures, and sewage sludges. The volatiles isolated and identified in significant quantities are carbon disulphide, carbonyl sulphide, methyl mercaptan, dimethyl sulphide, and dimethyl disulphide; however, the total amounts released from soils or decomposing organic materials under aerobic or waterlogged conditions appear to be very small (Barjac, 1952; Greenwood and Lees, 1956; Frederick *et al.*, 1957; Takai and Asami, 1962; Lewis and Papavizas, 1970; Banwart and Bremner, 1975a,b; 1976a,b; Bremner and Steele, 1978).

Carbon disulphide was produced from the decomposition of the sulphur-containing amino acids, cysteine and cystine; these two compounds may be the source of this gas in soils and manures (Banwart and Bremner, 1975b). When lanthionine and djenkolic acid were decomposed in soil carbonyl sulphide was produced (Banwart and Bremner, 1975b); this gas was also found when thiocyanates and isothiocyanates were metabolized (Bremner, 1977). Methyl mercaptan, dimethyl sulphide, and dimethyl disulphide seem to be produced principally from the decomposition of methionine and related compounds or from sulphonium compounds (Frederick *et al.*, 1957; Kadota and Ishida,

1972; Banwart and Bremner, 1975b; Salsbury and Merricks, 1975; Bremner, 1977).

Lovelock *et al.* (1972) found that living leaves of cotton, oak, spruce, and pine trees emitted from 2 to 43 $\times$ 10^{-12} g of dimethyl sulphide per g of oven-dried tissue per hour, while decaying leaves from the same plants emitted 10 to 100 times as much. Soils also exhaled dimethyl sulphide when incubated in air and the emission rates varied from 21 to 84 $\times$ 10^{-12} g^{-1} soil^{-1} (Lovelock *et al.*, 1972). Hitchcock (1977) used these figures, together with estimates of the global mass of plant materials and soils, to calculate the global production of dimethyl sulphide from these sources. She calculated that fresh leaves, senescent leaves, and soils could emit 0.01, 0.53 and 1.5 to 4.9 TgS year^{-1}, respectively. Hitchcock (1977) concluded that soil was a major source of dimethyl sulphide and that the global production of this gas was unlikely to exceed 5 TgS year^{-1}.

Aneja *et al.* (1979) estimated the emission rates of carbon disulphide and carbonyl sulphide emanating from salt marshes in North Carolina. Emission rates for carbon disulphide over the mud flat zone were less than 0.05 gS m^{-2} year^{-1}, and over a clipped marsh grass (*Spartina alterniflora*) area averaged 0.13 gS m^{-2} year^{-1} for all measurements, suggesting that carbon disulphide emanated from *S. alterniflora*. They calculated the global emission from marshes to be 0.07 TgS year^{-1}, assuming that all marshes emit uniformly and that the only species present was *S. alterniflora*. This they calculated to be <0.07% of the total biogenic emission of sulphur. In the same study it was found that carbonyl sulphide emission was not associated with the presence of marsh grass; the mean emissions of this gas from mud-flats and from the clipped marsh grass were 0.05 and 0.03 gS m^{-2} year^{-1}, respectively (Aneja *et al.*, 1979). They calculated that carbonyl sulphide emission from marshlands was responsible for only a small portion of the total biogenic sulphur emission.

D. Adsorption of sulphur compounds from the atmosphere by soil and their subsequent metabolism

Sulphur can be removed from the atmosphere by various mechanisms of which dry deposition on the soil surface and absorption by vegetation, soil, and stone are important. Sorption of certain atmospheric compounds by soil and their subsequent metabolism is believed to be an important mechanism for the removal of gaseous pollutants (Rasmussen *et al.*, 1975).

Sulphur Dioxide

Numerous workers have shown that soils can absorb sulphur dioxide (e.g. Vandecaveye *et al.*, 1936; Alway *et al.*, 1937; Fuhr *et al.*, 1948; Terraglio and

Manganelli, 1966; Seim, 1970; Abeles *et al.*, 1971; Payrissat and Beilke, 1975; Yee *et al.*, 1975) and it is now known that soils have the capacity rapidly to remove significant amounts of sulphur dioxide from the overlying gas (e.g. Smith *et al.*, 1973; Bremner and Banwart, 1976; Ghiorse and Alexander, 1976).

Sulphur dioxide was more rapidly absorbed by moist soils than by dry soils (Terraglio and Manganelli, 1966; Smith *et al.*, 1973) and there are reports that soil properties such as pH, texture, specific surface, organic matter content, ion-exchange capacity, and porosity affect the degree of absorption (Nakayama and Scott, 1962; Terraglio and Manganelli, 1966; Faller, 1968; Seim, 1970; Smith *et al.*, 1973; Yee *et al.*, 1975; Lockyer *et al.*, 1978).

There are conflicting reports on the role of micro-organisms in the sorption of sulphur dioxide by soil; Abeles *et al.* (1971) found that autoclaved soils sorbed less sulphur dioxide than untreated soils, but Ghiorse and Alexander (1976) and Smith *et al.* (1973) observed that steam sterilization had no effect on the rate of sorption, and Smith *et al.* (1973) found no effect on the amount sorbed. Microflora may, however, be involved in the removal of the sorption products.

Few estimates have been made on the amounts of sulphur dioxide that are absorbed by soils; generally deposition velocities for soils, defined as

$$v_g(\text{cm sec}^{-1}) = \frac{SO_2 \text{ deposited } (\mu g \, cm^{-2} \, sec^{-1})}{SO_2 \text{ concentration } (\mu g \, cm^{-3})}$$

have been found to be in the range 0.2–0.7 cm sec^{-1} (e.g. Spedding, 1969b; Owers and Powell, 1974; Garland, 1977; Lockyer *et al.*, 1978), and Smith *et al.* (1973) found that air-dry soils could absorb 1.1–15.3 and wet soils 9.3–66.8 mgSO$_2$ g^{-1} soil. Eriksson (1963) estimated that 25 TgS year^{-1} was absorbed by soil globally, while Junge (1963), Robinson and Robbins (1968), Kellogg *et al.* (1972), and Friend (1973) estimated dry deposition on land surfaces to be 10–20 TgS year^{-1}.

It has been known for many years that plants can sorb sulphur dioxide directly from the atmosphere and use it for their metabolism (e.g. Thomas and Hill, 1937; Thomas *et al.*, 1944; Fried, 1948; Olsen, 1957). Sulphur dioxide uptake by plants has been assessed by a number of different methods including micrometeorological techniques (e.g. Garland, 1977; Fowler, 1978; Galbally *et al.*, 1979), the use of $^{35}SO_2$ (Owers and Powell, 1974; Garland and Branson, 1977) and sulphur isotope abundances (Belot *et al.*, 1974; Krouse, 1977).

Surface wetness has a major influence on dry deposition rates (Fowler and Unsworth, 1974) and there appear to be diurnal and seasonal variations in uptake by grasses (Shepherd, 1974; Fowler, 1976, 1978). Uptake of sulphur dioxide occurs through the stomata, although there is evidence for some uptake on or through the cuticle (Belot *et al.*, 1974; Fowler, 1976, 1978; Garland and Branson, 1977; Garland, 1978).

Deposition velocities for grasses have been found to be in the range $0.3-1.2$ cm sec^{-1}, soybeans 1.25 and forests $0.1-2$ cm sec^{-1} (Fowler and Unsworth, 1974; Garland, 1977; Garland and Branson, 1977; Garland, 1978; Galbally *et al.*, 1979).

Fowler (1978) has estimated the dry deposition of sulphur dioxide on the agricultural area of Britain to be ~ 0.5 TgS year^{-1}, of which 0.14 TgS is deposited in soil in winter and the remainder is deposited on the vegetation. Galbally (1980) calculated the dry deposition of sulphur dioxide over Australia by dividing the continent into two areas; a desert and semi-arid region and a temperate and wet tropical region with deposition velocities of ~ 0.1 and 0.85 cm sec^{-1}, respectively. Assuming that the background concentration of sulphur dioxide was in the range $0.5-2$ μgSO$_2$ m^{-3} he estimated the dry deposition to be between 0.3 and 1 TgS year^{-1}. Estimates of the global uptake of sulphur dioxide by vegetation vary from 15 to 75 TgS year^{-1} (Eriksson, 1963; Junge, 1963; Robinson and Robbins, 1968; Kellogg *et al.*, 1972; Friend, 1973).

Sulphur dioxide can also be removed from the atmosphere by reaction with stone and an estimate of the global removal by this process is 4.5 TgS year^{-1} (Rasmussen *et al.*, 1975).

Reduced Sulphur Compounds

Soils also have the capacity to sorb hydrogen sulphide and volatile organic sulphur compounds such as methyl mercaptan, dimethyl sulphide, dimethyl disulphide, carbonyl sulphide and carbon disulphide (Fuhr *et al.*, 1948; Carlson and Gumerman, 1966; Carlson and Leiser, 1966; Kanivets, 1970; Smith *et al.*, 1973; Bremner and Banwart, 1976).

Both air-dry and moist soils will sorb dimethyl sulphide, dimethyl disulphide, carbonyl sulphide and carbon disulphide, but moist soils sorb larger amounts of these compounds. The capacity of soils for sorption of these gases is much smaller than their capacity for the sorption of sulphur dioxide, hydrogen sulphide, or methyl mercaptan (Bremner and Banwart, 1976; Bremner and Steele, 1978).

Bremner and Banwart (1976) working with sealed systems found that four soils from Iowa absorbed $23-442$ μg dimethyl sulphide, $101-306$ μg dimethyl disulphide, $14-104$ μg carbon disulphide, and $72-2340$ μg carbonyl sulphide per gram of soil from the gas phase initially containing 500 ppm of the gas. Smith *et al.* (1973) using the same system and 100 ppm of gas found that the same four soils absorbed $40.6-65.2$ mg hydrogen sulphide and $18-321$ mg of methyl mercaptan per gram of soil.

Micro-organisms are partly responsible for the sorption of dimethyl sulphide, dimethyl disulphide, carbonyl sulphide, and carbon disulphide from moist soils (Bremner and Banwart, 1976), but do not appear to be involved in the sorption of hydrogen sulphide or methyl mercaptan by soils (Smith *et al.*, 1973). When carbonyl sulphide is absorbed by moist soils, carbon disulphide is released in small amounts, and this appears to be due to the metabolism of carbonyl sulphide by micro-organisms (Bremner and Banwart, 1976).

Little information seems to be available for the uptake of reduced sulphur compounds by vegetation; however, Granat *et al.* (1976) suggests that uptake of hydrogen sulphide and dimethyl sulphide may be much slower than the uptake of sulphur dioxide because of the lower solubility of these gases in water.

3.2.4 Leaching, Runoff, and Erosion of Sulphur Compounds from Unamended Soil

Sulphur can be lost from soils by leaching with rain or irrigation water; the amount lost varies widely depending on the chemical and physical form of the sulphur, soil type, and properties, fertilization practices, cropping system, amount and distribution of rainfall, and temperature (see Harward and Reisenauer, 1966; Tisdale and Nelson, 1966; Jones and Woodmansee, 1979).

In most agricultural soils the bulk of the sulphur is present in complex organic molecules and there is no evidence for appreciable leaching of this organic sulphur. Sulphate appears to be the usual form leached and thus any factor which affects the transformation and movement of sulphate will affect the rate and amount of sulphur leached. Thus, reactions such as mineralization of organic sulphur, immobilization of sulphate, oxidation of sulphide, reduction of sulphate (section 3.2.3A), and adsorption of sulphate are important in promoting or preventing leaching.

Sulphate sorption capacity varies from soil to soil and also from layer to layer within the soil profile. It is affected by pH (retention increases as pH decreases), type of clay mineral (kaolinite type minerals retain more sulphate than the montmorillonite group; aluminium and iron hydrous oxides absorb strongly), and associated cations, and is concentration dependent (Harward and Reisenauer, 1966; Barrow, 1967b, 1975; Walker and Gregg, 1975).

More sulphur can be lost by leaching from bare soil than from soil under grass (Dutil and Ballip, 1979). Losses are reduced by cropping and are less with deep-rooted and perennial crops than with annual crops (Gachon, 1972).

Leaching losses have been determined by lysimeter studies and have been shown to vary from very small amounts to more than 310 kgS ha^{-1} year^{-1} (see Harward and Reisenauer, 1966; Baker *et al.*, 1975). For example, a study in New Zealand showed that, after the initial losses due to cultivation when the lysimeters were filled, the sulphur content of the soil actually increased rather than decreased (Muller, 1975). On the other hand, Kuhn and Weller (1977), working with two unfertilized mineral soils in Germany, found leaching losses of the order of 110 kgS ha^{-1} year^{-1}. Leaching there occurred mainly in the winter and to a lesser extent in the spring. A lysimeter study on unfertilized soil in Norway showed that the annual loss of sulphur in drainage water (30–40 kg ha^{-1}) was up to seven times higher than that supplied in rainwater; it was suggested that the balance came from sulphur absorbed from the air, weathering of inorganic sulphur, and mineralization of organic sulphur in soil (Ødelien, 1965).

Apart from the lysimeter studies, very few field studies on sulphate movement as a result of mineralization of soil organic matter have been made. Bromfield (1972) showed that mineralized sulphur moved down the soil profile but accumulated at depth.

Because leaching of sulphur varies so greatly with amount and distribution of rain, vegetation, soil type, etc., it is very difficult to estimate leaching losses from unamended soils on a global basis.

The extent of water erosion, or runoff, depends on the climate, topography, vegetation, and soil type (Baver, 1956) and it is well established that man's activities accelerate erosion. The average loss of soil due to erosion has been estimated at 22.4 tonnes ha^{-1} year^{-1} (Rhoades and Bernstein, 1971). Organic matter losses have been estimated to range from 378 to 1290 kg ha^{-1} (Barrows and Kilmer, 1963), while soluble sulphur and total sulphur have been found to vary from 12.9 to 29.7 and 47 to 113 kgS ha^{-1} year^{-1}, respectively (Stallings, 1957; Barrows and Kilmer, 1963).

Loss of sulphur due to wind erosion is discussed in Chapter 4.

3.3 THE EFFECT OF MAN ON THE SULPHUR CYCLE IN SOIL

3.3.1 Agricultural Practices and the Cycling of Soil Sulphur

Bringing native forest, bush or grassland under cultivation leads to appreciable loss of soil organic matter through mineralization processes (Allison, 1973). Such losses are invariably accompanied by corresponding losses of soil organic sulphur (Harper, 1959; Williams and Lipsett, 1961; Bettany *et al.*, 1973; McLaren and Swift, 1977).

These losses sometimes occur without appreciable change in the C : N : S ratio of the soil organic matter (Harper, 1959; Williams and Lipsett, 1961), but in some soils substantial changes in the ratio occur (McLaren and Swift, 1977; Bettany *et al.*, 1980). Following the clearing of native scrub—or grassland—and establishing clover pasture, large changes can occur in the C : S ratio of soils (Walker *et al.*, 1959; Jackman, 1964; Bettany *et al.*, 1979, 1980). In most cases, however, some changes in the chemical nature of the organic sulphur clearly occur since most of the mineralized sulphur appears to be derived from the carbon-bonded sulphur (Freney *et al.*, 1975; McLachlan and De Marco, 1975; McLaren and Swift, 1977).

A. Effect of fertilizer additions, and pasture production or cropping practices on soil sulphur

The losses in sulphur from surface horizons during cultivation generally take place whether or not fertilizer sulphur is applied. Table 3.10 summarizes a number of reports showing changes in soil sulphur as a result of cropping. In all cases a decrease in total soil sulphur had occurred even although, in some cases, fertilizer sulphur considerably in excess of that removed in crop produce had been applied. Sears *et al.* (1965) found that losses of sulphur from soil kept continuously under fallow exceeded the losses occurring under cropping and pasture. It is thus apparent that the losses of sulphur from the soil under cropping are largely due to the mineralization of organic sulphur to sulphate. Part of this may be recovered by the crop, but the remainder is leached to deeper horizons of the soil profile. In many cases this sulphate may be retained in the subsoil horizons by adsorption and can be recovered, at least in part, by deep-rooted species (Broomfield, 1972).

Effects of Liming

Addition of liming materials to soils may increase the availability of sulphur to plants and its potential for leaching in three ways.
(1) It stimulates the rate of mineralization of soil organic matter, and consequently the rate of mineralization of organic sulphur (Freney *et al.*, 1962; Williams and Steinbergs, 1964; Williams, 1967; Probert, 1976).
(2) By raising the soil pH it releases adsorbed sulphate from exchange sites, thereby favouring leaching and plant uptake.
(3) Liming materials frequently contain insoluble sulphates which can be released to soluble forms after reaction of the liming materials with the soil (Williams and Steinbergs, 1964).
Liming then enhances the uptake of sulphur by plants (Williams and Steinbergs, 1964), but it may also enhance the leaching of sulphate and could be expected to increase the rate of depletion of sulphur in the surface soil under cultivated systems.

Effect of Pasture

Organic matter frequently accumulates in the soil under permanent pasture until eventually an equilibrium level is reached (e.g. Richardson, 1938). There is thus a concurrent accumulation of organic sulphur. Accumulation of

Table 3.10 Loss of sulphur from the cultivated layer of soils under continuous cropping

Country	Virgin soil (μgS g^{-1})	Cropped soil (μgS g^{-1})	Average period of cropping (years)	Estimated soil sulphur loss (kgS ha^{-1})	Probable addition in fertilizers (kgS ha^{-1})	Estimated soil and fertilizer loss (kgS ha^{-1})	Estimated removal in crops (kgS ha^{-1})	References
Australia	128	92	50–60	50	90	140	35	Williams and Lipsett (1961)
Canada	497	308	65	345	Nil	345	230	Bettany et al. (1973)
USA								
Oklahoma Kansas	197	139	a	110	a	a	a	Harper (1959)
Humid regions	350	260	39	180	a	a	140	Swanson and Latshaw (1922)
Subhumid regions	440	420	32	40	a	a	110	Swanson and Latshaw (1922)
Semi-arid regions	280	250	25	60	a	a	85	Swanson and Latshaw (1922)
England	410	330	100	150	600	750	a	Cowling and Jones (1970)
Scotland	597	351	a	460	a	a	a	McLaren and Swift (1977)
New Zealand	340	307	3	60	Nil	60	90	Sears et al. (1965)
	340	318	3	30	250	280	190	Sears et al. (1965)

^{a}Not known.

organic sulphur is to be expected, especially in soils from which organic matter has been depleted by extensive cropping, and in soils naturally low in organic matter when improved pastures are established under regimes of regular fertilizer application. Substantial increases in the total sulphur content of the surface layers of soils treated in this way have been recorded (Donald and Williams, 1954; Hingston, 1959; Barrow, 1969a; Walker and Adams, 1959; Watson, 1969). An example of the accumulation of sulphur in soils under permanent subterranean clover pasture is given in Fig. 3.1.

In most well-drained soils nearly all of the sulphur in the surface horizons is accumulated in organic forms (Donald and Williams, 1954) and indeed, in many cases, that portion of the total sulphur input converted to organic forms is permanently retained, the rest being lost by leaching as sulphate. The accumulation of sulphur in the surface layers of soil under permanent pasture is thus influenced by several factors which include the rate of sulphur input, the intensity of leaching, level of pasture production, and the soil and environmental factors governing the rate of accumulation of soil organic mat-

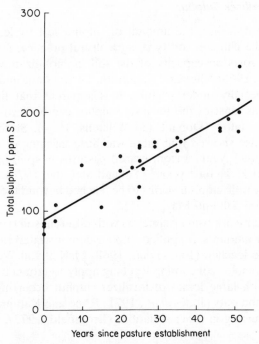

Fig. 3.1 The accumulation of sulphur in the surface 10 cm of soil under permanent sub-terranean clover pasture top-dressed regularly with single superphosphate ($S = 92.3 + 2.20t$; S = total sulphur in μgS g^{-1} soil, and t = time in years)

ter. Fertilizer and environmental sulphate leached from the surface layers is frequently not lost from the profile but is retained by adsorption in the subsoil horizons and can sometimes be utilized by deep-rooted species (Walker and Gregg, 1975).

In Australian pasture soils the total carbon-bonded sulphur fraction was found to account for 84% of the organic sulphur accumulated under subterranean clover pastures (Freney *et al.*, 1975). Of this only 42% was reducible by Raney nickel. Similarly, when organic sulphur was lost from red–brown earths under a cereal cropping system, 80% of the loss was from the carbon-bonded fraction (McLachlan and De Marco, 1975). Similar changes have been found by McLaren and Swift (1977). In incubation and pot culture studies (Freney *et al.*, 1975) it was found that all three fractions contributed available sulphur for plant uptake. Approximately 60% of the sulphur taken up by sorghum plants was derived from the carbon-bonded fraction and involved both Raney nickel and non-reducible fractions.

Leaching of Fertilizer Sulphur

As pointed out earlier, the amount of sulphur lost by leaching will vary depending on the climate, soil type, agricultural practice, fertilizer rate and form, sulphate sorption capacity of the soil, immobilization potential, etc. (e.g. Jones *et al.*, 1968; Rhue and Kamprath, 1973; Gregg and Goh, 1978). In some soils, especially under pasture, it is apparent that little leaching of sulphur occurs as most of the fertilizer sulphur can be accounted for in the surface organic matter (Donald and Williams, 1954). Studies using radioactive sulphur also showed that there was little leaching of the applied sulphate (<0.1 kg ha^{-1} year^{-1}) from grazed pastures in spite of apparent total sulphur losses of 25 kg ha^{-1} year^{-1} (Till and McCabe, 1976). At low rates of fertilization very little sulphur seems to be leached beyond the plant root zone (May *et al.*, 1968; Till and May, 1971).

However, other work using lysimeters with different soil types and pastures and much larger amounts of applied sulphur demonstrated that considerable sulphur could be leached (Jones *et al.*, 1968; McKell and Williams, 1960).

The same principles concerning leaching apply to cropped soils as well as pasture soils, with large losses of fertilizer sulphur occurring in some soils sown to wheat and oats (Jones *et al.*, 1971; Rhue and Kamprath, 1973), and small losses occurring in other situations (Bromfield, 1972).

Irrigated Agriculture

Irrigation and use of fertilizer on semi-arid and arid soils for the production of pastures or crops can have a marked effect on the quality of the drainage water percolating through or draining off the land. For example, Sylvester and Seabloom (1962) compared the sulphate content of the applied irrigation water with that of the surface and subsurface drainage water and found that the sulphate concentration had increased markedly, from 5.4 to 37 and 39 mg litre^{-1}, respectively. They estimated that 280 kg of sulphate per hectare per year were lost from the irrigated Yakima River valley in Washington (Sylvester and Seabloom, 1962; 1963). Similarly in the San Joaquin valley in California the subsurface drainage water had a concentration of 3600 mg of sulphate per litre compared with 170 for the surface drainage water (Walton, 1966).

Some of the sulphur in the drainage waters comes from sulphur applied in the fertilizer or absorbed from the atmosphere and some is dissolved from gypsum previously precipitated in the soil profile (Doneen, 1966).

B. Effect of grazing animals on the sulphur cycle

The increasing world population and rising standards of living in many parts of the world has brought about an increase in the consumption of animals products. Consequently, there have been large increases in animal numbers, animal excreta, and sulphur removal in animal products. The production of greater numbers of animals has required an increase in land sown to forage crops and because of their greater requirement for sulphur a greater use of sulphur fertilizers (Saalbach, 1965/66; Spencer, 1975). These various interactions have increased the amounts and rates of sulphur cycling within the soil–plant–animal system and the potential for increased losses from that system by gaseous loss to the atmosphere or by runoff and leaching to the hydrosphere. The problems, such as degradation of vegetation, soil erosion, and the depletion or silting up of water resources, which arise when native pastures are over-grazed have been well documented (Till, 1980).

Sulphur fertilization improves not only the quantity of pasture available to the grazing animal but also the quality of that feed; for example improvement in the protein quality and vitamin A content of alfalfa, and increases in the protein content of legumes and grasses have been recorded (Tisdale *et al.*, 1950; Fox *et al.*, 1964a; Saalbach, 1966; Blenkhorn, 1974) as a result of sulphur addition. Increases in the dietary intake of sulphur by animals has led to increases in wool growth and quality, meat production (e.g. Rendig and Weir, 1957; Wagnon *et al.*, 1958; Pund, 1969; Kennedy and Siebert, 1972; Morozov, 1972; Odynets *et al.*, 1972; Stepanov and Yakimchuk, 1973; Doyle and Moir, 1979), and milk production by lactating dairy cows (Bouchard and Conrad, 1973a,b).

When plant material or feed supplements are ingested by stock a small amount of sulphur is retained by the animal, but the bulk of the sulphur is returned to the soil in urine or faeces (Walker, 1957; Barrow, 1967c). Sulphur losses through the skin of sweating animals such as man and cattle may be important, while additional losses of organic sulphur must occur during the shedding of epithelial cells of the skin of all animals (Doyle and Moir, 1979). Possible losses during respiration or through the skin of sheep are not considered to be important, but considerable sulphur as hydrogen sulphide may be lost by belching during supplemental feeding (Doyle and Moir, 1979).

The proportion of sulphur excreted in the faeces or urine varies and seems to be related to the metabolic fate of the dietary or supplemental sulphur (Doyle and Moir, 1979). The proportion of the total sulphur excreted in the urine varies from 6 to ~90% (Warth and Krishnan, 1935; Walker, 1957; Barrow and Lambourne, 1962; Till, 1975; Doyle and Moir, 1979) with the proportion for a non-deficient pasture being about 50–60%. Work with ruminants has shown that the excretion of sulphur increases as the intake of sulphur increases (Bird, 1971; Bird and Hume, 1971; Bird and Moir, 1972; Langlands *et al.*, 1973; Kennedy *et al.*, 1975; Doyle and Moir, 1979).

Most of the sulphur in faeces is in organic combination although some inorganic sulphate is present (Walker, 1957; Bird, 1971; Bird and Hume, 1971; Doyle and Moir, 1979). The amount and form of sulphur excreted in urine appears to vary with sulphur intake, but inorganic sulphate seems to be the major form excreted. Considerable ester sulphate is also present (Walker, 1957; Bird, 1971; Bird and Hume, 1971; Kennedy *et al.*, 1975; Doyle and Moir, 1979) and this would probably be rapidly converted to inorganic sulphate in soil.

Thus the sulphur in urine is in a form which is immediately available to plants or in a form which is readily converted into an available form (Barrow, 1967c). Depending on the climate, soil type, etc., this available sulphur can also be readily leached. On the other hand, little of the sulphur in faeces is sulphate and the organic sulphur in faeces has to be mineralized before it is available for plant uptake or leaching. The rate of mineralization of sulphur depends on the concentration of sulphur in the faeces and it needs to be higher than that in plant material for mineralization to occur (Barrow, 1961). Even though faeces are mineralized more slowly than plant material, the passage of plant material through the animal, and the partitioning of the sulphur between urine and faeces in favour of urine, means that sulphur is mineralized more rapidly than if the plant material was returned directly to the soil (Barrow, 1961, 1967c; Till, 1975). The stocking rate will thus have a large effect on the proportion of the plant sulphur that is mineralized by the comparatively slow process of microbial breakdown of plant residues and that mineralized by the faster process in the grazing animal (Till, 1975).

However, faeces and urine are not uniformly distributed over a pasture (Petersen *et al.*, 1956; Lotero *et al.*, 1966) and there can be a transfer of sulphur from one part of the paddock to another. Sulphur ingested over a large area is voided on to a small area, thus concentrating the sulphur at that site. Losses by leaching can therefore be accentuated (Petersen *et al.*, 1956). Doak (1949) reported that a large part of the sulphur returned in a urine patch by sheep (at a concentration of $\sim$42.5 kgS ha^{-1} per urination) may be lost by leaching.

In addition to this random return of sulphur in excreta, sulphur can also be concentrated in one particular area of the field by the tendency of animals regularly to congregate in a specific area (camp-site), by the use of feedlots to fatten cattle, or by dairy farmers bringing their animals to a chosen area twice daily for milking. For example, Hilder (1964) in a study on the return of sulphur in faeces by grazing sheep found that the concentration of faeces increased towards the centre of the camp-site and that 34% and 22% of the total faeces were voided on areas which comprised only 10% and 3%, respectively of the total area. The distribution of urine is not known. Wrenford (1968) measured soil sulfate levels of $\sim$31 ppm in sheep camps compared to 7 ppm in grazed areas removed from the camps.

Till (1980) has estimated the amount of sulphur present in plant ($\sim$6 $\times$ 10^2 Tg) and animal pools on a global scale to be about 3 $\times$ 10^{-5}% of the total sulphur in the biogeochemical cycle.

Till (1980) also estimated the overall quantities of sulphur present and the major flows for the Australian pastoral industry. These are reproduced in Table 3.11. He estimated the total soil sulphur pool in the top 30 cm of soil to be 1.2 $\times$ 10^2 TgS assuming the mean concentration to be 75 μgS g^{-1}. The Australian pastoral industry supports about 150 $\times$ 10^6 sheep and 30 $\times$ 10^6 cattle on an area of approximately 400 $\times$ 10^6 ha of which only $\sim$30 $\times$ 10^6 ha are improved pasture. The stocking rate varies from 1 to 30 sheep ha^{-1} with an overall stocking rate of 1 sheep ha^{-1} assuming that one cattle beast is equivalent to eight sheep (Till, 1980). He calculated that for Australia as a whole, animals would contain $\sim$0.022 TgS and the sulphur removed in animal products would be 0.019 TgS year^{-1}, assuming that the average sheep weighed 40 kg and had a sulphur content of 0.14% S. As only 5–10% of the sulphur ingested is retained by animals, the amount excreted and available for uptake by pasture plants or leaching would be $\sim$0.016 TgS.

Volatiles

Recent work has shown that volatile sulphur compounds can be released to the atmosphere during the aerobic and anaerobic decomposition of manures (Burnett, 1969; Elliott and Travis, 1973; Banwart and Bremner, 1975a; Smith *et al.*, 1977). Cattle, sheep, swine, and poultry manures have been studied.

Banwart and Bremner (1975a) found that trace amounts of dimethyl sul-

Table 3.11 Sulphur budgets for grazed pastures (after Till, 1980)

| | Australian totals | | Typical values for sheep grazing systems | | | |
| | | | Native pasture | | Improved pasture | |
Component	Mass (tonnes)	Flux (tonnes year^{-1})	Mass (kg ha^{-1})	Flux (kg ha^{-1} year^{-1})	Mass (kg ha^{-1})	Flux (kg ha^{-1} year^{-1})
Inputs						
Atmosphere		2×10^5–4×10^5		<1		1–4
Fertilizer		1×10^5–2×10^5				25
Soil						
Total (0–30 cm)	1.2×10^8				1000	
Available (0–10 cm)	1.6×10^6		300		20	
Plants						
Total	0.8×10^6		2		12	
Production		1.5×10^6		3		50
Domestic animals						
Total	2.2×10^6		0.06		1.2	
Products		1.9×10^4		0.05		1.0
Intake		1.6×10^5		0.5		20
Microfauna and flora						
Total					0.6	
Production		—		—		35

phide were emitted from fresh poultry and sheep manure, but no emission could be detected from air-dried manures. They also found that all of their manure samples released hydrogen sulphide, methyl mercaptan, and dimethyl sulphide when incubated under anaerobic conditions, and that some samples also released dimethyl disulphide, carbonyl sulphide, and carbon disulphide. Most of the sulphur emitted was in the form of hydrogen sulphide and methyl mercaptan.

When homogenized manures were incubated in sealed bottles for 30 days the total amount of sulphur volatilized was between 0.02 and 0.53% of the total sulphur in the manure. Hydrogen sulphide and methyl mercaptan were the main gases evolved and together they accounted for 70–97% of the sulphur volatilized (Banwart and Bremner, 1975a). However, when the manures were incubated with a sandy soil under aerobic or waterlogged conditions no hydrogen sulphide was evolved (Bremner, 1977). Dimethyl disulphide and carbon disulphide were emitted from these mixtures under aerobic conditions but, with an anaerobic atmosphere, methyl mercaptan, dimethyl sulphide, and carbonyl sulphide were emitted in addition to the two disulphides (Bremner, 1977). The total emission amounted to 0.1–0.4% of the sulphur added. Other work by Banwart and Bremner (1976b) on the amounts of sulphur emitted from soils amended with animal manures, sewage sludges, and plant materials incubated under aerobic or waterlogged conditions showed that, on the average, less than 0.5% of the sulphur added was volatilized from manures or sewage sludges, whereas up to 3.3% was volatilized from plant materials. The range of values was quite large, 0–15.6%.

Volatile sulphur compounds are emitted from soil during the decomposition of certain pesticides. Moje *et al.* (1964) detected the emission of carbonyl sulphide from soil during the decomposition of nabam (disodium-ethylene-bis-dithiocarbamate) while Munnecke *et al.* (1962) reported the formation of carbon disulphide in soils treated with nabam, thiram, and zineb and methylisothiocyanate from soils receiving vapam and mylone.

Acidification of Soils

The importance of leguminous pastures and the sulphur fertilizer, super-phosphate, in improving soil fertility by increasing soil organic matter and plant-available nitrogen is well known (Russell, 1973). Unfortunately, growth of clover pastures also leads to a gradual increase in soil acidity. Decreases in soil pH under subterranean clover pastures have been reported on a wide range of soil types in southern Australia (Cook, 1939; Donald and Williams, 1954; Russell, 1960; Barrow, 1969a; Watson, 1969; Flemons and Siman, 1970; Kohn *et al.*, 1977), and Williams (1980) has calculated a relationship between pH and years of pasture growth over a 50-year period (equation 8)

$$\text{pH} = 5.95 - 0.905\,(1 - e^{-0.045n}) \tag{8}$$

where t = time in years During this period an average of ~55 kg super-phosphate was applied per hectare each year to the pastures. The increase in exchangeable hydrogen in the surface 10 cm of these soils due to pasture growth and fertilization is shown in Fig. 3.2 (C. H. Williams, unpublished).

In some pasture soils acidity has reached levels high enough to depress plant growth and, in some cases, to cause serious losses in crop or pasture production (Flemons and Siman, 1970; Williams and David, 1976; Osborne *et al.*, 1978; Cregan *et al.*, 1979). In most cases the problems are due to the development of manganese or aluminium toxicities (Williams, 1980), but can also be due to a decrease in molybdenum, calcium, potassium, phosphorus, and magnesium availability to plants, reduced nodulation, and increased fungal disease of the pasture plants.

The increase in acidity appears to be due, in part, to increases in the cation-exchange capacity of the soil as a result of the accumulation of soil organic matter (Williams and Donald, 1957; Russell, 1960) and cation depletion by increased leaching with rainwater (Williams, 1980).

Figure 3.1 shows the build-up in total sulphur, in the surface 10 cm of soils derived from granodiorite, due to pasture improvement at Binda, NSW. This build-up in sulphur, due to an average application of ~55 kg superphosphate $ha^{-1} year^{-1}$, reflects changes in total soil organic matter as the changes in sulphur were closely correlated with the increases in total soil nitrogen

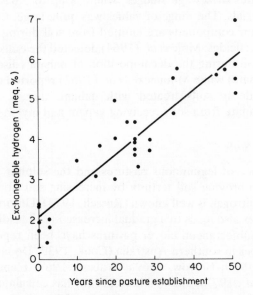

Fig. 3.2 The relationships between the period under subterranean clover pasture and the exchangeable hydrogen at pH 7.0 of the surface 10 cm of yellow podzolic soils

($r = 0.95$) and organic phosphorus ($r = 0.95$) (Williams, 1980); there were also corresponding changes in cation-exchange capacity ($r = 0.85$).

The accumulation of organic matter is mainly confined to the surface 10 cm of soil (Williams and Steinbergs, 1958), but the decrease in pH extends well below this depth. This indicates the importance of other factors such as cation depletion associated with increased leaching of nitrate following the input of legume fixed nitrogen (Helyar, 1976; Williams, 1980).

It seems unlikely that the acidity of the superphosphate itself was directly responsible for the increase in soil acidity. Long-term field experiments involving the application of large amounts of superphosphate to crops have generally shown little measurable effect on soil pH (Leeper, 1952; Williams and Lipsett, 1961; Piper and de Vries, 1964; Russell, 1973). However, addition of sulphur fortified superphosphates and other acid-forming sulphur fertilizers such as elemental sulphur and ammonium sulphate would be expected to increase soil acidity.

The rate at which pH decreases in soil under improved pasture depends upon several interacting factors such as the buffering capacity of the soil, the rate of build-up of organic matter and nitrogen, stocking rate, leaching intensity, soil type, and climate, and thus is likely to vary from soil to soil.

About half of the coastal areas and tablelands of New South Wales now have soils sufficiently acid for non-tolerant plants to be affected, and in the wheat belt on the western plains about one-sixth of the soils have become too acid for optimal growth. Soils in other areas of Australia that have a Mediterranean climate are also likely to be affected.

Approximately 75% of the improved pastures in southern Australia are less than 30 years old and it is, therefore, likely that the areas affected by acidity problems will increase in the future. Where legume-based pastures are grown in other parts of the world similar effect could be expected.

C. Removal of sulphur in crops and animal products

The amounts of sulphur removed in crops depends on the yield of the crop, the portion of the crop removed, and the treatment of the crop residues. Obviously, any factor which affects the yield of the crop, such as level of an essential nutrient, will affect removal. Some crops have a far greater requirement for sulphur and thus will remove more sulphur from soil during growth. Thus Spencer (1975) has grouped plants into three categories—high, moderate, and low—according to the magnitude of their needs in $kgS\ ha^{-1}$ for sulphur. Group I (high) contains the cruciferous forages (40–80), lucerne (30–70), and rapeseed (20–60); Group II includes (moderate) sugar-cane and coffee (20–40) clovers and grasses (10–40), and cotton (10–30); while Group III includes sugar-beet (15–25), cereal forages (10–20) cereal grains (5–20), and peanuts (5–10). An adequate supply of readily available sulphur is required for vegetable crops which have a short growing season.

Table 3.12   Sulphur in farm produce

Product		Tisdale and Nelson (1966)	Beaton and Fox (1971)	Biederbeck (1978)	C. H. Williams (unpublished)
			Sulphur in product (kg ha⁻¹)		
Grains	Corn	16[a]	49[b]	16–18[b]	16[a]
		11[c]			17[c]
	Wheat	6[a]	25[b]	14–17[b]	2–4[a]
		10[c]			3–6[c]
	Barley		28[b]		4[a]
					5[c]
	Oats	7[a]	22[b]		3[a]
		10[c]			5[c]
	Sorghum		43[b]		7[a]
	Rice	3[a]	20[b]		8–17[b]
Forage crops	Lucerne	34	34[b]	23–28	6[b]
	Clover	11	20[b]	17–22	6[b]
	Grass		18–56[b]	9–13	
Oil crops	Soybeans	11 (beans)	11 (beans)		
	Peanut	18 (nuts)	28[b]		
	Rapeseed		39[b]		
Vegetables	Potatoes	11 (tubers)	30 (tops and tubers)	8–11	2 (tubers)
	Cabbage		41 (heads)	22–43	4
	Turnip		43 (tops and roots)	28–29	
	Onions		28 (tops and bulbs)	22–23	
	Tomatoes	24 (fruit)			2
Miscellaneous	Sugar-cane		90[b]		20
	Sugar-beet		55 (tops and roots)	21–31	
	Cotton	2 (seed)	26[b]		3 (seed and lint)
	Tobacco	22 (leaves)	24[b]		
	Apples				2
Animal products	Milk				2
	Cattle beast				2
	Fat lambs				3
	Wool				1

[a] Grain.
[b] All above-ground parts.
[c] Stubble.

The figures for crop removal given in Table 3.12 are an indication of the rate at which the reserve supply in soil is being depleted.

Not all of the sulphur in a crop would necessarily be removed because in many cases only part of the crop plant is harvested, for example in many areas cereal straw is incorporated into the soil and in others it is burned *in situ*.

The amounts of sulphur removed in animal products, e.g. in wool production and beef or fat lamb production are very small compared with the amounts removed in plant products (Table 3.12), and play virtually no part in determining the requirements for available nutrients (Till, 1980).

D. Flooded rice production and the sulphur cycle

Sulphur is required for the nutrition of paddy rice as well as for upland rice, and sulphur deficiency has been found to reduce markedly rice yields and quality (Aiyar, 1945; Sen, 1938; Saran, 1949; Wang *et al.*, 1976a,b; Mamaril *et al.*, 1976; Ismunadji and Zulkarnaini, 1978; Wang, 1978; Blair *et al.*, 1979a,b).

Dramatic responses to sulphur fertilization by flooded rice have been recorded in the tropics (e.g. Wang *et al.*, 1976a,b) although excessive rates of application can decrease grain yield (e.g. Wang, 1978). Application of high rates of sulphur (>400 kgS ha^{-1}) caused burning of the tips and margins of the old leaves.

Not all of the sulphur added in fertilizer, irrigation water, or from the atmosphere is absorbed by the rice crop; some is leached and some may be reduced to sulphide in the anaerobic zone. Under anaerobic conditions sulphate can be reduced to sulphide by anaerobic bacteria (*Desulfovibrio* and *Desulfotomaculum*) which use sulphate as the terminal electron acceptor in respiration (Roy and Trudinger, 1970; Ponnamperuma, 1972). Under these flooded conditions sulphide can also be formed during the metabolism of the sulphur-containing amino acids, cysteine, cystine, and methionine (Barker, 1961).

Under flooded conditions the reduction of soil components is more or less sequential; after the disappearance of oxygen the sequence of reduction is nitrate, nitrite, manganic compounds, ferric compounds, and sulphate. Reduction of one component does not have to be completed before reduction of the next one can commence, although there are exceptions. Oxygen and nitrate must be depleted before ferric iron is reduced and sulphate is not reduced to sulphide whenever oxygen or nitrate are present. Sulphate can be found in waterlogged soils in the presence of reduced iron and reduced manganese. Redox potentials low enough for sulphate to be reduced do not generally occur until most of the reactable iron is in ferrous form. Therefore, whenever reactable iron and sulphur are present in a soil there will always be ferrous iron to react with and precipitate any sulphide formed when reduction processes are sufficiently intense to produce hydrogen sulphide (Patrick and Mikkelsen, 1971). More sulphide is produced when organic matter is added to a flooded soil (Yamane and Sato, 1961).

Although large amounts of hydrogen sulphide can be produced in waterlogged soils, the concentration of water-soluble hydrogen sulphide may be extremely low (Misra, 1938; Mortimer, 1941; Hutchinson, 1957) due to its reaction with ferrous iron and possibly other cations to form insoluble sul-phides (Ponnamperuma, 1972). When excessive amounts of sulphur are added to soils low in active iron, or other cations, some sulphide will remain in solution. This soluble sulphide can injure the roots of rice plants and cause the physiological disease termed 'Akiochi' or late summer disease (Mitsui, 1956; Tisdale and Nelson, 1966). This disease has been found to occur in sandy soils in Japan and Korea which are normally low in cations, but can also occur in soils as a result of cation depletion due to many years of continuous rice cultivation (Shioiri, 1943; Tisdale and Nelson, 1966).

Okajima and Takagi (1953, 1955, 1956) investigated the mechanisms of injury caused by hydrogen sulphide and found that the efficiency of nutrient absorption was impaired, the translocation of inorganic nutrients in the plant was inhibited, and the formation and translocation of carbohydrate was lowered. Other workers have also shown that hydrogen sulphide depressed nutrient absorption and yield of rice crops (Hashimoto *et al.*, 1948; Mitsui *et al.*, 1951; Yamane *et al.*, 1956; Baba, 1958; Takijima, 1963).

Suzuki and Shiga (1956) attempted to relate the production of hydrogen sulphide in soils with the severity of the disease. They found an average of <0.05, 0.23, and 0.54 mgS as hydrogen sulphide formed when 30 mg of S as sulphate per 100 g soil (dry basis) was incubated with normal, weakly de-graded, and strongly degraded soils, respectively. Yoneda (1958) and Shiga and Suzuki (1964) found that approximately 2.5 mgS was sufficient to kill rice plants.

There does seem to be some disagreement as to whether hydrogen sulphide is the real cause of Akiochi disease. However, many farmers have decided to reduce the risk by substituting ammonium chloride for ammonium sulphate in their fertilizer programme for rice (Tisdale and Nelson, 1966).

Soils which produce hydrogen sulphide when wet will become more acid as they are dried because of the oxidation of the sulphide to sulphuric acid.

E. Soil amendment with gypsum and other sulphur materials

Sulphur plays a number of roles in crop production; it is required as a plant nutrient, and it is used as a soil amendment, particularly on saline, alkali, saline–alkali, and calcareous soils. On these soils crops grow poorly, or not at all, because of the high salt content and the unavailability of nutrients or poor soil structure. These soils occur in many of the arid and semi-arid areas of the world which are used for irrigated agriculture. The main points of difference between these four classes of soil have been described by Richards (1954) and Stromberg and Tisdale (1979) and are summarized in Table 3.13. The arbitrary exchangeable sodium percentage (ESP) value of 15 which is used to differentiate soil classes (Table 3.13) has been modified by a number of people on the basis of morphological and experimental evidence. Bernstein (1974) proposed on ESP value of 10 for fine-textured, and 20 for coarse-

Table 3.13 Some properties of saline, saline–alkali, and calcareous soils

	Electrical conductivity of saturation extract (mS^a cm^{-1})	Exchangeable sodium percentage[b]	pH
Saline	>4	<15	<8.2
Alkali (or sodic)	<4	>15	>8.5
Saline–alkali	>4	>15	>8.5
Calcareous	<4	<10	7.5–8.2

[a]Millisiemens.
[b]Ratio of exchangeable sodium to cation-exchange capacity.

textured soils. Australian experience suggests that an ESP value of 15 or 10 is too high. Northcote and Skene (1972) suggested an ESP of 6 as a better boundary value and this was supported by Loveday and Pyle (1973). McIntyre (1979) presents results which show that 5 would be more relevant for certain plastic soils from eastern Australia. Exchangeable magnesium also affects the swelling and dispersion of soils (e.g. Bakker *et al.*, 1973; Bakker and Emerson, 1973; Emerson and Bakker, 1973; Emerson, 1977) and probably should be taken into account when setting a critical ESP value.

The main requirements for the reclamation of these soils are reduced salinity, improvement in physical characteristics (which impede water penetration, root development, and seedling emergence), lowering of pH (and improvement in nutrient availability), and removal of toxic amounts of boron (Loveday, 1975; Miyamoto *et al.*, 1975; Stromberg and Tisdale, 1979). On some alkaline and calcareous soils many crops suffer from nutrient deficiencies; under these conditions the plant cannot assimilate some nutrients such as phosphorus, iron, copper, manganese, and zinc. The availability of these nutrients is markedly increased by adjusting the pH from 8 to 7 (Stromberg and Tisdale, 1979).

Salinity is usually lowered by leaching; if the soil contains sufficient soluble calcium to replace sodium on the exchange sites no amendment aside from adequate drainable is required (Kelley, 1951; Loveday, 1975). In the absence of calcium, if sufficient swelling clay is present, a reduction in salt content by leaching will cause the clay to swell and disperse and water penetration will decrease. This can be overcome by addition of calcium to reduce the amounts of sodium on the clay. Gypsum is widely used for this purpose throughout the world (Tisdale, 1970).

To reclaim alkali and saline–alkali soils the excess sodium on the cation-exchange complex must be replaced by calcium. This can be done by applying gypsum, as above, or by applying acid or an acid-forming amendment to dissolve the calcium carbonate that is present in many alkali soils. The amendment must be followed by adequate leaching to remove the soluble salts (Overstreet *et al.*, 1951; McGeorge *et al.*, 1956; Tisdale, 1970; Miyamoto *et al.*, 1975; Miyamoto and Stroehlein, 1975; Stromberg and Tisdale, 1979).

Materials which have been used for this purpose include gypsum, elemental sulphur, sulphuric acid, sulphur dioxide, ammonium bisulphite, iron sulphate, ammonium polysulphide, and calcium polysulphide (Thomas, 1936; Thorup, 1972; Mohammed *et al.*, 1979; Stromberg and Tisdale, 1979). Many of these compounds have an acidifying effect on high pH soils; when the pH is lowered the availability to plants of phosphorus and micronutrients such as zinc, iron, and manganese is improved (e.g. McGeorge and Greene, 1935; Thorne, 1944; Mitchell *et al.*, 1952; Mortvedt and Giordano, 1973; Clement, 1978; Stromberg and Tisdale, 1979). The use of sulphuric acid on these soils has been studied in detail (Thorne, 1944; Miyamoto *et al.*, 1974; Ryan and Stroehlein, 1974), but little work has been done with other sulphur sources (Clement, 1978). Elemental sulphur must be oxidized to sulphuric acid before any reaction with calcium carbonate, soil phosphorus, or micronutrients can occur (Clement, 1978).

The choice of a soil amendment depends on the availability and cost of the amendment, the presence or absence of calcium carbonate in the soil, and soil temperature. Materials other than gypsum are generally required in smaller amounts, resulting in economies of storage, handling, and application. The acid-forming materials also have the dual ability of increasing water penetration and nutrient availability (Stromberg and Tisdale, 1979). Some of them, however, have undesirable properties.

The amount of material required for a response is dependent on the acid-forming properties of the compound used, the amount of calcium carbonate present in the soil, the crop species, and soil texture (Miyamoto *et al.*, 1975; Stromberg and Tisdale, 1979). It ranges from 2–6 tonnes ha^{-1} for improving moderately affected soils to 6–12 tonnes ha^{-1} for reclaiming severely affected soils (Miyamoto *et al.*, 1975).

Sodium-affected soils are a serious problem in agriculture (Nyborg, 1978). The full extent of salt-affected soils throughout the world is not accurately known so it is difficult to calculate the possible beneficial effects of amendment with sulphur-containing materials or the possible pollution of groundwaters which might result from these amendments. Approximately 15 million ha of land are affected to some degree by salt in the United States (Beaton and Fox, 1971). Solonetzic soils occupy about 5 million ha in the grass and parkland areas of Alberta and ~3 million ha in Saskatchewan and Manitoba (Cairns and Bowser, 1969; Toews, 1973; Cairns and Beaton, 1976), and 40 million ha in Asia, Africa, Europe, and Oceania (Bohn, 1976). Dr N. I. Bazilevich (pers. comm.) estimates that there are 47 million ha of soil affected by salt to a depth of 2 m in the European part of the USSR and 350 million ha in the whole of the USSR.

Results of laboratory column studies show that addition of sulphuric acid to the surface of arid soils followed by leaching with water is a more effective way of removing toxic amounts of boron than leaching with water alone (Prather and Rhoades, 1975).

There is also the possibility that sulphur could be used to lower high nitrate concentrations in soil water and sewage effluent, thereby reducing nitrate contamination of lakes and streams (Sikora and Keeney, 1976). Mann *et al.* (1972) found that denitrification rates in two soils were higher when sulphur was added to the soils; the nitrate was reduced to nitrogen by the action of the organism *Thiobacillus denitrificans* on the elemental sulphur.

Potentially all of the sulphur that is added as a soil amendment can be leached to contaminate the ground-water. Whether or not pollution occurs depends on the amount and type of acid or acid-forming compounds used, water management, and soil properties. In many areas the sulphur will be leached beyond the root zone but will remain in the soil profile.

According to Miyamoto (1977) 24.2 cm of water completely dissolved the calcium sulphate that had precipitated on the soil surface after acid application of <5 tonnes ha^{-1}, yielding soluble salts approximately equal to 100×10^3 eq ha^{-1}. The salts originally present in the same layer amounted to 126×10^3 eq ha^{-1}. Thus the acid application nearly doubled the salt load of the leaching water.

On the positive side, sulphur dioxide produced from the burning of coal or from the recovery of minerals by smelting could be used for the reclamation of salt-affected soils instead of being discharged into the atmosphere. In the south-western region of the United States there is a surplus production of sulphuric acid which lowers the pH. The pH of the natural precipitation, therefore, depends on the relative amounts of the various acidic and basic salt-affected soils. Further increases in the production of sulphuric acid are projected in the near future (McKee, 1969; Jones, 1972; Miyamoto, 1977).

3.3.2 Acid Rain

Rain-water in equilibrium with carbon dioxide at normal concentrations and pressures has a pH of 5.7 at 25 °C (Kramer, 1978). However, other substances which enter the atmosphere locally can raise or lower the pH. For example, in regions with calcareous soils, incorporation of dust can raise the pH of rain as will the volatilization of ammonia from grazing animals. Sulphur dioxide from sources such as volcanoes is oxidized in the atmosphere to sulphuric acid which lowers the pH. The pH of the natural precipitation, therefore, depends on the relative amounts of the various acidic and basic substances which enter the atmosphere and react with the rain-water (Kramer, 1978; Likens *et al.*, 1979).

Precipitation which fell before the Industrial Revolution, and is preserved in glaciers and continental ice-sheets, normally has a pH above 5. For example, Gjessing and Gjessing (1973) found a mean pH of 6.32 for a snow profile in Antarctica. Man's activities, particularly the burning of fossil fuels and the smelting of sulphide ores, has changed this. The oxides of sulphur and nit-rogen emitted during these practices are converted into strong acids in the atmosphere and returned to the ground in precipitation. These strong acids (sulphuric and nitric) completely dissociate in water and lower the pH to less than 5.6 (Likens *et al.*, 1979). On an annual basis rain and snow over large regions of the world are now from 5 to 30 times more acid than the rain from unpolluted areas. In large areas of the eastern United States, south-eastern Canada and western Europe the annual average pH of precipitation ranges from 4 to 4.5 (Likens, 1972; Hornstrom *et al.*, 1973; Cogbill and Likens, 1974; Likens *et al.*, 1979). With normal concentrations of CO_2 in the atmos-sphere the amount of sulphur required to form sufficient sulphuric acid to lower the pH of rainfall to 4.0 is only 1.5 ppmS (Terman, 1978). Sulphuric is the dominant species in acid precipitation, although nitric appears to be responsible for about 30% of the acidity.

The increase in acidity is accompanied by increases in sulphate concentration in the rain-water. Although sulphur produces visible damage to soil and vegetation close to emission sources (McGovern, 1972; McGovern and Bal-sillie, 1973; Hutchinson and Whitby, 1976) there are no direct toxic effects from sulphate on flora and fauna. In many areas of the world the sulphate is beneficial because it corrects a sulphur deficiency in the soil. Correction of the pollution problem by controlling the emission of sulphur dioxide will eventu-ally mean a greater requirement for sulphur fertilizers throughout the agricul-tural areas in Europe and North America.

The effects of acid precipitation on soils and vegetation may be insidious, and Malmer (1974) and McFee *et al.* (1976) suggest that periods of the order of a century may be required to cause irreversible effects. At the present time acute effects of acid precipitation on soils and vegetation have been clearly identified only under extreme conditions, for example very close to sources of air pollution (Overrein, 1978).

Acid rain may affect the soil environment by slightly intensifying the natural weathering processes which causes leaching of bases and development of acid soils. These processes result in dissolution of carbonates (calcite, dolomite), oxides (iron, aluminium, manganese), and aluminosilicates, adsorption of hydrogen ions on oxides, silicate minerals, and organic matter, and mobilization of metal ions (Kramer, 1978; Overrein, 1978). Deposition of acid rain causes exchange of hydrogen ions for calcium, magnesium, sodium, and potassium on clay minerals. Whether or not this will result in immediate marked acidification of soils depends on the buffering capacity of the soil. Calcareous soils show very little immediate effect from deposition of acid rain, and it has been concluded that the most adverse effects of acid precipitation are to be expected in soil types that are transitional between brown earths and podzols (Wiklander, 1973; Malmer, 1974). Overrein (1972) found that the pH of the precipitation had to be below 4 to make a significant effect on the leaching of calcium from soil and on the pH of ground-water.

Natural ecosystems are extremely complex with numerous interactions at all organizational levels; evaluating the effects of acid precipitation on these systems is difficult and expensive (Likens *et al.*, 1979). It is also very difficult to separate out the effect of sulphuric acid brought down in rainfall and that caused by dry deposition of sulphur dioxide. As far as pollution effects are concerned the actual causative agent is not important. Close to sources the bulk of the sulphur comes from direct absorption of sulphur dioxide by the soil or vegetation. This has been shown by setting out potted soils of known sulphur contents at a number of sites downwind from a source (Johansson, 1959; Cox, 1975; Siman and Jansson, 1976b; Nyborg *et al.*, 1977). More sulphur was absorbed from the air than was gained in rainfall (Johansson, 1959).

A number of workers have studied the spatial accumulation of sulphate in soils and the acidification of soils around smelters and other industrial concerns. They have found that the sulphate content of the soil falls off rapidly with distance from the source and that the acidification effect also decreases with distance (Katz *et al.*, 1939; Gorham and Gordan, 1960a,b; Gordan and Gorham, 1963; Bohn, 1972; Cox, 1975; Moss, 1975; Baker *et al.*, 1977). This reflects the similar trends observed in precipitation.

Vegetation may also be affected adversely by acid precipitation or by direct absorption of sulphur dioxide. Damage in the vicinity of emission sources has been observed for many years but, recently, concern has been expressed that damage to forests far from emission sources may also be occurring (Tamm, 1976).

3.3.3 Effect of Coal and Mineral Mining on Soils

The wastes from coal and heavy metal mining and treatment plants are primarily effluent water, chemicals, and tailings. The coarse residue from mining operations is usually piled high to form large dumps and the fine residues or slimes are usually pumped as a slurry to storage areas or dams where the solid materials settle out and the water can be evaporated or reused (Craze, 1980). In some mining areas these dumps are enormous; for example, at Magna, Utah, the tailings covered an area greater than 2000 ha and material was being added at the rate of 0.11 Tg day^{-1} (Nielson and Peterson, 1972). Solids dumps and slime dams are not only unsightly but are also sources of air and water pollution (Craze, 1980).

Chemical pollution as a result of mining operations is caused by exposing sulphide minerals to oxidation and leaching, resulting in the production of drainage waters that are extremely high in sulphate, acid, and heavy metals. Barton (1978) found that sulphate in the effluent from coal and zinc mine drainage, a uranium tailings pile and a copper smelting acid plant scrubber ranged from 490 to 63 000 ppm, while the pH of the effluents varied from 1.8 to 2.7. Some of the sulphide minerals involved in pollution caused by mining activities are pyrite (FeS_2), chalcopyrite ($CuFeS_2$), sphalerite (ZnS), and galena (PbS).

The sequence proposed for the oxidation of pyrite with the formation of acid and sulphate (Temple and Delchamps, 1953) has already been described in section 3.3.2.

In summary, the reactions involved are

$$2FeS_2 + 2H_2O + 7O_2 \rightarrow 2FeSO_4 + 2H_2SO_4 \tag{9}$$

$$4FeSO_4 + O_2 + 2H_2SO_4 \rightarrow 2Fe_2(SO_4)_3 + 2H_2O \tag{10}$$

$$Fe_2(SO_4)_3 + FeS_2 \rightarrow 3FeSO_4 + 2S \tag{11}$$

$$2S + 3O_2 + 2H_2O \rightarrow 2H_2SO_4 \tag{12}$$

Ferric sulphate is a powerful oxidant and thus chalcopyrite, sphalerite, and galena can be oxidized to form sulphuric acid along with soluble copper and zinc sulphates and insoluble lead sulphate (Craze, 1980).

$$CuFeS_2 + 2Fe_2(SO_4)_3 + 2H_2O + 3O_2 \rightarrow CuSO_4 + 5FeSO_4 + 2H_2SO_4 \tag{13}$$

$$2ZnS + 2Fe_2(SO_4)_3 + 2H_2O + 3O_2 \rightarrow 2ZnSO_4 + 4FeSO_4 + 2H_2SO_4 \tag{14}$$

$$2PbS + 2Fe_2(SO_4)_3 + 2H_2O + 3O_2 \rightarrow 2PbSO_4 + 4FeSO_4 + 2H_2SO_4 \tag{15}$$

All these reactions except (11) require an abundant supply of oxygen, and water is also required for most of them to proceed. Because of this, three zones have been observed in aged slime dumps; (1) an oxidized zone, out of which zinc, copper, and lead are removed; (2) a cemented hard-pan where due to oxygen depletion, iron oxides, copper, zinc, and lead are deposited; and (3) a reduced zone where the composition of the dump is unchanged and pHs are in the range of 6–7 (Boorman and Watson, 1976).

A number of factors in additon to oxygen and water are reported to have an affect on the overall dissolution rate of the minerals: e.g. pH, ferric ion concentration, catalytic agents such as bacteria or specific ions, and inhibitors such as bactericides or alkali (Hanna *et al.*, 1963).

Plants will not establish on these mine dumps or on areas affected by drainage waters from the mines or dumps because of the high salt contents or the increased solubility of toxic elements such as aluminium and manganese caused by increased acidity (Blevins *et al.*, 1970; Hoyt and Nyborg, 1972; Nyborg, 1978; Craze, 1980).

Many miles of streams and rivers, reservoirs, and large areas of land have been affected by mining activity.

3.4 SULPHUR BUDGET FOR THE PEDOSPHERE

It has not been possible to calculate the global content of soil sulphur because of lack of data on the sulphur content and mass of the many different soil types.

Bolin *et al.* (1979) presented data for total carbon contents of soil organic matter in world ecosystems. Using their data and C : S ratios of 200 : 1 for natural grassland and forest soils, 130 : 1 for peats, and 100 : 1 for cultivated soils (see above), we calculate the total organic sulphur content to be 1.1×10^4 TgS.

On the basis of annual mineralization of 1–3% of the soil organic matter, the conversion of soil organic sulphur to inorganic forms would amount to $1.1–3.3 \times 10^2$ TgS. Most of this sulphur will be released in the surface 10–15 cm of the profile and may be subject to leaching. However, leaching, should it occur, may not necessarily remove all of this sulphur from the soil, since this will be determined by factors such as rainfall intensity and adsorption reactions. Kilmer *et al.* (1974) estimates a leaching and erosion loss from arable soils of 3 kg ha^{-1} year^{-1} which would result in a global loss of 4.8 TgS.

A suggested sulphur budget for the pedosphere based on these calculations and published data is presented in Table 3.14. Clearly there are many gaps and inaccuracies and more research and data is essential.

Table 3.14 Sulphur budget for the pedosphere

Reservoirs		Total mass (TgS)	Reference
Soil organic matter		1.1×10^4	
Land plants		600	Friend (1973), Till (1980)
Biosphere		6×10^4	Wedepohl (1978)

Fluxes	Local (kg ha^{-1} year^{-1})	Global (TgS year^{-1})	Reference
Internal			
Mineralization of soil organic matter (1–3% year)		$1.1–3.3 \times 10^2$	
Cycled through plants		14	Till (1980)
Inputs			
Amendments for saline soils	$2–12 \times 10^3$		
Fertilizer		26	Friend (1973)
		29	Nriagu (1978)

Table 3.14 (continued)

Fluxes	Local (kg ha^{-1} year^{-1})	Global (TgS year^{-1})	Reference
Weathering			
Magma		5	Granat et al. (1976)
Igneous and sedimentary		33	Granat et al. (1976)
Pesticides (<1% of fertilizer)		<0.26	Coleman (1966)
Rainfall		43–86	Kellogg et al. (1972)
Dry deposition		10–20	Junge (1963) Robinson and Robbins (1968) Friend (1973)
Outputs			
Erosion	47–113		Barrows and Kilmer (1973)
Leaching	0–319		
Volatilization			
Dimethyl sulphide			
Fresh leaves		0.01	Hitchcock (1977)
Senescent leaves		0.53	
Soils		1.5–4.9	
Carbonyl sulphide			
Natural and agricultural fires		0.2–0.3	Crutzen et al. (1979)
Soils		0.5	Adams et al. (1979)

REFERENCES

Abeles, F. B., Cracker, L. E., Forrence, L. E., and Leather, G. R. (1971) Fate of air pollutants: removal of ethylene, sulfur dioxide, and nitrogen dioxide by soil. *Science* (Wash. DC), **173**, 194–916.

Adams, D. F., Farwell, S. O., Pack, M. R., and Bamesburger, W. L. (1979) Preliminary measurements of biogenic sulfur-containing gas emissions from soils. *J. Air Pollut. Control Assoc.*, **29**, 380–383.

Adams, F., and Hajak, B. F. (1978) Effects of solution sulfate, hydroxide, and potassium concentrations on the crystallization of alunite, basaluminite and gibbsite from dilute aluminum solutions. *Soil Sci.*, **126**, 169–173.

Adams, F., and Rawajfih, Z. (1977) Basaluminite and alunite: a possible cause of sulfate retention by acid soils. *Soil Sci. Soc. Am. J.*, **41**, 686–691.

Aderikhin, P. G., and Tikhova, E. P. (1969) Sulphur in chernozems and grey forest soils of the Central Chermozemic Zone. *Agrokhimiya*, No. 11, 121–127 [in Russian].

Ahmed, S., and Jha, K. K. (1969) Sulphur status of Bihar soils. *J. Indian Soc. Soil. Sci.*, **17**, 197–202.

Aiyar, S. P. (1945) A chlorosis of paddy (*Oryza sativa*) due to sulphate deficiency. *Curr. Sci.*, **14**, 10–11.

Alexander, M. (1961) *Introduction to Soil Microbiology.* Wiley, New York, 472 pp.

Allison, F. E. (1973) *Soil Organic Matter and its Role in Crop Production.* Elsevier, Amsterdam, 637 pp.

Almy, L. H., and James, L. H. (1926) A method for the study of the formation of volatile sulfur compounds by bacteria. *J. Bacteriol.*, **12**, 319–331.

Alway, F. J., Marsh, A. W., and Methley, W. J. (1937) Sufficiency of atmospheric sulfur for maximum crop yields. *Soil Sci. Soc. Am. Proc.*, **2**, 229–238.

Aneja, V. P., Overton, J. H., Cupitt, L. T., Durham, J. L., and Wilson, W. E. (1979) Carbon disulphide and carbonyl sulphide from biogenic sources and their contributions to the global sulphur cycle. *Nature (Lond.)*, **282**, 493–496.

Aylmore, L. A. G., Karim, M., and Quirk, J. P. (1967) Adsorption and desorption of sulfate ions by soil constituents. *Soil. Sci.*, **103**, 10–15.

Baba, I. (1958) Nutritional studies on the occurrence of *Helminthosporium* leaf spot and akiochi of rice plant. *Bull. Nat. Inst. Agric. Sci.*, Ser. D, **7**, 1–157.

Baker, J., Hocking, D., and Nyborg, M. (1977) Acidity of open and intercepted precipitation in forests and effects on forest soils in Alberta, Canada. *Water Air Soil Pollut.*, **7**, 449–460.

Baker, J. L., Campbell, K. L., Johnson, H. P., and Hanway, J. J. (1975) Nitrate phosphorus and sulfate in subsurface drainage water. *J. Environ. Qual.*, **4**, 406–412.

Bakker, A. C., and Emerson, W. W. (1973). The comparative effects of exchangeable calcium, magnesium, and sodium on some physical properties of red–brown earth subsoils. III. The permeability of Shepparton soil and comparison of methods. *Aust. J. Soil Res.*, **11**, 159–165.

Bakker, A. C., Emerson, W. W., and Oades, J. M. (1973) The comparative effects of exchangeable calcium, magnesium, and sodium on some physical properties of red–brown earth subsoils. I. Exchange reactions and water contents for dispersion of Shepparton soil. *Aust. J. Soil Res.*, **11**, 143–150.

Balyanin, M. I., and Evstigneeva, L. T. (1974) Sulphur content in chenozems of the Pre-Urals in the Tatar ASSR. In: Kosloskova, A. V. *Physico-chemical Properties of Soils and Fertilizer Application*, pp. 13–18. Izd. Kazan Univ., Kazan [in Russian]

Banwart, W. L., and Bremner, J. M. (1974) Gas chromatographic identification of sulfur gases in soil atmospheres. *Soil Biol. Biochem.*, **6**, 113–115.

Banwart, W. L., and Bremner, J. M. (1975a) Identification of sulfur gases evolved from animal manures. *J. Environ. Qual.*, **4**, 363–366.

Banwart, W. L., and Bremner, J. M. (1975b) Formation of volatile sulfur compounds by microbial decomposition of sulfur-containing amino acids in soils. *Soil Biol. Biochem.*, **7**, 359–364.

Banwart, W. L., and Bremner, J. M. (1976a) Volatilization of sulfur from unamended and sulfate-treated soils. *Soil Biol. Biochem.*, **8**, 19–22.

Banwart, W. L., and Bremner, J. M. (1976b) Evolution of volatile sulfur compounds from soils treated with sulfur-containing organic materials. *Soil Biol. Biochem.*, **8**, 439–443.

Barjac, H. de (1952) Metabolism of sulfur-containing amino acids, especially of methionine in the soil *Ann. Inst. Pasteur (Paris)*, **82**, 623–628.

Barker, H. A. (1961) Fermentations of nitrogenous organic compounds. In: Gunsalus, I. C. and Stanier, R. Y. (eds), *The Bacteria*, Vol. 2. Academic Press, New York, pp. 151–207.

Barrow, N. J. (1960) A comparison of the mineralization of nitrogen and of sulfur from decomposing organic materials. *Aust. J. Agric. Res.*, **11**, 960–969.

Barrow, N. J. (1961) Studies on mineralization of sulfur from soil organic matter. *Aust. J. Agric. Res.*, **12**, 306–319.

Barrow, N. J. (1966) The residual value of the phosphorus and sulphur components of superphosphate on some Western Australian soils. *Aust. J. Exp. Agric. Anim. Husb.*, **6**, 9–16.

Barrow, N. J. (1967a) Studies on extraction and availability to plants of adsorbed plus soluble sulfate. *Soil Sci.*, **104**, 242–249.

Barrow, N. J. (1967b) Studies on the adsorption of sulfate by soils. *Soil Sci.*, **104**, 342–349.

Barrow, N. J. (1967c) Some aspects of the effects of grazing on the nutrition of pastures. *J. Aust. Inst. Agric. Sci.*, **33**, 254–262.

Barrow, N. J. (1969a) The accumulation of soil organic matter under pasture and its effect on soil properties. *Aust. J. Exp. Agric. Anim. Husb.*, **9**, 437–444.

Barrow, N. J. (1969b) Effects of adsorption of sulfate by soils on the amount of sulfate present and its availability to plants. *Soil Sci.*, **108**, 193–201.

Barrow, N. J. (1975) Reactions of fertilizer sulphate in soils. In: McLachlan, K. D. (ed.), *Sulphur in Australasian Agriculture*. Sydney University Press, Sydney, pp. 50–57.

Barrow, N. J., and Lambourne, L. J. (1962) Partition of excreted nitrogen, sulphur and phosphorus between the faeces and urine of sheep being fed pasture. *Aust. J. Agric. Res.*, **13**, 461–471.

Barrow, N. J., and Spencer, K. (1959) Lateral movement of fertilizer sulphur. *J. Aust. Inst. Agric. Sci.*, **25**, 208–209.

Barrow, N. J., Spencer, K., and McArthur, W. M. (1969) Effects of rainfall and parent material on the ability of soils to adsorb sulfate. *Soil Sci.*, **108**, 120–126.

Barrows, H. L., and Kilmer, V. J. (1963) Plant nutrient losses from soils by water erosion. *Adv. Agron.*, **15**, 303–315.

Barton, P. (1978) The acid mine drainage. In: Nriagu, J. O. (ed.), *Sulfur in the Environment*, Pt II. Wiley, Chichester, pp. 313–358.

Bassett, H., and Goodwin, T. H. (1949) The basic aluminum sulfates. *J. Chem. Soc. (Lond.)*, 2239–2279.

Baver, L. D. (1956) *Soil Physics*. Wiley, New York, 489 pp.

Beaton, J. D., and Fox, R. L. (1971) Production, marketing and use of sulfur products. In: Olson, R. A., Army, T. J., Hanway, J. J. and Kilmer, V. J. (eds), *Fertilizer Technology and Use*. Soil Sci. Soc. Am., Madison, Wisconsin, pp. 335–379.

Beattie, J. A., and Haldane, A. D. (1958) The occurrence of palygorskite and barytes in certain parna soils of the Murrumbidgee region New South Wales. *Aust. J. Sci.*, **20**, 274–275.

Begheijn, L. Th., van Breemen, N., and Velthorst, E. J. (1978) Analysis of sulfur compounds in acid sulfate soils and other recent marine soils. *Commun. Soil Sci. Plant Anal.*, **9**, 873–882.

Belot, Y., Bourreau, J. C., Dubois, M. L., and Pauly, C. S. (1974) Use of sulfur-34 to measure the absorption rate of sulfur dioxide on the leaves of plants. Isotope ratios as pollutant source and behaviour indicators. *Proc. Symp.*, 18–22 Nov. 1974 Vienna (IAEA – SM-191-18), 403–416 [in French].

Berg, W. A., and Thomas, G. W. (1959) Anion elution patterns from soils and soil clays. *Soil Sci. Soc. Am. Proc.*, **23**, 348–350.

Bernstein, L. (1974) Crop growth and salinity. In: Schilfgaarde, J. van (ed.), *Drainage for Agriculture*, Agronomy 17. Amer. Soc. Agron., Madison, pp. 39–54.

Bettany, J. R., Saggar, S., and Stewart, J. W. B. (1980) Comparison of the amounts and forms of sulfur in soil organic matter fractions after 65 years of cultivation. *Soil Sci. Soc. Am. J.*, **44**, 70–75.

Bettany, J. R., Stewart, J. W. B., and Halstead, E. H. (1973) Sulfur fractions and carbon, nitrogen, and sulfur relationships in grassland, forest and associated transitional soils. *Soil Sci. Soc. Am. Proc.*, **37**, 915–918.

Bettany, J. R., Stewart, J. W. B., and Saggar, S. (1979) The nature and forms of sulfur in organic matter fractions of soils selected along an environmental gradient. *Soil Sci. Soc. Am. J.*, **43**, 981–985.

Bhan, C., and Tripathi, B. R. (1973) The forms and contents of sulphur in some soils of U.P. *J. Indian Soc. Soil Sci.*, **21**, 499–504.

Bhardwaj, S. P., and Pathak, A. N. (1969) Fractionation of sulphur in soils of Uttah Pradesh and its relation to phosphorus. *J. Indian Soc. Soil Sci.*, **17**, 285–289.

Biederbeck, V. O. (1978) Soil organic sulfur and fertility. In: Schnitzer, M. and Khan, S. U. (eds), *Soil Organic Matter*. Elsevier, Amsterdam, pp. 273–310.

Birch, H. F. (1960a) Nitrification in soils after different periods of dryness. *Plant Soil*, **12**, 81–96.

Birch, H. F. (1960b) Soil drying and soil fertility. *Trop. Agric.*, **37**, 3–10.

Bird, P. R. (1971) Sulphur metabolism and excretion studies in ruminants. II. Organic and inorganic sulphur excretion by sheep after intraruminal or intraduodenal infusions of sodium sulphate. *Aust. J. Biol. Sci.* **24**, 1329–1339.

Bird, P. R., and Hume, I. D. (1971) Sulphur metabolism and excretion studies in ruminants. IV. Cystine and sulphate effects upon the flow of sulphur from the rumen and upon sulphur excretion by sheep. *Aust. J. Agric. Res.*, **22**, 443–452.

Bird, P. R., and Moir, R. J. (1972) Sulphur metabolism and excretion studies in ruminants. VIII. Methionine degradation and utilization in sheep when infused into the rumen or abomasum. *Aust. J. Biol. Sci.*. **25**, 835–848.

Blair, G. J. (1971) The sulphur cycle. *J. Aust. Inst. Agric. Sci.*, **37**, 113–121.

Blair, G. J., Mamaril, C. P., Pangerang Umar, A., Momuat, E. O., and Momuat, C. (1979a) Sulfur nutrition of rice. I. A survey of soils of South Sulawesi, Indonesia. *Agron. J.*, **71**, 473–477.

Blair, G. J., Momuat, E. O., and Mamaril, C. P. (1979b) Sulfur nutrition of rice. II. Effect of source and rate of S on growth and yield under flooded conditions. *Agron. J.*, **71**, 477–480.

Blenkhorn, H. D. (1974) Effect of sulphur and magnesium on silage corn in Ontario. *Sulphur Inst. J.*, **10**, 10–12.

Blevins, R. L., Bailey, H. H., and Ballard, G. E. (1970) The effect of acid mine water on floodplain soils in the western Kentucky coalfields. *Soil Sci.*, **110**, 191–196.

Bloomfield, C. (1969) Sulphate reduction in waterlogged soils. *J. Soil. Sci.*, **20**, 207–221.

Bohn, H. L. (1972) Soil absorption of air pollutants. *J. Environ. Qual.*, **1**, 372–377.

Bohn, H. L. (1976) Estimate of organic carbon in world soils. *Soil Sci. Soc. Am. J.*, **40**, 468–470.

Bolin, B., Degens, E. T., Duvigneaud, P., and Kempe, S. (1979) The global biogeochemical carbon cycle. In: Bolin, B., Degens, E. T., Kempe, S. and Ketner, P. (eds), *The Global Carbon Cycle*. Wiley, Chichester, pp. 1–56.

Boorman, R. S., and Watson, D. M. (1976) Chemical processes in abandoned sulphide tailings dumps and environmental implication for Northeastern New Brunswick, *CIM. Bulletin*, **69**, 86–96.

Bouchard, R., and Conrad, H. R. (1973a) Sulfur requirement of lactating dairy cows. I. Sulfur balance and dietary supplementation. *J. Dairy Sci.*, **56**, 1276–1282.

Bouchard, R., and Conrad, H. R. (1973b) Sulfur requirement of lactating dairy cows. II. Utilization of sulfates, molasses, and lignin-sulfonate. *J. Dairy Sci.*, **56**, 1429–1434.

Bremner, J. M. (1977) Role of organic matter in volalization of sulfur and nitrogen from soils. In; *Proc. Symp. Soil Organic Matter Studies*, Vol. II. Braunschweig. Federal Republic of Germany 1976. International Atomic Energy Agency, Vienna, pp. 229–240.

Bremner, J. M., and Banwart, W. L. (1974) Identifying volatile S compounds by gas chromatography. *Sulphur Inst. J.*, **10**, 6–9.

Bremner, J. M., and Banwart, W. L. (1976) Sorption of sulfur gases by soils. *Soil Biol. Biochem.*, **8**, 79–83.

Bremner, J. M., and Steele, C. G. (1978) Role of microorganisms in the atmospheric sulfur cycle. *Adv. Microb. Ecol.*, **2**, 155–201,

Bromfield, A. R. (1972) Sulphur in Northern Nigeria soils. 1. The effects of cultivation and fertilizers on total S and sulphate patterns in soil profiles. *J. Agric. Sci.*, **78**, 465–470.

Brümmer, G., Grunwaldt, H.-S., and Schroeder, D. (1971) Contributions to the genesis and classification of marsh soils. III. Contents, oxidation states and bonding mechanisms of sulphur in polder soils. *Z. Pflanzenernaehr. Dueng Bodenk D.*, **129**, 92–108.

Buchner, A. (1958) The sulphur supply for agriculture in West Germany. *Landwirtsch. Forsch.*, **11**, 79–92 [in German].

Burbano, O. H., and Blasco, L. M. (1975) Volcanic soils of Nicaragua. II. Sulphur distribution and content. *Turrialba*, **25**, 429–435.

Burnett, W. E. (1969) Air pollution from animal wastes. Determination of malodors by gas chromatographic and organoleptic techniques. *Environ. Sci. Technol.*, **3**, 744–749.

Cadle, R. D. (1975) The sulfur cycle. In: Parker, H. S. (ed.), *Sulphur in the Environment*. Missouri Botanical Garden and Union Electric Company, St Louis, Missouri, pp. 1–13.

Cairns, R. R., and Beaton, J. D. (1976) Improving a solonetzic soil by nitrogen–sulphur materials. *Sulphur Inst. J.*, **12**, 10–12.

Cairns, R. R., and Bowser, W. E. (1969) *Solonetzic Soils and their Management*. Canada Dept. Agr. Pub. 1391. Ottawa, Ontario, 24 pp.

Carlson, D. A., and Gumerman, R. C. (1966) Hydrogen sulfide and methyl mercaptan removal with soil columns. In: *Proceedings of the 21st Industrial Waste Conference*. Purdue University, Lafayette, Indiana, pp. 177–191.

Carlson, D. A., and Leiser, C. P. (1966) Soil beds for the control of sewage odors. *J. Water Pollut. Control Fed.*, **38**, 829–840.

Chae, Y. M., and Lowe, L. E. (1980) Distribution of lipid sulphur and total lipids in soils of British Columbia. *Can. J. Soil. Sci.*, **60**, 633–640.

Chang, M. L., and Thomas, G. W. (1963) A suggested mechanism for sulfate adsorption by soils. *Soil Sci. Soc. Am. Proc.*, **27**, 281–283.

Chao, T. T., Harward, M. E., and Fang, S. C. (1962a) Movement of S35 tagged sulfate through soil columns. *Soil Sci. Soc. Am. Proc.*, **26**, 27–37.

Chao, T. T., Harward, M. E., and Fang, S. C. (1962b) Adsorption and desorption phenomena of sulfate ions in soils. *Soil Sci. Soc. Am. Proc.*, **26**, 234–237.

Chao, T. T., Harward, M. E., and Fang, S. C. (1962c). Soil constituents and properties in the adsorption of sulfate ions. *Soil Sci.*, **94**, 276–283.

Chao, T. T., Harward, M. E., and Fang, S. C. (1963) Cationic effects on sulfate adsorption by soils. *Soil Sci. Soc. Am. Proc.* **27**, 35–38.

Chao, T. T., Harward, M. E., and Fang, S. C. (1964) Iron and aluminium coatings in relation to sulfate adsorption characteristics of soils. *Soil Sci. Soc. Am. Proc.*, **28**, 632–635.

Chaudhry, I. A., and Cornfield, A. H. (1967a) Effect of moisture content during incubation of soil treated with organic materials on changes in sulphate and sulphide levels. *J. Sci. Food Agric.*, **18**, 38–40.

Chaudhry, I. A., and Cornfield, A. H. (1967b) Effect of temperature of incubation on sulphate levels in aerobic and sulphide levels in anaerobic soils. *J. Sci. Food Agric.*, **18**, 82–84.

Chow, W. T., and Ng, S. K. (1969) A preliminary study on acid sulphate soils in West Malaysia. *Malays. Agric. J.*, **47**, 253–267.

Clark, J. S., Gobin, C. A., and Sprout, P. N. (1961) Yellow mottles in some poorly drained soils of the lower Fraser Valley, British Columbia. *Can. J. Soil Sci.*, **41**, 218–227.

Clement, L. (1978) Sulphur increases availability of phosphorus in calcareous soils. *Sulphur in Agric.*, **2**, 9–12.

Cogbill, C. V., and Likens, G. E. (1974) Acid precipitation in northeastern United States. *Water Resour. Res.*, **10**, 1133–1137.

Coleman, R. (1966) The importance of sulfur as a plant nutrient in world crop production. *Soil Sci.*, **101**, 230–239.

Cook, L. J. (1939) Further results secured in 'top dressing' poor south-eastern pasture lands with phosphatic fertilizers. *J. Agric. South Aust.*, **42**, 791–865.

Cooper, P. J. M. (1972) Arylsulphatase activity in northern Nigerian soils. *Soil Biol. Biochem.*, **4**, 333–337.

Cormis, L. de (1968a) Contribution à l'étude de l'absorption du soufre par les plantes soumises à une atmosphère contenant due dioxyde de soufre. *Ann. Physiol. Veg. Univ. Brux.*, **10**, 99–112.

Cormis, L. de (1968b) Dégagement d'hydrogène sulfuré par des plantes soumises à une atmosphère contenant de l'anhydride sulfureux. *C. R. Hebd. Séances Acad. Sci.*, *Ser D*, **226**, 683–685.

Cowling, D. W., and Jones, L. H. P. (1970). Deficiency in soil sulfur supplies for perennial ryegrass in England. *Soil Sci.*, **110**, 340–348.

Cox, G. L. (1975) Effects of smelter emissions on the soils of the Sudbury area. M.Sc. Thesis, University of Guelph, Guelph, Ontario.

Craze, B. (1980) Problems on the revegetation of mine spoil dumps. In: Freney, J. R. and Nicolson, A. J. (eds), *Sulfur in Australia*. Australian Academy of Science, Canberra, pp. 256–264.

Cregan, P. D., Sykes, J. A., and Dymock, A. J. (1979) Pasture improvement and soil acidification. *Agric. Gaz. N. S. W.*, **90**, 33–35.

Crutzen, P. J., Heidt, L. E., Krasnec, J. P., Pollock, W. H., and Seiler, W. (1979) Biomass burning as a source of atmospheric gases CO, H_2, N_2O, NO, CH_3Cl and COS. *Nature (Lond.)*, **282**, 253–256.

Dabin, B. (1972) Preliminary results of a survey of the sulphur contents of the soils of tropical Africa. Int. Symp. Sulphur in Agriculture. *Ann. Agron.*, **72**, 113–136 [in French].

Davidesco, D., and Palovski, G. (1965) Considerations on the sulfur content of soils in the Rumanian People's Republic. *Agrochimica*, **9**, 178–182.

Delver, P. (1962) Properties of saline soils in Iraq. *Neth. J. Agric. Sci.*, **10**, 194–210.

Doak, B. W. (1949) The fate of nitrogenous constituents in animals excreta when voided on soil. In: *Plant and Animal Nutrition in Relation to Soil and Climatic Factors*. British Commonwealth Scientific Specialist Conference, Australia, pp. 418–423.

Donald, C. M., and Williams, C. H. (1954) Fertility and productivity of a podzolic soil as influenced by subterranean clover (*Trifolium subterraneum* L.) and super-phosphate. *Aust. J. Agric. Res.*, **5**, 664–687.

Doneen, L. D. (1966) Effect of soil salinity and nitrates on tile drainage in San Joaquin Valley, California. University of California, Davis, Dept. Water Sci. Eng., Paper 4002, 48 pp.

Doyle, P. T. and Moir, R. J. (1979) Sulfur and methionine metabolism in sheep. III. Excretion and retention of dietary and supplemented sulfur and production responses to intraruminal infusions of DL-methionine. *Aust. J. Agric. Res.*, **30**, 1185–1196.

Dutil, P., and Ballip, J. L. (1979) Biochemical oxidation of different forms of sulphur in chalk soils of Champagne. *C.R. Hebd. Séances Acad. Agric. Fr.*, **65**, 370–375.

Elliott, L. F., and Travis, T. A. (1973) Detection of carbonyl sulfide and other gases emanating from beef cattle manure. *Soil Sci. Soc. Am. Proc.*, **37**, 600–702.

Emerson, W. W. (1977) Physical properties and structure. In: Russell, J. S. and Greacen, E. L. (eds), *Soil Factors in Crop Production in a Semi-Arid Environment*. University of Queensland Press, Brisbane, pp. 78–104.

Emerson, W. W., and Bakker, A. C. (1973) The comparative effects of exchangeable calcium, magnesium and sodium on some physical properties of red–brown earth subsoils. II. The spontaneous dispersion of aggregates in water. *Aust. J. Soil. Res.*, **11**, 151–157.

Ensminger, L. E. (1954) Some factors affecting the adsorption of sulfate by Alabama soils. *Soil Sci. Soc. Am. Proc.*, **18**, 259–264.

Eriksson, E. (1963) The yearly circulation of sulfur in nature. *J. Geophys. Res.*, **68**, 4001–4008.

Evans, C. A., and Rost, C. O. (1945) Total organic sulfur and humus sulfur of Minnesota soils. *Soil Sci.*, **59**, 125–136.

Faller, N. N. (1968) Der Schwefeldioxydgehalt der Luft als Koomponente der Schwefelversorgung der Pflanze. Diss. Doktorgrades, Inst. Pflanzenernährung, Justus Liebig Univ., Giessen, 120 pp.

Fedorov, M. V. (1954) *Soil Microbiology*. Izdatel'stvo Sovetskaya Nauka [in Russian].

Fleming, J. F., and Alexander, L. T. (1961) Sulfur acidity in South Carolina tidal marsh soils. *Soil Sci. Soc. Am. Proc.*, **25**, 94–95.

Flemons, K., and Siman, A. (1970) Goulburn lucerne failures linked with induced manganese toxicity. *Agric. Gaz. N.S.W.*, **81**, 664–687.

Fowler, D. (1976) Uptake of sulfur dioxide by crops and soil. Ph.D. Thesis, University of Nottingham.

Fowler, D. (1978) Dry deposition of SO_2 on agricultural crops. *Atmos. Environ.*, **12**, 369–373.

Fowler, D., and Unsworth, M. H. (1974) Dry deposition of sulphur dioxide on wheat. *Nature (Lond.)* **249**, 389–390.

Fox, R. L., Flowerday, A. D., Hosterman, F. W., Rhoades, H. F., and Olson, R. A. (1964a) Sulfur fertilizers for alfalfa production in Nebraska. *Nebr. Agric. Exp. Stn. Bull. 214*, 37 pp.

Fox, R. L., Olson, R. A., and Rhoades, H. F. (1964b) Evaluating the sulfur status of soils by plant and soil tests. *Soil Sci. Soc. Am. Proc.*, **28**, 243–246.

Frederick, L. R., Starkey, R. L., and Segal, W. (1957) Decomposability of some organic sulfur compounds in soil. *Soil Sci. Soc. Am. Proc.*, **21**, 287–292.

Freney, J. R. (1958) Determination of water-soluble sulfate in soils. *Soil Sci.*, **86**, 241–244.

Freney, J. R. (1961) Some observations on the nature of organic sulphur compounds in soil. *Aust. J. Agric. Res.* **12**, 424–432.

Freney, J. R., Barrow, N. J., and Spencer, K. (1962) A review of certain aspects of sulphur as a soil constituent and plant nutrient. *Plant Soil*, **17**, 295–308.

Freney, J. R., Melville, G. E., and Williams, C. H. (1969) Extraction, chemical nature, and properties of soil organic sulphur. *J. Sci. Food Agric.*, **20**, 440–445.

Freney, J. R., Melville, G. E., and Williams, C. H. (1970) The determination of carbon bonded sulfur in soil. *Soil Sci.*, **109**, 310–318.

Freney, J. R., Melville, G. E., and Williams, C. H. (1971) Organic sulphur fractions labelled by addition of ^{35}S-sulphate to soil. *Soil Biol. Biochem.*, **3**, 133–141.

Freney, J. R., Melville, G. E., and Williams, C. H. (1975) Soil organic matter fractions as sources of plant-available sulphur. *Soil Biol. Biochem.*, **7**, 217–221.

Freney, J. R., and Spencer, K. (1960) Soil sulphate changes in the presence and absence of growing plants. *Aust. J. Agric. Res.*, **11**, 339–345.

Freney, J. R., Stevenson, F. J., and Beavers, A. H. (1972) Sulfur-containing amino acids in soil hydrolysates. *Soil Sci.*, **114**, 468–476.

Freney, J. R., and Williams, C. H. (1980) Forms and amounts of sulfur in soils. In: Freney, J. R. and Nicolson, A. J. (eds), *Sulfur in Australia*. Australian Academy of Science, Canberra, pp. 170–175.

Fried, M. (1948) The absorption of sulfur dioxide by plants as shown by the use of radioactive sulfur. *Soil Sci. Soc. Am. Proc.*, **13**, 135–138.

Friend, J. P. (1973) The global sulfur cycle. In: Rasool, S. I. (ed.), *Chemistry of the Lower Atmosphere*. Plenum Press, New York, pp. 177–201.

Fuhr, I., Bransford, A. V., and Silver, S. D. (1948) Sorption of fumigant vapors by soil. *Science* (Wash. D.C.), **107**, 274–275.

Gachon, L. (1972) Sulphur losses by drainage. Intern. Symp. on Sulphur in Agriculture. *Ann. Agron.*, **72**, 11–21.

Galbally, I. E. (1980) Sulfur dioxide uptake at the Earth's surface. In: Freney, J. R. and Nicolson, A. J. (eds.), *Sulfur in Australia*. Australian Academy of Science, Canberra, pp. 71–74.

Galbally, I. E., Garland, J. A., and Wilson, M. J. G. (1979) Sulphur uptake from the atmosphere by forest and farmland. *Nature (Lond.)*, **280**, 49–50.

Gallagher, P. A. (1969) The effect of sulphur in fertilizers, rainwater and soils on crop nutrition. *Sci. Proc. R. Dublin Soc.*, **2B**, 191–204.

Galoppini, C. (1964) Lo zolfo nei depositi alluvionali della Toscana e la sua distribuzione in alcuni terreni tipici. *5^e Symp. Intern. Agrochimica*, Palermo, 225–233.

Garland, J. A. (1977) The dry deposition of sulphur dioxide to land and water surfaces. *Proc. R. Soc. (Lond.)* Ser. A, **354**, 245–268.

Garland, J. A. (1978) Dry and wet removal of sulphur from the atmosphere. *Atmos. Environ.*, **12**, 349–362.

Garland, J. A., and Branson, J. R. (1977) The deposition of sulphur dioxide to pine forest assessed by a radioactive method. *Tellus*, **29**, 445–454.

Ghiorse, W. C., and Alexander, M. (1976) Effect of microorganisms on the sorption and fate of sulfur dioxide and nitrogen dioxide in soil. *J. Environ. Qual.*, **5**, 227–230.

Gillman, G. P. (1973) Studies on some deep sandy soils in Cape York Peninsula, North Queensland. 3. Losses of applied phosphorus and sulphur. *Aust. J. Exp. Agric. Anim. Husb.*, **13**, 418–422.

Gjessing, Y. T., and Gjessing, E. T. (1973) Chemical composition of an Antarctic snow profile. *Vatten*, **3**, 233–237.

Gordan, A. G., and Gorham, E. (1963) Ecological aspects of air pollution from an iron-sintering plant at Wawa, Ontario. *Can. J. Bot.*, **41**, 1063–1078.

Gorham, E., and Gordan, A. G. (1960a) Some effects of smelter pollution N.E. of Falconbridge, Ontario. *Can. J. Bot.*, **38**, 307–312.

Gorham, E., and Gordan, A. G. (1960b) Influence of smelter fumes upon surrounding vegetation. *Can. J. Bot.*, **38**, 477–487.

Granat, L., Rodhe, H., and Hallberg, R. O. (1976) The global sulphur cycle. In: Svensson, B. H. and Söderlund, R. (eds), *Nitrogen, Phosphorus and Sulphur–Global Cycles*. SCOPE Report 7. *Ecol. Bull (Stockholm)*, **22**, 89–134.

Grant, R. C. N., Hughes, E. W., Moerman, J., and Colzer, D. T. (1964) Soil chemistry. In: *Ann. Rep. Agric. Research Council Central Africa*, pp. 15–21.

Green, V. E. Jr. (1957) The culture of rice on organic soils – a world survey. *Agron. J.*, **49**, 468–472.

Greenwood, D. J., and Lees, H. (1956) Studies on the decomposition of amino acids in soils. 1. A preliminary survey of techniques. *Plant Soil*, **7**, 253–262.

Gregg, P. E. H., and Goh, K. M. (1978) Field studies with radioactive sulphur-labelled gypsum fertilizer. I. Soil factors affecting the movement of fertilizer sulphur. *N.Z. J. Agric. Res.*, **21**, 593–601.

Gulimov, S., and Mukhanova, V. L. (1976) Content of sulphur and its forms in irrigated soils of Uzbekistan. *Pochvovedenie, No. 11*, 28–32 [in Russian].

Hanna, G. P. Jr., Lucas, J. R., Randles, C. I., Smith, E. E., and Brant, R. A. (1963) Acid mine drainage research potentialities. *J. Water Pollut. Control Fed.*, **35**, 275–296.

Haque, I., and Walmsley, D. (1972) Incubation studies on mineralization of organic sulphur and organic nitrogen. *Plant Soil*, **37**, 255–264.

Haque, I., and Walmsley, D. (1973) Adsorption and desorption of sulphate in some soils of the West Indies. *Geoderma*, **9**, 269–278.

Haque, I., and Walmsley, D. (1974a). Movement of sulphate in two Caribbean soils. *Plant Soil*, **40**, 145–152.

Haque, I., and Walmsley, D. (1974b) Sulphur investigations in some West Indian soils. *Trop. Agric.*, **51**, 253–263.

Harmsen, G. W., Quispel, A., and Otzen, D. (1954) Observations on the formation and oxidation of pyrite in the soil. *Plant Soil*, **5**, 324–348.

Harper, H. J. (1959) Sulfur content of Oklahoma soils, rainfall and atmosphere. *Okla. Agric. Exp. Stn. Bull. B-536*, 18 pp.

Hart, M. G. R. (1959) Sulphur oxidation in tidal mangrove soils of Sierra Leone. *Plant Soil*, **11**, 215–236.

Harter, R. D., and McLean, E. O. (1965) The effect of moisture level and incubation time on the chemical equilibria of a Toledo clay loam soil. *Agron. J.*, **57**, 583–588.

Harward, M. E., Chao, T. T., and Fang, S. C. (1962) The sulfur status and sulfur supplying power of Oregon soils. *Agron. J.*, **54**, 101–106.

Harward, M. E., and Reisenauer, H. M. (1966) Reactions and movement of inorganic soil sulfur. *Soil Sci.*, **101**, 326–335.

Hasan, S. M., Fox, R. L., and Boyd, C. C. (1970) Solubility and availability of sorbed sulfate in Hawaiian soils. *Soil Sci. Soc. Am. Proc.*, **34**, 897–901.

Hashimoto, H., Koda, T., and Mitsui, S. (1948) Effect of the addition of base and silica on degraded paddy soils. *J. Sci. Soil Manure, Tokyo*, **19**, 61–62.

Helyar, K. R. (1976) Nitrogen cycling and soil acidification. *J. Aust. Inst. Agric. Sci.*, **42**, 217–221.

Hesse, P. R. (1957) Sulphur and nitrogen changes in forest soils of East Africa. *Plant Soil*, **9**, 86–96.

Hilder, E. J. (1964) The distribution of plant nutrients by sheep at pasture. *Proc. Aust. Soc. Anim. Prod.*, **5**, 241–248.

Hingston, F. J. (1959) The loss of applied phosphorus and sulphur from soils under pasture in W. A. *J. Aust. Inst. Agric. Sci.*, **25**, 209–213.

Hitchcock, D. R. (1977) Biogenic contributions to atmospheric sulfate levels. In: Cecil, L. K. (ed.), *Proceedings of the Second National Conference on Complete*

WateReuse. Water's Interface with Energy Air and Solids. American Institute of Chemical Engineers, New York, pp. 291–310.

Hollingworth, S. E., and Banister, F. A. (1952) Basaluminite and hydrobasaluminite; two new minerals from Northamptonshire. *Mineral Mag.*, **29**, 1–17.

Hörnstrom, E., Ekström, C., Miller, U., and Dickson, W. (1973) Effects of the acidification of lakes in the Swedish west coast region. *Statens Naturvårdsverk*, 1973: **7**, Vallingby, Sweden, 97 pp.

Houghton, C., and Rose, F. A. (1976) Liberation of sulfate from sulfate esters by soils. *Appl. Environ. Microbiol.*, **31**, 969–976.

Hoyt, P. B., and Nyborg, M. (1972) Use of dilute calcium chloride for the extraction of plant-available aluminium and manganese from acid soil. *Can. J. Soil Sci.*, **52**, 163–167.

Hsu, P. H., and Bates, T. F. (1964) Formation of X-ray amorphous and crystalline aluminium hydroxides. *Mineral. Mag.*, **33**, 749–768.

Hutchinson, G. E. (1957) *A Treatise on Limnology*, vol. 1. Wiley, New York, 1015 pp.

Hutchinson, T. C., and Whitby, L. M. (1976) The effects of acid rainfall and heavy metal particulates on a boreal forest ecosystem near the Sudbury smelting region of Canada. U.S. Department of Agriculture, Forest Service. *Tech. Rept. NE-23*, pp. 745–766.

Ismunadji, M., and Zulkarnaini, I. (1978) Sulphur deficiency of lowland rice in Indonesia. *Sulphur Agric.*, **2**, 17–19.

Ivanov, M. V. (1968) *Microbiological Processes in the Formation of Sulfur Deposits.* Israel Program for Scientific Translations, 298 pp.

Jackman, R. H. (1964) Accumulation of organic matter in some New Zealand soils under permanent pasture. I. Patterns of change of organic carbon, nitrogen, sulphur, and phosphorus. *N.Z. J. Agric. Res.*, **7**, 445–471.

Jenny, H. (1941) *Factors of Soil Formation.* McGraw-Hill, New York, 278 pp.

Johansson, O. (1959) On sulfur problems in Swedish agriculture. *K. Lantbrukshogsk. Ann.*, **25**, 57–169.

Jones, H. R. (1972) *Pollution Control in the Nonferrous Metals Industry.* Noyes Data Corp., Park Ridge, NJ, 201 pp.

Jones, L. H. P., Cowling, D. W., and Lockyer, D. R. (1972) Plant-available and -extractable sulfur in some soils of England and Wales. *Soil Sci.*, **114**, 104–114.

Jones, M. B., Martin, W. E., and Williams, W. A. (1968) Behavior of sulfate sulfur and elemental sulfur in three California soils in lysimeters. *Soil Sci. Soc. Am. Proc.*, **32**, 535–540.

Jones, M. B., Williams, W. A., and Martin, W. E. (1971) Effect of waterlogging and organic matter on the loss of applied sulfur. *Soil Sci. Soc. Am. Proc.*, **35**, 343–346.

Jones, M. B., and Woodmansee, R. G. (1979) Biogeochemical cycling in annual grassland ecosystems. *Bot. Rev.*, **45**, 111–114.

Jorgensen, B. B., Hansen, M. H., and Ingvarson, K. (1978) Sulfate reduction in coastal sediments and the release of H_2S to the atmosphere. In: Krumbein, W. E. (ed.), *Environmental Biogeochemistry and Geomicrobiology*, vol. 1. Ann Arbor Science, Michigan, pp. 245–253.

Junge, C. E. (1963) Sulfur in the atmosphere. *J. Geophys. Res.*, **68**, 3975–3976.

Kadota, H., and Ishida, Y. (1972) Production of volatile sulfur compounds by microorganisms. *Annu. Rev. Microbiol.*, **26**, 127–138.

Kamprath, E. J., Nelson, W. L., and Fitts, J. W. (1956) The effect of pH, sulfate and phosphate concentrations on the adsorption of sulfate by soils. *Soil Sci. Soc. Am. Proc.*, **20**, 463–466.

Kanivets, V. I. (1970) Reaction of hydrogen, methane, and hydrogen sulfide with the mineral part of the soil. *Pochvovedenie*, No. 5, 52–59.

Kanwar, J. S., and Takkar, P. N. (1964) Distribution of sulphur forms in tea soils of the Punjab. *J. Res. Punjab Agric. Univ.*, **1**, 1–15.

Katz, M., Wyatt, F. A., and Atkinson, H. J. (1939) The hydrogen ion concentration, base-exchange capacity and sulfate content of soils. In: *Effects of SO$_2$ on Vegetation* (National Research Council, Report No. 815), Ch. 5. NRC, Ottawa, pp. 131–164.

Kelley, W. P. (1951) *Alkali Soils, their Formation, Properties, and Reclamation*, Reinhold, New York, 176 pp.

Kellogg, W. W., Cadle, R. D., Allen, E. R., Lazrus, A. L., and Martell, E. A. (1972) The sulfur cycle. *Science* (Wash. D.C.), **175**, 587–596.

Kelly, D. P. (1972) Transformations of sulphur and its compounds in soils. *Symp. Internat. sur le Soufre in Agric.* Versailles (1970), pp. 217–232.

Kennedy, P. M., and Siebert, B. D. (1972) The utilization of spear grass (*Heteropogon contortus*). II. The influence of sulphur on energy intake and rumen and blood parameters in cattle and sheep. *Aust. J. Agric. Res.*, **23**, 45–56.

Kennedy, P. M., Williams, E. R., and Siebert, B. D. (1975) Sulfate recycling and metabolism in sheep and cattle. *Aust. J. Biol. Sci.*, **28**, 31–42.

Kilmer, V. J., Gilliam, J. W., Lutz, J. F., Joyce, R. T., and Eklund, C. D. (1974) Nutrient loss from fertilized grassed watersheds in western North Carolina. *J. Environ. Qual.*, **3**, 214–219.

Kodama, H., and Singh, S. S. (1972) Hydroxy aluminum sulfate–montmorillonite complex. *Can. J. Soil Sci.*, **52**, 208–218.

Kohn, G. D., Osborne, G. J., Batten, G. D., Smith, A. N., and Lill, W. J. (1977) The effect of topdressed superphosphate on changes in nitrogen : carbon : sulphur : phosphorus and pH on a red earth soil during a long term grazing experiment. *Aust. J. Soil Res.*, **15**, 147–158.

Kondo, M. (1923) The production of mercaptan from l-cystine by bacteria. *Biochem. Z.*, **136**, 198–202.

Korkman, J. (1973) Sulphur status of Finnish cultivated soils. *Maataloustiet Aikak.*, **45**, 121–215.

Kovda, V. A. (1973) *The Principles of Pedology. General Theory of Soil Formation*, vol. 2. Nauka, Moscow, pp. 421–425 [in Russian].

Kramer, J. R. (1978) Acid precipitation. In: Nriagu, J. O. (ed.), *Sulfur in the Environment*. Wiley, New York, pp. 325–370.

Krouse, H. R. (1977) Sulphur isotope abundance elucidate uptake of atmospheric sulphur emissions by vegetation. *Nature (Lond.)*, **265**, 45–46.

Krupskii, N. K., and Mamontova, E. G. (1974) Sulphur in typical thick chernozems of the Ukrainian SSR. *Trans. 10th Int. Congress Soil Sci.*, **II**, 168–173 [in Russian].

Krupskii, N. K., Mamontova, E. G., and Batsula, A. A. (1971) Sulfur content in humic and fulvic acids of some Ukranian soils. *Pochvovedenie*, No. 10, 37–41 [in Russian].

Kuhn, H., and Weller, H. (1977) Six-year studies on sulphur inputs from precipitation and losses by leaching (in lysimeters). *Z. Pflanzenernaehr. Dueng Bodenk D.*, **140**, 431–440 [in German].

Langlands, J. P., Sutherland, H. A. M., and Playne, M. J. (1973) Sulfur as a nutrient for Merino sheep. 2. The utilization of sulphur in forage diets. *Br. J. Nutr.*, **30**, 537–543.

Laurence, R. C. N., Gibbons, R. W., and Young, C. T. (1976) Changes in the yield, protein, oil and maturity of groundnut cultivars with application of sulphur fertilizers and fungicides. *J. Agric. Sci.*, **86**, 245–250.

Leeper, G. W. (1952) *Introduction to Soil Science*. 2nd edn, Melbourne University Press, Melbourne, 222 pp.

Lewis, J. A., and Papavizas, G. C. (1970) Evolution of volatile sulfur-containing compounds from decomposition of crucifers in soils. *Soil Biol. Biochem.*, **2**, 239–246.

Li, P., and Caldwell, A. C. (1966) The oxidation of elemental sulfur in soil. *Soil Sci. Soc. Am. Proc.*, **30**, 370–372.

Lichtenwalner, D. C., Flenner, A. L., and Gordon, N. E. (1923) Adsorption and replacement of plant food in colloidal oxides of iron and aluminum. *Soil Sci.*, **15**, 157–165.

Likens, G. E. (1972) *The Chemistry of Precipitation in the Central Finger Lakes Region.* Cornell University, Water Resource and Marine Science Center Tech. Rept. T50, Ithaca, N.Y., 47 pp.

Likens, G. E., Wright, R. F., Galloway, J. N., and Butler, T. J. (1979) Acid rain. *Sci. Am.*, **241**, 39–47.

Lipsett, J., and Williams, C. H. (1971) The sulphur status of wheat in red-brown earths in southern New South Wales. *Aust. J. Exp. Agric. Anim. Husb.*, **11**, 59–63.

Little, R. C. (1957) Sulphur in soils. II. Determination of the total sulphur content of soil. *J. Sci. Food Agric.*, **8**, 271–279.

Lockyer, D. R., Cowling, D. W., and Fenlon, J. S. (1978) Laboratory measurements of dry deposition of sulphur dioxide on to several soils from England and Wales. *J. Sci. Food Agric.*, **29**, 739–746.

Lotero, J., Woodhouse, W. W. Jr., and Petersen, R. G. (1966) Local effect on fertility of urine voided by grazing cattle. *Agron. J.*, **58**, 262–265.

Loveday, J. (1975) The use of sulphur and its compounds in soil amendment. In: McLachlan, K. D. (ed.), *Sulphur in Australasian Agriculture.* Sydney University Press, Sydney, pp. 163–171.

Loveday, J., and Pyle, J. (1973) The Emerson dispersion test and its relation to hydraulic conductivity. Aust. CSIRO Div. Soils, Tech. Paper. No. 15.

Lovelock, J. E., Maggs, R. J., and Rasmussen, R. A. (1972) Atmospheric dimethyl sulphide and the natural sulphur cycle. *Nature (Lond.)*, **237**, 452–453.

Lowe, L. E. (1964) An approach to the study of the sulfur status of soils and its application to selected Quebec soils. *Can. J. Soil Sci.*, **44**, 176–179.

Lowe, L. E. (1965) Sulfur fractions of selected Alberta soil profiles of the chernozemic and podzolic orders. *Can. J. Soil Sci.*, **45**, 297–303.

Lowe, L. E. (1969) Sulfur fractions of selected Alberta profiles of the gleysolic order. *Can. J. Soil Sci.*, **49**, 375–381.

Lowe, L. E., and De Long, W. A. (1963) Carbon bonded sulphur in selected Quebec soils. *Can. J. Soil Sci.*, **43**, 151–155.

McClung, A. C., De Freitas, L. M. M., and Lott, W. L. (1959) Analyses of several Brazilian soils in relation to plant responses to sulphur. *Soil Sci. Soc. Am. Proc.*, **23**, 221–224.

McFee, W. W., Kelly, J. M., and Beck, R. H. (1976) Acid precipitation effects on soils in the humid temperature zone. US Department of Agriculture, Forest Service. *Tech. Rept. NE-23*, pp. 725–736.

McGeorge, W. T., Breazeale, E. L., and Abbott, J. L. (1956) Polysulfides as soil conditioners. *Ariz. Agric. Exp. Stn Bull. 131*, 3–29.

McGeorge, W. T., and Green, R. A. (1935) Oxidation of sulfur in Arizona soils and its effect on soil properties. *Ariz. Agric. Exp. Stn Bull. 59*, 299–325.

McGovern, P. C. (1972) *Sulphur Dioxide Levels and Environmental Studies in the Sudbury Area during 1971.* Ministry of the Environment, Air Quality Branch, Sudbury, Ontario, 33 pp.

McGovern, P. C., and Balsillie, D. (1973) *Sulphur Dioxide (1972) – Heavy Metal (1971) Levels and Vegetative Effects in the Sudbury area.* Ministry of the Environment, Air Quality Branch, Sudbury, Ontario, 50 pp.

McIntyre, D. S. (1979) Exchangeable sodium, subplasticity and hydraulic conductivity of some Australian soils. *Aust. J. Soil Res.*, **17**, 115–120.

McKee, A. G. (1969) *Systems Study for Control of Emissions. Primary Nonferrous*

Smelting Industry. Final report US National Technical Information Service PB184884. Washington, DC, 190 pp.

McKell, C. M., and Williams, W. A. (1960) A lysimeter study of sulfur fertilization of an annual-range soil. *J. Range Manage.*, **13**, 113–117.

Mackenzie, A. F., De Long, W. A., and Ghanem, I. S. (1967) Total sulfur, acetate-extractable sulfur and isotopically exchangeable sulfate in some eastern Canadian soils. *Plant Soil*, **27**, 408–414.

McLachlan, K. D., and De Marco, D. G. (1975) Changes in soil sulphur fractions with fertilizer additions and cropping treatments. *Aust. J. Soil Res.*, **13**, 169–176.

McLaren, R. G., and Swift, R. S. (1977) Changes in soil organic sulphur fractions due to the long term cultivation of soils. *J. Soil Sci.*, **28**, 445–453.

Malmer, N. (1974). *On the Effects on Water, Soil and Vegetation of an increasing Atmospheric Supply of Sulphur*. National Swedish Environmental Protection Board SNVPM 402E, Solna, 125 pp.

Mamaril, C. P., Pangerang Umar, A., Manwan, I., and Momuat, C. J. S. (1976) *Sulphur Response of Lowland Rice in South Sulawesi, Indonesia*. Contrib. Cent. Res. Inst. Agric. Bogor, 22 pp.

Mann, L. D., Focht, D. D., Joseph, H. A., and Stolzy, L. H. (1972) Increased denitrification in soils by additions of sulfur as an energy source. *J. Environ. Qual.*, **1**, 329–332.

Massoumi, A., and Cornfield, A. H. (1964) Total sulphur and water-soluble sulphate contents of soils and their relation to other soil properties. *J. Sci. Food Agric.*, **15**, 623–625.

Massoumi, A., and Cornfield, A. H. (1965) Sulphate levels in soil of varying pH during incubation with organic materials. *J. Sci. Food Agric.*, **16**, 565–568.

May, P. F., Till, A. R., and Downes, A. M. (1968) Nutrient cycling in grazed pastures I. A preliminary investigation of the use of (^{35}S) gypsum. *Aust. J. Agric. Res.*, **19**, 531–543.

Metson, A. J., and Blackmore, L. C. (1978) Sulphate retention by New Zealand soils in relation to the competitive effect of phosphate. *N.Z. J. Agric. Res.*, **21**, 243–253.

Misra, R. D. (1938) Edaphic factors in the distribution of aquatic plants in English lakes. *J. Ecol.*, **26**, 411–451.

Mitchell, J., Dehn, J. E., and Dion, H. G. (1952) The effect of small additions of elemental sulphur on the availability of phosphate fertilizers. *Sci. Agric.*, **32**, 311–316.

Mitsui, S. (1956) *Inorganic Nutrition, Fertilization and Soil Amelioration of Lowland Rice*, 3rd edn, Yokendo, Tokyo, Japan, 107 pp.

Mitsui, S., Aso, S., and Kumazawa, K. (1951) Dynamic studies on nutrient uptake by crop plant. Pt 1. The nutrient uptake of rice root as influenced by hydrogen sulfide. *J. Sci. Soil Manure, Tokyo*, **22**, 46–52.

Miyamoto, S. (1977) Predicting effects of sulfuric acid on qualities of irrigation and drainage waters in calcareous soils. *J. Environ. Qual.*, **6**, 12–18.

Miyamoto, S., Bohn, H. L., Ryan, J., and Yee, M. S. (1974) Effect of sulfuric acid and sulfur dioxide on the aggregate stability of calcareous soils. *Soil Sci.*, **118**, 299–303.

Miyamoto, S., Ryan, J., and Stroehlein, J. L. (1975) Potentially beneficial uses of sulfuric acid in south western agriculture. *J. Environ. Qual.*, **4**, 431–437.

Miyamoto, S., and Stroehlein, J. L. (1975) Sulfuric acid for increasing water penetration into some Arizona soils. *Prog. Agric. Ariz.*, **27**, 13–16.

Mohammed, E. T. Y., Letey, J., and Branson, R. (1979) Sulphur compounds in water treatment – effect on infiltration rate. *Sulphur Agric.*, **3**, 7–11.

Moje, W., Munnecke, D. E., and Richardson, L. T. (1964) Carbonyl sulphide, a volatile fungitoxicant from nabam in soil. *Nature (Lond.)*, **202**, 831–832.

Moormann, F. R. (1963) Acid sulfate soils (cat-clays) of the tropics. *Soil Sci.*, **95**, 271–275.

Morozov, A. Z. (1972) Effects of sulfur feed supplements on the productivity of ewes. *Tr. Alma-At. Zoovet. Inst.*, **24**, 185–187.

Mortimer, C. H. (1941) The exchange of dissolved substances between mud and water in lakes. *J. Ecol.*, **29**, 280–329.

Mortvedt, J. J., and Giordano, P. M. (1973) Grain sorghum response to iron in a ferrous sulfate–ammonium thiosulfate–ammonium polyphosphate suspension. *Soil Sci. Soc. Am. Proc.*, **37**, 951–955.

Moser, U. S., and Olson, R. V. (1952) Sulfur oxidation in four soils as influenced by moisture tension and bacteria. *Soil Sci.*, **76**, 251–257.

Moss, M. R. (1975) Spatial patterns of sulphur accumulation by vegetation and soils around industrial centres. *J. Biogeography*, **2**, 205–222.

Muller, F. B. (1975) Sulphur received in rainfall and leached from a yellow brown loam. *N.Z. J. Sci.*, **18**, 243–252.

Munnecke, D. E., Domsch, K. H., and Eckert, J. W. (1962) Fungicidal activity of air passed through columns of soil treated with fungicides. *Phytopathology*, **52**, 1298–1306.

Nakayama, F., and Scott, A. D. (1962) Gas sorption by soils and clay minerals 1. Solubility of oxygen in moist materials. *Soil Sci.*, **94**, 106–110.

Neller, J. R. (1959) Extractable sulfate sulfur in soils of Florida in relation to amount of clay in the profile. *Soil Sci., Soc. Am. Proc.*, **23**, 346–348.

Neptune, A. M. L., Tabatabai, M. A., and Hanway, J. J. (1975) Sulfur fractions and carbon–nitrogen–phosphorus–sulfur relationships in some Brazilian and Iowa soils. *Soil Sci. Soc. Am. Proc.*, **39**, 51–55.

Nicolson, A. J. (1970) Soil sulphur balance studies in the presence and absence of growing plants. *Soil Sci.*, **109**, 345–350.

Nielson, R. F., and Peterson, H. B. (1972) Treatment of mine tailings to promote vegetative stabilization. *Utah Agric. Exp. Stn. Bull.* **485**, 22 pp.

Nikolov, N. (1964) Total sulphur content of the basic soil types of Bulgaria. *Rozteniev Nauki*, **1**, 23–30 [in Bulgarian].

Northcote, K. H., and Skene, J. K. M. (1972) *Australian Soils with Saline and Sodic Properties*. CSIRO Aust. Soil Publ. No. 27, 62 pp.

Nriagu, J. O. (1978) Production and uses of sulfur. In: Nriagu, J. O. (ed.), *Sulfur in the Environment*, Pt I. Wiley, New York, pp. 1–21.

Nyborg, M. (1978) Sulfur pollution and soils. In: Nriagu, J. O. (ed.), *Sulfur in the Environment*, Pt II. Wiley, New York, pp. 359–390.

Nyborg, M., Crepin, J., Hocking, D., and Baker, J. (1977) Effect of sulphur dioxide on precipitation and on the sulphur content and acidity of soils in Alberta, Canada. *Water Air Soil Pollut*, **7**, 439–448.

Ødelien, M. (1965) Investigations on the leaching of sulphate from soil. *Forsk. Fors. Landbruket*, **16**, 39–76.

Odynets, R. N., Perelygina, V. S., and Mossorova, R. (1972) Use of sulfur preparations in the feeding of sheep. *Izv. Akad. Nauk. Kirg. SSR*, Ser. Biol. Nauk, No. 5, 46–52 [in Russian].

Ogoleva, V. P., and Vershinina, G. A. (1976) Sulphur content and its distribution in soils of the Volgogrod Region. *Agrokhimiya*, No. 3, 89–91 [in Russian].

Okajima, H., and Takagi, S. (1953) Physiological behavior of hydrogen sulfide in the rice plant. Part 1. Effect of hydrogen sulfide on the absorption of nutrients. *Sci. Rep. Res. Inst. Tôhoku Univ.*, Ser. D, **5**, 21–31.

Okajima, H., and Takagi, S. (1955) Physiological behaviour of hydrogen sulfide in the rice plant. Part 2. Effect of hydrogen sulfide on the content of nutrients in the rice plant. *Sci. Rep. Res. Inst. Tôhoku Univ. Ser. D*, **6**, 89–99.

Okajima, H., and Takagi, S. (1956) Physiological behaviour of hydrogen sulfide in the rice plant. Part 4. Effect of hydrogen sulfide on the distribution of radioactive P^{32} in the rice plant. *Sci. Rep. Res. Inst. Tôhoku Univ, Ser. D*, **7**, 107–113.

Oke, O. L. (1967) The sulphur status of Nigerian soils. *J. Indian Soc. Soil. Sci.*, **15**, 207–208.

Olivero, C. (1960) Reserves of organic-sulphur compounds in Italian soils. *Ann. Stn Chim. Agrar. Sper. Roma*, Ser. 3, No. 179, 1–5 [in Italian].

Olsen, R. A. (1957) Absorption of sulphur dioxide from the atmosphere by cotton plants. *Soil Sci.*, **84**, 107–111.

Osborne, G. J., Wright, W. A., and Sykes, J. (1978). Increasing soil acidity threatens farming system. *Agric. Gaz. N.S.W.*, **89**, 21.

Overrein, L. N. (1972) Sulfur pollution patterns observed; leaching of calcium in forest soil determined. *Ambio*, **1**, 145–147.

Overrein, L. N. (1978) Changes in soil productivity through acidification. *Trans. 11th Congress Intern. Soil Sci. Soc.*, **3**, 260–278.

Overstreet, R., Martin, J. C., and King, H. M. (1951) Gypsum, sulfur and sulfuric acid for reclaiming an alkali soil of the Fresno Series. *Hilgardia*, **21**, 113–127.

Owers, M. J., and Powell, A. W. (1974) Deposition velocity of sulphur dioxide on land and water surfaces using a ^{35}S tracer method. *Atmos. Environ.*, **8**, 63–68.

Pallotta, U., Sandri, G., and Rossi, N. (1964) Sulphur in the soils of Emilia, *5ᵉ Symp. Intern. Agrochimica.* Palermo, 276–290 [in Italian].

Patrick, W. H., and Mikkelsen, D. S. (1971) Plant nutrient behaviour in flooded soil. In: Olson, R. A., Army, T. J., Hanway, J. J., and Kilmer, V. J. (eds), *Fertilizer Technology and Use.* Soil Sci. Soc. Am., Madison, pp. 187–215.

Paul, E. A., and Schmidt, E. L. (1961) Formation of free amino acids in rhizosphere and nonrhizosphere soil. *Soil Sci. Soc. Am. Proc.*, **25**, 259–362.

Payrissat, M., and Beilke, S. (1975) Laboratory measurements of the uptake of sulphur dioxide by different European soils. *Atmos. Environ.*, **9**, 211–217.

Petersen, R. G., Lucas, H. L., and Woodhouse, W. W. Jr. (1956) Distribution of excreta by freely grazing cattle and its effect on pasture fertility 1. Excretal distribution. *Agron. J.*, **48**, 440–444.

Piper, C. S., and de Vries, M. P. C. (1964) The residual value of superphosphate on a red–brown earth in South Australia. *Aust. J. Agric. Res.*, **15**, 234–272.

Ponnamperuma, F. N. (1972) The chemistry of submerged soils. *Adv. Agron.*, **24**, 29–96.

Prather, R. J., and Rhoades, J. D. (1975) Sulphuric acid as an amendment for reclaiming soils high in boron. *Agron. Abs.*, **121**.

Probert, M. E. (1974) The sulphur status of some North Queensland soils. In: McLachlan, K. D. (ed.), *Handbook on Sulphur in Australian Agriculture.* CSIRO, Melbourne, pp. 64–68.

Probert, M. E. (1976) Studies of 'available' and isotopically exchangeable sulphur in some North Queensland soils. *Plant Soil*, **45**, 461–475.

Probert, M. E. (1977) The distribution of sulphur and carbon–nitrogen–sulphur relationships in some north Queensland soils. Aust CSIRO Div. Soils Tech. Paper No. 31, 20 pp.

Probert, M. E. (1980) Effect of soil forming processes on the distribution of sulfur in soils. In: Freney, J. R. and Nicolson, A. J. (eds), *Sulfur in Australia*, Australian Academy of Science, Canberra, pp. 158–169.

Pund, W. A. (1969) Sulphur improves urea-treated corn silage. *Sulphur Inst. J.*, **5**, 7–9.

Putnam, H. D., and Schmidt, E. L. (1959) Studies on the free amino acid fractions of soils. *Soil Sci.*, **87**, 22–27.

Ralph, B. J. (1980) Weathering of sulfur in rocks. In: Freney, J. R. and Nicolson, A. J. (eds), *Sulfur in Australia*. Australian Academy of Science, Canberra, pp. 146–157.

Rasmussen, K. H., Taheri, M., and Kabel, R. L. (1975) Global emissions and natural processes for removal of gaseous pollutants. *Water Air Soil Pollut.*, **4**, 33–64.

Reddy, C. S., and Mehta, B. V. (1970) Relationship of carbon, nitrogen and sulphur in Gujarat soils. *Indian J. Agric. Sci.*, **40**, 630–633.

Rehm, G. W., and Caldwell, A. C. (1968) Sulfur supplying capacity of soils and the relationship to soil type. *Soil Sci.*, **105**, 355–361.

Rendig, V. V., and Weir, W. C. (1957) Evaluation by lamb feeding tests of alfalfa hay grown on a low-sulphur soil. *J. Anim. Sci.*, **16**, 451–461.

Rhoades, J. D., and Bernstein, L. (1971) Chemical, physical and biological characteristics of irrigation and soil water. In: Ciaccio, L. L. (ed.), *Water and Water Pollution Handbook*, vol. 1. Marcel Dekker, New York, pp. 141–222.

Rhue, R. D., and Kamprath, E. J. (1973) Leaching losses of sulphur during winter months when applied as gypsum, elemental S or prilled S. *Agron. J.*, **65**, 603–605.

Richards, L. A. (1954) *Diagnosis and Improvement of Saline and Alkali Soils*. Handbook No. 60, US Department of Agriculture, 160 pp.

Richardson, H. L. (1938) The nitrogen cycle in grassland soils with especial reference to the Rothamsted Park grass experiment. *J. Agric. Sci.*, **28**, 73–121.

Roberts, B. R. (1974) Foliar absorption of atmospheric SO_2 by woody plants. *Environ. Pollut.*, **7**, 133–140.

Robinson, E., and Robbins, R. C. (1968) *Sources, Abundance and Fate of Gaseous Atmospheric Pollutants*. Final Report, Project PR-6755, Stanford Research Institute, Menlo Park, California, 110 pp.

Roy, A. B., and Trudinger, P. A. (1970) *The Biochemistry of Inorganic Compounds of Sulphur*. Cambridge University Press, Cambridge, UK, 400 pp.

Ruhal, D S., and Paliwal, K. V. (1978) Status and distribution of sulphur in soils of Rajasthan. *J. Indian Soc. Soil Sci.*, **26**, 352–358.

Russell, E. J. (1973) *Soil Conditions and Plant Growth*, 10th edn. Longman, London, 849 pp.

Russell, J. S. (1960) Soil fertility changes in the long term experimental plots at Kybybolite, South Australia. 1. Changes in pH, total nitrogen, organic carbon and bulk density. *Aust. J. Agric. Res.*, **11**, 902–926.

Ryan, J., and Stroehlein, J. L. (1974) Use of sulfuric acid on phosphorus deficient Arizona soils. *Prog. Agric. Ariz.*, **25**, 11–13.

Saalbach, E. (1965/66) The influence of sulfur on the yield of forage crops in West Germany. *Sulphur Inst. J.*, **1**, 7–9.

Saalbach, E. (1966) Sulphur fertilization and protein quality. *Sulphur Inst. J.*, **2**, 2–5.

Sachdev, M. S., and Chhabra, P. (1974) Transformations of ^{35}S-labelled sulfate in aerobic and flooded soil conditions. *Plant Soil*, **41**, 335–341.

Salsbury, R. L., and Merricks, D. L. (1975) Production of methanethiol and dimethyl sulfide by rumen microorganisms. *Plant Soil*, **43**, 191–209.

Saran, A. B. (1949) Some observations on an obscure disease of paddy, *Oryza sativa*. *Curr. Sci.*, **18**, 378–379.

Schalscha, E. B., Estrada, C., and Galindo, G. G. (1972) Sulphur status of some volcanic ash derived soils in Chile. *Agrochimica*, **16**, 77–82.

Schwertmann, U. (1961) The occurrence and origin of jarosite (maibolt) in marshy soil. *Naturwissenschaften*, **48**, 159–160.

Scott, N. M. (1976) Sulphate contents and sorption in Scottish soils. *J. Sci. Food Agric.*, **27**, 367–372.

Scott, N. M., and Anderson, G. (1976) Sulphur, carbon and nitrogen contents of organic fractions from acetylacetone extracts of soils. *J. Soil. Sci.*, **27**, 324–330.

Sears, P. D., Goodall, V. C., and Jackman, R. H. (1965) Pasture growth and soil fertility. IX. Repeated cropping of a soil previously under high-quality permanent pasture. *N.Z. J. Agric. Res.*, **8**, 497–510.

Seim, E. C. (1970) Sulfur dioxide absorption by soil. Ph.D. Thesis. University of Minnesota, 138 pp.

Sen, A. T. (1938) Further experiments on the occurrence of depressed yellow patch of paddy in the Mandalay farm. *Burma Dept. Agric. Rept. 1937–1938*, pp. 35–36.

Shedd, O. M. (1928) Influence of sulfur and gypsum on the solubility of potassium in soils and on the quantity of this element removed by certain plants. *Soil Sci.*, **22**, 335–354.

Shepherd, J. G. (1974) Measurements of the direct deposition of sulphur dioxide onto grass and water by the profile method. *Atmos. Environ.*, **8**, 69–74.

Shiga, H., and Suzuki, S. (1964) Studies on the behavior of hydrogen sulfide in water-logged soils. Part 8. Influence of free hydrogen sulfide in soils on the growth of rice plants. *Bull. Chugoku Nat. Agric. Exp. Stn.*, Ser. **A10**, 113–130.

Shioiri, M. (1943) *Chemistry of Paddy Soils*. Japanese Agricultural Society, Dai Nihon Nokai, Tokyo, 64 pp.

Shkonde, E. I. (1957) The role of sulphur in plant nutrition. *Dokl. Vses Akad. SKH. Nauk*, No. 2, 22–25 [in Russian].

Shukla, U. C., and Gheyi, A. K. (1971) Sulphur status of some Rajasthan soils. *Indian J. Agric. Sci.*, **41**, 247–253.

Sikora, L. J., and Keeney, D. R. (1976) Evaluation of a sulfur–*Thiobacillus denitrificans* nitrate removal system. *J. Environ. Qual.*, **5**, 298–303.

Siman, G., and Jansson, S. L. (1976a) Sulphur exchange between soil and atmosphere, with special attention to sulphur release directly to the atmosphere. 1. Formation of gaseous sulphur compounds in soil. *Swed. J. Agric. Res.*, **6**, 37–45.

Siman, G., and Jansson, S. L. (1967b) Sulphur exchange between soil and atmosphere, with special attention to sulphur release directly to the atmosphere. 2. The role of vegetation in sulphur exchange between soil and atmosphere. *Swed. J. Agric. Res.*, **6**, 135–144.

Simon-Sylvestre, G. (1965) Annual evolution of sulphur in soil compared to that of nitrogen. *C.R. Hebd. Séances Acad. Agric. Fr.*, 426–431 [in French].

Simon-Sylvestre, G. (1967a) Observations on the annual cycle of sulphur in soil under a moderately wet climate. *Agrochimica*, **12**, 60–68 [in French].

Simon-Sylvestre, G. (1967b) New observations on the annual cycle of sulphur and nitrogen in soil. *C.R. Hebd. Séances Acad. Agric. Fr.*, 90–96 [in French].

Simon-Sylvestre, G. (1969a) Soluble sulphates in soil. *Ann. Agron.*, **20**, 435–447 [in French].

Simon-Sylvestre, G. (1969b) First results of a survey on the total sulphur content of arable soils in France. *Ann. Agron.*, **20**, 609–625 [in French].

Simon-Sylvestre, G. (1972) Sulphur in soils – Its evolution. Int. Symp. on Sulphur in Agriculture, Versailles, 3–4 Dec. 1970. *Ann. Agron.*, **72**, 181–199 [in French].

Singh, B. R., Uriyo, A. P., and Kilasara, M. (1979) Sorption of sulphate and distribution of total sulphate and mineralisable sulphur in some tropical soil profiles in Tanzania. *J. Sci. Food Agric.*, **30**, 8–14.

Singh, S. S. (1967) Sulfate ions and ion activity product (Al) $(OH)^3$ in Wyoming suspensions. *Soil Sci.*, **104**, 433–438.

Singh, S. S., and Brydon, J. E. (1967) Precipitation of aluminium by calcium hydroxide in the presence of Wyoming bentonite and sulfate ions. *Soil Sci.*, **103**, 162–167.

Singh, S. S., and Brydon, J. E. (1969) Solubility of basic aluminium sulfates at equilibrium in solution and in the presence of montmorillonite. *Soil Sci.*, **107**, 12–16.

Sklodovski, P. (1969) Sulphur distribution in the profiles of some soil types in Poland. *Rocz. Glebozn*, **19**, 99–119 [in Polish].

Smith, K. A., Bremner, J. M., and Tabatabai, M. A. (1973) Sorption of gaseous atmospheric pollutants by soils. *Soil Sci.*, **116**, 313–319.

Smith, M. S., Francis, A. J., and Duxbury, J. M. (1977) Collection and analysis of organic gases from natural ecosystems: application to poultry manure. *Environ. Sci. Technol.*, **11**, 51–55.

Smittenberg, J., Harmsen, G. W., Quispel, A., and Otzen, D. (1951) Rapid methods for determining different types of sulphur compounds in soil. *Plant Soil*, **3**, 353–360.

Sowden, F. J. (1955) Estimation of amino acids in soil hydrolysates by the Moore and Stein method. *Soil Sci.*, **80**, 181–188.

Sowden, F. J. (1956) Distribution of amino acids in selected horizons of soil profiles. *Soil Sci.*, **82**, 491–496.

Sowden, F. J. (1958) The forms of nitrogen in the organic matter of different horizons of soil profiles. *Can. J. Soil Sci.*, **38**, 147–154.

Spáleny, J. (1977) Sulphate transformation to hydrogen sulphide in spruce seedlings. *Plant Soil*, **48**, 557–563.

Spedding, D. J. (1969a) Uptake of sulphur dioxide by barley plants at low sulphur dioxide concentrations. *Nature (Lond.)*, **224**, 1229–1231.

Spedding, D. J. (1969b) Sulphur dioxide uptake by limestone. *Atmos. Environ.*, **3**, 683.

Spek, J. van der (1950) Katteklei. *Versl. Landbouwkd. Onderz.*, **56**(2), 1–40.

Spencer, K. (1975) Sulphur requirements of plants. In: McLachlan, K. D. (ed.), *Sulphur in Australasian Agriculture*. Sydney University Press, Sydney, pp. 98–109.

Stace, H. C. T., Hubble, G. D., Brewer, R., Northcote, K. H., Sleeman, J. R., Mulcahy. M. J., and Hallsworth, E. G. (1968) *A Handbook of Australian Soils*. Rellim, Glenside, SA, 435 pp.

Stallings, J. H. (1957) *Soil Conservation*. Prentice-Hall, Englewood Cliffs, NJ. 575 pp.

Starkey, R. L. (1966) Oxidation and reduction of sulfur compounds in soils. *Soil. Sci.*, **101**, 297–306.

Stepanov, D. G., and Yakimchuk, E. F. (1973) Effect of sulfur containing compounds on the live weight and wool productivity of Ascanian fine wooled ewe lambs. *Trudy Kishinev. SKH. Inst.*, **113**, 72–78 [in Russian].

Stevenson, F. J. (1956) Isolation and identification of some amino compounds in soils. *Soil Sci. Soc. Am. Proc.*, **20**, 201–208.

Stewart, B. A., Porter, L. K., and Viets, F. G. Jr. (1966a) Effect of sulfur content of straws on rates of decomposition and plant growth. *Soil Sci. Soc. Am. Proc.*, **30**, 355–358.

Stewart, B. A., Porter, L. K., and Viets, F. G. Jr. (1966b) Sulfur requirements for decomposition of cellulose and glucose in soil. *Soil Sci. Soc. Am. Proc.* **30**, 453–456.

Stewart, B. A., and Whitfield, C. J. (1965) Effect of crop residue soil temperature and sulfur on the growth of winter wheat. *Soil Sci. Soc. Am. Proc.*, **29**, 752–755.

Stotzky, G., and Norman, A. G. (1961) Factors limiting microbial activities in soil. II. The effect of sulfur. *Arch. Mikrobiol.*, **40**, 370–382.

Stromberg, L. K., and Tisdale, S. L. (1979) Treating irrigated arid-land soils with acid-forming sulphur compounds. *Tech. Bull. 24*. The Sulphur Institute, Washington, 26 pp.

Suzuki, S., and Shiga, H. (1956) Studies on physical and chemical characteristics of Akiochi paddy soils. Part 2. Relation between production of free hydrogen sulfide and Akiochi degree. *Bull. Chugoku Nat. Agric. Exp. Stn*, **3**, 69–80.

Swaby, R. J., and Fedel, R. (1973) Microbial production of sulphate and sulphide in some Australian soils. *Soil Biol. Biochem.*, **5**, 773–781.

Swanson, C. O., and Latshaw, W. L. (1922) Sulfur as an important fertility element. *Soil Sci.*, **14**, 421–430.

Sylvester, R. O., and Seabloom, R. W. (1962) *A Study on the Character and Significance of Irrigation Return Flows in the Yakima River Basin.* Univ. Washington. Dept. Civil Engineering, 104 pp.

Sylvester, R. O., and Seabloom, R. W. (1963) Quality and significance of irrigation return flow. *J. Irrig. Drain. Div. Am. Soc. Civ. Engrs*, **89**, 1–27.

Tabatabai, M. A., and Bremner, J. M. (1972a) Forms of sulfur, and carbon, nitrogen and sulfur relationships in Iowa soils. *Soil Sci.*, **114**, 380–386.

Tabatabai, M. A., and Bremner, J. M. (1972b) Distribution of total and available sulfur in selected soils and soil profiles. *Agron. J.*, **64**, 40–44.

Takai, Y., and Asami, T. (1962). Formation of methyl mercaptan in paddy soils. 1. *Soil Sci. Pl. Nutr.*, **8**, 40–44.

Takijima, Y. (1963) Studies on behavior of the growth inhibiting substances in paddy soils with special reference to the occurrence of root damage in the peaty paddy field. *Bull. Natn. Inst. Agric. Sci. Ser. B.*, **13**, 117–252.

Tamiya, N. (1951a) Aerobic decomposition of cystine by *Escherichia coli*. I. *J. Chem. Soc. Japan*, **72**, 118–121.

Tamiya, N. (1951b) Aerobic decomposition of cystine by *Escherichia coli*. II. *J. Chem. Soc. Japan*, **72**, 121–124.

Tamm, C. F. (1976) Acid precipitation – biological effects on soil and on forest vegetation. *Ambio* **5**, 235–238.

Tarr, H. L. A. (1933a) Anaerobic decomposition of 1-cystine by washed cells of *Proteus vulgaris*. *Biochem. J.*, **27**, 759–763.

Tarr, H. L. A. (1933b) Enzymic formation of hydrogen sulfide by certain heterotrophic bacteria. *Biochem. J.*, **27**, 1869–1874.

Temple, K. L., and Delchamps, E. W. (1953) Autotrophic bacteria and the formation of acid in bituminous coal mines. *Appl. Microbiol.*, **1**, 255–258.

Terman, G. L. (1978) Atmospheric sulphur—the agronomic aspects. *Tech. Bull. No. 23.* The Sulphur Institute, Washington, 15 pp.

Terraglio, F. P., and Manganelli, R. M. (1966) The influence of moisture on the adsorption of atmospheric sulfur dioxide by soil. *Int. J. Air Water Pollut.*, **10**, 783–791.

Thomas, E. E. (1936) Reclamation of black-alkali soils with various kinds of sulfur. *Hilgardia*, **19**, 127–142.

Thomas, M. D., Hendricks, R. H., and Hill, G. R. (1944) Some chemical reactions of sulfur dioxide after absorption by alfalfa and sugar beets. *Plant Physiol.*, **19**, 212–226.

Thomas, M. D., and Hill, G. R. (1937) Relation of sulfur dioxide in the atmosphere to photosynthesis and respiration of alfalfa. *Plant Physiol.*, **12**, 309–383.

Thorne, D. W. (1944) The use of acidifying materials on calcareous soils. *J. Am. Soc. Agron.*, **36**, 815–828.

Thorup, J. T. (1972) Soil sulphur application: a new approach. *Sulphur Inst. J.*, **8**, 16–17.

Till, A. R. (1975) Sulphur cycling in grazed pastures. In: McLachlan, K. D. (ed.), *Sulphur in Australasian Agriculture.* Sydney University Press, Sydney, pp. 68–75.

Till, A. R. (1980) Sulphur cycling in soil–plant–animal systems. In: Freney, J. R. and Nicolson, A. J. (eds), *Sulfur in Australia.* Australian Academy of Science, Canberra, pp. 204–217.

Till, A. R., and McCabe, T. P. (1976) Sulfur leaching and lysimeter characterization. *Soil Sci.*, **121**, 44–47.

Till, A. R., and May, P. F. (1971) Nutrient cycling in grazed pastures. IV. The fate of sulphur-35 following its application to a small area in a grazed pasture. *Aust. J. Agric. Res.*, **22**, 391–400.

Tisdale, S. L. (1970) The use of sulphur compounds in irrigated aridland agriculture. *Sulphur Inst. J.*, **6**, 2–7.

Tisdale, S. L., Davis, R. L., Kingsley, A. F., and Mertz, E. T. (1950) Methionine and cystine content of two strains of alfalfa as influenced by different concentrations of the sulphate ion. *Agron. J.*, **42**, 221–225.

Tisdale, S. L., and Nelson, W. L. (1966) *Soil Fertility and Fertilizers*. Macmillan, New York, 694 pp.

Tiwari, R. C., and Ram, M. (1973) Distribution of different forms of sulphur in soils of Varanasi. *Bhartiya Krishi Anusandhan Patrika*, **1**, 33–38 [in Hindustani].

Toews, B. (1973) Deep plowing solonetzic soils in Alberta. *Proc. 17th Manitoba Soil Sci. Meeting*, Winnipeg, Manitoba, 5–6 Dec., pp. 107–111.

Toxopeus, M. R. J. (1970) Sulphur deficiency in grassland on young soils from volcanic ash and its possible prediction by laboratory tests. *Proc. XI Intern. Grassl. Cong.*, 345–346.

Trudinger, P. A. (1975) The biogeochemistry of sulphur. In: McLachlan, K. D. (ed.), *Sulphur in Australasian Agriculture*, Sydney University Press, Sydney, pp. 11–20.

Tsuji, T. (1975) Sulphur supplying power of Japanese grassland soils. *JARQ*, **9**, 142–147.

Turco, R. P., Whitten, R. C., Toon, O. B., Pollack, J. B., and Hamill, P. (1980) OCS, stratospheric aerosols and climate. *Nature (Lond.)*, **283**, 283–286.

Val'nikov, I. U. (1970) Forms of sulfur in forest–steppe soils of the Tatar ASSR and role of sulfur in soil fertility. *Agrokhimiya, No. 2*, 60–64 [in Russian].

Val'nikov, I. U., Grishin, P. V., and Lomako, E. I. (1971) Sulphur in compact Chuvash chernozems. *Agrokhimiya*, No. 8, 93–96 [in Russian].

Val'nikov, I. U., and Mishin, A. M. (1974) The forms of sulphur in soils of the central Volga. *Agrokhimiya*, No. 12, 112–118 [in Russian].

Vámos, R. (1964) The release of hydrogen sulphide from mud. *J. Soil Sci.*, **15**, 103–109.

Vandecaveye, S. C., Horner, G. M., and Keaton, C. M. (1936). Unproductivenesss of certain orchard soils as related to lead arsenate spray accumulations. *Soil Sci.*, **42**, 203–215.

Venkateswarlu, J., Subbiah, B. V., and Tamhane, R. V. (1969) Vertical distribution of forms of sulphur in selected rice soils of India. *Indian J. Agric. Sci.*, **39**, 426–431.

Virmani, S. M., and Kanwar, J. S. (1971) Distribution of forms of sulphur in six soil profiles of north-East India. *J. Indian Soc. Soil Sci.*, **19**, 73–77.

Vishniac, W., and Santer, M. (1957) The thiobacilli. *Bacteriol. Rev.*, **21**, 195–213.

Vitolins, M. I., and Swaby, R. J. (1969) Activity of sulphur-oxidising microorganisms in some Australian soils. *Aust. J. Soil Res.*, **7**, 171–183.

Wagnon, K. A., Bentley, J. R., and Green, L. R. (1958) Steer gains on annual-plant range pastures fertilized with sulphur. *J. Range Manage.*, **11**, 177–182.

Walker, P. H. (1972) Seasonal and stratigraphic controls in coastal flood plain soils. *Aust. J. Soil Res.*, **10**, 127–142.

Walker, T. W. (1957) The sulfur cycle in grassland soils. *J. Br. Grassl. Soc.*, **12**, 10–18.

Walker, T. W., and Adams, A. F. R. (1958) Studies on soil organic matter. I. Influence of phosphorus content of parent materials on accumulations of carbon, nitrogen, sulfur and organic phosphorus in grassland soils. *Soil Sci.*, **85**, 307–318.

Walker, T. W., and Adams, A. F. R. (1959) Studies on soil organic matter: 2. Influence of increased leaching at various stages of weathering on levels of carbon, nitrogen, sulfur and organic and total phosphorus. *Soil Sci.*, **87**, 1–10.

Walker, T. W., and Gregg, P. E. H. (1975) The occurrence of sulphur deficiency in

New Zealand. In: McLachlan, K. D. (ed.), *Sulphur in Australasian Agriculture*. Sydney University Press, Sydney, pp. 145–153.

Walker, T. W., Thapa, B. K., and Adams, A. F. R. (1959) Studies on soil organic matter: 3. Accumulation of carbon, nitrogen, sulfur, organic and total phosphorus in improved grassland soils. *Soil Sci.*, **87**, 135–140.

Walton, G. (1966) Agricultural waste water; treatment and recycling for agricultural waste. *Proc. Symp. on Agric. Waste Water*, Davis, Univ. Calif., 6–8 Apr., 1966, pp. 273–281 (Calif. Univ. Water Resources Center. Rep. 10).

Wang, C. H. (1978) Sulphur fertilization of rice. *Sulphur Agric.*, **2**, 13–16.

Wang, C. H., Liem, T. H., and Mikkelsen, D. S. (1976a) Sulfur deficiency—a limiting factor in rice production in the Lower Amazon basin. I. Development of sulfur deficiency as a limiting factor for rice production. *IRI Bull. 47*, New York, 46 pp.

Wang, C. H., Liem, T. H., and Mikkelsen, D. S. (1976b) Sulfur deficiency—a limiting factor in rice production in the Lower Amazon basin II. Sulfur requirement for rice production. *IRI Bull. 48*, New York, 38 pp.

Warth, F. J., and Krishnan, T. S. (1935) Sulfur and sulfate balance experiments with sheep. *Indian J. Vet. Sci. Anim. Husb.*, **5**, 319–331.

Watson, E. R. (1969) The influence of subterranean clover pastures on soil fertility. III. The effect of applied phosphorus and sulphur. *Aust. J. Agric. Res.*, **20**, 447–456.

Watson, K. A. (1964) Fertilizers in Northern Nigeria. Current Utilization and Recommendations for their Use. *Samaru Res. Bull.* 38, 20 pp.

Wedepohl, K. H. (1978) *Handbook of Geochemistry*. Springer-Verlag, Berlin.

Weir, R. G. (1975) The oxidation of elemental sulphur and sulphides in soil. In: McLachlan, K. D. (ed.), *Sulphur in Australasian Agriculture*. Sydney University Press, Sydney, pp. 40–49.

Whitehead, D. C. (1964) Soil and plant-nutrition aspects of the sulphur cycle. *Soils Fert.*, **27**, 1–8.

Widdowson, J. P., and Hanway, J. J. (1974) Available sulfur status of some representative Iowa soils, *Iowa Agric. Exp. Stn Res. Bull.*, No. 579, 714–736.

Wiklander, L. (1973) The acidification of soil by acid precipitation. *Grundforbattring*, **26**, 155–164.

Wiklander, L., and Hallgren, G. (1949) Studies on Gyttja soils. I. Distribution of different sulfur and phosphorus forms and of iron manganese and calcium carbonate in a profile from Kungsängen. *K. Lantbrukshogsk. Ann.*, **16**, 811–827.

Williams, C. (1975a) The distribution of sulphur in the soils and herbage of North West Pembrokeshire. *J. Agric. Sci.*, **84**, 445–452.

Williams, C. H. (1967) Some factors affecting the mineralization of organic sulphur in soils. *Plant Soil*, **26**, 205–223.

Williams, C. H. (1968) Seasonal fluctuations in mineral sulphur under subterranean clover pasture in southern New South Wales. *Aust. J. Soil Res.*, **6**, 131–139.

Williams, C. H. (1974) The chemical nature of sulphur in some New South Wales soils. In: McLachlan, K. D. (ed.), *Handbook on Sulphur in Australian Agriculture*. CSIRO, Melbourne, pp. 16–23.

Williams, C. H. (1975b) The chemical nature of sulphur compounds in soils. In: McLachlan, K. D. (ed.), *Sulphur in Australasian Agriculture*, Sydney University Press, Sydney, pp. 21–30.

Williams, C. H. (1980) Soil acidification under clover pasture. *J. Aust. Inst. Agric. Sci.*, **20**, 561–567.

Williams, C. H., and David, D. J. (1976) Effects of pasture improvement with subterranean clover and superphosphate on the availability of trace metals to plants. *Aust. J. Soil Res.*, **14**, 85–93.

Williams, C. H., and Donald, C..M. (1957) Changes in organic matter and pH in a podzolic soil as influenced by subterranean clover and superphosphate. *Aust. J. Agric. Res.*, **8**, 179–189.

Williams, C. H., and Lipsett, J. (1961) Fertility changes in soils cultivated for wheat in southern New South Wales. *Aust. J. Agric. Res.*, **12**, 612–629.

Williams, C. H., and Steinbergs, A. (1958). Sulphur and phosphorus in some eastern Australian soils. *Aust. J. Agric. Res.*, **9**, 483–491.

Williams, C. H., and Steinbergs, A. (1959) Soil sulphur fractions as chemical indices of available sulphur in some Australian soils. *Aust. J. Agric. Res.*, **10**, 340–352.

Williams, C. H., and Steinbergs, A. (1962) The evaluation of plant-available sulphur in soils: I. *Plant Soil*, **17**, 279–294.

Williams, C. H., and Steinbergs, A. (1964) The evaluation of plant available sulfur in soils. II. *Plant Soil*, **21**, 50–62.

Williams, C. H., Williams, E. G., and Scott, N. M. (1960) Carbon, nitrogen sulphur and phosphorus in some Scottish soils. *J. Soil Sci.*, **11**, 334–346.

Wilson, L. G., Bressan, R. A., and Filner, P. (1978) Light-dependent emission of hydrogen sulfide from plants. *Plant Physiol.*, **61**, 184–189.

Wolt, J. D., and Adams, F. (1979) The release of sulfate from soil-applied basalumi-nite and alunite. *Soil Sci. Soc. Am. J.*, **43**, 118–121.

Wrenford, M. (1968) Nutritional deficiencies of white clover *Trifolium repens* in improved pastures of New England. M. Rur. Sci. Thesis, Univ. of New England, 175 pp.

Yamane, I., and Sato, I. (1961) Metabolism in muck paddy soil. Part 3. Role of soil organic matter in the evolution of free hydrogen sulfide in water-logged soil. *Sci. Rep. Res. Inst Tôhoku Univ., Ser.* **D12**, 73–86.

Yamane, I., Usami, T., Ikeda, K., and Omukai, S. (1956) Influence of root damage by hydrogen sulfide on the rice growth. *Bull. Tôhoku Nat. Agric. Exp. Str.*, **10**, 134–155.

Yee, M. S., Bohn, H. L., and Miyamoto, S. (1975) Sorption of sulfur dioxide by calcareous soils. *Soil Sci. Soc. Am. Proc.*, **39**, 268–270.

Yoneda, S. (1958) Salt damage and soil (3). *Agr. and Hort.*, **33**, 1337–1342.

Young, A. (1969) Present rate of land erosion. *Nature (Lond.)*, **224**, 851–852.

ZoBell, C. E. (1963) Organic geochemistry of sulfur. In: Breger, I. A. (ed.), *Organic Geochemistry*. Macmillan, New York, pp. 543–578.

The Global Biogeochemical Sulphur Cycle
Edited by M. V. Ivanov and J. R. Freney
© 1983 Scientific Committee on Problems of the Environment (SCOPE)

CHAPTER 4
The Atmospheric Sulphur Cycle

A. G. RYABOSHAPKO

4.1 INTRODUCTION

Compared to other geospheres, the atmosphere is a very mobile system in which the predominating processes occur during no more than a few days. The slowest processes, which take months or years, are the transfer of air between hemispheres and the stratosphere–troposphere exchange. The high mobility of the system causes high rates of transfer and redistribution of sulphur compounds in spite of relatively small average concentrations in the atmosphere. In this sense the atmosphere differs sharply from the other reservoirs.

The atmospheric part of the global sulphur cycle has during recent years been a focus of attention since it is undergoing drastic changes due to the industrial activity of man. As will be shown below, the anthropogenic sources of sulphur are comparable in size with the natural global emissions to the atmosphere, and in some regions of the globe the anthropogenic sources may be more than 10 times higher than natural sources.

Eriksson (1960) made the first systematic study of the atmospheric sulphur cycle. Some of his estimates are still valid, while others have been substantially revised. One of the latest publications on the global sulphur cycle is the report of SCOPE made by the Swedish scientists, Granat *et al.* (1976). We shall attempt to improve some of the estimates using relatively inaccessible data from Soviet investigators as well as results published in world literature during the period 1976–80.

4.2 CONCENTRATION AND MASS OF SULPHUR COMPOUNDS IN THE ATMOSPHERE

We shall review the literature on concentrations of three basic forms of sulphur: reduced sulphur compounds, sulphur dioxide, and sulphate in the troposphere. Since these concentrations may vary over a wide range depend-

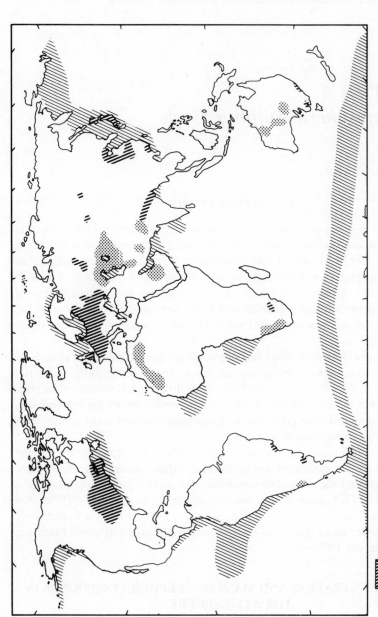

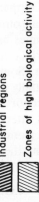

Industrial regions

Zones of high biological activity

Zones of aeolian weathering of sulphates

ing on location, we shall group data separately for oceanic, continental, and industrial and urban regions. Similar reviews were undertaken by Granat *et al.* (1976), Georgii (1978), and Meszaros (1978b). These regions are marked in Fig. 4.1, the derivation of which will be discussed later. The stratospheric sulphate layer will be considered separately. To standardize units throughout the text, we shall express concentration as the mass of sulphur per unit volume of air under normal conditions (273 K, 1 bar) $\mu gS\ m^{-3}$, the sulphur content in individual reservoirs as TgS, and fluxes as TgS year^{-1}, (1 Tg = 10^{12} g).

4.2.1 Concentration of Reduced Sulphur Compounds

Reduced (sulphide) sulphur is present in the atmosphere in several compounds. The simplest (and at the same time the most common) are hydrogen sulphide H_2S, dimethyl sulphide CH_3SCH_3, carbon disulphide CS_2, and carbonyl sulphide OCS. Other more complicated compounds of sulphide sulphur occur in the atmosphere, e.g. mercaptans and dimethyl disulphide. We shall consider only those compounds which contribute substantially to sulphur fluxes between reservoirs or whose atmospheric content is relatively large.

In the oxidizing conditions of the atmosphere, hydrogen sulphide and dimethyl sulphide are chemically unstable and degrade quickly; we shall describe these as short-lived reduced sulphur compounds. Carbon disulphide and carbonyl sulphide are more resistant to oxidation; we shall describe them as reduced sulphur compounds with a long residence time.

Reduced sulphur compounds enter the atmosphere as a result of biological and decomposition processes both on land and in water and to some extent through volcanic and anthropogenic activity. Reduced sulphur compounds occur in rather low concentrations in the atmosphere except for narrow tidal zones along coasts where the local concentrations are so high that under certain weather conditions they can be detected by their characteristic odour.

Table 4.1 contains data on concentrations of both classes of reduced sulphur compounds in the atmosphere over different regions. (Note that the long-lived reduced sulphur compounds are less variable in space.)

Direct measurements of reduced sulphur compounds have been made only in recent years and they are not yet sufficiently numerous to elucidate the time and space variations.

Over ocean regions with high bioproductivity, i.e. in upwelling areas and near coasts, the concentration of reduced sulphur compounds in the atmosphere can be substantially higher than over the open ocean (e.g. the direct observations of Jaeschke *et al.* (1979) in coastal regions of the North Sea). From the limited data available (Slatt *et al.*, 1978; Maroulis and Bandy, 1977; Graedel, 1979) we assume that over the open regions of the world's ocean, and in polar areas, the concentration of hydrogen sulphide is

Table 4.1 Reduced sulphur in the atmosphere

Type of atmosphere	Region	Compound	Concentration (μgS m^{-3})		Reference	Comments
			Range	Mean		
Oceanic	Central Atlantic	H_2S	0.007–0.07	0.049	Slatt et al. (1978)	Direct measurements
	Eastern coast of the USA, wind from ocean	DMS		0.083 ± 0.003	Maroulis and Bandy (1977)	Direct measurements[a]
Continental, including industrial regions	Central USA	H_2S	0.03–0.09		Natusch et al. (1972)	
	Central USA	H_2S	0.17–0.49	0.32	Breeding et al. (1973)	
	Central Europe	H_2S		1.0	Georgii (1978)	
	England	H_2S	0.14–0.54		Smith et al. (1961)	
	West Germany, flood areas, bogs	H_2S	0.1–1.03		Jaeschke et al. (1978)	
	West Germany, area between Mainz and Wiesbaden	H_2S	0.33–1.55		Jaeschke et al. (1978)	Possible anthropogenic effect
	West Germany, flood area	H_2S		0.2	Jaeschke et al. (1978)	October
				0.4	Jaeschke et al. (1978)	March
				1.0	Jaeschke et al. (1978)	May
				2.0	Jaeschke et al. (1978)	July
	Northern West Germany			0.05	Jaeschke et al. (1978)	Altitude 4 km

Location	Compound	Concentration	Range	Reference	Notes
		0.11		Jaeschke et al. (1978)	Altitude 2 km
	H_2S		0.14–0.37	Jaeschke et al. (1978)	Altitude 1 km
Northern sea coast	H_2S		0.47–1.04	Jaeschke et al. (1978)	Surface layer
	H_2S		0.09–0.36	Jaeschke et al. (1979)	Direct measurements
Urban					
New York	H_2S		1.7–5.4	Jacobs et al. (1957)	
Industrial region near steel mills and coke industry	H_2S		1.4–85	Smith et al. (1961)	
Whole atmosphere					
South and central Pacific, western region of North America	COS	0.73 ± 0.09		Torres et al. (1980)	Direct measurements. Uniform distribution with altitude up to the tropopause
Philadelphia, USA	COS	0.76 ± 0.04		Torres et al. (1980)	
Philadelphia, USA	COS	0.62 ± 0.08		Maroulis et al. (1977)	
Wyoming, USA	COS	0.65 ± 0.04		Maroulis et al. (1977)	
Oklahoma, USA	COS	0.73 ± 0.05		Maroulis et al. (1977)	
England	COS	0.73	0.57–0.80	Sandalls and Penkett (1977)	
Stratosphere over the USA	COS		0.37–0.75	Inn et al. (1979)	Altitude 15.2 km
	COS		0.25–0.70	Inn et al. (1979)	Altitude 21.3 km
	COS		0.02–0.025	Inn et al. (1979)	Altitude 31.2 km
England	CS_2	0.27	0.10–0.53	Sandalls and Penkett (1977)	

[a]Over the open ocean concentrations may be approximately half this value (Graedel, 1979).

0.05 ± 0.03 μg m^{-3}, and the concentration of dimethyl sulphide and other reduced compounds with short residence time is 0.04 ± 0.03 μgS m^{-3}, making a total of 0.09 ± 0.06 μgS m^{-3}. For zones of a high bioproductivity we suggest a total concentration of 0.3 ± 0.2 μgS m^{-3}, even if there are no measurements to support this estimate.

Practically nothing is known about the vertical distribution of hydrogen sulphide, dimethyl sulphide, and other reduced sulphur compounds with short residence time in the atmosphere over the ocean. Based on inversion height data, Ryaboshapko *et al.* (1978) assumed their distribution to be uniform in the layer up to 2 km. Jaeschke *et al.* (1978) showed that hydrogen sulphide is found mainly in the lower 2 km layer. Model calculations of Omstedt and Rodhe (1978) suggest that the height of this layer should not exceed 1500 m and most likely is probably about 1 km (at the most probable hydrogen sulphide residence time in the atmosphere). We assume the average mixing height of reduced sulphur compounds in the atmosphere over oceans is 1.5 ± 0.5 km. We recall that the mixing height of a substance in the atmosphere is defined as

$$H = \frac{\int_0^\infty c(h)\, \mathrm{d}h}{c_0}$$

where $c(h)$ and c_0 are the concentrations at altitude h and directly at the surface, respectively.

The hydrogen sulphide concentration over continents has been measured frequently; however, the accuracy of methods employed is not high, and the values obtained may be highly erroneous (Cadle 1975b). Recently a reliable method has been developed (Jaeschke and Haunold, 1977) which makes it possible to determine hydrogen sulphide concentrations within the ppb range. Measurements made by Natusch *et al.* (1972) and Breeding *et al.* (1973) in the unpolluted atmosphere of central USA showed a relatively small variability with values in the range of a few tenths of a μgS m^{-3}. Smith *et al.* (1961) give hydrogen sulphide concentration values of 0.1 to 0.5 μgS m^{-3} for industrially developed regions of England and Wales. Concentrations in the surface air layer over land measured by Georgii (1978) appeared to be high (1 μg m^{-3}); however, even in this case a rapid decrease in concentration to 0.2 μg m^{-3} at 1 km altitude and a further slow decrease to 0.1 μg m^{-3} at 3 km were observed. Based on these data, Georgii (1978) finds it impossible to neglect anthropogenic hydrogen sulphide emission.

Values given in Table 4.1 suggest that the space distribution of reduced sulphur compounds with short residence time in the atmosphere over continents is quite irregular. All measurements were made in the temperate latitudes during warm seasons when an intensive hydrogen sulphide emission from biological processes may be expected. As shown by Georgii (1978), the hydrogen sulphide concentrations over marshes and littorals decrease sharply with a decrease in temperature. This was vividly demonstrated by the seasonal data of Jaeschke *et al.* (1978) who showed that the near-surface hydrogen sulphide concentrations over flooded zones in summer were higher than autumn ones by an order of magnitude. Nothing is known about the concentrations of reduced sulphur compounds such as dimethyl sulphide and dimethyl disulphide over continents, though Francis *et al.* (1975) have shown that soils emit these gases into the atmosphere.

Summarizing, we assume that the average concentration of sulphur in the form of reduced compounds with short residence time in the atmosphere over continents is 0.2 ± 0.1 μgS m^{-3}, and (as for the oceanic atmosphere) the scale height is 1.5 ± 0.5 km.

There are very few regular observations of hydrogen sulphide concentrations in the urban atmosphere over extended periods of time. We cannot use the results of occasional measurements since the concentration of pollutants in the urban atmosphere can vary with time over a wide range. The data of Jacobs *et al.* (1957) suggest an average concentration of reduced sulphur with short residence time in the urban atmosphere of 3 ± 1.5 μg m^{-3}. (Any error in this estimate will not substantially affect the calculation of the total content of reduced sulphur in the atmosphere.) For the urban atmosphere, a mixing height for reduced sulphur compounds with short residence time is assumed to be 0.5 ± 0.3 km.

Data obtained by Torres *et al.* (1980), Maroulis *et al.* (1977), and Sandalls and Penkett (1977) allow us to make a reliable estimate for the mean tropospheric concentration of carbonyl sulphide. Torres *et al.* (1980) measured the carbonyl sulphide concentration from an aircraft over the Pacific Ocean in the Southern Hemisphere to latitude 57°S and over the Pacific Ocean and North America in the Northern Hemisphere to latitude 70°N, at different heights from the near-surface layer to the tropopause. The measurements showed the constancy of concentrations both with latitude and height in the troposphere. Inn *et al.* (1979) obtained similar results to those of Torres *et al.* (1980) in the vicinity of the tropopause. Earlier analyses of samples taken in different regions of the USA (Maroulis *et al.*, 1977) and England (Sandells and Penkett, 1977) are in good agreement with the data of Torres *et al.* (1980). These data indicate a mean carbonyl sulphide concentration of 0.7 ± 0.1 μgS m^{-3}.

Less information is available on the carbon disulphide concentration of the atmosphere. From the data of Sandalls and Penkett (1977) and Cox and Sheppard (1980) it is assumed that the mean carbon disulphide concentration is 0.2 ± 0.1 μgS m^{-3}.

Carbonyl sulphide and carbon disulphide are believed to have a long residence time and direct measurements suggest that the volume concentration of these compounds in the troposphere is constant. Thus, taking into account the mean height of the tropopause (14 km) and variations in pressure with height, we obtain a mixing height for carbonyl sulphide and carbon disulphide of 6.5 ± 1 km.

4.2.2 Concentration of Sulphur Dioxide

Since sulphur dioxide emissions into the atmosphere are localized in space (thermal power plants, volcanoes), rather high sulphur dioxide concentrations (up to several mgS m^{-3}) can be detected near the emission sources. Concentrations measured far from the sources are several orders of magnitude lower and make up a tenth or even a hundredth of 1 μgS m^{-3}. Also, sulphur dioxide concentrations vary with time, i.e. daily, seasonally and annually. Table 4.2 summarizes data on the concentration of sulphur dioxide in the atmosphere of polluted and unpolluted regions.

If the data summarized in Table 4.2 are considered in chronological order, there appears to be a decrease with time in the measured background concentrations. This may be due to the development of more reliable techniques for the analysis of sulphur dioxide and to the measurement of background levels in the 'pure' regions of the world.

In the early measurements (1950–60), the detection limit was often higher than the measured sulphur dioxide concentration. As shown by Granat *et al.* (1976), during a series of 37 measurements conducted by Fisher *et al.* (1968), concentrations in 21 of the samples were lower than the detection limit (~0.5 μgS m^{-3}). Relatively high concentrations were observed by Lodge *et al.* (1974) in the air over the Panama coast and Barbados Islands. This might be explained by the emission of reduced sulphur compounds and their subsequent oxidation in the atmosphere over the coastal zone. The possibility of such an effect was noted by a number of authors, for example Nguyen *et al.* (1974) and Georgii (1978) (see below).

Extensive investigations on the content of sulphur-containing compounds in the atmosphere over the Atlantic Ocean were carried out by researchers from the Federal Republic of Germany on board the research vessel *Meteor*. Georgii (1970) employed a highly sensitive method to obtain the distribution with latitude along 30°W and showed that maximum sulphur dioxide concentrations occurred near 40°N (average 2 μgS m^{-3}, varying up to 3.5 μgS m^{-3}). Farther south, concentrations fell to 0.25–0.5 μgS m^{-3} at latitude 10–20°N and were beyond the detection limit of the method in the region of the equator. Georgi attributed this distribution to the latitudinal transport of polluted air masses from the USA. Georgii (1978) estimated the average concentration over the North Atlantic Ocean to be 0.5 μgS m^{-3}; over the Central Atlantic to be 0.25–0.5 μgS m^{-3}, and over the tropical zone of the Atlantic Ocean to be 0.15 μgS m^{-3}. We consider it rather risky to assume an influence of anthropogenic sulphur dioxide emissions over the whole of the North Atlantic Ocean. On board the *Meteor*, Büchen and Georgii (1971)

Table 4.2 Sulphur dioxide concentrations in the atmosphere

Type of atmosphere	Region	Concentration (μgS m^{-3})		Reference	Comments
		Range	Mean		
Oceanic and polar	Hawaiian Islands		0.5	Junge (1957)	
	Pacific Ocean	0.5–1.6	1.5	Lodge et al. (1960)	
	Antarctica	0.5–2.0		Cadle et al. (1968)	
	Antarctica	0.00–0.25		Fisher et al. (1968)	
	Faeroe Islands		0.12	Büchen and Georgii (1971)	
	Caribbean Sea	0.6–1.8	1.3	Lodge et al. (1974)	
	Indian and Pacific oceans	0.02–0.45	0.06	Nguyen et al. (1974)	
	Subantarctic regions	0.025–0.25	0.09	Nguyen et al. (1974)	
	Faeroe Islands		0.10 ± 0.05	Prahm et al. (1976)	
	North Atlantic		0.5	Georgii (1978)	Constant concentrations with height
	Central Atlantic		0.25–0.5	Georgii (1978)	
	Atlantic, tropical zone		0.15	Georgii (1978)	
	Pacific, central part	0.02–0.06	0.035	Ryaboshapko et al. (1978)	
	Pacific and Bering Sea, northern part	0.025–0.35	0.12	This report	
	Greenland	0.03–1.0	0.21	Flyger et al. (1976)	Aircraft observations
	North Atlantic	0.025–1.5	0.3	Flyger et al. (1976)	Aircraft observations
	Southern Pacific, layer over water surface		0.081 ± 0.026	Maroulis et al. (1980)	22 samples, aircraft observations
	Southern Pacific, free troposphere		0.12 ± 0.04	Maroulis et al. (1980)	100 samples, aircraft observations
Continental	Panama	0.45–1.6		Lodge et al. (1974)	
	Brasil		0.45	Lodge et al. (1974)	
	Africa (tropical)		3.05	Delmas et al. (1978)	

Table 4.2 continued

Type of atmosphere	Region	Concentration (μgS m^{-3})		Reference	Comments
		Range	Mean		
	Colorado, USA	0.25–1.0		Georgii (1970)	Aircraft observations
	North America, western part, surface layer		0.16 ± 0.11	Maroulis et al. (1980)	Aircraft observations, 20 samples
	North America, western part, free troposphere		0.23 ± 0.14	Maroulis et al. (1980)	Aircraft observations, 59 samples
	Northern Europe	0.07–0.11		Georgii (1978)	Altitude 6–13 km Aircraft observations
	USSR, northern Caucasus	0.17–0.31	0.12	ICM (1973)	Altitude 2.4 km Mountains
	USSR, northern Caucasus	0.01–0.8	0.095	Gedraitis (1979)	Altitude 2 km Mountains
	USSR, Borovoe, Kazakhstan	0.15–0.4	0.2	Rovinsky et al. (1980)	Summer time
	Hungary, Trans-Danube part	50–100	17	Varkonyi (1974)	Automobile survey
	England			Garland and Branson (1976)	
Industrial regions	Central Europe		10	Georgii (1978)	In winter
	Central Europe		5	Georgii (1978)	In summer
	Central Europe	5–10		Jost (1970)	Aircraft observations
	Central Europe		5	OECD (1977)	Average value for north-western Europe ($\sim$3 × 10^{12} m^2) based on measurements 1973–4

Region	Range	Value	Reference	Notes
Ukraine	0.9–20	6	This report	Episodical measurements, summer 1976
Lithuanian SSR	0.25–1.0		Shopauskas et al. (1974)	
Western and eastern Europe		7	EMEP (1980)	Average value for north-western and Central Europe based on many measurements 1977–79
USSR, Baltic Sea coast	0.02–4.7	0.63	Gedraitis (1979)	141 samples 1972–76
USA, Saint Louis region	0.00–35.7	7.5	Breeding et al. (1976)	
USA, central part	0.7–6.7	4.2	Breeding et al. (1973)	
USA, north eastern states		5	Altshuller (1973)	Average value based on measurements at 4 points
Urban				
USA, north eastern		33	Altshuller (1973)	Average for 148 cities for 1964–68
USA, western cities		11	Altshuller (1973)	Averages for 60 cities for 1964–68
Cities of the USA		13	CAPITA (1978)	Average annual concentrations in 1974
Los Angeles		6.2	Hidy et al. (1978)	Data for 1974–75
New York		21.5	Hidy et al. (1978)	Data for 1974–75
Cities of Japan		95	EAJ (1974)	
Stratosphere		0.05	Georgii (1978)	Flights over Europe. Exchange with troposphere possible

obtained an average sulphur dioxide concentration of 0.12 μgS m^{-3} in the region of the Faeroe Islands, while Prahm *et al.* (1976) reported 0.035 μgS m^{-3} for the same region. Evidently, the area measurably affected by anthropogenic emissions occupies only the western part of the North Atlantic (probably 1000 km from the North American coast) and a minor strip west of Great Britain. Under certain meteorological conditions, however, a plume of sulphur dioxide can be episodically observed at more considerable distances. This was confirmed by numerous (about 200) observations in the region of Japan (Korolev and Ryaboshapko, 1979). They showed that at distances of 100–200 km from the Japanese coast, sulphur dioxide concentrations in the plume from industrial regions can amount to 5 μgS m^{-3}. At about 500 km, they drop to 0.5–1 μgS m^{-3}, and at distances of more than 1000 km the contribution of anthropogenic sulphur dioxide in the plume transported from Japan is only 0.1 μgS m^{-3}.

Flyger *et al.* (1976) found almost the same sulphur dioxide concentrations over the North Atlantic and Greenland using the isotope dilution method. No relation was observed between concentration and height (up to 5 km).

Ngyen *et al.* (1974) measured the concentrations of sulphur-containing compounds in the atmosphere of the Southern Hemisphere. They showed that sulphur dioxide concentrations varied over the range 0.02–0.45 μgS m^{-3}. No significant difference was observed between the atmosphere in the region of 40°S and that of the sub-Antarctic zone. Most measurements fell within the range 0.02–0.1 μgS m^{-3}, and only in a few cases were higher concentrations, up to 0.45 μgS m^{-3}, detected. These authors suggested that the sulphur dioxide came from natural emissions of reduced sulphur in some oceanic areas. In 1974 investigations were conducted from the Soviet research vessel *A. I. Voejkov* in the trade-wind zone of the Pacific (20–30°N latitude) and near the Aleutian ridge (Ryaboshapko *et al.*, 1978). In both regions the concentrations were essentially the same, with the average value from 12 samples being 0.035 ± 0.02 μgS m^{-3}. The trajectory analysis showed that in all 12 cases the air masses did not contact land in the five days before sampling. During the joint Soviet–American expedition on board the Soviet research vessel *Volna* in 1977, sulphur dioxide concentrations of 0.025–0.35 μgS m^{-3}, with an average value of 0.12 μgS m^{-3}, were determined in the northern part of the Pacific (>50°N) and in the Bering Sea.

Maroulis *et al.* (1980) determined the sulphur dioxide concentration over the Pacific Ocean in the Southern Hemisphere, and over the Pacific Ocean and western part of North America in the Northern Hemisphere, i.e. from 57°S to 70°N and from the boundary layer to the tropopause. They showed that the concentration in the Northern Hemisphere was somewhat higher than in the Southern Hemisphere, and that the sulphur dioxide concentrations in the free troposphere were everywhere higher than in the boundary layer.

Summarizing, in the atmosphere over the open oceanic regions characterized by low biological productivity, the sulphur dioxide concentration is about 0.10 ± 0.03 μgS m^{-3}. In upwelling zones, in the ocean regions of high biological productivity, a concentration of 0.2 ± 0.1 μg m^{-3} can be taken as average.

The profile of the sulphur dioxide concentration in the oceanic atmosphere (studied by Flyger *et al.* 1976, Gravenhorst 1975, and Maroulis *et al.* 1980 from aircraft) should be determined by its sources and sinks. Since the ocean surface cannot be a direct source of sulphur dioxide and serves only as a sink, it is reasonable to assume that there should be some increase in concentration with height in the lower atmosphere. If the precursors of sulphur dioxide are hydrogen sulphide, dimethyl sulphide, and other reduced sulphur compounds, then the source should be in the lower troposphere; after a small increase in concentration in this layer, there should be a decrease in the sulphur dioxide concentration higher up. Maroulis *et al.* (1980) believed that carbonyl sulphide is the precursor of sulphur dioxide: if this is the case, there should be a uniform distribution of sulphur dioxide throughout the troposphere. The problem of the sulphur dioxide precursor in the free troposphere still needs to be resolved (Rodhe 1981). However, from the data available, and if we assume that the mixing ratio of sulphur dioxide (the ratio between sulphur dioxide and air volumes) is constant throughout the troposphere, then, we have a mixing height of 6.5 ± 1 km. The increased sulphur dioxide concentrations in zones of higher biological productivity should have a smaller mixing height, which we assume to be 4.5 ± 1 km.

We have less information on sulphur dioxide concentrations in unpolluted (clean) continental atmospheres than in the atmosphere over the isolated regions of the ocean and Antarctica. The majority of continental observations carried out in Europe and the USA do not characterize a clean continental atmosphere because of the considerable contribution from anthropogenic sources. We have no data on the sulphur dioxide concentration in the atmosphere over large regions of Asia and Australia. Meszaros (1978b) gives a value of 3.05 μgS m^{-3} for Central Africa. This is not in line with data on sulphur dioxide concentrations in the continental atmosphere summarized in Table 4.2. Lodge *et al.* (1974) give values in the range 0.5–1.5 μgS m^{-3} for the atmosphere of tropical forests in Panama and Brazil. Georgii's (1970) data obtained during aircraft observations in the clean atmosphere in the western region of the USA are somewhat lower. In 1973 in the Caucasus a comparison of methods was made for the analysis of pollutants in background air; the value of 0.12 μgS m^{-3} given in Table 4.2 is the average of all the values obtained by various techniques during one week of comparisons. Extremely interesting routine data were obtained in the Borovoe region of Eurasia (Rovinsky *et al.*, 1980) from 1976 to 1978. This region is characterized by the absence of local industrial sources of sulphur dioxide (at least during summer). We believe that the average summer concentration of

0.2 μgS m^{-3} represents the clean continental atmosphere. Maroulis *et al.* 1980 present data on the sulphur dioxide concentration in the boundary layer of the atmosphere and in the free troposphere over the western part of North America. As in the case of the oceanic atmosphere, the authors noted some increase in the sulphur dioxide concentration with height. This agrees with Georgii's (1978) data obtained in the upper troposphere over Europe.

Thus the average sulphur dioxide concentration for the clean continental atmosphere is assumed to be 0.2 ± 0.1 μgS m^{-3} and the mixing height 6.5 ± 1 km.

The areas of regional atmospheric pollution cover considerable parts of Europe and North America. Furthermore, practically the whole of Japan, the densely populated regions of China, India, the South African Republic, etc. can also be characterized as regionally polluted areas. Owing to a number of international and national programmes on regional monitoring of atmospheric pollutants, much information has been published recently on the atmospheric distribution of sulphur dioxide with space and time in industrial regions and surrounding areas. Measurements made episodically over a national reserve in Lithuania, with no sources of anthropogenic emission of any importance in the radius of several tens of kilometres (Shopauskas *et al.* 1974) suggest that the rather low concentration of 0.25–1.0 μgS m^{-3} most likely characterizes the continental European background. Similar results were obtained by Gedraitis (1979) from routine measurements in the same region. Much higher concentrations obtained by Varkonyi (1974), from an automobile survey in an industrialized region of Hungary, were characteristic of a high extent of urbanization. When assessing average values of sulphur dioxide concentrations in the atmosphere of industrial regions (Table 4.2), one has to determine whether the results are representative of the region, i.e. the duration of observations has to be taken into account. In this regard, the data from the long-term programmes of OECD (1977), EMEP (1980), and Altshuller (1973) should be representative of the regions studied. All the other values of regional sulphur dioxide concentrations summarized in Table 4.2 fall within a narrow range. From these data we assume the average sulphur dioxide concentration in the atmosphere of industrial regions to be 5 ± 2 μgS m^{-3}.

The vertical distribution of sulphur dioxide in the atmosphere of industrial regions was studied by a number of authors. Meszaros (1978b), referring to investigations by Varhelyi, gives a value of 0.6 km for Hungarian conditions. Special investigations aimed at the estimation of an average mixing height for sulphur dioxide in Europe were conducted within the OECD (1977) programme. From 23 profiles an average scale height of 1220 m was obtained. The same value was measured by Garland and Branson (1976) and a similar value (1100 m) was obtained by Smith and Jeffrey (1975) in observations made over England during aircraft surveys. We consider the value

$H = 600$ m to be an underestimate, since it was obtained for urban atmospheric conditions rather than regional. In our calculations we shall assume a mixing height of 1.2 ± 0.2 km for the regional atmosphere.

The sulphur dioxide concentration in urban atmospheres is measured continuously in many of the large cities of the world. The results of three reviews summarizing such data are given in Table 4.2. Concentrations vary with time, and over a wide range from several μgS m^{-3} to several mgS m^{-3}. One can distinguish short-term variations caused mainly by meteorological conditions and seasonal variations connected with the increase in emissions in certain seasons. Since the late 1960s, average annual concentrations in many towns have decreased as a result of the centralization of heating systems, construction of high stacks, and use of cleaner fuels. Average concentrations vary considerably from town to town depending on their size, level of industrialization, type of fuel used, etc. There is considerable uncertainty in estimates of the average concentration in the urban atmosphere for a specific country and especially for the world as a whole. In our estimates we assume the average concentration to be 20 ± 10 μgS m^{-3}.

In various countries both episodic and systematic investigations of the vertical distribution of sulphur dioxide in the urban atmosphere have been performed using helicopters, aircraft, towers, and masts. From the data of Goroshko *et al.* (1968) we assume the average scale height of sulphur dioxide over towns to be 0.6 ± 0.2 km.

Investigators from the Federal Republic of Germany measured the sulphur dioxide concentration in the lower stratosphere (Georgii, 1978) and found it to be 0.05 μgS m^{-3}. So far this is the only measurement in this region and we cannot judge how representative it is for the whole stratosphere, especially as Georgii noted that intensive exchange between the stratosphere and troposphere was occurring during his observations.

4.2.3 Concentration of Sulphate

Sulphate can be added to the atmosphere by various mechanisms: sea sprays, oxidation of sulphur dioxide in the atmosphere, aeolian soil weathering, etc., (Junge, 1963a). Differences in these mechanisms lead to varying physical and chemical characteristics of the sulphates in the atmosphere. In regions of intensive anthropogenic sulphur dioxide emissions, sulphates are formed mainly as a result of sulphur dioxide oxidation, and according to Whitby (1978), Friedlander (1978), and others, practically all of these sulphates are airborne particles from 0.1 to 1 μm in size. Sea-water sulphates enter the atmosphere in sea spray. Gravenhorst (1975) showed that these sulphates are predominantly particles ranging from 0.5 to 10 μm in diameter, although most are within the range 2 to 5 μm. Similar data were obtained by Levkov *et al.* (1975) from Cuba. It appears that sulphates from aeolian weathering occur in larger particles; Andreev and Lavrinenko (1968) showed that in a dusty atmosphere, especially during dust-storms, the majority of the sulphate was found on large aerosol particles.

The chemical composition of the sulphate-containing substances depends on their origin. If the sulphates are formed in the atmosphere due to the oxidation of sulphur dioxide, the most probable species formed are ammonium sulphate $(NH_4)_2SO_4$ and its acid forms—$(NH_4)_3H(SO_4)_2$, NH_4HSO_4 (Brosset, 1978). Sulphate can also occur in the atmosphere in the form of sulphuric acid (OECD, 1977). Sea sulphates are associated with the basic sea-salt cations: Na, Ca, Mg, etc. According to Khusanov *et al.* (1974), $CaSO_4$ is the most probable form in aeolian dusts sampled over Central Asia.

Data on sulphate concentrations in the atmosphere of different regions are presented in Table 4.3. Here we separate the marine sulphate contribution by the sulphate : sodium ratio, as was done previously by Granat *et al.*, 1976. Nguyen *et al.* (1974) determined the sulphate content in the atmosphere over various regions of the world's ocean from Antarctica to 50°N in the Atlantic Ocean. The value obtained by them over the Mediterranean Sea (Table 4.3) appears to be anomalously high compared with their other observations. It is difficult to hypothesize any specific distinguishing characteristic of the Mediterranean Sea from the world oceans except for the possible anthropogenic sulphate concentration in this region. The rest of the data do not vary significantly from the average value of 0.56 $\mu gS\ m^{-3}$. Gravenhorst (1975, 1978) showed that excess sulphate accounted for 70–80% of the total sulphate, the concentrations of which varied around the mean value of 0.35 $\mu gS\ m^{-3}$. Both Büchen and Georgii (1971) and Gravenhorst (1978) observed increased sulphate concentrations in air masses from the Sahara. The same result was observed by Winkler (1975) who studied the sulphur content in submicron particles. According to Winkler (1975) and Gravenhorst (1975), small particles are predominantly excess sulphur. Prahm *et al.* (1976) analysed four samples from clean air of the North Atlantic Ocean which gave an average concentration of 0.25 $\mu gS\ m^{-3}$ and an excess sulphur contribution of 56%, which is in agreement with the above results.

Investigations performed by Gillette and Blifford (1971) and Flyger *et al.* (1976) show that sulphate is found predominantly in the lower atmosphere, i.e. up to 2–3 km above the ocean.

Summarizing, we assume that the average concentration of sulphate over the ocean is $0.5 \pm 0.15\ \mu gS\ m^{-3}$, which includes $0.35 \pm 0.15\ \mu gS\ m^{-3}$ of excess sulphur. The mixing height of sulphate over the ocean is assumed to be 2.5 ± 0.5 km.

Direct measurements of sulphate in the surface continental atmosphere are less numerous than in the oceanic atmosphere. Gillette and Blifford (1971) found the sulphate concentrations in the surface layer of the atmosphere over

Table 4.3 Sulphate concentrations in the atmosphere

Type of atmosphere	Region	Total (μgS m^{-3})		Excess (μgS m^{-3})		Reference	Comments
		Range	Average	Range	Average		
Oceanic and polar	Antarctic coast	0.17–0.9	0.52			Nguyen et al. (1974)	
	Subantarctica	0.3–0.9	0.56			Nguyen et al. (1974)	
	Pacific Ocean, 20–40°S lat.		0.38			Nguyen et al. (1974)	
	Atlantic Ocean 8–50°N lat.	0.3–1.78	0.78			Nguyen et al. (1974)	
	Mediterranean Sea	6.0–13.2	2.81			Nguyen et al. (1974)	
	Central Atlantic	0.3–0.7				Büchen and Georgii (1971)	
	Central Atlantic	1.0–1.3				Büchen and Georgii (1971)	Wind from Africa
	North Atlantic	0.1–3.0	0.23	0.1–2.9	0.13	Gravenhorst (1975)	
	Central Atlantic		0.37		0.32	Gravenhorst (1975)	
	Central Atlantic		0.4		0.3 ± 0.17	Gravenhorst (1978)	
	Central Atlantic		0.93		0.77	Gravenhorst (1978)	Wind from Africa
	North Atlantic	0.18–0.32	0.25	0.03–0.23	0.14	Prahm et al. (1976)	
	Central Atlantic	0.2–2.6	0.9		0.6	Petrenchuk (1979)	
	Pacific Ocean surface layer	0.14–1.4				Huebert and Lazrus (1980)	
	Atlantic Ocean		0.2		0.2	Winkler (1975)	Aitken nuclei, r ~ 0.2 μm
	Atlantic Ocean		0.46		0.46		Transport from Sahara
	Pacific coast		0.23		0.2	Gillette and Blifford (1971)	Water surface layer
	Pacific coast		0.1		0.09	Gillette and Blifford (1971)	Altitude 2 km
	Pacific coast		0.05		0.05	Gillette and Blifford (1971)	Altitude 5 km

Table 4.3 (continued)

Type of atmosphere	Region	Total (μgS m^{-3})		Excess (μgS m^{-3})		Reference	Comments
		Range	Average	Range	Average		
Continental	North Atlantic	0.03–0.63	0.122			Flyger et al. (1976)	Altitude 4 km
	Greenland	0.03–0.22	0.089			Flyger et al. (1976)	Altitude 4 km
	All continents				0.17	Junge (1965)	Estimate
	All continents				0.5	Friend (1973)	Estimate
	Bolivia		0.13		0.12	Boueres et al. (1977)	Altitude 5.2 km
	Bolivia		0.1		0.09	Adams et al. (1977)	Altitude 5.2 km
	Salvador		0.55		0.5	Guzman (1977)	Altitude 1.8 km
	South America		0.057			Lawson and Winchester (1978)	Particles ~1 μm
	Central Africa	0.5–0.6	0.77			Delmas et al. (1978)	
	Sudan		1.05			Penkett et al. (1979)	Transport from North
	Sudan		0.67			Penkett et al. (1979)	Transport from South
	Alaska, Barrow	Max. 1.7				Rahn and McCaffrey (1979)	Possible anthropogenic impact
	West Canada	0.83–1.6	1.1			Chung (1978)	
	East Canada		0.8			Shaw (1979)	Clean air masses
	USA, western part		0.3		0.26	Gillette and Blifford (1971)	Surface layer
	USA, western part		0.2		0.17	Gillette and Blifford (1971)	Altitude 2 km
	USA, western part		0.05		0.05	Gillette and Blifford (1971)	Altitude 5 km
	North America		0.07 ± 0.057			Lezberg et al. (1979)	Below the tropopause
	North America		0.17 ± 0.057			Lezberg et al. (1979)	Above the tropopause
	Continental boundary layer	0.085–0.17				Huebert and Lazrus (1980)	

Location	Range	Value		Reference	Remarks
Troposphere above boundary layer	0.043–0.26			Huebert and Lazrus (1980)	
USSR, Middle Asia		11.2		Andreev and Lavrinenko (1968)	Excluding dust-storms
USSR, Middle Asia		28.0		Andreev and Lavrinenko (1968)	During dust-storms at 300 m
USSR, Middle Asia		9.5		Khusanov et al. (1974)	
USA, desert in south-west	Min. 0.03	0.25		Hoffer et al. (1979)	Excluding dust-storms
Over industrial regions USA, eastern part		2.7		Altshuller (1973)	Obtained at 14 points, 1965–68
USA, western part		0.87		Altshuller (1973)	Obtained at 15 points, 1965–68
USA, eastern part		3.4		EPA (1975)	Characteristic of area $\sim 2.5 \times 10^{12}$ m^2
USA, western part		1.5		EPA (1975)	Characteristic of area $\sim 2.5 \times 10^{12}$ m^2
USA, north-eastern part	1.3–3.5	2.7		Hidy et al. (1979)	Surface layer
USA, north-eastern part	1.6–6.0	3.1		Hidy et al. (1979)	0.15–1.5 km layer
USA industrial regions		3.3		CAPITA (1978)	Max. in summer
South-eastern Canada	1.6–4.6	3.3		Chung (1978)	April–May 1975
North-western Europe			1.75	OECD (1977)	Average for area 3×10^{12} m^2, 1973–74
North-western and Central Europe			2.5	EMEP (1980)	Average for area 5×10^{12} m^2, 1977–79
Urban USA, eastern cities	3.3–8.0			EPA (1975)	All big US cities
USA, western cities		2.7		EPA (1975)	All big US cities
USA, cities		4.2		Altshuller (1976)	22 cities, 1963–72
Nagoya		4.8		Kadowaki (1976)	1973–74
Budapest		2.4		Meszaros (1978b)	1974–76
Toronto		2.4		Heidorn (1978)	Data for 8 points, 1976

California to be 0.3 μgS m^{-3}. As in the oceanic atmosphere, concentrations fell with height to 0.2 μgS m^{-3} at an altitude of 2 km and to 0.05 μgS m^{-3} at altitudes over 5 km. This agrees well with the data obtained by Lezberg *et al.* (1979) at the level of the tropopause. Recently, Boueres *et al.* (1977) and Adams *et al.* (1977) reported sulphate concentrations at an altitude of 5200 m over South America. These data allow us to estimate a mean height for the distribution of sulphate in the continental atmosphere, but unfortunately they give no idea of its concentration directly over the South American continent. Guzman (1977) found concentrations of 0.55 μgS m^{-3} for Central America (Salvador) at an altitude of 1800 m. In these instances, the contribution of sea-water sulphur was no more than 10%. Meszaros (1978b) reported measurements in the surface atmosphere over Africa performed by Delmas *et al.* (1978) of 0.5–0.6 μg m^{-3}. There is a good agreement between the values obtained by Rahn and McCaffrey (1979), Chung (1978), and Shaw (1979) for North America. For South America, Lawson and Winchester (1978) reported a value which was lower by an order of magnitude, but they analysed the fine fraction of aerosol only. From the limited data available at the time, Junge (1965) inferred the average concentration of sulphate in the atmosphere over continents to be 0.17 μgS m^{-3}, while Friend (1973) estimated it to be 1.5 μgS m^{-3}.

Observations in dusty regions should be considered separately. The Soviet investigators Andreev and Lavrinenko (1968) and Khusanov *et al.* (1974) made measurements in Middle Asia in a region characterized by intensive aeolian weathering. In that region the soils are mainly sands, serozems, and saline soils. Saline soils with sulphate concentrations of up to 5% are known to have a loose structure and are capable of producing dust at wind speeds of 3–4 m sec^{-1} (Zakharov, 1965). The average sulphate concentration of the atmosphere near Tashkent is 11.2 μgS m^{-3} according to Andreev and Lavrinenko (1968), and 10 μgS m^{-3} according to Khusanov *et al.* (1974) Both groups noticed that, during dust-storms, concentrations increase drastically and reach almost 30 μgS m^{-3}. This confirms the aeolian origin of sulphate in the atmosphere of this region.

These measurements were made during a warm season. During rainy seasons, or when the soil is covered with snow there is practically no aeolian weathering. Therefore, the data of Andreev and Lavrinenko (1968) and Khusanov *et al.* (1974) do not reflect the average annual conditions of the region. We can, however, single out specific regions of intensive aeolian weathering where sulphate concentrations are much higher than the average continental value.

The data obtained in other arid zones of the world differ greatly from the results above. Penkett *et al.* (1979) reported values of 1 μgS m^{-3} for East Africa. Still lower concentrations were found by Hoffer *et al.* (1979) for deserts in the south-western USA. Increased sulphate concentrations were observed more than once in the atmosphere during the transport of aerosols from Sahara to the Atlantic (Gravenhorst, 1978; Büchen and Georgii, 1971; Winkler, 1975), and abnormally high deposition of sulphate from the atmosphere was observed in the arid zones of Australia (Hingston and Gailitis, 1976).

From a consideration of the above data we can assume that the average sulphate concentration over continents is $0.6 \pm 0.2 \ \mu gS \ m^{-3}$, and the mixing height of the layer is 2.5 ± 0.5 km. Taking into account seasonal variations, we may assume an average annual value for regions with intensive aeolian weathering of sulphate-containing soils to be $4 \pm 2 \ \mu gS \ m^{-3}$. Since the aeolian dust consists mainly of large particles, it would be expected that they would not rise high into the atmosphere. Observations by the author, made from an aircraft in Middle Asia, showed that the uppermost boundary of the dusty air was usually at an altitude of 1–1.5 km. During dust-storms, however, aeolian particles may rise to heights greater than 6 km. Let us assume the mixing height of aeolian sulphates to be 1.5 ± 0.5 km.

The increased sulphate concentrations in the atmosphere of industrial regions come from anthropogenic emissions and subsequent oxidation of sulphur dioxide. Levels of sulphate sulphur depend on the extent of industrialization and the sulphur content of the fuels used. For instance, in Europe, the central regions of England, Germany, and Poland are distinguished by high sulphate concentrations. Altshuller, (1973) and Perhac (1978) showed a clear-cut difference in concentrations between the western and eastern regions of the USA. Concentrations of $1–1.3 \ \mu gS \ m^{-3}$ over the western states are approximately the same as those of a clean continental atmosphere. Over the eastern, more industrialized, states the average concentration was $3.3–5.0 \ \mu gS \ m^{-3}$. According to OECD (1977), the average concentration of sulphate over all of western Europe was $1.7 \ \mu gS \ m^{-3}$. Based on the data in Table 4.3, we shall assume $3 \pm 0.5 \ \mu gS \ m^{-3}$ as an average concentration of sulphate in industrial regions.

Our estimate of the sulphate mixing height in the regional atmosphere based on the data of OECD (1977), Rodhe (1972b), Jost (1974), Bolin and Persson (1975), Meszaros (1978a), and Georgii (1978) is 1.5 ± 0.3 km.

Since the transformation of anthropogenic sulphur dioxide to sulphates takes several hours or even days, most of the sulphate is formed at distances remote from the source. Therefore, the urban concentrations of sulphate (i.e. near the sources of sulphur dioxide emissions) do not differ substantially from the regional average. From the data given in Table 4.3 we can assume an average concentration of sulphate in the urban atmosphere of $4 \pm 1 \ \mu gS \ m^{-3}$. The mixing height of sulphate in the urban atmosphere is taken to be 1.5 ± 0.3 km.

4.2.4 Total Sulphur Content of the Atmosphere

In previous sections we estimated average concentrations and mixing heights of various sulphur compounds in the atmosphere of six different zones: oceanic without the influence of marine biota, oceanic with high bioproductivity, clean continental, dusty industrial and urban regions. The area of each of these zones is required for the determination of the sulphur content in each.

A high bioproductivity is characteristic of rather small regions of the world's ocean; these are the areas where organic matter, phosphorus, nitrogen, and other nutrients are available in sufficient amounts. This situation occurs in regions of upwelling and along the continental shelf. The area of the world's ocean with a high bioproductivity comprises 15% of the entire oceanic surface, or 55×10^{12} m^2 (Stepanov, 1974). The other 305×10^{12} m^2 is referred to as the zone without marine biota or open ocean.

Regional atmospheric pollution is characteristic of almost the whole of Europe and the major part of the USA. In Japan, the zone of regional pollution covers the whole territory and spreads far over the ocean. Other countries of Asia, Africa, and South America are characterized by insignificant areas of regional pollution. Detailed estimates of areas of regional pollution are given in Table 4.4.

The area characterized by urban levels of atmospheric pollution is assumed to be 1×10^{12} m^2. Therefore, the area of regional pollution accounts for 11×10^2 m^2. The area of all the deserts and semi-deserts in the world is 50×10^{12} m^2, and 9.5×10^{12} m^2 are occupied by areas in which the upper soil horizon is salinized (Kovda and Sabolsz, 1980). More than half of this salinization is the chloride–sulphate type. From this we assume the area with a high content of aeolian sulphates in the atmosphere is 5×10^{12} m^2. Therefore, the clean continental atmosphere accounts for an area of 133×10^{12} m^2.

Table 4.5 summarizes data on sulphur concentrations for reduced compounds with short and long residence times, sulphur dioxide, total and excess

Table 4.4   Areas of industrial regions in various countries and continents

Country or continent	Area (10^{12} m^2)
Europe and Asian part of USSR	5.2
USA and South-east Canada	3.9
Japan	0.8
China	1.0
India	0.4
Australia	0.1
Africa	0.2
South America	0.15
Total	11.75

Table 4.3 Concentrations, mixing heights and contents of various forms of sulphur in the atmosphere

Type of atmosphere	Area (10^{12} m^2)	Form	Concentration (μgS m^{-3})	Mixing height (km)	Content (TgS) Range	Content (TgS) Mean
Oceanic, without biota effect	305	S^a	0.09 ± 0.06	1.5 ± 0.5	0.01–0.09	0.04
		SO$_2$	0.10 ± 0.03	6.5 ± 1.0	0.12–0.30	0.20
		SO$_4^{2-}$ total	0.5 ± 0.15	2.5 ± 0.5	0.21–0.60	0.38
		SO$_4^{2-}$ excess	0.35 ± 0.15	2.5 ± 0.5	0.12–0.46	0.27
Oceanic, with biota effect	55	S^a	0.3 ± 0.2	1.5 ± 0.5	0.005–0.055	0.025
		SO$_2$	0.2 ± 0.1	4.5 ± 1.0	0.02–0.09	0.05
		SO$_4^{2-}$ total	0.5 ± 0.15	2.5 ± 0.5	0.04–0.10	0.07
		SO$_4^{2-}$ excess	0.35 ± 0.15	2.5 ± 0.5	0.02–0.08	0.05
Continental clean	133	S^a	0.2 ± 0.1	1.5 ± 0.5	0.01–0.08	0.04
		SO$_2$	0.2 ± 0.1	6.5 ± 1.0	0.07–0.30	0.17
		SO$_4^{2-}$	0.6 ± 0.2	2.5 ± 0.5	0.11–0.32	0.20
Industrial region	11	S^a	0.2 ± 0.1	1.5 ± 0.5	0.00–0.01	0.005
		SO$_2$	5.0 ± 2.0	1.2 ± 0.2	0.03–0.11	0.07
		SO$_4^{2-}$	3.0 ± 0.5	1.5 ± 0.3	0.03–0.07	0.05
Continental dusty	5	S^a	0.2 ± 0.1	1.5 ± 0.5	0.00–0.005	0.002
		SO$_2$	0.2 ± 0.1	6.5 ± 1.0	0.00–0.01	0.005
		SO$_4^{2-}$	4.0 ± 2.0	1.5 ± 0.5	0.01–0.06	0.03
Urban	1	S^a	3.0 ± 1.5	1.0 ± 0.3	0.00–0.01	0.005
		SO$_2$	20.0 ± 10.0	0.6 ± 0.2	0.005–0.02	0.01
		SO$_4^{2-}$	4.0 ± 1.0	1.5 ± 0.3	0.00–0.01	0.005
		Total sulphur with short residence time			0.68–2.24	1.355
		Including excess sulphates			0.14–0.54	0.32
Troposphere	510	COS	0.7 ± 0.1	6.5 ± 1.0	1.68–3.06	2.32
		CS$_2$	0.2 ± 0.1	6.5 ± 1.0	0.28–1.15	0.66
		Total sulphur with long residence time			1.96–4.21	2.98
		Total sulphur in troposphere			2.64–6.45	4.335

^{a}Reduced sulphur with short residence time.

Table 4.6 Sulphur contents (TgS) of the troposphere according to different authors

Reduced sulphur with short residence time	SO₂	Total SO₄²⁻	Excess SO₄²⁻	Total sulphur with short residence time	Reduced sulphur with long residence time	Reference
0.99	0.52	0.25	0.57	1.76		Granat et al. (1976)
0.08–0.75	0.45	0.80		1.3–2.0		Friend (1973)
0.11	0.50	0.74	0.32	1.35	2.98	Meszaros (1978b)
0.03–0.25	0.25–0.83	0.4–1.16	0.14–0.54	0.68–2.24	1.96–4.21	Present work

sulphates in individual parts of the atmosphere, mixing heights, and areas of zones. It also gives the sulphur content of different compounds in separate parts of the atmosphere and the troposphere as a whole.

Table 4.6 shows that the estimates of the amount of sulphur in the troposphere agree reasonably well with those of other authors. Only the estimates of Friend (1973) for sulphur with short residence time and sulphate sulphur seem to be out of line.

4.2.5 Stratospheric Sulphate Layer

The existence of an aerosol layer in the stratosphere was suspected at the end of the last century on the basis of a light-scattering effect. In the last 20 years, several attempts have been made to measure directly the concentrations of stratospheric airborne particles and to determine their chemical composition. Studies conducted by Junge and Manson (1961) and Rosen (1964) showed that in contrast to the regularly decreasing profiles of Aitken nuclei, the profile of larger particles ($r > 0.1$ μm) shows a maximum concentration in the lower stratosphere. Beginning at the tropopause, the concentration of particles increases with height, reaches a maximum at an altitude of 20 km, and then drops quickly; the maximum concentration is approximately four times higher than that at the tropopause level. Data obtained from a number of aircraft and balloons showed a low range in concentration of such particles over most of the globe, from 70°N to 60°S (Junge, 1965).

Sampling with an impactor, Junge (1963b) showed that the mean radius of particles is approximately 0.15 μm and the distribution of particles by size has a clear upper limit of 1–2 μm. The concentration of smaller particles increases rapidly as the radius decreases and its change may be expressed by the relation:

$$dN/d(\log r) = cr^{-3.5}$$

where c is a constant and r varies over the range 0.1–2 μm.

From analyses of the chemical composition of particles sampled in this layer, Junge et al. (1961) found that the sulphate in them amounts to about 90%. Thus it is called the sulphate layer (or sometimes the Junge layer). Friend (1966) found that these particles may contain persulphates in addition to sulphates. It has been shown recently (Farlow et al., 1978) that the sulphur chemistry in the stratosphere is closely connected with the nitrogen chemistry and that compounds such as $NOHSO_4$, $NOHS_2O_7$, $(NH_4)_4SO_4$ and $(NH_4)_2S_2O_8$ are formed in the Junge layer. Even in the first chemical analysis of stratospheric aerosol, it was found that the number of cations detected was insufficient to neutralize all sulphate ions. This implies the presence of free sulphuric acid in the stratosphere.

Cadle (1975a) used filters of high efficiency to collect particles up to sizes of Aitken nuclei in aircraft flights at an altitude of about 18 km over the USA and the tropical areas of Central America. He found that the concentrations of sulphur varied from 0.025 to 0.12 μgS m^{-3}.

At present it is thought that the formation of the sulphate layer is governed by two mechanisms: (1) the diffusion of sulphur-containing gases to the stratosphere followed by their oxidation to sulphates (Junge, 1963b; Lazrus *et al.*, 1971); and (2) the release of volcanic gases and sulphates during powerful eruptions directly into the stratosphere. The existence of the first mechanism is suggested by the absence of sufficient cations to neutralize all of the sulphates. Free sulphuric acid can form only *in situ* as a result of the oxidation of sulphur-containing gases. According to Junge, Cadle, and Lazrus *et al.*, the sulphur cycle in the stratosphere can be described as follows: gaseous sulphur compounds diffuse from the troposphere into the lower stratosphere and are oxidized there to sulphur trioxide by atomic oxygen in a trimolecular reaction. The molecule is hydrated and transformed into sulphuric acid. The process of coagulation then causes droplets to grow; by the time they are of radius 1–2 μm, the particles are withdrawn from the stratosphere.

There has been uncertainty for a long time about which gaseous sulphur compounds could exist for long enough in the troposphere to penetrate through the tropopause to the stratosphere in appreciable quantity. The participation of sulphur dioxide, hydrogen sulphide, dimethyl sulphide, and dimethyl disulphide is questionable because of the short residence time of these compounds in the troposphere and the long period of troposphere–stratosphere exchange (Reiter, 1975). Crutzen (1976) suggested carbonyl sulphide and carbon disulphide as possible candidates. According to his calculations, the carbonyl sulphide flux to the stratosphere amounts to 0.05 TgS year^{-1}.

The role of carbonyl sulphide in the formation of stratospheric sulphates is confirmed by the data of Inn *et al.* (1979) who measured a rapid decrease in the mixing ratio with height in the stratosphere, indicating rapid oxidation of carbonyl sulphide.

The validity of the second mechanism for stratospheric sulphate formation is based on the increase in sulphate concentration in the stratosphere after the eruption of Agung in 1963. A comparison of the data obtained by Junge before the eruption with that of Cadle *et al.* (1975a) obtained in the period 1969–70 indicates a 30-fold increase in the concentration. According to Junge (1963b), the residence time of particles in the sulphate layer of the stratosphere is 0.5 year. Cadle's (1975a) estimates are 1 month at the tropopause and 1–2 years at an altitude of 20 km. Russell *et al.* (1976) considered it to be 8 months based on lidar measurements of the optical thickness of the stratospheric aerosol layer formed after the Fuego eruption. Even if we take the maximum value to be 2 years, then by the time of Cadle's flight in 1969, only 5% of the stratospheric sulphates resulted from the Agung eruption should remain. The most reasonable explanation for the discrepancy between

Table 4.7 Sulphate in the stratosphere

Year of measurement	Sampling height (km)	Region	Concentration (μgS m^{-3})[a]	Reference
1960	20	Global distribution	0.001–0.004[b]	Junge (1965)
1969	17–18	Northern Hemisphere, 0–30°N	0.06	Cadle and Grams (1975)
1969	17–18	Northern Hemisphere, 30–48°N	0.10	Cadle and Grams (1975)
1970–71	17–19	Northern Hemisphere, 0–30°N	0.02	Lazrus et al. (1971)
1970–71	17–18	Northern Hemisphere, 30–35°N	0.05	Lazrus et al. (1971)
1962–71	15–19	Northern Hemisphere	0.08–0.15	Castleman et al. (1974)
1962–71	15–19	Southern Hemisphere	0.05	Castleman et al. (1974)
1976	10–25	Global distribution	0.015–0.1	Lazrus et al. (1979)

[a]Per unit volume of stratospheric air.
[b]Underestimates because of low particle trapping efficiency.

the concentrations measured in 1960 and 1969 was given by Cadle (1975a) who pointed out the difference in efficiency of sampling devices used by Junge (impactor) and Cadle (filter). We do not reject the possible injection of large quantities of sulphur into the stratosphere during intensive eruptions, but practically nothing remains in the stratosphere from the Agung eruption. From 1963 to the present, at least seven eruptions have occurred which have been accompanied by the release of volcanic products into the stratosphere (Cronin, 1971; Fedotov *et al.*, 1976). Cadle (1975a), for instance, estimated the mean volcanic emission to the stratosphere to be ~0.14 TgS year^{-1}.

Table 4.7 summarizes the data on sulphate concentrations with height in the Junge layer obtained by different authors during the last 20 years. When Junge's data, which are considered to be underestimates, are not taken into account, the data of the other authors varies within an order of magnitude.

It is difficult to draw any definite conclusions on the sulphur content of the stratosphere. Junge reported a value of 0.01–0.03 Tg of sulphur, which is based on underestimated concentrations. Granat *et al.* (1976), using the data of Castleman *et al.* (1974), estimated the sulphur content in the stratosphere of the Northern Hemisphere to be 0.2 Tg and that in the Southern Hemisphere to be 0.1 Tg. Lazrus *et al.* (1979) give an estimate of ~0.6 TgS for the spring of 1975 after the eruption of Fuego. These estimates seem reasonable but, to obtain more accurate data, one should know the vertical concentration profiles at various latitudes. This is especially so in view of the influence of the altitude of the tropopause on the chemical composition and total amount of sulphates. If the unexpectedly high content of sulphur dioxide in the lower stratosphere of 0.18 μgS m^{-3} (Jaeschke *et al.*, 1976) is representative, then we should assume that the total sulphur content in the stratosphere can reach 0.5 Tg. This constitutes an appreciable amount in relation to the sulphur content in the troposphere. However, the sulphur flux through the tropopause is rather small compared to the tropospheric fluxes of 0.5–1 TgS year^{-1} if the average residence time of sulphur in the stratosphere is 0.5–1 year.

The small value for this natural flux is indicative of its sensitivity to change with anthropogenic factors (Crutzen, 1976). Hofmann and Rosen (1980) have measured a sharp increase in the concentration of aerosol particles in the Junge layer. They point out that during recent years the content of particles has increased annually by 9 ± 2%. They consider anthropogenic emissions of carbonyl sulphide and carbon disulphide to be the possible cause of this phenomenon. This suggests that reduced sulphur compounds with long residence times could affect climate.

4.3   SOURCES OF ATMOSPHERIC SULPHUR

Sulphur enters into the atmosphere from other geospheres via a number of biotic and abiotic processes. Some of these fluxes are continuous in time, e.g.

sulphate flux from sea salts, while others are of a sporadic character, e.g. emissions of sulphur compounds during volcanic eruptions. The rate of sulphur input by biological processes depends on the temperature of the environment and, therefore, for temperate and high latitudes it undergoes seasonal variations. These fluxes are unaffected by human activity and therefore their values may be regarded as being constant during hundreds to thousands of years. During the last century, man's activity has made a considerable impact on the global geochemistry of sulphur and on its atmospheric cycle in particular.

Man-made effects on the environment are diversified and therefore it is often difficult to draw a distinction between natural and anthropogenic emissions. For example, as a result of agricultural activity, man has drastically changed the ecology of vegetation which in turn has affected the rates of sulphur input into the atmosphere by decomposition of organic matter on land.

As a result of natural and anthropogenic processes, various compounds of sulphur can enter the atmosphere. Biogenic sulphur is represented usually by its reduced compounds from which we single out hydrogen sulphide, dimethyl sulphide, carbonyl sulphide, and carbon disulphide. A great variety of sulphur compounds, e.g. reduced compounds, sulphur dioxide, and sulphates originate from anthropogenic activity. Here we consider the main fluxes influencing the global atmospheric sulphur balance.

4.3.1 Biogenic Emission from Coastal Regions and the Open Ocean

Gaseous reduced sulphur compounds can be formed by microbiological reduction of sulphates and various organic matter decomposition processes. Estimates of the oceanic and continental biogenic sulphur flux into the atmosphere vary from 34 TgS year^{-1} (Granat *et al.*, 1976) to 267 TgS year^{-1} (Eriksson, 1963). This large discrepancy in estimates is attributed to the fact that none of the authors had factual material on emissions of hydrogen sulphide. Nor did they consider data on the intensity of microbial processes in different ecosystems which may assist in the estimation of the possible upper limit of emission.

As can be seen from Tables 6.11 and 6.17, the most active reduction of sulphate occurs in periodically flooded and shallow parts of sea basins, especially in parts with considerable organic matter. Many investigators have commented on the odour of hydrogen sulphide over shallow parts of marine bays and coasts of saline lakes (Kuznetsov and Romanenko, 1968; Chukhrov *et al.*, 1975; Jørgensen *et al.*, 1978).

However, the emission of hydrogen sulphide into the atmosphere depends not only on the rate of its production but also on the rate of its removal by oxidation and pyrite formation. Hydrogen sulphide emission was comprehensively studied by Danish ecologists in Jutland on shallow sediments (Jørgensen *et al.*, 1978; Hansen *et al.*, 1978), who used plexiglas boxes equipped with measuring devices to estimate the emission. Sulphate reduction occurs in the sediments under study; in the daytime, part of the hydrogen sulphide formed is oxidized by photosynthetic bacteria in the surface sludge film and part is oxidized by oxygen emitted during photosynthesis by green plants in the water. Toward the end of the day, the activity of both groups of photosynthetic organisms falls and hydrogen sulphide increases in concentration, firstly in the water and then in the atmosphere close to the water. The maximum emission into the atmosphere is observed at night. After sunrise, the photosynthetic processes resume, the pH of the water increases, free oxygen appears in the water, and emissions to the atmosphere cease.

For the sandy sediments of Limfjorden containing 0.5–1.0% of organic matter in the form of *Zostera marina* detritus, the annual emission of hydrogen sulphide was estimated as 18 gS m^{-2}. In sediments of the small lagoon, Kano Vig, to the north of Aarhus, rich in the organic matter of decomposing algae, this value amounts to 450 gS m^{-2} (Hansen *et al.*, 1978).

The question arises as to whether these individual data can form the basis for calculating the emission of biogenic hydrogen sulphide into the atmosphere. We believe that they can be used to estimate the upper limit of the flux of biogenic sulphur, since the data obtained by Jørgensen *et al.* (1978) and Hansen *et al.* (1978) on the intensity of sulphate reduction agree well with the results of other investigators obtained on the coasts of the Baltic, Caspian, and Azov seas (see Tables 6.11 and 6.17).

Assuming, as did Karo (1956), that the length of the world's coastline is of the order of 5×10^5 km, and that the mean width of the littoral zone is 100 m , we obtain a value for the area of shallow sediments of $50 \times 10^3 \text{ km}^2$.

Assuming that 50% of these sediments are sands and 50% are highly enriched with organic matter or decaying algae, then the maximum total emission of hydrogen sulphide from such sediments is about 10 TgS year^{-1}.

Aneja *et al.* (1979a,b) determined experimentally the magnitude of the emission of reduced sulphur compounds (dimethyl sulphide, hydrogen sulphide, carbonyl sulphide, and carbon disulphide) from grass-covered saline marshes. The sulphur flux was mainly in the form of dimethyl sulphide and amounted to $\sim 0.66 \text{ gS m}^{-2} \text{ year}^{-1}$.

As the total area of marshes is $3.8 \times 10^5 \text{ km}^2$ (Aneja *et al.* 1979b) the global emission from marshes may be of the order of 0.25 TgS year^{-1}.

To obtain more reliable information on biogenic sulphur emissions into the atmosphere, it will be necessary to determine the flux in different geographic zones. It seems reasonable at present to assume a value for the flux of reduced sulphur in littoral (coastal) zones of the world's ocean of $5 \pm$ TgS year^{-1}.

In section 4.2.1, the amount of sulphur in the form of its reduced compounds in the clean maritime regions was estimated to be equal to 0.04 Tg. If we assume an average residence time of reduced sulphur compounds in the atmosphere to be one day (section 4.4.1) then the flux of these compounds from the open ocean will be ~15 TgS year^{-1}. In view of the uncertain concentration of reduced sulphur in the atmosphere over oceans, this flux is assumed to be 15 ± 15 TgS year^{-1}.

Thus, the total flux of reduced sulphur from the oceans is 20 ± 20 TgS year^{-1}. The great uncertainty of this estimate reflects the inadequacy of our knowledge. Nguyen et al. (1978), using the dimethyl sulphide concentration in the surface layer of sea-water, estimated this emission to the atmosphere to be 27 TgS year^{-1}. Maroulis and Bandy (1977), from direct measurements of dimethyl sulphide on the Atlantic coast, concluded that its emission from the world's oceans should not exceed 2 TgS year^{-1}. Graedel (1979) reviewing this work estimated the total emission of hydrogen sulphide and dimethyl sulphide by the world's oceans to be 30 TgS year^{-1}, and Hitchcock (1975) reported the dimethyl sulphide emission by algae to be 0.05 TgS year^{-1}. The range of these values indicates the uncertainty of our knowledge and this is reflected in our error estimate.

4.3.2 Biogenic Emission from Land

Formation of volatile sulphur compounds occurs mainly under the anaerobic conditions found in marshes (Bremner and Steele, 1978) and microorganisms play a leading role in this process. Emissions are likely to vary, therefore, with the temperature and moisture status of the environment and with the availability of nutrients.

The direct estimation of the biogenic flux from land is rather difficult since its value may vary in space and time by several orders of magnitude. Emission of reduced sulphur seems to be very low in arid regions, but in tropical marshlands it may be rather high. Field investigations of the emission of reduced sulphur by Farwell et al. (1979) showed that soil emitted mainly hydrogen sulphide along with carbon disulphide and carbonyl sulphide and negligible amounts of dimethyl sulphide, dimethyl disulphide, and methyl mercaptan. Bremner and Steele (1978), however, found that more was emitted as dimethyl sulphide, dimethyl disulphide, and methyl mercaptan: hydrogen sulphide was completely absorbed in the soil and transformed to metal sulphides (predominantly FeS). It also seemed likely that soil was the source of carbonyl sulphide.

We shall attempt to estimate the biogenic flux indirectly. As seen from Table 4.5 the content of sulphur with short residence time in the atmosphere over continents (except for the urban atmosphere) is estimated to be about 0.045 Tg.

If reduced sulphur with short residence time is assumed to have a lifetime of one day (see section 4.4.1), its flux into the atmosphere of continents should be 16 TgS year^{-1}. Considering the uncertainty of the estimate of reduced sulphur with short residence time in this reservoir, the flux may be within the range 3.5–30 TgS year^{-1}. Hitchcock (1975) estimated the flux of reduced sulphur into the continental atmosphere to be 2–5.4 TgS year^{-1}, which agrees with the lower limit of our estimate. Rodhe and Isaksen (1980) estimate the total flux of reduced sulphur with short residence time into the continental and oceanic atmosphere to be, at most, 40 TgS year^{-1}.

In conclusion we shall mention the estimate of Adams *et al.* (1979) of the emission of carbonyl sulphide from soil; 0.23 TgS year^{-1}. It is also noteworthy that carbonyl sulphide (and sulphur dioxide) may be formed during forest-fires, but we have made no estimate of the magnitude of this flux.

4.3.3 Aeolian Weathering of Sulphates from the Continental Surface

Almost all investigators of the global atmospheric sulphur cycle ignore this flux (Eriksson, 1963; Junge, 1965; Kellogg *et al.*, 1972; Friend, 1973) or consider it to be insignificant (Granat *et al.*, 1976). The calculation of Granat *et al.* (1976) was based on the estimate of Butcher and Charlson (1972) for the total emission of soil dust into the atmosphere (200 Tg year^{-1}). Granat *et al.* (1976) assumed that only one-third of this amount is submicron particles which are transferred over considerable distances, and they further assumed that the sulphur content of dust approximates that of weathered rocks (0.33%). Thus Granat *et al.* (1976) estimated the aeolian sulphur emission into the atmosphere to be 0.2 TgS year^{-1}, which is a negligible quantity when compared with the anthropogenic emission. At the same time they underline the possibility of a considerable contribution by wind erosion to the sulphur flux into the atmosphere at regional and local levels. There are difficulties in the estimation of aeolian sulphur emission similar to those found when estimating volcanic emissions. In both cases the emission varies significantly with space and time. We attempt to re-estimate the amount of aeolian sulphur emission based on observations of dust fluxes into the atmosphere, the sulphur content in the dust layer of soil and in the dust proper, and from the dust deposition in arid zones, etc.

The dust produced by wind erosion rises to the atmosphere in the arid zones of continents. The annual isohyet for 400 mm coincides reasonably well with the boundary of formation of dust-storms (Zakharov, 1965). The total area of these zones amounts to more than 37×10^6 km^2, of which 15.2×10^6 km^2 is in Eurasia, 13.7×10^6 km^2 in Africa, 4.2×10^6 km^2 in Australia, 3.6×10^6 km^2 in North America, and 0.2×10^6 km^2 in South America. Naturally, as dust-storms are occasional events, the dusty zone occupies substantially less area at certain times. There is no clear-cut definition of a dust-storm, and this impedes the interpretation of reports on dust-storms in different regions of the world. Dust-storms may occupy areas from several tens to millions of square kilometres. They vary also with time: local dust-storms last for less than an hour, but sometimes they may continue for two days and more. Berg (1947) noted that in Middle Asia in the Osh region there was only one day without dust during the whole summer of 1913. Depending on the colour of the weathered soil, we recognize black, red (brown, yellow), and white storms (Nalivkin, 1969). In the first case, chernozem soils are eroded; in the second, the soils are characteristically coloured loams, sands and clay; and in the third case when saline soils are eroded, the salts colour the dust particles white. Usually salts from saline soils contain considerable quantities of sulphates.

Estimates of dust fluxes into the atmosphere vary over a wide range and the situation is impeded by the fact that there is no clear agreement of the upper limit of the sizes of dust particles.

Table 4.8 summarizes the data on global rates of dust emissions into the atmosphere. We think that the majority of these estimates are not in accord with observed fluxes for aeolian material on to the earth's surface. For example, Kravchenko (1959) pointed out that 200 Tg of dust fall annually in the Volgograd and Saratov regions of the USSR (total area 0.2×10^6 km²). However, it should be noted that experimental methods for measuring dust deposition are inaccurate.

If we assume that such rates of deposition of aeolian material apply in the minimum case to one-third of the arid zones and in the maximum case to the whole area of the arid zones (37×10^6 km²), then we obtain a rate of aeolian weathering of 12 000 to 37 000 Tg year⁻¹. It is possible that a substantial fraction of this flux is of a local character, and is therefore not quite comparable to some of the other estimates. According to Matveev *et al.* (1976), an average value for deposition of dry dust on the territories of Stavropol and Krasnodar regions and Kalmyk ASSR is 75 km⁻² year⁻¹. This is a minimum value since the wash-out of dust with precipitation has not been taken into account. Using this rate and assuming the area of the global arid zone to be

Table 4.8 Flux of aeolian dust into the atmosphere

Flux (Tg year⁻¹)	Comments	Reference
200		Butcher and Charlson (1972)
500	All sizes	Peterson and Junge (1971)
200	$<5 \mu m$	Peterson and Junge (1971)
7–365		Hidy and Brock (1971)
200 ± 100		Joseph *et al.* (1973)
8000	$<20 \mu m$	Petrenchuk (1979)
15–75	Only in sub-Aral region	Grigoriev and Kondratiev (1979)
15	Only in Ukraine, 1928	Doskach and Trushkovsky (1963)
60–200	Only from Sahara	Morales (1979)
1000–3000		Present work

$12-37 \times 10^6 \text{ km}^2$, the rate of aeolian weathering is estimated to be $1000-3000$ Tg year^{-1}. The higher estimate (compared to others in Table 4.8) is reinforced by the data of Grigoriev and Kondratiev (1979) on dust emissions in the sub-Aral region which makes up no more than 1% of the area of arid zones of the globe. Petrenchuk (1979) estimated the dust emission into the atmosphere at the global level from the terrigenous material in precipitation (particles $<20 \mu$m) to be 8000 Tg year^{-1}. In subsequent work we will use the value obtained from the data of Matveev *et al.* (1976) of $1000-3000$ Tg year^{-1}. We find it unreasonable to consider only submicron fractions of dust, as was done by Granat *et al.* (1976) since larger particles are also transferred to distances of 1000 km or more.

Table 4.9 summarizes the data on the sulphur content in the dust layer of soil and in the aerosols of arid zones. The data show that saline soils can make a major contribution to the sulphur content of aerosols. These soils are abundant in arid zones (Gerasimov, 1959; Capot-Rey, 1953) and, according to the work of Kovda and Sabolsz (1980), they cover $9.5 \times 10^6 \text{ km}^2$. Here a clear-cut fractionation of sulphur with depth is observed. This is due to the fact that underground waters rise to the surface, evaporate, and enrich the uppermost layer of soil with salts containing sulphates. A conservative estimate is that the sulphur content of dust is within the range 0.3–0.7%. (The minimum value coincides with the estimate of Granat *et al.* (1976) for the average sulphur content of weathered rocks.) Considering the possible total emission of aeolian dust and its sulphur content, we estimate the aeolian emission to be $3-21$ TgS year^{-1}.

Zverev (1968) calculated the average deposition of sulphate in arid zones of the USSR. He singled out an arid zone with less than 377 mm of precipitation per year (area $3 \times 10^6 \text{ km}^2$) and a dry zone with less than 149 mm of precipitation per year (area $2.45 \times 10^6 \text{ km}^2$). The rate of dry deposition of sulphate was 1.4 gS m^{-2} year^{-1} in the first zone and 1.8 gS m^{-2} year^{-1} in the second.

The major part of this sulphur is of aeolian origin. Even if we assume that only half is due to aeolian emission we obtain a value for the sulphur flux in the USSR of 4.7 TgS year^{-1}. If the rate of sulphate deposition in arid zones of the USSR is characteristic of all arid zones of the world, we estimate the global aeolian emission to be 32 TgS year^{-1}.

The significance of aeolian emission in the input of sulphur into the atmosphere is illustrated by two white dust-storms in the region of the Caspian Sea on April 11–12 and 18–22, 1955 (Kravchenko, 1959). Routes of transfer and zones of dustfall are shown in Fig. 4.2. During April 11–12 the dust-haze was accompanied by the formation of an ash-grey film ~ 0.1 mm thick over the territories of Kalmyk ASSR and south of the Astrakhan region. Analyses of the sedimented material showed that water-soluble salts accounted for 47.4%, and consisted mainly of sulphates (90.6%). About 25 kg of sodium sulphate

Table 4.9 Sulphur concentration of soil dust layer and aerosols

Region	Sample	Concentration (% S)	Reference
Middle Asia	Sandy soil	0.009	Andreev and Lavrinenko (1968)
	Grey soil	0.1	Andreev and Lavrinenko (1968)
	Saline soil	1.87	Andreev and Lavrinenko (1968)
	Aerosol	0.7	Andreev and Lavrinenko (1968)
Ustyurt	Hard pan crust	0.12	Gerasimov (1959)
Kopet-Dag	Hard pan crust	0.23	Gerasimov (1959)
Gobi Desert	Coarse gravel gibber	0.7	Gerasimov (1959)
Sahara	Dense salt crust	5.67	Gerasimov (1959)
Uzbek SSR	Soil samples, mean	1.57	Khusanov et al. (1974)
Kalmyk ASSR	Aerosol during white dust-storm (transported from Aral–Caspian depression)	11.5	Kravchenko (1959)
Krasnodar	Sedimented dry dust	0.97	Matveev et al. (1976)
Novocherkassk	Sedimented dry dust	1.64	Matveev et al. (1976)
Salsk	Sedimented dry dust	4.59	Matveev et al. (1976)
Otkaznoe	Sedimented dry dust	1.50	Matveev et al. (1976)
Sahara	Aerosol	0.8	Calculated from data of Gravenhorst (1978)

fell on each hectare near Elista in 35 h. As seen from Fig. 4.2, Elista was at the very edge of the zone of dust fall-out during April 11–12. Undoubtedly, near the region of dust formation, the intensity of fall-out might be higher by an order of magnitude. Let us assume an average fall-out of 100 kg of sodium sulphate on a hectare or 2.25 gS m^{-2}. Assuming the area of sedimentation to be ~4 × 10^{11} m^2, then the amount of sulphur entering the atmosphere with aeolian dust during this storm was 1 Tg.

The white dust-storm during April 18–22 was larger. In this case the salt-haze covered the Astrakhan, Volgograd, Voronezh, and Saratov regions and the zone of dustfall extended more than 1500 km from the place of formation (Fig. 4.2). In the Volga region, the thickness of salt sediment was 1–2 mm, and over the remainder of the territory it was 0.5 mm. In some regions there were sediments 2–4 mm in thickness. The upper limit for the distribution of the dust-cloud was 3–4 km. Assuming an average thickness of sediment of 1 mm, then at a sediment density of 0.3 g cm^{-3} we obtain a total dustfall of 0.3 kg m^{-2}. Since this is the same region as that of the dust-storm during April 11–12, we may assume that the chemical composition of the dust is

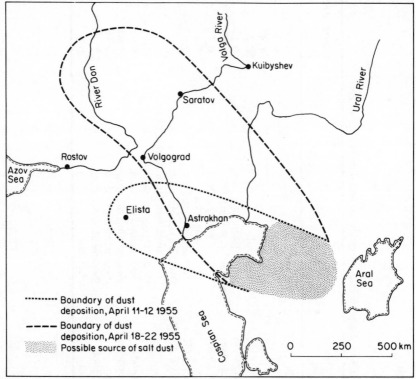

Fig. 4.2   Areas of dust deposition during the white dust-storms in April 1955

identical to that discussed previously. The area of fall-out, according to Fig. 4.2, amounts to $\sim 8 \times 10^{11}$ m^2. Therefore (after allowing for chemical composition), on April 18–22 1955, more than 30 Tg sulphur entered the atmosphere and were transferred a considerable distance.

Similar events happened in 1933 in the USA when huge amounts of sulphur were transferred more than 1000 km from the region between the Sierra Nevada and the Rocky Mountains (Fett, 1961), and in 1950 in the Lower Volga (Zamorsky, 1952). Unfortunately, in these cases no data are available on the chemical composition of the sulphur-rich dust or the density of the sediments.

Undoubtedly these events are rare. Their singularity consists not in the amount of salt dust entering the atmosphere but in the unusual stability of the atmospheric transport process which carried the salt-haze into regions where it does not usually occur. Local dust-storms with high salt contents very often occur in the Aralo-Caspian depression (Grigoriev and Kondratiev, 1979; Romanov, 1961). We conclude that at the global level, the previously estimated maximal value for the aeolian emission of sulphur (21 Tg year^{-1}) is closer to the true value than the minimal one (3 Tg year^{-1}).

In conclusion we estimate the sulphur flux with dust from the Sahara. A number of investigators (Büchen and Georgii, 1971; Gravenhorst, 1975, 1978) noted the increased concentrations of sulphate in air masses from the Sahara. Gravenhorst (1978) found that the sulphur content in these is ~ 0.5 μg m^{-3} higher than in clean Atlantic air. Assuming that the width of the dust-layer is 1000 km, that the average altitude of the layer is 2.5 km, and that the mean wind speed is 5 m sec^{-1}, and taking into account that half of the time the wind blows from the Sahara towards the Atlantic, the annual flux of sulphur from the Sahara to the Atlantic should be ~ 0.1 TgS year^{-1}. This is less than half of the estimate of Granat *et al.* (1976) for the global aeolian emission of sulphur. It should be noted that we did not estimate the emission but the flux of sulphur carried at a large distance from the source. Therefore, the estimate of Granat *et al.* (1976) is undoubtedly too low. Glaccum and Prospero (1980) present a hypothesis which conflicts with our conclusions. They suggest that gypsum is formed during the transport of Saharan dust by reaction of sea-salt calcite with sulphur dioxide. It is possible to improve the estimate of global aeolian emission by generalization of the experimental data for all arid zones of the globe. For the present we assume that the aeolian sulphur flux into the atmosphere is 20 ± 10 Tg year^{-1}.

4.3.4 Emission of Sea salt Sulphate from the Ocean

Significant amounts of sea salt containing sulphate enter the atmosphere as sea sprays. There are several quite contradictory estimates of this emission. Eriksson's (1960) estimate of 44 TgS year^{-1} is the most commonly cited figure, whereas the lowest estimate from the data of Bruevich and Ivanenkov (1971) is 130 Tg year^{-1} and from Grabovsky's (1956) work it is 700 Tg year^{-1}. Recalculation of Selezneva's data (1977) gives a value of 300 Tg year^{-1}.

Eriksson's (1960) estimate was based on the assumption that the ratio of sulphates in sea salts redeposited into the ocean and transferred to continents is the same as that in moisture evaporated over the ocean (i.e. 9 : 1). In our opinion, such an approach underestimates the magnitude of marine sulphate emission, since it is unreasonable to believe that these ratios should be the same for water vapour and aerosols. This becomes evident if we take into account the fact that the residence time of water vapour in the atmosphere is ~10 days (Junge, 1972), whereas that of aerosol particles containing sea salt is undoubtedly less. Of great importance in this case is the assumed upper limit for the sizes of salt particles.

The size of particles corresponding to the deposition rate in the atmosphere of ~10 cm sec^{-1} is assumed as a conditional limit. The residence time of such particles appears to be several hours; this is sufficient for their transport over considerable distances, and in particular, for their transfer to the continental atmosphere. Large particles differ from small ones not only in atmospheric residence time but also in composition (see below).

At present there are two views on the contribution of sea salt to the composition of atmospheric sulphates over both the ocean and land. The first stems from the known weight ratio between cations and anions in sea-water, which is practically invariable in any region of the world's ocean. It is assumed that this ratio does not change during the transfer of salts from the ocean to the atmosphere. The contribution of sea-water sulphates in any sample can be calculated from the relations $(SO_4^{2-})/(Cl^-)$, $SO_4^{2-})/(Na^+)$, and $(SO_4^{2-})/(Mg^{2+})$, in sea water via the formulae:

$$(SO_4^{2-})_{ex} = (SO_4^{2-})_{tot} - 0.14 \, (Cl^-)$$
$$(SO_4^{2-})_{ex} = (SO_4^{2-})_{tot} - 0.25 \, (Na^+)$$
$$(SO_4^{2-})_{ex} = (SO_4^{2-})_{tot} - 1.89 \, (Mg^{2+})$$

where $(SO_4^{2-})_{ex}$ is the quantity of 'excess' sulphate formed in the sample from non-marine sources and $(SO_4^{2-})_{tot}$ is the total sulphate in the sample. According to Junge (1965), the use of the ratio $(SO_4^{2-})/(Na^+) = 0.25$ is preferable since aerosol products may lack chloride due to its volatility. Many investigators use the ratio of magnesium to sulphur in sea-water. Such calculations presume the absence of any fractionation between the basic ions of sea salt on formation of sea-salt aerosols and assume that chloride, sodium, and magnesium ions are exclusively of marine origin.

Excess sulphates are found everywhere in the atmosphere when these relationships are used. According to the first view the excess sulphate formation over the ocean is usually explained by biogenic processes and emission of gaseous sulphur compounds from the ocean into the atmosphere followed by their oxidation to sulphates. Gravenhorst (1975) explains the availability of excess sulphates in the clean oceanic atmosphere by their formation in gas-phase reactions rather than by fractionation of chlorine and sulphate ions. To support his idea he adduces the fact that 'excess' sulphates are present as smaller aerosols (0.1–1 μm) than sea sulphates (0.5–10 μm).

The second view is based on the mechanism of ion fractionation during transfer of salts into the atmosphere. Bruevich and Kulik (1967) noted that during the evaporation of sea-water and the transfer of atmospheric precipitation from the ocean to land and their further transformation, the concentration of sulphates increases compared to chlorine. According to Livingstone (1963) the mean $(SO_4^{2-})/(Cl^-)$ ratio for river runoff not exposed to anthropogenic influence is 0.82. In rain-water over the Pacific and Indian oceans it is within the range 0.39–0.63, in Antarctic snow it is 0.80, and in rain-water over the Black Sea it is equal to 0.42. All of these ratios exceed the value of 0.14 characteristic of sea-water. From an analysis of the above data, Bruevich and Kulik (1967) concluded that sulphates are transferred predominantly to aerosols while chlorides remain in sea-water. Continental river runoff is rich in sulphate and compensates for this preferential loss from the ocean.

Changes in the ion ratio on transition from the ocean to the atmosphere are also characteristic of the other components of sea-water. To explain the fractionation, hypotheses have been put forward based on differences in atomic weights, ionic energy, ion hydration radii, and differences in surface and volume concentrations of sea salts. However, there is still no satisfactory explanation for the mechanism of fractionation. From analyses of many ions in sea salts, Korzh (1976) showed that the fractionation occurs while passing the ocean–atmosphere interface and the ratio of ions in the atmosphere is connected with their ratio in sea-water by the following empirical relation:

$$\left[\frac{C_A}{C_{Cl}}\right]_{atm} = \left[\frac{C_A}{C_{Cl}}\right]_{ocean}^{2/3}$$

where C_A and C_{Cl} are molar concentrations of any ion designated A and the chlorine ion, respectively. According to Korzh (1976), the contribution of sea salt sulphates to the composition of atmospheric aerosols should be calculated from the formulae:

$$(SO_4^{2-})_{ex} = (SO_4^{2-})_{tot} - 0.38\ (Cl^-)$$
$$(SO_4^{2-})_{ex} = (SO_4^{2-})_{tot} - 0.67\ (Na^+)$$
$$(SO_4^{2-})_{ex} = (SO_4^{2-})_{tot} - 2.71\ (Mg^{2+})$$

It should be noted that these formulae are not universal since they contradict many experimental results. The contradiction resides in the fact that the calculated contribution of sea sulphates may exceed the experimentally determined quantity of sulphates; i.e. the coefficients of 0.38, 0.67, and 2.71 are overestimated. This is observed when aerosol, precipitation, and dry deposition are sampled close to the coast. Discussion of these mechanisms has continued for many years and the problem is still far from solution (Granat *et al.*, 1976).

Changes in the composition of atmospheric moisture over continents and further changes in the composition of river runoff are indicative of the extent of continental participation. Korzh (1971) estimates that 15.9% of the sulphate in precipitation over the USSR is of oceanic origin, based on the assumptions that 100% of the chloride in precipitation over the USSR comes from the ocean and that the sulphate of oceanic origin can be calculated using the sulphate to chloride ratio observed in the ocean. Korzh (1971) also calculates, using identical assumptions about the chloride, that 16% of the sulphate in world river runoff is of oceanic origin. He further calculates that 28.1% of all salts in world river runoff are of oceanic origin. Other workers have estimated this latter component to be 14% (Sugavara 1964), 10% (Poldervaart 1957), and 6% (Clarke 1924). After considering these figures, we suggest that Korzh's (1971) estimate of 16% may be an upper limit for the percentage of sulphur of oceanic origin in world river runoff.

It seems reasonable to assume that oceanic sulphates form 10% of the total sulphates of river-water. Then, taking into account the sulphur content of river runoff (60 Tg year^{-1}), the amount of oceanic sulphur in river runoff returning to the ocean amounts to 6 Tg year^{-1}.

4.3.5 Anthropogenic Emission

The main human activities resulting in sulphur emissions to the atmosphere are the combustion of fossil fuels for the production of energy, smelting of ferrous and non-ferrous ores, oil processing, and production of sulphuric acid. Sulphur occurs in all fossil fuels, but its content varies widely. The 'purest' fuel with regard to sulphur is natural gas; on average the sulphur concentration is 0.05% S (by weight), although there are sources, for example, in Alberta, Canada which contain almost 80% of hydrogen sulphide in gas. In African oils (Algeria and Nigeria) the sulphur concentration is usually 0.3–0.5% by weight, whereas in Venezuelan oils it exceeds 5% (Santa-Olalla, 1973). The sulphur content of coals averages 2.2% (by weight) throughout the world (Bhatia, 1978) and may vary from fractions of one percent to 5% for some coals from Donbass and Ruhr deposits, and reach 8% or more in Kizelovsk coals, USSR.

In natural gases and oils, sulphur exists as hydrogen sulphide and organic compounds whereas in coals it is present as organic compounds, pyrites, and sulphate. The sulphate concentration in coals is not high, usually within

0.1–0.2% by weight. On combustion, organic sulphur and pyrite are oxidized to sulphur dioxide (and partially to sulphur trioxide) and together with flue gases are released into the atmosphere. It is generally accepted that 95% of the sulphur in fuel is released into the atmosphere on combustion (Kellogg *et al.*, 1972), and 96% of this is in the form of sulphur dioxide, the remaining 4% being sulphur trioxide (Kiyoura *et al.*, 1970).

In ores of non-ferrous metals, sulphur exists in the sulphide form (pyrites). The sulphur concentration in some pyrites reaches 45% (dry weight). During smelting, predominantly of copper, zinc, lead, and nickel, sulphide sulphur is oxidized to sulphur dioxide. If it is not utilized for other processes, it is emitted into the atmosphere.

The contributions of various industries to sulphur dioxide emissions differ depending on the country, reflecting the general extent of industrialization, the development of individual industries, the predominance of some types of fuel balances, etc. For example, Canada is characterized by a high level of emission from smelting of non-ferrous metals (65% of total emission; Brown, 1973). The sulphur emissions from large industrialized countries are characterized by a predominant contribution from energy production based on coal and oil as fuels. Table 4.10 shows the sources of anthropogenic sulphur emissions for the USA, England, the Federal Republic of Germany, and the USSR, from which it is evident that the emissions come mainly from energy production and metallurgical processes in each of these countries.

Anthropogenic emission is not constant with time. Its changes are mainly conditioned by the uneven consumption of fuels for heat and electrical energy. Three types of periodic fluctuations can be singled out: daily, weekly, and seasonal. The daily fluctuations of emission in the USA are about 20% (Lavery *et al.*, 1980). Weekly fluctuations are characterized by the marked decline in emission on Saturdays and Sundays, due to the decrease in the load of thermoelectric power stations. The seasonal variations are characteristic of Europe, with maximum emission in winter due to the heating of buildings and the prolonged use of electricity for illumination during the long hours of darkness. According to Barnes (1976), the ratio between winter and summer sulphur emissions is 1.6 : 1 in England and Wales; OECD (1977) estimated this ratio from maximum winter and minimum summer emissions for western Europe as ~2. An interesting change has been observed in the USA where maximum winter emissions occurred in the 1950s. Since 1960, the summer emissions have increased annually by 5.8% and winter emissions by 2.8%. This is due to the increase in energy consumption for air-conditioning during summer and it has resulted in maximum summer emissions in a number of regions of the USA (CAPITA, 1978).

Estimates of the global anthropogenic sulphur emission have been made by a number of authors, e.g. Robinson and Robbins (1968) and Cullis and Hirschler (1980). These estimates were based on statistical data for coal

Table 4.10 Source of anthropogenic sulphur dioxide emission in industrialized countries (TgS year⁻¹)

Type of production	USA Environmental Quality (1972); EPA (1975); Roderick (1975)		Great Britain Fjeld and Ottar (1975); Reay (1973)		West Germany Fjeld and Ottar (1975)		USSR Brodsky (1977); Solomatina (1977)	
Energy	10.65	(55.1)[a]	1.53	(53.0)	1.35	(58.5)	7.2	(52.8)
Petroleum treatment and refinery	1.65	(8.6)	0.2	(6.9)	0.15	(5.8)	0.3	(2.1)
Ferrous metallurgy	4.05	(20.8)	0.25	(7.8)	0.25	(10.8)	1.3	(9.7)
Non-ferrous metallurgy							2.25	(16.5)
Industrial processes	0.90	(4.7)	0.51	(17.8)	0.55	(24.9)	0.65	(4.8)
Coal processing			0.10	(3.0)			0.5	(3.7)
Others	2.1	(10.8)	0.3	(11.5)			1.4	(10.4)
Total	19.35	(100)	2.9	(100)	2.3	(100)	13.6	(100)

[a]Figures in parentheses denote %.

production, petroleum production and processing, metal smelting, and on emission factors per production unit. Rodhe (personal communication) has reviewed the independent estimates of emission and has prepared a diagram showing the increase in sulphur emission on a global scale during the last 120 years (Fig. 4.3). The global anthropogenic emission of sulphur in 1980 was estimated to be 110 TgS (Cullis and Hirschler, 1980). During combustion of fuels part of the sulphur may be emitted in the form of sulphate; the percentage depends on the ash content of the fuel and may vary from 2% for liquid fuels to 50% for lignites and shales with high ash content (Danilova and Dergachyov, 1977). From the data of Cullis and Hirschler (1980) and Danilova and Dergachyov (1977) it may be assumed that the anthropogenic flux of sulphur dioxide is 98 TgS year^{-1} and that of sulphate is 12 TgS year^{-1}, giving a total of 110 TgS year^{-1}. We consider that an uncertainty of ±15% should be assigned to these estimates.

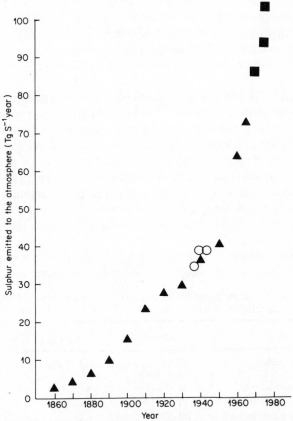

Fig. 4.3 Change with time in the global emission of sulphur from anthropogenic sources. ■ Cullis and Hirshler, 1980; ○ Katz, 1956; ▲ Robinson and Robbins, 1968

It is difficult to estimate the emissions of hydrogen sulphide and other reduced sulphur compounds into the atmosphere. There is one opinion that this emission is insignificant, because the range of hydrogen sulphide concentrations is even in space and has no time fluctuations. However, Georgii (1978), in considering measurements of hydrogen sulphide concentrations over the Federal Republic of Germany, concluded that the contribution of anthropogenic hydrogen sulphide to the total anthropogenic sulphur emission may be significant in certain regions. This conclusion was supported by Jaeschke *et al.* (1978) and Smith *et al.* (1961). In his comprehensive work devoted to the emission of organic sulphides from industrial sources, Bhatia (1978) singles out the following basic sources of reduced sulphur: gasification and enrichment of coal, wood working, oil processing, some synthetic material production, kraft-paper production, etc. Unfortunately, Bhatia (1978) presents no estimates of emissions except for the emissions of hydrogen sulphide from paper-mills which vary within the range 0.64–4.1 kgH$_2$S per t of dry pulp. More concrete estimations were made by Sitting (1975) (see Table 4.11). The data of Kalyuzhnyi (1961) and Jarzebski *et al.* (1971) on hydrogen sulphide emission from the ferrous industry and coke firing were used in the preparation of these estimates, as was the data of Kornel (1980) on paper production.

The value given in Table 4.11 is probably an underestimate of the total amount of anthropogenic emission of reduced sulphur into the atmosphere. Several processes have not been taken into account which must make a contribution to this flux, for example, production of plastics and synthetic fibres and oil processing. The possibility of significant emissions of hydrogen sulphide from fuel combustion has been pointed out by Sitting (1975), but no attempt has been made to quantify this flux. For global estimates, we shall assume the amount of anthropogenic flux of reduced sulphur with short residence time to be 3 ± 2 TgS year^{-1} which is in accord with the estimate of Rasmussen *et al.* (1975).

Anthropogenic emissions of carbonyl sulphide and carbon disulphide possibly occur in a number of industrial processes, but there are no reliable

Table 4.11 Anthropogenic emission of short-lived reduced sulphur in 1980

Industry	Compound	Emission (TgS year^{-1})
Paper-mills	Mercaptans H$_2$S, DMS	0.1–0.65
Coking	H$_2$S	0.07–0.7
Pig-iron smelting	H$_2$S	0.1–0.5
Cement production	H$_2$S	1.2
Total		1.5–3.05

estimates. According to Crutzen *et al.* 1979, carbonyl sulphide emission during biomass combustion, including naturally occurring fires, amounts to 0.24 TgS year^{-1} on a global scale. Hoffman and Rosen (1980) estimated the anthropogenic emission of carbon disulphide in the USA in 1973 to be 0.085 TgS year^{-1}. By 1980 this value may reach 0.1 TgS year^{-1}. Assuming that the carbon disulphide emission in the USA accounts for one-third of the global emission, we obtain a value of 0.3 TgS year^{-1} as a global estimate. In view of the paucity of data no attempt has been made to estimate the global anthropogenic emissions of reduced sulphur compounds with long atmospheric residence times.

4.4 PROCESSES OF SULPHUR REMOVAL FROM THE ATMOSPHERE

In this section we include all processes resulting in the elimination of a given sulphur species from the atmosphere, even if sulphur is not removed from the atmosphere in the process. Therefore, chemical transformations of sulphur dioxide and reduced sulphur compounds in the atmosphere are included. Where possible, we shall generalize observational data concerning the intensity of the various processes.

4.4.1 Oxidation of Reduced Sulphur Compounds in the Atmosphere

The atmosphere is a system with oxidative properties, and therefore all reactions involving valence changes lead to the formation of sulphate. Oxidation of reduced sulphur compounds may occur in gas-phase reactions, on the surface of solid particles, and in the liquid phase of cloud drops. The chemistry of these reactions has been described in detail by Heicklen (1976), Davis and Klauberg (1975), and Harrison *et al.* (1976). Here we shall concentrate on the estimation of rates of the most important chemical reactions of reduced sulphur resulting in its removal from the atmosphere.

The most important reactions for the gas-phase oxidation of reduced sulphur compounds appear to be those involving the hydroxyl (OH) radical. Measurements of these reaction rates from the work of Cox and Sheppard (1980) are presented in Table 4.12. We assume that the 24 h average concentration of OH in the lower troposphere is of the order of 10^6 molecules cm^{-3}. We use this to calculate the residence times due to gas-phase oxidation for the reduced sulphur compounds in the lower atmosphere and suggest that an average residence time of the order of one day is appropriate for the short-lived reduced sulphur compounds.

This is consistent with model estimates by Junge (1972) and Rodhe and Isaksen (1980). Somewhat unexpected in this connection are the results of Hitchcock *et al.* (1980) who found indications of a rapid formation of aerosol sulphuric acid near a natural source of reduced sulphur. This might possibly

Table 4.12 Reaction rates of reduced sulphur compounds with OH radicals in the troposphere (Cox and Sheppard, 1980)

Sulphur compound	Reaction rate constant (cm^3 molecule^{-1} sec^{-1})
DMDS	$(2.33 \pm 0.8) \times 10^{-10}$
MS	$(9.04 \pm 0.85) \times 10^{-11}$
DMS	$(9.1 \pm 1.4) \times 10^{-12}$
H_2S	$(5.0 \pm 0.3) \times 10^{-12}$
CS_2	$(0.43 \pm 0.16) \times 10^{-12}$
OCS	4×10^{-14}

be explained by the fact that the sulphuric acid precursor in this case was dimethyl disulphide or methyl mercaptan with residence time less than a day (Cox and Sheppard, 1980) (Table 4.12).

In accordance with the model suggested by Maroulis *et al.* (1980) concerning carbon disulphide oxidation, the final products may be sulphur dioxide and carbonyl sulphide. Carbonyl sulphide in turn is oxidized to sulphur dioxide. Let us assume average residence times for carbon disulphide of 70 days (Cox and Sheppard, 1980) and for carbonyl sulphide of 500 days (Kurylo, 1978). From the concentrations and residence times we can calculate the amount of sulphur dioxide formed by oxidation. This amounts to 1.5 TgS year^{-1} over land and 3.5 TgS year^{-1} over the ocean, giving a total production of 5 TgS year^{-1}.

4.4.2 Sulphur Dioxide Oxidation in the Atmosphere

Sulphur dioxide may be oxidized in the atmosphere in the gas phase, on the surface of soil particles, and in the liquid phase of droplets in clouds and fog. It is difficult to say which of these mechanisms makes the greatest contribution to the chemical transformation of sulphur dioxide at the global level, though the majority of investigators are now tilting the balance in favour of gas-phase oxidation. The kinetic aspects of sulphur dioxide oxidation in the atmosphere were discussed by Möller (1980).

The rates of various gas-phase oxidation reactions (Cox and Sheppard 1980) along with the likely atmospheric concentrations of the reactants, leads us to suggest that the reaction with OH is the primary one and reactions with HO_2 and CH_3O_2 are secondary. (The rate constant for the reaction of SO_2 with OH is $0.72 \pm 0.16 \times 10^{-12}$ cm^3 molecules^{-1} sec^{-1}, Cox and Sheppard 1980). It should be appreciated that the atmospheric OH concentration probably varies by orders of magnitude with cloudiness, time of day, season, and latitude (i.e. intensity of solar radiation). Direct measurements of OH have not been made on any extensive basis in the lower troposphere, but model calculations are readily available (e.g. Altshuller, 1979). As a consequence of this OH variation, we can expect the rate of sulphur dioxide oxidation to vary greatly depending on cloudiness, time of day, season, and latitude.

Direct observations have shown that the rate of chemical transformation of sulphur dioxide depends on the length of time that the polluted air masses have been present in the atmosphere. In a smoke plume near a source, the rate of sulphur dioxide chemical transformation is high and it decreases as the polluted air is carried away and diluted. In the plume, oxidation reactions on the surfaces of particles and catalytic and non-catalytic oxidation in the liquid phase of droplets are of importance.

Table 4.13 summarizes the rates of chemical transformation of sulphur dioxide under different conditions. These are expressed as rate constants, assuming first-order kinetics. The data are separated into three groups depending on the level of atmospheric pollution: background, regional, urban. When considering Table 4.13, proper allowance must be made for change in rates of sulphur dioxide oxidation to sulphates with time of day (Husar *et al.*, 1978) and season (Calvert *et al.* 1978). In this connection one should be careful when using experimental results obtained mainly in the day time and during warm seasons to estimate average results. The rate of chemical transformations in plumes may depend on the type of fuel burnt. According to Homolya and Fortune (1978) oxidation of sulphur dioxide to sulphate occurs 5–10 times faster during the combustion of oil than during coal burning. The authors suggest that this is caused by the catalytic action of vanadium and nickel, the content of which is substantially higher in oil than in coal. These data contradict those of Dlugi and Jordan (1979) who observed an extremely high rate of sulphate formation on the surfaces of coal-ash particles. They estimated the ability of ash particles to remove sulphur dioxide to be 0.07 and 0.095 $mgSO_2$ mg^{-1} ash at 30% and 60% relative humidity, respectively. This process could be of importance in the immediate vicinity of combustion sources. In the liquid phase of water drops, the oxidation of sulphurous acid to sulphuric acid occurs rather quickly, and the reaction therein may be of a catalytic (Penkett *et al.*, 1979) or non-catalytic (Hegg and Hobbs, 1979) character. In the latter case H_2O_2 and O_3 may be the dominant oxidants. Oxidation of sulphur dioxide usually yields sulphuric acid. The average rate of sulphur dioxide oxidation in the liquid phase is determined not only by the droplet chemistry and the rate at which sulphur dioxide can diffuse into the droplet, but also by the time taken for the sulphur dioxide molecule, after release into the atmosphere, to reach a cloud.

After considering the date of McMahon *et al.* (1976) and Calvert *et al.* (1978) we assume that the average oxidation rate of SO_2 in the oceanic and clean continental atmosphere is $\sim$0.0033 h^{-1}. Liquid-phase oxidation was not considered; therefore this estimate must be regarded as an upper limit. The oxidation rate increases substantially under conditions of regional atmospheric pollution. From a consideration of the results of McMahon *et al.* (1976), Eliassen (1978), Rodhe (1978), and Hidy *et al.* (1978), we assume the average oxidation rate of sulphur dioxide in the regional polluted atmosphere is 0.015 h^{-1}.

Table 4.13 Sulphur dioxide transformation rates in the atmosphere, h^{-1}

Type of pollution	Conditions	Rate constant	Reference
Background	Clean atmosphere over continents	0.01	McMahon et al. (1976)
	Average tropospheric conditions		
	January	0.001	Calvert et al. (1978)
	July	0.002	Calvert et al. (1978)
	Over ocean	0.033	Tsunogai (1971)
Regional	Non-urban regions	0.02	McMahon et al. (1976)
	Average European conditions	0.007	Eliassen and Saltbones (1975)
	Average European conditions	0.012	Eliassen (1978)
	Average European conditions	0.0036	Fisher (1978)
	Average European conditions	0.0125	Rodhe (1978)
	North-eastern USA	0.015	Hidy et al. (1978)
	Transfer of plume to 400 km		
	in daytime	0.01–0.04	Husar et al. (1978)
	at night	0.005	Husar et al. (1978)
	Transfer of plume to 300 km	0.02–0.03	Gillani (1978)
	Transfer of polluted air masses $3 < t < 40$ h	$(4 + 0.9t)^{-1}$	Korolev and Ryaboshapko (1979)
	Long-range transport of plume from electric power stations	0.074	Davis et al. (1974)

	Description	Value	Reference
Regional	Average conditions of industrial regions	0.02–0.08	Miller (1978)
	First day of plume transfer	0.14 ± 0.04	Alkezweeny and Powel (1977)
	Second day of plume transfer	0.10 ± 0.02	Alkezweeny and Powel (1977)
	18 h transport from Great Britain	Max. 0.01	Smith and Jeffrey (1975)
	Transfer of plume for 24 h (av.)	0.0025	Roberts and Williams (1979)
	(at night)	0.00	
	At 0°C	0.02	Ronneau and Snappe-Jacob (1978)
	At 20°C	0.04	Ronneau and Snappe-Jacob (1978)
Urban	Polluted atmosphere	0.01–0.1	Cox (1974)
	First hour of transport	1.83	Flyger and Fenger (1976)
	First 2–4 h of transport	2.0	Flyger et al. (1976)
	0–10 min of transport	0.18–3.7	McMahon et al. (1976)
	Distance from source ~20 km	0.067–0.1	McMahon et al. (1976)
	Los Angeles: photochemical smog	0.1	Henry and Hidy (1979)
	High humidity, low photochemical activity	0.04	Henry and Hidy (1979)
	Plume of electric power plant, 50 km (about 100 min of transport)	0.02	Forrest et al. (1979a)
	Plume of electric power plant, in light at noon	0.03	Davis et al. (1979)
	Average for 24 h	0.01	Davis et al. (1979)
	Transport from city during 8 h	Max. 0.04	Forrest et al. (1979b)
	Transport of polluted air masses from city	0.042	Chang (1979)
	Transport of polluted air masses from Budapest	0.1	Horvath and Bonis (1980)

4.4.3   Removal of Sulphur Compounds by Precipitation

It is customary to divide the removal of sulphur compounds from the atmosphere by precipitation into two processes: uptake into raindrops within clouds (in-cloud scavenging) and uptake into raindrops as they are falling to the ground beneath the clouds (sub-cloud scavenging).

It is difficult to distinguish between in-cloud and sub-cloud scavenging. Sulphur removal from the atmosphere by sub-cloud scavenging is greater in regions of intensive atmospheric pollution where most of the sulphur is found beneath the cloud layer of the atmosphere. Unfortunately, information on rates of in-cloud scavenging and the content of sulphur compounds in cloud-water and fog is meagre. The most extensive investigations of the chemical composition of cloud-water have been made in the Soviet Union by Petrenchuk (1979). He showed that, in the eastern region of the European part of the USSR, the concentration of sulphate sulphur in sub-inversion clouds reached 15 mgS litre^{-1}. The sulphate concentration in precipitation in this region was 3 mgS litre^{-1}. This suggests that rain drops have a high absorbing capacity and that sulphates may be concentrated in cloud-water during partial evaporation of cloud-drops. Rather high concentrations (1.5 mgS litre^{-1}) were observed in cloud-water of the clean marine atmosphere by Lazrus *et al.* (1970), thereby indicating a high efficiency of in-cloud scavenging. Junge (1965) commented on the very high (up to 17 mgS litre^{-1}) concentrations of sulphates in fog-water. These data are confirmed by Kirkaite *et al.* (1974) who summarized a great deal of data on the chemical composition of fog-water and showed that a high concentration of sulphates in fog-water is a widespread phenomenon. For example, they showed that in Lithuania, concentrations of sulphate in fog-water varied over the range 10–27 mgS litre^{-1}, while in rain-water they were 0.2–5 mg litre^{-1}. When mountain peaks are within clouds the deposition of sulphur in fog droplets may be rather high.

There are many studies of scavenging of sulphur dioxide and sulphates from the sub-cloud layer. Mechanisms of scavenging of sulphur dioxide and sulphates differ from each other: concentration and surface area of drops are decisive factors for sulphates; for sulphur dioxide, they also include the pH of rain- or cloud-water (Hill and Adamowicz, 1977). Sulphur dioxide scavenging and oxidation in the water phase have been described in detail by Adamowicz (1979) and Overton *et al.* (1979), who studied the effect of drop size pH of solution, ammonia, carbon dioxide, and ozone concentration in air, and the time taken for drops to fall to the ground, on sulphur dioxide oxidation in the liquid phase.

We may assume that the quantity of substance scavenged depends on the intensity of precipitation and its content in the sub-cloud layer, and that the rate of scavenging obeys first-order kinetics, where m is sulphate or sulphur dioxide mass in an air column of the sub-cloud layer, and λ is a proportionality

$$\frac{dm}{dt} = -\lambda m,$$

factor (often called the Langmuir factor) which depends on precipitation intensity. Naturally, as far as averge conditions of any region are concerned, not only the intensity of rains but the frequency of occurrence of precipitation should be taken into account. In this case, the average residence time with regard to scavenging or its reciprocal value, the scavenging factor K_{scav}, is given by

$$\tau_{scav.} = \frac{1}{K_{scav.}} = \frac{1}{f \cdot \lambda} + \tau_d$$

where f is the fraction of time when precipitation is occurring and τ_d is the time from an arbitrary moment during a dry period to the start of precipitation. Here, τ_d and f are dependent on the climatological characteristics of each region (Rodhe and Grandell, 1972).

Tables 4.14 and 4.15 summarize published data on λ and $\tau_{scav.}$ for sulphates and sulphur dioxide. They show that the uncertainty of our knowledge of this factor is quite high. When the precipitation intensity is 1 mm h^{-1} the estimates of λ_{SO_2} vary from 10^{-5} to more than 2.5×10^{-4}. An even wider range is found for sulphates, from 4×10^{-6} to 10^{-4}. If we take $f = 0.1$ for Central Europe, then typical λ values will lead to average residence times for sulphur dioxide

Table 4.14   Langmuir factors and mean residence times for scavenging of sulphate

Conditions	Langmuir factor (sec^{-1})	Mean residence time (h)	Reference
Europe, summer		100–300	Rodhe and Grandell (1972)
Europe, winter		30–80	Rodhe and Grandell (1972)
	$3 \times 10^{-5} I$		McMahon *et al.* (1976)
$I^a = 1$	3×10^{-5}		Chamberlain (1955)
5 μm particles	$1.6 \times 10^{-5} I^{0.8}$		Englemann (1965)
0.5 μm particles	10^{-6}		Slinn and Hales (1970)
	4×10^{-6}		Esmen (1972)
	8×10^{-5}		Makhon'ko (1967)
Europe, average conditions	10^{-4}		Fisher (1975)
Europe, average conditions	10^{-4}		Garland (1978)
Europe, average conditions		70	Rodhe (1978)
Europe, average conditions	$10^{-4} \sqrt{I}$		Scriven and Fisher (1975)

$^a I$ = intensity of precipitation (mm h^{-1}).

Table 4.15　Langmuir factors and mean residence times for sulphur dioxide removal in precipitation

Conditions	Langmuir factor (sec^{-1})	Mean residence time (h)	Reference
Europe, average conditions	$10^{-4}\sqrt{I}$	28	Scriven and Fisher (1975)
	10^{-4}		Makhon'ko (1967)
$I^a = 1$	$(1.2-6.2) \times 10^{-5}$		Okita (1972)
$I = 10$	$(8.4-44) \times 10^{-5}$		Okita (1972)
$I = 0.52$	2.53×10^{-4}		Högström (1974)
	1.58×10^{-4}		Enger and Högström (1979)
$I = 15$	7.4×10^{-4}		Beilke (1969)
	$6.0 \times 10^{-5} I$		McMahon *et al.* (1976)
$I = 1$	10^{-4}		Chamberlain (1960)
$I = 10$	3.5×10^{-4}		Chamberlain (1960)
Europe, average conditions	4×10^{-5}		Eliassen (1978)
Europe, average conditions	10^{-5}		Garland (1978)
Europe, average conditions	10^{-4}		Fisher (1978)
Europe, average conditions	4×10^{-5}	100	Rodhe (1978)
England	$3 \times 10^{-5} I$		Maul (1978)

^aI = intensity of precipitation (mm h^{-1}).

and sulphates, with regard to scavenging, of 60–130 and 70–200 h, respectively.

A number of authors have attempted to determine the chemical form of sulphur in newly fallen rain. Table 4.16 summarizes the results of these experiments.

From Table 4.16 it follows that about 80% of the sulphur in precipitation is present as sulphates. However, this does not mean that sulphates and sulphur dioxide are washed out by precipitation in this ratio, since during the time taken for drops to fall part of the sulphur dioxide is oxidized to sulphates.

Table 4.16　Proportion of sulphate and sulphite in raindrops reaching the ground

Sulphate (%)	Sulphite (%)	Reference
87–100	0–13	Högström (1979)
80–90	10–20	Hales and Dana (1979)
72	28	Davies (1979)—over city
83	17	—at distance from cities
	7 ± 4	Gravenhorst *et al.* (1980)

4.4.4 Sulphur Concentration in Precipitation

Traditionally, the sulphur content of precipitation is determined by sampling the rain-water and subsequent laboratory analysis, although some attempts have been made to use instruments based on ion-selective electrodes to determine sulphate concentrations during rainy periods. The procedures used for sampling and analysis are of extreme importance for a correct interpretation of results. More than 10 years ago it was widespread practice to use permanently open funnels, though in some cases samplers were open only during the collection of precipitation. Later, samplers with automatically opening caps were widely used. Collectors which are permanently open collect not only precipitation but also dry aerosol matter. According to Martin and Barber (1978) in East England, the amount of sulphate in precipitation is 2.4 times that collected during dry weather. In arid zones with intensive dust formation, dry deposition is substantially higher (Matveev *et al.*, 1976). When generalizing experimental data we shall use information gathered with both constantly open and automatically opening samplers. However, the sulphur content in precipitation estimated by the first technique will be taken as an upper limit.

In the present section, as in section 4.2.3, we shall attempt to distinguish the contribution by particles of sea salt using the sulphur : sodium and sulphur : chloride ratios characteristic of sea-water. This is valid for samples taken over the ocean or in coastal zones. In continental conditions such a calculation may yield erroneous estimates due to possible terrigenous contribution of sodium and increased volatility of chlorine (Junge, 1965). In arid zones this technique is absolutely unsuitable since the soils may contain large quantities of sodium and chlorine.

We shall now summarize data on sulphur in precipitation and estimate the sulphur washout from the atmosphere on a global scale. Table 4.17 presents published data on the sulphur content in precipitation of polar and Alpine regions. As expected, the concentration of sulphur in the precipitation of polar regions is very low. In Antarctica, no appreciable increase in concentration is observed in ice layers chronologically associated with the industrial epoch (last 40 years). In Greenland, however, the concentration has increased significantly during recent years. According to Koide and Goldberg (1971), the ice formed before 1940 contained 0.027 mgS litre^{-1}, while the seasonal snow (determined in the mid 1960s) contained 0.063 mgS litre^{-1}. Delmas (1979) found that during recent decades the concentration in Greenland ice increased from 0.03 to 0.05 mgS litre^{-1}. The sulphur concentration in Antarctic ice and snow is of the same order of magnitude as that in Greenland, varying from 0.02 to 0.12 mg litre^{-1}. From a consideration of the Antarctic and pre-industrial Greenland data, we may assume that the minimal concentration of non-anthropogenic sulphur in precipitation is ~0.05 mgS litre^{-1} in these regions. In all the other regions of the world, sulphur concentrations in precipitation will be higher.

Interesting data have been presented by Vilenskii and Koroleva (1973) on the intensity of sulphur deposition on the surface of the ice-shield of Antarctica. They showed that near the coast the deposition amounts to 0.04 g m^{-2} year^{-1} and that it decreases significantly with distance inland, due to the decrease in sulphur concentration in precipitation and the amount of precipitation in the heart of the continent. At a distance of 1000 km from the coast, the intensity of sulphur deposition was only 0.003 gS m^{-2} year^{-1}. The authors consider that the major part of this flux is dry deposition. In all Greenland and Antarctic data, the sulphur content reflects both fall-out with precipitation and dry deposition. When considering the global atmospheric sulphur cycle, the sulphur deposited in precipitation in polar regions may be neglected because of the low sulphur concentrations in precipitation and the insignificant amount of precipitation, especially in the continental regions of Antarctica.

In Table 4.17 we included data from the Caucasus glaciers (altitude 2.5–3.5 km). Matveev's data (1964) on Elbrus give substantially lower values than the later data. Matveev *et al.* (1976) obtained concentrations an order of magnitude higher (0.6 mg litre^{-1}) for Mt Cheget (~10 km from the sampling site on Elbrus and ~700 m lower) from routine samples taken by automatic samplers. The data of other authors are in good agreement (0.3–0.47 mg litre^{-1}) with the Mt Cheget concentration. These concentrations, an order of magnitude higher than those in polar regions, reflect the considerable impact of anthropogenic and natural constituents on the sulphur content of precipitation in this region.

Few data have been obtained routinely on the sulphur content of precipitation in oceanic regions far enough from land to exclude the possible admixture of dry deposition in the samples. The available information is summarized in Table 4.18. The data of Chukhrov *et al.* (1977) and Andreev and Rozhkova (1971) are not entirely of oceanic origin since the sampling sites are located on continental coasts and the precipitation chemistry is undoubtedly affected by some land trajectories. In addition, Andreev and Rozhkova's (1971) samples were taken using constantly opened samplers; therefore their data reflect the maximum possible values for concentration and fall-out with precipitation in oceanic regions. The data of Eriksson (1957), Tsunogai *et al.* (1972, 1975), and Junge (1965) were obtained from episodic observations and cannot characterize average annual conditions in sampling regions. However, episodic sampling of individual precipitation minimizes the impact of dry deposition, and thus these results give more representative values for sulphur concentration in precipitation.

From the above, we find that the range of possible values for total sulphur in precipitation at sea-level is from 0.3 to 1.0 mgS litre^{-1} and for excess sulphur, 0.1–0.5 mgS litre^{-1}. We assume an average value for total sulphur in precipitation over oceans of 0.5 ± 0.15 mgS litre^{-1} and for excess sulphur,

Table 4.17 Sulphur in precipitation, snow, and ice of polar and alpine regions

Region	Total sulphur (mg litre⁻¹)	Excess sulphur (mg litre⁻¹)	Element for calculation of excess S	Comments	Reference
Greenland	0.083			Ice of 1915–57, average	Junge (1960)
Greenland	0.063	0.061	Cl	Old snow during 1960–65	Koide and Goldberg (1971)
Greenland	0.027	0.024	Cl	Ice of 1300–1940	Koide and Goldberg (1971)
Antarctica	0.026	0.023	Cl	Old snow	Koide and Goldberg (1971)
Greenland	0.027			For last 800 years	Cragin et al. (1974)
Antarctica, Bird station	0.02			For last 1000 years	Cragin et al. (1974)
Antarctica	0.024			Old snow, 2–1070 km off shore	Delmas and Boutron (1978)
Greenland	0.03	0.027	Na	Ice of pre-industrial period	Delmas (1979)
Greenland	0.05	0.047	Na	Ice of pre-industrial period	Delmas (1979)
Greenland	0.016 ± 0.002	0.015	Na, Cl	Ice of pre-industrial period	Busenberg and Langway (1979)
Greenland	0.040 ± 0.009	0.039	Na, Cl	Ice of 1974–75	Busenberg and Langway (1979)
Antarctica, Vostok station	0.04			Annual precipitation	Doronin (1975)
Antarctic coast	0.12			Old snow	Vilenskii and Koroleva (1973)
Antarctica, Mirnyi–Vostok route	0.066			Old snow	Matveev (1961)
Caucasus, Mt. Elbrus 3700 m	0.06			Recent snow	Matveev (1964)
Caucasus, Mt. Cheget 3000 m	0.6			Fresh precipitation	Matveev et al. (1976)
Caucasus, glaciers	0.3			Seasonal snow	Supatashvili (1970)
Caucasus, glaciers	0.47			Fresh precipitation	Pkhalagava (1963)

Table 4.18   Sulphur in precipitation over oceanic regions

Region	Total sulphur (mg litre^{-1})	Excess sulphur (mg litre^{-1})	Element for calculation of excess S	Comments	Reference
Hawaiian Islands	0.64	0.19	Na	Many samples on one day, 10 m above sea-level	Eriksson (1957)
Pacific Islands	0.55	0.47	Na	One station, one day 80 m above sea-level	Tsunogai et al. (1972)
Sakhalin	0.23			Episodical snow sampling	Chukhrov et al. (1977)
Sakhalin	1.88	1.65		Routine sampling of dry and wet deposition	Andreev and Rozhkova (1971)
Okhotsk Sea coast	1.43	0.81		Routine sampling of dry and wet deposition	Andreev and Rozhkova (1971)
Petropavlovsk–Kamchatsky	1.95	1.73		Routine sampling of dry and wet deposition	Andreev and Rozhkova (1971)
Vladivostok	1.56	1.43		Routine sampling of dry and wet deposition	Andreev and Rozhkova (1971)
Bermuda Islands	0.71	0.13	Cl	Long-term observations	Junge (1963a)
USA, Florida coast	0.33	0.22			Junge (1963a)
USA, western coast	1.13	0.08			Junge (1963a)
Newfoundland	0.72	0.31			Junge (1963a)
Japan coast	0.45	0.41		Data of 1946	Koyama et al. (1965)
Northern Atlantic, central part	0.8	0.1	Na	Sampling on board ships Samples with high salt content and those related to transport from industrial regions of N. America are excluded	Nyberg (1977)

0.2 ± 0.1 mgS litre^{-1}. The intensity of precipitation (Korzun *et al.*, 1974) varies over an order of magnitude. The value of 1280 mm year^{-1} is taken as an average. From these we calculate the total sulphur flux with precipitation from the atmosphere over oceans (area 3.6×10^{14} m^2) to be 230 ± 70 TgS year^{-1}, and the excess sulphur to be 90 ± 45 TgS year^{-1}.

To estimate the sulphur flux with precipitation to the surface in clean continental regions, we use European data from the pre-industrial period and modern data obtained in isolated regions of other continents. Table 4.19 summarizes the concentrations and deposition for different continents. It is evident that, in Eurasia, concentrations vary widely from 0.3 to 4.7 mgS litre^{-1}. The high values are characteristic of zones of intensive dust formation (Matveev *et al.* 1976; Khasanov and Rakhmatullina, 1969). The data of Matveev *et al.* (1976) are especially interesting since they were obtained on a routine basis using automatic samplers which exclude dry deposition. The data of Drozdova *et al.* (1964), Andreev and Rozhkova (1971), and PGO (1970) were obtained by permanently opened samplers and therefore give the sum of dry deposition and precipitation. We consider that they provide an upper limit of possible concentrations, viz. 1.3 mgS litre^{-1}. Data of Vityn (1911), Votintsev (1954), and Granat *et al.* (1976) vary over the range 0.3–0.7 mgS litre^{-1}. The estimated sulphur concentration for Eurasia can be taken as 0.7 ± 0.2 mg litre^{-1}. Using this value and an average rate of precipitation over all continents of 400 mm year^{-1}, we calculate the intensity of fall-out to be 0.28 ± 0.08 gS m^{-2} year^{-1}.

In Table 4.19 the data of Whelpdale (1978b) and Junge (1965) for North America present a range of values similar to those over Eurasia. Thus for this continent we assume that the concentration is 0.7 ± 0.2 mgS litre^{-1} and wet deposition is 0.28 ± 0.08 gS m^{-2} year^{-1}. Unfortunately, we have no information on the chemical composition of precipitation in South America.

The data on Australia and Africa are quite variable. Overall, values of wet deposition for these continents are lower than those for Eurasia and North America. It is noteworthy, however, that arid zones in Australia and to a lesser extent in Africa occupy considerable areas.

Based on Table 4.19 and the above analysis, we shall assume that the average sulphur concentration in precipitation over the clean regions of continents is 0.7 ± 0.2 mgS litre^{-1}, and the intensity of wet deposition is 0.28 ± 0.08 gS m^{-2} year^{-1}. Taking into account the area of 'clean' regions of continents (133×10^{12} m^2) this estimate gives the sulphur flux with precipitation over 'clean' regions as 37 ± 10 TgS year^{-1}. For arid zones, we assume an average concentration of 3.0 ± 1.5 mgS litre^{-1}. With an average intensity of precipitation of 200 mm year^{-1} and an area of 5×10^{12} m^2, we calculate the sulphur flux with precipitation for arid zones to be 3.0 ± 1.5 TgS year^{-1}.

The chemical composition of precipitation in industrial regions has been studied quite intensively in nearly all developed countries. However, in many

Table 4.19 Sulphur concentration in precipitation and deposition rates over continents

Region	Concentration (mgS litre⁻¹)	Amount deposited (gS m⁻² year⁻¹)	Comments	Reference
Eurasia				
Smolensk region	0.33	0.22	Data for 1909–10	Vityn (1911)
Kuibyshev region	0.35	0.19		Vityn (1911)
Trans-Baikal region	0.32			Votintsev (1954)
Komi ASSR	1.2			Drozdova et al. (1964)
Khabarovsk territory	1.4		1957–62	Andreev and Rozhkova (1971)
Irkutsk region	0.83	0.33	1963–67	PGO (1970)
Central Yakutia	1.2	0.3	1962–63	PGO (1970)
Kalmyk ASSR	2.32	1.07	1962–65	Matveev et al. (1976)
Stavropol territory	2.13	0.92	1966–70	Matveev et al. (1976)
Magadan region	0.57		Dry deposition excluded	Chukhrov et al. (1977)
Trans-Baikal region	1.24		Episodic snow in 1975–76	Chukhrov et al. (1977)
Kazakhstan	1.07		Episodic snow in 1973–74	Chukhrov et al. (1977)
Aral Sea coast	4.7	0.49	Episodic snow in 1969–70	Khasanov and Rakhmatullina (1969)
Fergana valley	4.3	0.76	Dry and wet deposition in 1963	Khasanov and Rakhmatullina (1969)
Tashkent region	3.4	0.90		Khasanov and Rakhmatullina (1969)

Northern Sweden	0.4	0.16	Calculated value	Granat et al. (1976)
Great Britain	1.0	0.8	Excess sulphate calculated, 1881–87	Miller (1905)
America				
North-western USA	0.23–0.5		1955–56	Junge (1965)
Clean regions of North America	1.0–2.0		Excess sulphate calculated	Whelpdale (1978b)
Africa				
Nigeria	0.13	0.11		Bromfield (1974)
Uganda, Kampala	0.6	0.8		Visser (1961)
Zaire		0.28	Excess sulphate calculated	Eriksson (1966)
Kiyuhu		0.1	Excess sulphate calculated	Hesse (1957)
Equatorial eastern Africa	0.25		Monthly sampling at 8 points 1974–78. Excess sulphate calculated	Rodhe et al. (1981)
Australasia				
Australia		0.2	Excess sulphate calculated	Eriksson (1960)
Australia		0.33	Excess sulphate calculated	Drover (1960)
Western Australia		0.1	Offshore regions	Hingston and Gailitis (1976)
Western Australia		0.09	Clean continental regions	Hingston and Gailitis (1976)
Western Australia		0.6	Dusty regions	Granat et al. (1976)

Table 4.20 Sulphur concentration in precipitation and deposition rates over industrial regions

Region	Concentration (mgS litre^{-1})	Amount deposited (gS m^{-2} year^{-1})	Comments	Reference
Lithuanian SSR	1.5	2.7	Sum of dry and wet deposition	Shopauskas et al. (1976)
Moscow region	7.1		Episodic snow sampling	Chukhrov et al. (1977)
Moscow region		3.1	Sum of dry and wet deposition.	Fomin et al. (1977)
USSR, central European part	2.6		Continuous observation 1968–76 Sum of dry and wet deposition, 1958–62	Drozdva et al. (1964)
Minsk region	2.6	1.6	Sum of dry and wet deposition	PGO (1970)
Smolensk region	2.1	1.3	Data for 1962–1965	PGO (1970)
Central England	3.3	1.8		Prince and Ross (1972)
Eastern England	2.5	0.94	Data for 1976	Martin and Barber (1978)
North-western Europe	1.3	1.0	Data for 1973–74, numerous sampling points	OECD (1977)
Western and central Europe	1.75		Data for 1977–79, numerous sampling points	EMEP (1980)
USA, north-eastern part	0.9		Data for 1965–73	Likens and Bormann (1974)
USA, north-eastern part	0.7–1.0		Data for 1955–56	Junge (1965)
South-eastern Canada	1.0–1.7		Data for 1976	Whelpdale (1978a)
Nagoya region, Japan	1.5		Data for 1959	Koyama et al. (1965)

cases separate values for dry and wet deposition are not available. In this connection, the most reliable of all the data presented in Table 4.20 are those of OECD (1977) and EMEP (1980). Values for concentration and deposition obtained in the USSR may be considered to be the upper limits. Assuming the precipitation in industrial regions to be 600 mm year^{-1} and the average sulphur concentration in precipitation to be 1.5 ± 0.3 mgS litre^{-1}, we calculate the intensity of sulphur deposition as 0.9 ± 0.2 gS m^{-2} year^{-1}. We have estimated the area of industrial regions, including cities, to be 12×10^{12} m^2, and thus the sulphur flux with precipitation on industrial regions is 11 ± 2 TgS year^{-1}.

4.4.5 Uptake of Sulphur Dioxide by Underlying Surfaces

Dry deposition cannot be measured in a deposit gauge. It is, therefore, customary to infer deposition from other types of measurements. The most widely used and fundamental assumption about gas uptake at the earth's surface is that the rate of uptake per unit area of surface, U_{SO_2}, is proportional to the gas concentration in contact with the surface. The constant of proportionality is the inverse of a surface resistance, r_s. This resistance is assumed to be constant for a given surface in a given condition, but varies according to the physical, chemical, and biological properties of the surface, e.g. from water to bare soil to forest. When the gas concentration, C_{SO_2} is measured some distance above the surface (as it always is), then there is an additional resistance to gas uptake due to the air between the level of the concentration measurement and the surface, r_a. This resistance varies according to the intensity of turbulent mixing in the atmosphere and can be very high when turbulence is low, e.g. during a nocturnal inversion. These two resistances can be added. The inverse of their sum is called the deposition velocity, V_{SO_2} which relates the uptake at the surface to the concentration at a fixed height for a specific rate of atmospheric mixing according to the expression

$$U_{SO_2} = C_{SO_2}/(r_s + r_a) = V_{SO_2} \times C_{SO_2}$$

In practice some ill-defined 'average' conditions are assumed (see Galbally 1974). As mentioned above, the physical, chemical, and biological properties of the absorbing surface are important in determining uptake rates. The acid nature of sulphur dioxide and its high solubility lead to extremely rapid absorption by humid surfaces and surfaces with alkaline properties. For example, calcareous surfaces (under the same conditions) will absorb sulphur dioxide more rapidly than acidic surfaces, and moist foliage is a better absorber than dry foliage, etc. Brimblecombe (1978) pointed out that dew covering the foliage may serve to increase the uptake of sulphur dioxide.

Table 4.21   Rates of sulphur dioxide uptake by underlying surfaces

Surface	Uptake rate ($cm\ sec^{-1}$)	Method of measurement	Reference
Grass, 0.1 m high	0.5		
Shrubs, 1 m high	0.7		
Trees, 10 m high	0.8		
Calcareous soil	0.3–1.0		
Acidic soil, dry	0.4	Generalized data to 1977	*Proceedings of the International Symposium on Sulphur in the Atmosphere*, September 1977, Dubrovnik, Yugoslavia
Acidic soil, humid	0.6		
Dry snow	0.1		
Water or wet snow	1.0		
Natural landscape	0.8		
Urban landscape	0.7		
Wet grass	0.5		
Moderately humid grass	0.17	Box experiment on natural landscape	Milne *et al.* (1979)
Dry grass	0.11		
Surface of a semi-desert	0.1		
Farm lands	0.6	Eddy correlation	Galbally *et al.* (1979)
Forest	0.2		

Shepherd (1974) has observed seasonal variability in deposition velocities at one site, and both Husar *et al.* (1978) and Wys *et al.* (1978) reported diurnal variations in deposition velocities with values an order of magnitude lower at night than in the daytime. These latter results are in keeping with the night-time closing of stomata and the decrease in turbulence that frequently occurs near the earth's surface. These factors, both seasonal and diurnal, should be taken into account when estimating average uptake rates, since field determinations of absorption rates are usually performed for limited periods and usually during the daytime, and the results do not reflect mean seasonal and diurnal conditions.

Recent data, including the review of all data up to 1977 made at the Dubrovnik meeting, are presented in Table 4.21. Considering the factors discussed so far, we assume $V_{SO_2} = 0.8 \pm 0.2$ cm sec^{-1} for water surfaces.

It is more difficult to determine an average value for continental conditions with various types of underlying surfaces ranging from snow to humid tropical forests. Possible V_{SO_2} values lies within the range 0.1–2 cm sec^{-1}. For northern regions with lasting winter, and arid zones, the value will be low, 0.1–0.3 cm sec^{-1}. For southern regions, especially with humid climates, the values are high, 1–2 cm sec^{-1}. Consequently, we assume $V_{SO_2} = 0.6 \pm 0.2$ cm sec^{-1} as an average value for continental conditions. Industrial regions are mainly located in moderate latitudes and are characterized by snow in winter. Therefore, for industrial and urban regions, we shall assume a V_{SO_2} equal to 0.5 ± 0.2 cm sec^{-1}.

In Table 4.22 the sulphur fluxes from the atmosphere due to sulphur dioxide absorption by underlying surfaces are given. Areas of the various regions and sulphur dioxide concentrations in the atmospheres of these regions are taken from Table 4.5. The calculation shows that 11 ± 6 TgS year^{-1} is taken up by the surface of the ocean and 17 ± 10 TgS year^{-1} is sorbed by the land surface.

Table 4.22 Sulphur flux from the atmosphere due to sulphur dioxide uptake at the surface

Zone	Area $(10^{12}$ m$^2)$	Concentration $(\mu\mathrm{gS\ m}^{-3})$	Uptake (cm sec^{-1})	Flux (TgS year^{-1})
Oceanic zone without biota effect	305	0.1 ± 0.03	0.8 ± 0.2	8 ± 4
Oceanic zone with biota effect	55	0.2 ± 0.1	0.8 ± 0.2	3 ± 2
Continental	138	0.2 ± 0.1	0.6 ± 0.2	5 ± 3
Industrial–regional	11	5 ± 2	0.5 ± 0.15	9 ± 5
Urban	1	20 ± 10	0.5 ± 0.15	3 ± 2
Total				28 ± 16

4.4.6   Dry Deposition of Sulphates

As noted above in section 4.2.3, sulphates occur in particles of different size depending on their origin; 0.1–1 μm for sulphates formed in gas-phase reactions, 0.5–10 μm for sulphates of sea salts, and 1–100 μm for sulphates of aeolian origin. For particles of submicron sizes, sedimentation may be neglected and the rate of dry deposition is determined by the turbulence of the atmosphere and the properties of underlying surfaces. For large particles, e.g. particles of a aeolian origin, sedimentation is the dominating factor.

Where the particles are sufficiently small so that their sedimentation rate is negligibly small compared with the rates of turbulent motion of the atmosphere the theory of uptake of particles by underlying surfaces is similar to gases (discussed in section 4.4.5). There are, of course, differences in the transfer processes in the immediate vicinity of the surface and there are conflicting experimental data which suggest that we do not fully understand these processes. For the present, a simple approach using average deposition velocities is the best available.

The dry deposition of sulphates present as submicron particles in a layer of uniform distribution and constant height can be described by the exponential law

$$c(t) = c(0)\exp\left(\frac{V_{SO_4^{2-}} \times t}{H}\right) = c(0)\exp(-K_{SO_4^{2-}} \times t)$$

where $c(0)$ and $c(t)$ are sulphate concentrations at time 0 and t, respectively; $V_{SO_4^{2-}}$ is the linear rate of dry deposition of sulphate and H is the height of the uniform distribution. For large particles, the picture is more complex, but in our rather rough calculations we shall use constant values of the dry deposition velocity, $V_{SO_4^{2-}}$.

Table 4.23 summarizes the estimates of the rate of dry deposition of sulphates given by different authors. All estimates refer to sulphates formed from gas-phase oxidation of sulphur dioxide. From this data we assume a rate of dry deposition of sulphates of 0.2 cm sec^{-1} (range 0.1–0.5 cm sec^{-1}) in clean continental regions.

Sulphates existing in the marine atmosphere are represented by a mixture of particle sizes including larger particles than those formed over continents in gas-phase reactions. Naturally, the average rate of dry deposition of sea sulphates should be higher than the value of 0.2 cm sec^{-1}. Our estimate of the $V_{SO_4^{2-}}$ value for sulphates of sea salts is 0.5 ± 0.2 cm sec^{-1}, and that for excess sulphates over the ocean is 0.2 cm sec^{-1} (range 0.1–0.5 cm sec^{-1}).

When evaluating the deposition rates of sulphates of aeolian origin, we used an average residence time of dust-clouds in dry weather of one to two days (Zakharov, 1965). Thus the coefficient of dry deposition of aeolian sulphates, defined as the inverse magnitude of the average residence time of dust particles in a cloud, is $K_{SO_4^{2-}} = (0.6–1.2) \times 10^{-5}$ sec^{-1}.

Table 4.23 Dry deposition rates for sulphate sulphur

Size or type of particles	Rate (cm sec^{-1})	Comments	Reference
0.05–0.1 μm	0.1–1.0		Wesely *et al.* (1977)
0.1–1.0 μm	0.01		Clough (1973)
0.5 μm	0.1		Esmen and Corn (1971)
Anthropogenic sulphate	0.2 ± 0.16	Deposition on water surface	Sievering *et al.* (1979)
Anthropogenic sulphate	1.4 ± 0.4	Deposition on land surface near St. Louis, USA	Everett *et al.* (1979)
Anthropogenic sulphate	0.47	Hungary, average conditions	Meszaros and Varhelyi (1975)
Anthropogenic sulphate	0.1	United Kingdom, average conditions	Garland (1978)
Anthropogenic sulphate	0.2–0.5	USA	Rodhe (1978)

In section 4.2.3 we estimated the average altitude of the mixed layer of aeolian sulphates to be 1500 ± 500 m. Hence, for aeolian sulphates the calculated value of $V_{SO_4^{2-}}$ is 1.5 ± 0.9 cm sec^{-1}.

Table 4.24 presents data on the sulphur flux from the atmosphere via dry deposition of sulphates. Areas of different zones and concentrations for these zones are taken from Table 4.5.

The calculation shows that 17 TgS year^{-1} (range 3–48) is deposited dry into the ocean. The total dry deposition of sulphate on land amounts to 16 TgS year^{-1} (range 5–47); half of this is deposited in zones of intensive aeolian weathering.

Table 4.24 Removal of sulphur from the atmosphere by dry deposition of sulphate

Zone	Zone area (10^{12} m^2)	Concentration (μgS m^{-3})	Rate of deposition, (cm sec^{-1})	Sulphur flux (TgS year^{-1}) Average	Range
Oceanic					
excess sulphate	360	0.35 ± 0.15	0.2 (0.1–0.5)	8	2–28
marine sulphate	360	0.15 ± 0.10	0.5 (0.3–0.7)	9	1–20
Clean continental	133	0.6 ± 0.2	0.2 (0.1–0.5)	5	2–17
Dusty continental	5	4.0 ± 2.0	1.5 (0.6–2.4)	9	2–23
Industrial regions	11	3.0 ± 0.5	0.2 (0.1–0.5) }	2	} 1–7
Urban	1	4.0 ± 1.0	0.2 (0.1–0.5) }		
Total				33	8–95

4.5 ATMOSPHERIC BALANCE OF SULPHUR

In this section we collate the estimates of sulphur in the various regions of the atmosphere, the fluxes, and the rates of individual processes obtained in the previous sections. The balance is based on kinetics of the first order and time changes are neglected. This enables us to determine the missing values for fluxes or rates. The balance is compiled for each of the different regions of the atmosphere but, to facilitate the structure, the reservoirs of urban and industrial atmospheres are combined.

4.5.1 Sulphur Balance in the Atmosphere of Industrial Regions

In this section the atmospheres of all industrial regions are treated collectively, but it must be realized that many regions, spatially disconnected, exist. The balance for individual regions may differ in many respects from the generalized structure presented here, see for example those constructed for North Europe (Rodhe, 1972a), Europe (Meszaros *et al.*, 1978), north-eastern USA, and south-eastern Canada (CAPITA, 1978; Galloway and Whelpdale, 1980) and for the Donbass industrial complex in the USSR (Lysak and Ryaboshapko, 1978).

Figure 4.4 illustrates the atmospheric sulphur balance for some major

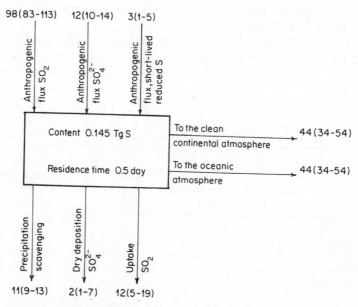

Fig. 4.4 Global sulphur balance in the atmosphere of industrial regions (TgS year^{-1})

industrial regions of the world. The input of sulphur is entirely anthropogenic and consists of sulphur dioxide and reduced sulphur compounds with short residence time (113 TgS year^{-1}; range 94–132). Removal processes are scavenging by precipitation, dry deposition of sulphates, and sulphur dioxide absorption by underlying surface (25 TgS year^{-1}; range 15–39). The uncertainty of the values for removal processes is 50%. From the balance it follows that horizontal advection from the reservoir is 88 TgS year^{-1} (uncertainty 25%). This sulphur enters both the continental and oceanic atmospheres. Let us assume the fluxes of anthropogenic sulphur to continental and oceanic atmospheres are each 44 ± 10 TgS year^{-1}. The sulphur content in the industrial regional atmosphere is 0.145 Tg (Table 4.5); given the total flux we obtain the average residence time for this reservoir of ~12 h.

4.5.2 Sulphur Balance in the Dusty Continental Atmosphere

Figure 4.5 depicts the sulphur balance for a dusty continental atmosphere. The only input here is the aeolian weathering of sulphur (20 ± 10 TgS year^{-1}). Removal processes are scavenging by precipitation and dry deposition of sulphate (12 TgS year^{-1}; range 3.5–27.5). The uncertainty of removal rates is >65%. To maintain a balanced condition it follows that the sulphur flux from the given reservoir is 8 TgS year^{-1}. This flux is directed towards the clean continental atmosphere. The residence time in this reservoir is ~14 h.

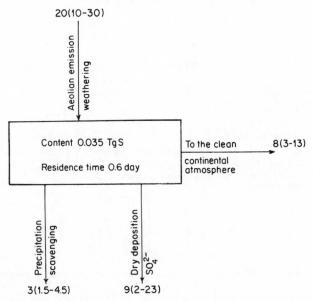

Fig. 4.5 Sulphur balance in the dusty continental atmosphere (TgS year^{-1})

4.5.3 Sulphur Balance for the Clean Continental Atmosphere

Figure 4.6 shows the sulphur budget for the clean continental atmosphere. Inputs are volcanic emissions, natural fluxes of sulphur with short residence time, sulphur dioxide formation (or H_2S) or oxidation of reduced sulphur with long residence time,[*] emissions from industrial-regional and dusty continental sources and sulphur flux from the ocean atmosphere. The last item is estimated in the following way. Aerosols are transferred into the continental atmosphere with the proportion of sea-salt sulphur to excess sulphur of 3 : 7 (see Table 4.5). The flux of sulphates of marine origin amounts to 6 ± 3 TgS and, from this, the flux of excess sulphates from oceanic to continental atmospheres is calculated to be 14 ± 7 TgS year^{-1}. Therefore, the total flux of sulphate sulphur from the oceanic atmosphere amounts to 20 ± 10 TgS year^{-1}. Volcanic emission was estimated in Chapter 2 to be 28 ± 14 TgS year^{-1}. It should be noted that zones of volcanic activity are grouped along the edges of continental shields. Hence, it is reasonable to assume that one half of the volcanic emission (14 ± 7 TgS year^{-1}) will be directed towards the clean continental atmosphere, and the other half towards the oceanic one. The flux resulting from oxidation of reduced sulphur is calculated from the total sulphur in the form of carbonyl sulphide and carbon disulphide in the atmosphere over continents and the residence times of these compounds, and amounts to 1.5 ± 0.5 TgS year^{-1}.

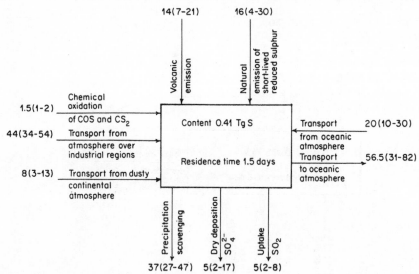

Fig. 4.6   Global sulphur balance for the clean continental atmosphere (TgS year^{-1})

[*]We neglected this flux in the reservoir of industrial regions and dusty continental atmospheres and attributed it to the reservoir of the clean continental atmosphere.

The sum of the inputs is 103.5 TgS year^{-1} (59–150) with an uncertainty of 45%. Removal processes are as follows: scavenging of sulphur by precipitation, sulphur dioxide absorption, and dry deposition of sulphate. The sum of these outputs is 47 TgS year^{-1} (range 31–72, uncertainty 45%). From balance calculations the flux to oceanic atmosphere should be 56.5 TgS year^{-1} (range 31–82, uncertainty 45%).

The residence time for sulphur in the clean continental atmosphere can be calculated from the total sulphur present (0.41 Tg, Table 4.5) and the flux value (103.5 TgS year^{-1}) to be 1.5 days ((0.41×365)/103.5 = 1.5 days).

4.5.4 Sulphur Balance in the Oceanic Atmosphere

Figure 4.7 shows the excess sulphur budget in the oceanic atmosphere. In compiling this budget use was made of average (i.e. the most probable) values of inputs and outputs. Our aim was to see whether excess sulphates over the oceans are the products of gas-phase reactions or due to fractionation of sea-salt ions on transition from the dissolved state in the hydrosphere to the aerosol state in the atmosphere. Input items here are the estimated volcanic emission, the natural emission of reduced sulphur with short residence time, and transfer of anthropogenic sulphur from the industrial atmosphere.

The flux due to carbonyl sulphide and carbon disulphide in oceanic atmospheres, estimated from their contents and residence times, is

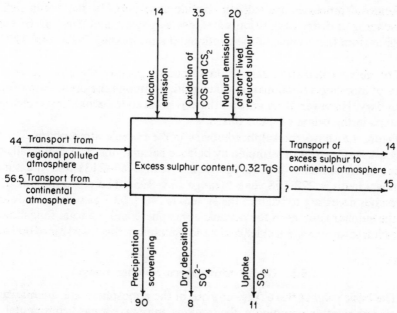

Fig. 4.7 Excess sulphur balance in the oceanic atmosphere (TgS year^{-1})

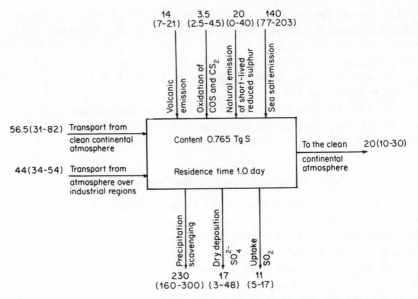

Fig. 4.8   Global sulphur balance in the oceanic atmosphere (TgS year^{-1})

3.5 ± 1 TgS year^{-1}. Therefore, the most probable input items amount to 138 TgS year^{-1}.

Removal processes are sulphur dioxide absorption by the ocean surface, scavenging and dry deposition of excess sulphates, and the flux of excess sulphur from the oceanic to the continental atmosphere. These total 123 TgS year^{-1}.

The above calculation cannot prove unambiguously the absence or presence of macro-ion fractionation on transition through the ocean–atmosphere boundary. However, it seems that there is no need to invoke any such theory to explain the origin of these fluxes.

Figure 4.8 shows the sulphur balance in the oceanic atmosphere. The flux of sea-salt sulphur is determined by balance calculations. The sum of all other input items amounts to 138 TgS year^{-1} (range 63–213) and the sum of removal items is 278 TgS year^{-1} (range 178–395). Therefore the flux of sea sulphates necessary to balance the system is 140 ± 63 TgS year^{-1}. Knowledge of the sulphur content in the oceanic atmosphere and the total flux allows us to estimate an average residence time for sulphur in this reservoir of one day.

4.5.5   Global Atmospheric Sulphur Budget

The basic sulphur fluxes into and out of the atmosphere are summarized in this section after combining the regions into two areas: continental and oceanic.

From Fig. 4.9, it follows that the total flux of sulphur into the atmosphere over land is 164.5 TgS year^{-1} and the efflux is 84 TgS year^{-1}, while over the ocean, the emission is 77.5 TgS year^{-1} and removal totals 258 TgS^{-1}. As the atmosphere as a whole should be a balanced system, if we assume that the sulphur flux from oceanic to continental atmosphere is 20 TgS year^{-1}, then

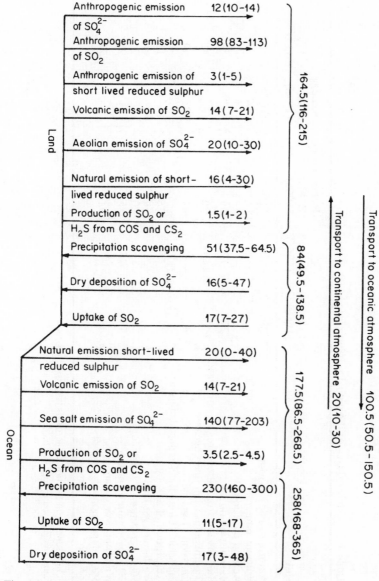

Fig. 4.9   Global atmospheric sulphur budget (TgS year^{-1})

Table 4:25 Major fluxes of the global sulphur cycle

Flux	Eriksson (1960)	Junge (1965)	Robinson and Robbins (1968)	Kellogg et al. (1972)	Friend (1973)	Granat et al. (1976)	Present work
Anthropogenic emission	39	40	70	50	65	65	113
Volcanic emission	—	—	—	1.5	2	3	28
Aeolian emission	—	—	—	—	—	0.2[a]	20
Bioemission on land	77	70	68	⎱ 90	58	—	16
Bioemission from littoral regions of oceans	⎱ 190	⎱ 160	⎱ 30		⎱ 48	5	5
Bioemissions from open oceans				⎰		27	15
Emission of sea salt	44	—[b]	44	43	44	44	140
Emission of long-lived reduced sulphur						—	5
Absorption of SO_2 by land	⎱ 77	⎱ 70	26	15	15	28	17
Wash-out over land			70	86	86	43	51
SO_4^{2-} deposition of land	57	55	20	10	20	—	16
SO_2 absorption by ocean	70	15	25	⎱ 72	25	10	11
Wash-out over ocean	⎱ 146	70	⎱ 71		⎱ 71	63	230
SO_4^{2-} deposition over ocean		60[b]		⎰		—	17

[a] Only submicron fraction of aerosols is considered.
[b] Junge calculated excess sulphate.

the reverse flux from the continental to oceanic atmosphere must equal 100.5 TgS year^{-1}.

Earlier, in section 4.2.4, we estimated the total amount of sulphur with short residence time to be 1.35 TgS. Now, given the value of the total sulphur flux or efflux as 342 Tg year^{-1}, we may estimate the average residence time of sulphur in the atmosphere to be $\cong$ 1.5 days.

In conclusion we compare the calculated global sulphur budget in the atmosphere with budgets obtained by the other authors (Table 4.25). This table shows that our budget agrees well with those of other authors. The major differences are the fluxes of volcanic activity, aeolian weathering, and emission of sea salts. In our opinion, volcanic emission was underestimated by other authors. Usually the volcanic emission estimated was associated with periodical eruptions, while the constant emission associated with fumaroles on drift fields and craters of resting volcanos was disregarded.

The ignoring of aeolian emission by other authors is reasonable since the transfer of aeolian sulphates is not global in nature. For this reason Granat *et al.* (1976) considered only the submicron fraction of aeolian dust, and estimated this flux to be only 0.2 TgS year^{-1}. For large regions, however, aeolian weathering is the major source of sulphur into the atmosphere and, therefore, we included it in the general scheme.

In our version of the atmospheric sulphur cycle, sulphur emission from the ocean together with sea salts is greatly enhanced compared to the values given by our predecessors. Almost all earlier workers used Eriksson's estimate without critical evaluation. The estimate of 140 TgS year^{-1} seems to us more probable than 44 TgS year^{-1}. This is evident from a comparison with the well-established value for sulphur washout over the oceans of 230 TgS year^{-1}.

4.5.6 The Global Sulphur Balance in the Pre-industrial Era

As experimental data on sulphur concentrations and sulphur deposition levels pertaining to the last century are not available, some assumptions must be made to estimate the pre-industrial sulphur balance. The anthropogenic fluxes of sulphur dioxide, sulphate and hydrogen sulphide should be negligible and are not considered in the balance. The volcanic emission should be unchanged and so we can use the estimate of 14 TgS year^{-1} for the emission to atmosphere over land and 14 TgS year^{-1} for that over the ocean. Aeolian emission may have undergone some changes during the last century, a view which is supported by an increased influx of salts to the atmosphere in the Aral Sea region due to increased agricultural and other human activities. However, quantitative estimates of any such change are not available and thus the flux of aeolian sulphates is considered to be unchanged (20 TgS year^{-1}). The same argument is used for the natural flux of short-lived reduced sulphur compounds; as no information is available to suggest that this flux was different prior to the industrial period its present value is used in the balance; viz. 16 TgS year^{-1}. The intensity of sulphur dioxide (or H$_2$S) formation from carbonyl sulphide and carbon disulphide during the pre-industrial period should be less important than it is now because some of the carbonyl sulphide and carbon disulphide emission is of anthropogenic origin. We assume that this flux in the pre-industrial period was 1 TgS year^{-1} over land, and 3 TgS year^{-1} over the ocean. Biogenic emission of short-lived reduced sulphur compounds and the emission of sea-salt sulphur should be unchanged, and thus the respective fluxes of 20 TgS year^{-1} and 140 TgS year^{-1} are used in the balance calculations. In addition the advective flux from the oceanic to the continental atmosphere should be unchanged at 20 TgS year^{-1}.

Assessment of the variation in runoff processes appears to be more difficult. One may assume that the sulphur dioxide concentration has not changed over the clean continental regions which means that sulphur dioxide uptake by underlying surfaces may be assessed as 5 TgS year^{-1} (Table 4.22). Dry deposition of sulphate in dusty regions (9 TgS year^{-1}) should remain the same since we have no evidence for a significant change in aeolian emission. With the assumption that without anthropogenic sources the concentration of sulphate in the continental atmosphere was 0.4 gS m^{-3} we can calculate the intensity of sulphate removal by dry deposition from the clean continental atmosphere as 4 TgS year^{-1}. Thus the total removal of sulphate by dry deposition over land amounts to 13 TgS year^{-1}. To assess the magnitude of sulphur scavenging by precipitation from the clean atmosphere, we make use of the facts that the anthropogenic component is now 43% of the total input (Fig. 4.6), and in the pre-industrial period the regions with clean continental atmosphere covered 145 × 10^{12} m^2 (Table 4.5). Thus the magnitude of sulphur wash-out over unpolluted regions in the past was probably of the order of 23 TgS year^{-1}. The intensity of sulphur scavenging in areas with dusty atmospheres should remain unchanged, i.e. 3 TgS year^{-1}. Therefore the flux of sulphur to land in the pre-industrial period should have been 26 TgS year^{-1}.

In summary, it may be concluded that in the pre-industrial period the total influx of sulphur to the atmosphere over land was 71 TgS year^{-1} and the output was 44 TgS year^{-1}. Therefore, the advective flux of sulphur from the continental to the oceanic atmosphere should have been 27 TgS year^{-1}. As the total supply of sulphur from the ocean amounts to 177 TgS year^{-1} the total flux from the atmosphere to the ocean must have been 184 TgS year^{-1}.

Sulphur dioxide uptake by the oceanic surface should have remained constant at 11 TgS year^{-1}, and it is assumed that the relationship between the intensities of sulphur dioxide uptake, dry deposition, and wash-out of sulphates in the pre-industrial and present-day periods remained the same. If so, then wet deposition over the ocean was 161 TgS year^{-1} and dry deposition was 12 TgS year^{-1} in the pre-industrial period.

Figure 4.10 presents the global sulphur balance for the atmosphere in the pre-industrial era. A comparison of Figs. 4.9 and 4.10 shows the marked changes that have taken place in the global atmospheric sulphur cycle under the impact of man. In the pre-industrial era the total flux of sulphur to the atmosphere appears to have been of the order of 228 TgS year^{-1}; the present value is 50% greater. Also in the former period the continental and oceanic components of the atmosphere were approximately balanced, whereas at the present time the advection of sulphur from continental to oceanic atmospheres is five times the flux in the reverse direction.

The changes which have occurred are especially marked in the atmosphere

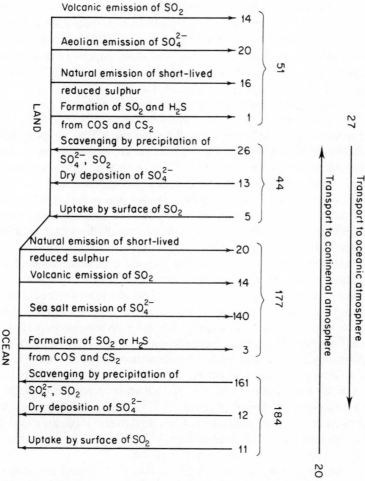

Fig. 4.10   Global atmospheric balance for the pre-industrial era (TgS year^{-1})

over land; anthropogenic emission represents about 70% of the total emission over land, and may account for more than 95% of the sulphur emitted into the atmosphere of industrial regions. Most of the anthropogenic emission is represented by sulphur dioxide which can be oxidized to sulphuric acid. As these two compounds are extremely reactive they have a greater impact on the environment compared to natural compounds such as sulphates of marine or aeolian origin.

4.6   THE ISOTOPIC COMPOSITION OF SULPHUR IN THE ATMOSPHERE

The isotopic composition of sulphur has been used with some success to determine the sources of sulphur emissions to the atmosphere. Two of the early studies (Östlund, 1959; Jensen and Nakai, 1961) were devoted to the source of sulphate in rain and snow, and later papers (e.g. Cortecci and Longinelli, 1970; Dequasi and Grey, 1970; Rabinovich, 1971; Holt *et al.*, 1972; Grey and Jensen, 1972; Castleman *et al.*, 1974; Ludwig, 1976; Chukhrov *et al.*, 1977) give examples which show how sulphur isotopic analysis can be applied to determine local sources of emission. These studies (some of which are summarized in Table 4.26) show that the $\delta^{34}S$ values of atmospheric precipitation and aerosols vary over a wide range, and it has been concluded that rain-water sulphate is depleted in ^{34}S with respect to sea-water sulphate. This effect is more pronounced in industrial areas (compare the results for industrial versus non-industrial regions; Jensen and Nakai, 1961).

It is evident that several sources may affect the isotopic composition and thus several examples are discussed in detail to show how this technique can be used in source identification.

Systematic investigations of sulphur dioxide concentration and isotopic composition in the air near sour-gas processing plants in Alberta, Canada, have been carried out by Krouse (1974). The sour-gas wells of Alberta are used to produce elemental sulphur; part of the hydrogen sulphide is oxidized to sulphur dioxide which is reacted with more hydrogen sulphide to yield elemental sulphur. During the process more than 1.2×10^9 gS day^{-1}, mostly as sulphur dioxide, are discharged into the atmosphere (Krouse, 1977). Data obtained since 1971 show that the sulphur oxides in the atmosphere have $\delta^{34}S$ values which range from +5 to +30‰, which is in the same range as the hydrogen sulphide processed (Krouse, 1977). Krouse (1974) concluded that emissions from different plants do not mix well over a large area and that the isotopic composition of the sulphur dioxide in the air depends on the type of raw material processed and the wind direction.

One example of the variation in $\delta^{34}S$ values and sulphur concentration in the air of the Ram River region is given in Fig. 4.11. The histogram of $\delta^{34}S$ values, which are related to the sulphur dioxide concentrations, has two

Table 4.26   Isotopic composition of sulphur in precipitation from different regions

Region	Isotopic composition ($\delta^{34}S$, %oo)		Reference
	Range	Mean	
Italy	-1 to $+3$	$+2.0$	Cortecci and Longinelli (1970)
Japan			
industrial regions	$+3.2$ to $+7.3$	$+6.0$	Jensen and Nakai (1961)
non-industrial regions	$+12.3$ to $+15.6$	$+14.0$	Jensen and Nakai (1961)
Sakhalin and Vladivostok region	$+4.2$ to $+8.6$	$+6.9$	Chukhrov *et al.* (1977)
Magadan region	$+3.9$ to $+7.6$	$+5.8$	Chukhrov *et al.* (1977)
Siberia	$+3.0$ to $+21.6$	$+10.6$	Chukhrov *et al.* (1977)
Kazakhstan			
steppe regions	$+2.1$ to $+3.7$	$+3.2$	
mountain regions	$+3.6$ to $+4.6$	$+4.1$	
Kirgizia, mountains	$+5.6$ to $+8.6$	$+6.9$	Chukhrov *et al.* (1977)
Tadzhikistan, mountains	$+3.8$ to $+12.9$	$+7.2$	Chukhrov *et al.* (1977)
Caucasus, mountains near near Elbrus	$+0.7$ to $+2.4$	$+0.7$ $+10.0$	Chukhrov *et al.* (1977) Chukhrov *et al.* (1977)
Kola Peninsula	$+3.8$ to $+6.2$	$+4.8$	Chukhrov *et al.* (1977)
Moscow and Novgorod regions	$+1.6$ to $+5.9$	$+4.0$	Chukhrov *et al.* (1977)
Urals	$+5.1$ to $+5.7$	$+5.4$	Gavrishin and Rabinovich (1971)
Rostov region	$+2.8$ to $+11.3$	$+7.1$	Rabinovich (1971)
Sweden	$+3.2$ to $+8.2$		Jensen and Nakai (1961)

peaks. For high sulphur dioxide concentrations the predominant $\delta^{34}S$ value is close to $+20$%oo which was typical of the emissions from the natural-gas processing plants. However, during a period when atmospheric concentrations were low, $\delta^{34}S$ values differed, often considerably, from those of the industrial source suggesting the existence of another source. The second peak, with value $+10$%oo, was considered to be due to a biological source.

Significant anthropogenic sulphur is observed in the atmosphere near ore-processing plants. The results of such a study for Salt Lake City (Table 4.27; Dequasi and Grey, 1970; Grey and Jensen, 1972) show a distinct difference in the isotopic composition of the atmosphere sampled before and during a strike by the smelter plant workers. During normal plant operation the isotopic composition of the sulphur in air was similar to that in the smelter plume. Studies made during the strike showed isotopic values within the range of biogenic sulphur. However, different values were obtained for the atmospheres on different sides of the study area. Results from one side seemed to be due predominantly to biogenic sulphur, but those on the other side sug-

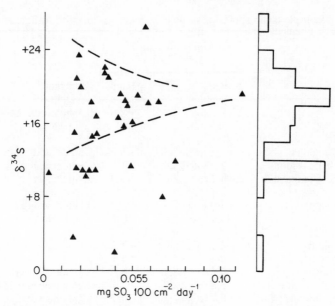

Fig. 4.11   Content and isotopic composition of sulphur in air over Ram River region, Alberta (Krouse, 1974)

Table 4.27   Isotopic composition for sources of atmospheric sulphur in the Salt Lake City area (Grey and Jensen, 1972)

Source	Isotopic composition ($\delta^{34}S$, ‰)		
	Range	Mean	Standard deviation
Automobile exhaust	+12.1 to +17.0	+15.1	
Crude oil		+16.6	
Great Salt Lake	+10.3 to +17.0	+15.3	0.5
Streams	+4.3 to +11.1	+9.6	0.8
Biogenic sulphur	+0.5 to +8.7	+5.3	1.7
Precipitation			
normal	−1.5 to +5.3	+2.2	1.0
during strike	+4.7 to +6.5	+6.0	
control[a]	+8.0 to +10.2	+9.0	0.3
Air samples			
normal	−1.0 to +3.1	+1.3	0.3
during strike	+5.0 to +7.5	+6.0	0.4
control[a]	+8.0 to +10.0	+9.0	0.2
Smelter emission	−3.8 to +3.4	+1.0	0.3
Copper ore	−4.3 to +2.4	0.0	0.6

[a]Control samples were collected far from the local sources of anthropogenic origin.

gested a mixture of biogenic sulphur (~90%) with refinery effluents and automobile exhausts (~10%).

As mentioned above, the precipitation from industrial regions of Japan had $\delta^{34}S$ values which averaged $+6^0/_{00}$ (Table 4.26; Jensen and Nakai 1961). Usually, the sulphur pollution of industrial countries is associated with the consumption of coal, but the widely differing $\delta^{34}S$ values of the precipitation and Japanese coal (average $+18^0/_{00}$) suggest another source. This source may be the large quantity of Persian Gulf oil ($\delta^{34}S = -5^0/_{00}$) used in Japan (Thode and Monster, 1970) which provides more than 60% of the total anthropogenic emissions to the atmosphere (Ryaboshapko and Erdman, 1978).

The data presented suggest an intensive local effect of anthropogenic sources on atmospheric sulphur, but an accurate assessment of the anthropogenic contribution can be made only if information on the isotopic composition of all pollution sources and the amounts of sulphur emitted to the atmosphere and deposited in precipitation is available.

Sulphur emission by natural processes such as aeolian transport, sea-salt spray, and biological activity has also been studied to explain the widely differing $\delta^{34}S$ values in precipitation, e.g. the very low values found for the Caucasus Mountains ($+0.7$ to $+2.4$) which cannot be attributed solely to anthropogenic effects (Table 4.26).

Rabinovich (1971) analysed 62 samples of rain and snow from the Rostov region and found that the $\delta^{34}S$ values for sulphate ($\sim+7^0/_{00}$) did not depend on the type of precipitation, the time of fall-out, or the direction of air movement. He concluded from the closeness of the $\delta^{34}S$ values for precipitation and soils that the sulphate in the lower layers of the atmosphere was principally of aeolian origin. Chukhrov *et al.* (1977) took the opposite view and suggested that the closeness of the values was due to the fact that the soil sulphur was derived from the atmosphere. This viewpoint would not explain the difference in isotopic composition of the precipitation from Siberia ($+10.5^0/_{00}$), Kazakhstan ($+3.2^0/_{00}$), or the mountainous regions of the Caucasus and the Pamirs.

In an attempt to understand the processes whereby marine aerosols often have a greater sulphur : chlorine ratio than that found in sea-water, Ludwig (1976) analysed samples of air taken from the San Francisco Gulf region for sulphur isotope composition. His results show that unpolluted marine air had sulphur isotope ratios that were significantly lower than those for samples subject to urban pollution. The low isotope ratio of marine air could not result from a mixture of sea-water sulphate and pollutant sulphur because these have higher isotope ratios (Fig. 4.12; Ludwig, 1976), and thus it is possible that the marine air aerosol material was of bacteriogenic origin. However, until more information is available on the isotopic composition of biogenic sulphur emission these results cannot be used to determine the importance of this source.

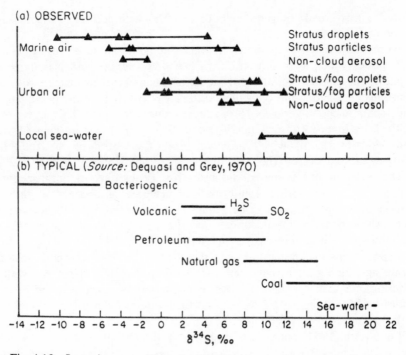

Fig. 4.12   Isotopic composition of sulphur in air over the San Francisco coast (after Ludwig, 1976)

Another complex problem is the determination of the contribution to the atmosphere of biogenic sulphur from continents. At present no data are available on the isotopic composition of sulphur produced during sulphate reduction in soils, swamps, and freshwater bodies.

ACKNOWLEDGEMENTS

The author gratefully acknowledges the advice and assistance of H. Rodhe, I. E. Galbally, and J. P. Friend during the preparation of this chapter, and the literature search made by D. I. Grigoryan.

REFERENCES

Adamowicz, R. F. (1979) A model for the reversible washout of sulfur dioxide, ammonia and carbon dioxide from a polluted atmosphere and the production of sulfates in raindrops. *Atmos. Environ.*, **13**, 105–121.

Adams, D. D., Farwell, S. O., Pack, M. R., and Bomsberger, W. L. (1979) Preliminary measurements of biogenic sulfur-containing gas emissions from soils. *J. Air Pollut. Control Assoc.*, **29**, 380–383.

Adams, F., Dams, R., Guzman, L., and Winchester, J. W. (1977) Background aerosol composition on Chacaltaya mountain, Bolivia. *Atmos. Environ.*, **11**, 629–634.

Alkezweeny, A. J., and Powel, D. C. (1977) Estimation of transformation rate of SO_2 to SO_4^{2-} from atmospheric concentration data. *Atmos. Environ.*, **11**, 179–182.

Altshuller, A. P. (1973) Atmospheric sulfur dioxide and sulfate distribution of concentration at urban and nonurban sites in United States. *Environ. Sci. and Technol.*, **7**, 709–712.

Altshuller, A. P. (1976) Regional transport and transformation of sulphur dioxide to sulphates in the US. *J. Air Pollut. Control. Assoc.*, **26**, 318–324.

Altshuller, A. P. (1979) Model predictions of the rates of homogenous oxidation of sulfur dioxide to sulfate in the troposphere. *Atmos. Environ.*, **13**, 1653–1661.

Andreev, B. G., and Lavrinenko, R. F. (1968) Some data on the chemical composition of atmospheric aerosols in the Middle Asia. *Meteorol. Gidrol.*, No. 4, 63–69 [in Russian].

Andreev, B. G., and Rozhkova, N. G. (1971) Study of chemical composition of precipitation over the Far East. In: *Synoptic Meteorology and Aerology*, Vol. 32, Hydrometeoizdat, Leningrad, pp. 213–227 [in Russian].

Aneja, V. P., Overton, J. H., Cupitt, L. T., Durham, J. L., and Wilson, W. E. (1979a) Direct measurements of emission rates of some atmospheric biogenic sulphur compounds. *Tellus*, **31**, 174–178.

Aneja, V. P., Overton, J. H., Cupitt, L. T., Durham, J. L., and Wilson, W. E. (1979b) Carbon disulphide and carbonyl sulphide from biogenic sources and their contributions to the global sulphur cycle. *Nature (Lond).*, **282**, 493–496.

Bach, W. (1976) Global air pollution and climatic change. *Rev. Geophys. Space Phys.*, **14**, 429–474.

Barnes, R. A. (1976) Long-term mean concentrations of atmospheric smoke and sulphur dioxide in country areas of England and Wales. *Atmos. Environ.*, **10**, 619–631.

Beilke, S. (1969) Neue Ergebnisse über das Auswaschen atmosphärischer Spurengase und Aerosole. *Ann. Meteorologie*, No. 4, 122–125.

Berg, L. S. (1947) *Climate and Life*. Geographizdat, Moscow, 242 pp. [in Russian].

Bhatia, S. P. (1978) Organosulfur emissions from industrial sources. In: Nriagu, J. O. (ed.), *Sulfur in the Environment*, Pt. 1, Wiley, New York, pp. 51–83.

Bolin, B., and Persson, C. (1975) Regional dispersion and deposition of atmospheric pollutants with particular application to sulfur pollution over western Europe. *Tellus*, **27**, 281–310.

Boueres, L. C. S., Adams, F. C., Winchester, J. W., Orsini, C. Q., Nelson, J. W., Dahill, T. A., and Lawson, D. R. (1977) Sulphur and heavy metals in South American urban and nonurban atmospheres. *Special Environmental Report*, No. 10, P. II, WMO No. 460, Geneva, pp. 102–108.

Breeding, R. J., Klonis, H. B., Lodge, J. P. Jr., Pate, J. B., Sheesley, D. C., Englert, T. R., and Sears, D. R. (1976) Measurements of atmospheric pollutants in the St. Louis area. *Atmos. Environ.*, **10**, 181–194.

Breeding, R. J., Lodge, J. P. Jr., Pate, J. B., Sheesley, D. C., Klonis, H. B., Fogle, B. T., Anderson, J. A., Englert, T. R., Haagenson, P. L., McBeth, R. B., Morris, A. L., Pouge, R., and Wartburg, A. F. (1973) Background trace gas concentrations in the central United States. *J. Geophys. Res.*, **78**, 7057–7064.

Bremner, J. M., and Steele, C. G. (1978) Role of microorganisms in the atmospheric sulfur cycle. *Adv. Microb. Ecol.*, **2**, 155–201.

Brimblecombe, P. (1978) 'Dew' as a sink for sulfur dioxide. *Tellus*, **30**, 151–157.

Brodsky, Yu. N. (1977) Sources of sulfur dioxide emission to the atmosphere. In: *Rationing and Control of Industrial Emissions to the Atmosphere*. Hydrometeoizdat, Leningrad, pp. 34–36 [in Russian].

Bromfield, A. R. (1974) The deposition of sulphur in the rainwater in Northern Nigeria. *Tellus*, **26**, 408–411.

Brosset, C. (1978) Water-soluble sulphur compounds in aerosols. *Atmos. Environ.*, **12**, 25–38.

Brown, D. (1973) National inventory of air pollutants. *Can. Paint and Finish.*, **47**, 29.

Bruevich, S. V., and Ivanenkov, V. N. (1971) Problems of the chemical balance in world oceans. *Okeanologiya*, **9**, No. 5, 835–841.

Breuvich, S. V., and Kulik, E. Z. (1967) Variation of the salt composition of oceanic water during its transfer to the atmosphere. *Dokl. Akad. Nauk SSSR*, **175**, No. 3, 697–699 [in Russian].

Büchen, M., and Georgii, H.-W. (1971) Ein Beitrag zum atmosphärischen Schwefelhaushalt über dem Atlantik. *'Meteor' Forschungsergeb.*, Reihe **B7**, 71–77.

Busenberg, E., and Langway, Ch. C. Jr. (1979) Levels of ammonium, sulfate, chloride, calcium, and sodium in snow and ice from southern Greenland. *J. Geophys. Res.*, **84**, 1705–1709.

Butcher, S. S., and Charlson, R. J. (1972) *An Introduction to Air Chemistry*. Academic Press, New York, 241 pp.

Cadle, R. D. (1975a) Volcanic emissions of halides and sulphur compounds to the troposphere and stratosphere. *J. Geophys. Res.*, **80**, 1650–1652.

Cadle, R. D. (1975b) The sulphur cycle. In: Parker, H. S. (ed.), *Sulphur in the Environment*. Missouri Botanical Garden and Union Electric Co. St. Louis, Missouri, pp. 1–13.

Cadle, R. D., Fisher, W. H., Frank, E. R., and Lodge, J. P. (1968) Particles in the Antarctic atmosphere. *J. Atmos. Sci.*, **25**, 100–103.

Cadle, R. D., and Grams, G. W. (1975) Stratospheric aerosol particles and their optical properties. *Rev. Space Phys.*, **13**, 475–501.

Calvert, J. G., Su, F., Bottenheim, J. W., and Strausz, O. P. (1978) Mechanism of the homogeneous oxidation of sulfur dioxide in the troposphere. *Atmos. Environ.*, **12**, 197–226.

CAPITA (1978) *Budget of Man-made Sulphur, Nitrogen and Hydrogen over the Eastern US and SE Canada Airshed*. Centre for Air Pollution Impact and Trend Analysis (CAPITA), Washington University, St. Louis N 63130, 38 pp.

Capot-Rey, R. (1953) *The French Sahara*. Presses Universitaires de France, 564 pp. [in French].

Castleman, A. W. Jr., Munkelwitz, H. R., and Manowitz, B. (1974) Isotopic studies of the sulfur component of the stratospheric aerosol layer. *Tellus*, **26**, 222–234.

Chamberlain, A. C. (1955) Aspects of travel and deposition of aerosol and vapour clouds. AERE, HP/R 1261.

Chamberlain, A. C. (1960) Aspects of the deposition of radioactive and other gases and particles. *J. Air Pollut.*, **3**, 63–88.

Chang, T. Y. (1979) Estimate of the conversion rate of SO_4^{2-} from the Da Vinci flight data. *Atmos. Environ.*, **13**, 1663–1664.

Chukhrov, F. V., Churikov, V. S., Ermilova, L. P., and Nosik, L. P. (1977) Isotope composition of atmospheric sulfur and its possible evolution in the history of the Earth. *Izv. Akad. Nauk SSSR, Ser. Geol.*, No. 7, 5–13 [in Russian].

Chukhrov, F. V., Ermilova, L. P., Churikov, V. S., and Nosik, L. P. (1975) On biogeochemistry of sulfur isotopes. *Izv. Akad. Nauk SSSR, Ser. Geol.*, No. 8, 32–48 [in Russian].

Chung, Y. S. (1978) The distribution of atmospheric sulfates in Canada and its relationship to long-range transport of air pollutants. *Atmos. Environ.*, **12**, 1471–1480.

Clarke, F. W. (1924) The Data of Geochemistry. *Government Printing Office*, Washington, 841 pp.

Clough, W. S. (1973) Transport of particles to surfaces *J. Aerosol Sci.*, **4**, 227–234.

Cortecci, G., and Longinelli, A. (1970) Isotopic composition of sulfate in rain water, Pisa, Italy. *Earth Planet. Sci. Lett.*, **8**, 36–40.

Cox, R. A. (1974) Particle formation from homogenous reactions of sulphur dioxide and nitrogen dioxide. *Tellus*, **26**, 235–240.

Cox, R. A., and Sheppard, D. (1980) Reactions of OH radicals with gaseous sulphur compounds. *Nature (Lond.)*, **284**, 330–331.

Cragin, J. H., Herron, M. M., Langway, C. C. Jr., and Klouda, G. (1974) Inter-hemispheric comparison of changes in the composition of atmospheric precipitation during the late Cenozoic era. Paper presented at ICSU/SCOR/SCAR Conference on Polar oceans, Montreal, May 1974.

Cronin, J. E. (1971) Recent volcanism and the stratosphere. *Science* (Wash. D.C.), **172**, 847–849.

Crutzen, P. J. (1976) The possible importance of COS for the sulfate layer of the stratosphere. *Geophys. Res. Lett.*, **3**, 73–76.

Crutzen, P. J., Heidt, L. E., Krasnec, J. P., Pollock, W. H., and Seiler, W. (1979) Biomass burning as a source of atmosphere gases CO, H_2, N_2O, NO, CH_3Cl and COS. *Nature (Lond.)*, **282**, 253–256.

Cullis, C. F., and Hirschler, M. M. (1980) Atmospheric sulfur: natural and man-made sources. *Atmos. Environ.*, **14**, 1263–1278.

Danilova, S. T., and Dergachyova, N. F. (1977) Assessment of emission to the atmosphere from thermal power stations. In: *Rate Setting and Control of Industrial Emissions to the Atmosphere*. Hydrometeoizdat, Leningrad, pp. 37–43.

Davey, T. R. A. (1973). The sulphur problem and the non-ferrous metal industries in Australia. Pt. 1. Contributory factors. *Process Chem. Eng.*, **26**, 20–27.

Davies, T. D. (1979) Dissolved sulfur dioxide and sulfate in urban and rural precipitation (Norfolk, U.K.). *Atmos. Environ.*, **13**, 1275–1285.

Davis, D. D., and Klauberg, G. (1975) Atmospheric gasphase oxidation mechanisms for the molecule SO_2. *Int. J. Chem. Kinet.*, **7**, 543–556. Discussion: 641–652.

Davis, D. D., Smith, G., and Klauberg, G. (1974) Trace gas analysis of power plant plumes via aircraft measurements: O_3, NO_x and SO_2 chemistry. *Science* (Wash. D.C.), **186**, 733–736.

Davis, D. D., Heaps, W., Philen, D., and McGee, T. (1979) Boundary layer measurements of the OH radical in the vicinity of an isolated power plant plume: SO_2 and NO_2 chemical conversion times. *Atmos. Environ.*, **13**, 1197–1203.

Delmas, R. (1979) Sulphate in polar snow and ice. Proc. Intern. Symposium on Sulphur Emissions and the Environment, London, 8–10 May 1979, Society of Chemical Industry (UK), pp.72–76.

Delmas, R., and Boutron, C. (1978) Sulphate in Antarctic snow: spatio-temporal distribution. *Atmos. Environ.*, **12**, 723–728.

Delmas, R., Baudet, J., and Servant, J. (1978) Mise en évidence des sources naturelles de sulfate en milieu tropical humide. *Tellus*, **30**, 158–168 [in French].

Dequasi, H. L., and Grey, D. C. (1970) Stable isotopes applied to pollution studies. *Amer. Lab. Dec.*, 19–27.

Dlugi, R., and Jordan, S. (1979) Formation of sulfuric acid and its sulfates on aerosol particles by catalytic reactions. *J. Aerosol Sci.*, **10**, 190–191.

Doronin, A. N. (1975) Chemical composition of the surface snow in the area of Vostok station and along the route Mirny–Vostok. *Informatsionny Bull. Sovetskoi Antarkticheskoi Ekspeditsii*, **91**, 62–68 [in Russian].

Doskach, A. G., and Trushkovsky, A. A. (1963) Dust storms in the southern part of the Russian plain. In: *Dust Storms and their Prevention.* Izv. Akad. Nauk SSSR, Moscow, pp. 5–30. [in Russian].

Drover, D. P. (1960) Accession of sulfur in the rainwater at Perth and Nedlands, Western Australia. *J.R. Soc. West. Aust.*, **43**, 81–82.

Drozdova, V. M., Petrenchuk, O. P., Selezneva, E. S., and Svistov, P. F. (1964) *Chemical Composition of the Atmospheric Fall-out over the European Part of the USSR.* Hydrometeoizdat, Leningrad, 242 pp. [in Russian].

EAJ (1974) *Quality of the Environment in Japan.* Environment Agency of Japan, Tokyo, 269 pp.

Eliassen, A. (1978) The OECD study of long range transport of air pollutants: long range transport modelling. *Atmos. Environ.*, **12**, 479–487.

Eliassen, A., and Saltbones, J. (1975) Decay and transformation rates of SO_2, as estimated from emission data, trajectories and measured air concentrations. *Atmos. Environ.*, **9**, 425–429.

EMEP (1980) *Cooperative Programme for Monitoring and Evaluation of the Long-range Transmission of Air Pollutants in Europe.* Summary Report Chem. Coordinating Centre for the First Phase of EMEP. Norwegian Inst. Air. Res. EMEP/CCC Report No. 4, 45 pp.

Enger, L., and Högström, U. (1979) Dispersion and wet deposition of sulfur from a power plant plume. *Atmos. Environ.*, **13**, 797–810.

Englemann, R. J. (1965) *The Calculation of Precipitation Scavenging.* USAEC Report BNWL-77, Battelle-Northwest Laboratory.

Environmental Quality, (1972) *The Third Annual Report of the Council on Environmental Quality.* Washinton. 6 pp.

EPA (1975) Position paper on regulation of atmospheric sulfates. US EPA-450/2-75-007, P.I.-XIX, pp. 1–88.

Eriksson, E. (1957) The chemical composition of Hawaiian rainfall. *Tellus*, **9**, 509–520.

Eriksson, E. (1960) The yearly circulation of chloride and sulphur in nature. *Tellus*, **12**, 63–109.

Eriksson, E. (1963) The yearly circulation of sulphur in nature. *J. Geophys. Res.*, **68**, 4001–4008.

Eriksson, E. (1966) Air and precipitation as sources of nutrients. In: Linser, H. (ed.), *Handbuch der Pflanzenernährung und Düngung.* Springer-Verlag, Vienna, pp. 774–792.

Esmen, N. R. (1972) *Particle Retention Efficiency of Scavenging Rainfall.* Div. Water, Air and Waste Chemistry, Am. Chem. Soc. Boston, Massachusetts.

Esmen, N. A., and Corn, M. (1971) Residence time of particles in urban air. *Atmos. Environ.*, **5**, 571–578.

Everett, R. G., Hicks, B. B., Berg, W. W., and Winchester, J. W. (1979) An analysis of particulate sulfur and lead gradient data collected at Argonne National Laboratory. *Atmos. Environ.*, **13**, 931–934.

Farlow, N. H., Snetsinger, K. F., Hayes, D. M., Lem, H. Y., and Topper, B. M. (1978) Nitrogen–sulfur compounds in stratospheric aerosols. *J. Geophys. Res.*, **83**, 6207–6211.

Farwell, S. O., Sherrard, A. E., Pack, M. R., and Adams, D. F. (1979) Sulfur compounds volatilized from soil at different moisture content. *Soil Biol. Biochem.*, **11**, 411–415.

Fedotov, S. A., Khrenov, A. P., and Chirkov, A. M. (1976) Great fissure eruption of the Tolbachinsky volcano in the Kamchatka in 1975. *Dokl. Akad. Nauk SSSR*, **228**, No. 5, 1193–1196 [in Russian].

Fett, W. (1961) *Atmospheric Dust.* Mir, Moscow. 336 pp. [in Russian].

Fisher, B. E. A. (1975) The long range transport of sulphur dioxide. *Atmos. Environ.*, 9, 1063–1070.

Fisher, B. E. A. (1978) The calculation of long term sulphur deposition in Europe. *Atmos. Environ.*, 12, 489–501.

Fisher, W. H., Lodge, J. P., Wartburg, A. F. and Pate, J. B. (1968) Estimation of some atmospheric trace gases in Antarctica. *Environ. Sci. Technol.*, 2, 464–466.

Fjeld, B., and Ottar, B. (1975) Draft report on the experience gained in determination of sources of SO_2 emission. UN EEC, Document ENN/WP.1/R.11, Geneva, 29 January, 1975.

Flyger, H., and Fenger, J. (1976) Conversion of sulphur dioxide in the atmosphere. *Fresenius Z. Anal. Chem.*, 282, 297–300.

Flyger, H., Heidam, N. Z., Hansen, K., Megaw, W. J., Walter, E. G., and Hogan, A. W. (1976) The background level of the summer tropospheric aerosol, sulphur dioxide and ozone over Greenland and the North Atlantic ocean. *J. Aerosol. Sci.*, 7, 103–140.

Fomin, P. I., Fomina, O. G., and Lazareva, R. P. (1977) Sulfur flux from the atmosphere in the Moscow Region. *Khim. Sel'sk. Khoz.*, No. 6, 29–31 [in Russian].

Forrest, J., Garber, R., and Newman, L. (1979a) Formation of sulfate, ammonium and nitrate in an oil-fired power plant plume. *Atmos. Environ.*, 13, 1287–1297.

Forrest, J., Schwartz, S. E., and Newman, L. (1979b) Conversion of sulfur dioxide to sulfate during the Da Vinci flights. *Atmos. Environ.*, 13, 157–167.

Francis, A. J., Duxbury, J. M., and Alexander, M. (1975) Formation of volatile organic products in soils under anaerobiosis. II. Metabolism of amino-acids. *Soil Biol. Biochem.*, 7, 51–56.

Friedlander, S. K. (1978) A review of the dynamics of sulphate containing aerosols. *Atmos. Environ.*, 12, 187–195.

Friend, J. P. (1966) Properties of the stratospheric aerosol. *Tellus*, 18, 465–473.

Friend, J. P. (1973) The global sulfur cycle. In Rasool, S. I. (ed.), *Chemistry of the Lower Atmosphere.* Plenum Press, New York, pp. 177–201.

Galbally, I. E. (1974) Gas transfer near the Earth's surface. *Adv. Geophys.*, 18B, 329–339.

Galbally, I. E., Garland, J. A., and Wilson, M. J. G. (1979) Sulfur uptake from the atmosphere by forest and farmland. *Nature (Lond.)*, 280, 49–50.

Galloway, J. N., and Whelpdale, D. M. (1980) An atmospheric sulfur budget for eastern North America. *Atmos. Environ.*, 14, 409–418.

Garland, J. A. (1978) Dry and wet removal of sulphur from the atmosphere. *Atmos. Environ.*, 12, 349–362.

Garland, J. A., and Branson, J. R. (1976) The mixing height and mass balance of SO_2 in the atmosphere above Great Britain. *Atmos. Environ.*, 10, 353–362.

Gavrishin, A. I., and Rabinovich, A. L. (1971) Possibility of use of sulfate sulfur isotope composition in natural waters as prospecting index of the pyrite ore formation in the Middle Urals. *Geokhimiya*, 7, 873–875 [in Russian].

Gedraitis, B. I. (1979) Measurement of the SO_2 background concentrations in the atmospheric air in regions remote from industrial centers. In: *Protection of the Atmosphere from Pollution.* Akad. Nauk Lith. SSSR, Institute of Physics, Mokslas, Vilnius, No. 5, pp. 78–84 [in Russian].

Georgii, H. W. (1970) Contribution to the atmospheric sulphur budget. *J. Geophys. Res.*, 75, 2365–2371.

Georgii, H. W. (1978) Large scale spatial and temporal distribution of sulfur compounds. *Atmos. Environ.*, 12, 681–690.

Gerasimov, I. P. (1959) *Essays on Physical Geography of Foreign Countries*. Geografizdat, Moscow, 358 pp. [in Russian].

Gillani, N. V. (1978) Project MISTT: mesoscale plume modeling of the dispersion transformation and ground removal of SO_2. *Atmos. Environ.*, **12**, 569–588.

Gillette, D. A., and Blifford, I. H. (1971) Composition of tropospheric aerosols as a function of altitude. *J. Atmos. Sci.*, **28**, 1199–1210.

Glaccum, R. A., and Prospero, J. H. (1980) Saharan aerosols over the tropical North Atlantic–mineralogy. *Mar. Geol.*, **37**, 295–321.

Goroshko, B. B., Zaitsev, A. S., and Nazarenko, V. Ya. (1968) Methodic problems of atmosphere pollution investigation using helicopter. In: *Atmospheric Diffusion and Air Pollution*. Hydrometeoizdat, Leningrad, pp. 85–92 [in Russian].

Grabovsky, R. I. (1956) *Atmospheric Nuclei of Condensation*. Hydrometeoizdat, Leningrad, 164 pp. [in Russian].

Graedel, T. E. (1979) Reduced sulfur emission from the open oceans. *Geophys. Res. Lett.*, **6**, 329–331.

Granat, L., Rodhe, H., and Hallberg, R. O. (1976) The global sulphur cycle. In: Svenson, B. H. and Söderlund, R. (eds), *Nitrogen, Phosphorus and Sulphur–Global Cycles*. SCOPE Report 7, *Ecol. Bull. (Stockholm)*, **22**, 89–134.

Gravenhorst, G. (1975) The sulphate component in aerosol samples over the North Atlantic. *'Meteor' Forschungsergeb.*, Ser. **B10**, 22–31.

Gravenhorst, G. (1978) Maritime sulfate over the North Atlantic. *Atmos. Environ.*, **12**, 707–713.

Gravenhorst, G., Beilke, S., Betz, M., and Georgii, H.-W. (1980) Sulfur dioxide absorbed in rain water. In: Hutchinson, T. C., and Havas, M. (eds), *Effects of Acid Precipitation on Terrestrial Ecosystems*. Plenum Press, New York, pp. 41–55.

Grey, D. C., and Jensen, M. L. (1972) Bacteriogenic sulphur in air pollution. *Science* (Wash. D.C.), **177**, 1099–1100.

Grigoriev, A. A., and Kondratiev, K. Ya. (1979) Atmospheric dust. Observations from outer space. *Nauka Zhizn*, No. 6, 88–92 [in Russian].

Guzman, G. T. L. (1977) Estation basica de contaminacion ambicutal en El Cerro Verde-El Salvador, primeros resultados. *Special Environmental Report*, No. 10, Pt. II, pp. 70–77. WMO, No. 460, Geneva.

Hales, J. M., and Dana, M. T. (1979) Regional-scale deposition of sulfur dioxide by precipitation scavenging. *Atmos. Environ.*, **13**, 1121–1132.

Hansen, M. H., Ingvorsen, K., and Jorgensen, B. B. (1978) Mechanisms of hydrogen sulfide release from coastal marine sediments to the atmosphere. *Limnol. Oceanogr.*, **23**, 68–76.

Harrison, H., Larson, T. V., and Hobbs, P. V. (1976) Oxidation of sulfur dioxide in the atmosphere: a review. In: *Int. Conf. Environ. Sensing and Assess.*, vol. 2. Las Vegas, Nev., 1975, New York, pp. 23-1/1–23-1/7.

Hegg, D. A., and Hobbs, P. V. (1979) The homogeneous oxidation of sulfur dioxide in cloud droplets. *Atmos. Environ.*, **13**, 981–987.

Heicklen, J. (1976) *Atmospheric Chemistry*. Academic Press, New York, XIV, 406 pp.

Heidorn, K. C. (1978) Sulfate and nitrate in total suspended particulate in Ontario. *J. Air Pollut. Control Assoc.*, **28**, 803–806.

Henry, R. C., and Hidy, G. M. (1979) Multivariate analysis of particulate sulfate and other air quality variables by principal components. Pt. 1. Annual data from Los Angeles and New York. *Atmos. Environ.*, **13**, 1581–1596.

Hesse, P. R. (1957) Sulphur and nitrogen changes in forest soils of East Africa. *Plant Soil*, **9**, 86–96.

Hidy, G. M., and Brock, I. P. (1971) An assessment of the global sources of tropospheric aerosols. In: *Proceedings of the 2nd International Clean Air Congress*. Academic Press, New York, pp. 1088–1097.

Hidy, G. M., Mueller, P. K., and Tong, E. Y. (1978) Spatial and temporal distributions of airborne sulfate in parts of the United States. *Atmos. Environ.*, **12**, 735–752.

Hidy, G. M., Mueller, P. K., Lavery, T. F., and Warren, K. K. (1979) Assessment of regional air pollution over the eastern United States: results from the sulfate regional experiment (SURE). In: *WMO Symposium on the Long-range Transport of Pollutants and its Relation to General Circulation Including Stratospheric/Tropospheric Exchange Processes.* Sofia, 1979, WMO, No. 538, Geneva, pp. 65–76.

Hill, F. B., and Adamowicz, R. F. (1977) A model for rain composition and the washout of sulphur dioxide. *Atmos. Environ.*, **11**, 917–927.

Hingston, F. J., and Gailitis, V. (1976) The geographic variation of salt precipitated over Western Australia. *Aust. J. Soil Res.*, **14**, 319–335.

Hitchcock, D. R. (1975) Dimethyl sulfide emissions to the global atmosphere. *Chemosphere*, **4**, 137–138.

Hitchcock, D. R., Spiller, L. L., and Wilson, W. E. (1980) Sulfuric acid aerosols and HCl release in coastal atmospheres: evidence of rapid formation of sulfuric acid particulates. *Atmos. Environ.*, **14**, 165–182.

Hoffer, T., Kliwer, J., and Moyer, I. (1979) Sulfate concentrations in the Southwestern Desert of the United States. *Atmos. Environ.*, **13**, 619–627.

Hofmann, D. I., and Rosen, I. M. (1980) Stratospheric sulfuric acid layer: evidence for an anthropogenic component. *Science* (Wash. DC) **208**, 1368–1370.

Högström, U. (1974) Wet fallout of sulphurous pollutants emitted from a city during rain and snow. *Atmos. Environ.*, **8**, 1291–1303.

Högström, U. (1979) Initial dry deposition and type of source in relation to long distance transport of air pollutants. *Atmos. Environ.*, **13**, 295–301.

Holt, B. D., Engelkemeir, A. G., and Venters, A. (1972) Variations of sulfur isotope ratios in samples of water and air near Chicago. *Environ. Sci. Technol.*, **6**, 338–341.

Homolya, J. B., and Fortune, C. R. (1978) The measurement of the sulfuric acid and sulfate content of particulate matter resulting from the combustion of coal and oil. *Atmos. Environ.*, **12**, 2511–2514.

Horvath, L., and Bonis, K. (1980) An attempt to estimate the rate constant of sulfur dioxide sulfate conversion in the urban plume of Budapest. *Idojaras*, **84**, 190–195.

Huebert, B. J., and Lazrus, A. L. (1980) Bulk composition of aerosols in the remote troposphere. *J. Geophys. Res.*, **85**, 7337–7344.

Husar, R. B., Patterson, D. E., Husar, J. D., Gillani, N. V., and Wilson, W. E. Jr. (1978) Sulfur budget of a power plant plume. *Atmos. Environ.*, **12**, 549–568.

ICM (1973) *International Comparison of Methods and Instruments of the Atmospheric Pollution Control at Stations of Background Measurements.* WMO. GUGMS, SSSR, 69 pp.

Inn, E. C. V., Vedder, I. F., Tyson, B. J., and O'Hara, D. (1979) COS in the stratosphere. *Geophys. Res. Lett.* **6**, 191–193.

Jacobs, M. B., Braverman, M. M., and Hochheiser, S. (1957) Ultra-microdetermination of sulphides in air. *Anal. Chem.*, **29**, 1349–1351.

Jaeschke, W., and Haunold, W. (1977) New methods and first results of measuring atmospheric H_2S and SO_2 in the ppb range. *Special Environmental Report*, No. 10, pp. 193–198. WMO, No. 460, Geneva.

Jaeschke, W., Schmitt, R., and Georgii, H. W. (1976) Preliminary results of stratospheric SO_2-measurements. *Geophys. Res. Lett.*, **3**, 517–519.

Jaeschke, W., Claude, H., Herrmann, J., and Vogler, D. (1979) Messung und verteilung von schwefelwasserstoff in reiner und verunreinigter atmosphäre. *Staub-Reinhalt. Luft*, **39**, 174–177 [in German].

Jaeschke, W., Georgii, H.-W., Claude, H., and Malewski, H. (1978) Contributions of H_2S to the atmospheric sulfur cycle. *Pure Appl. Geophys.*, **116**, 465–475.

Jarzebski, S., Grabowski, I., and Kapala, I. (1971) Principles of determination of the exhaust indices of harmful substances into atmospheric air during industrial processes. Part IV. The process of coal coking. Polish Academy of Sciences, Wroclaw–Warsaw, p. 66 [in Polish].

Jensen, M. L., and Nakai, N. (1961) Sources and isotopic composition of atmospheric sulphur. *Science* (Wash. DC), **134**, 2102–2104.

Jørgensen, B. B., Hansen, M. H., and Ingvorsen, K. (1978) Sulfate reduction in coastal sediments and the release of H_2S to the atmosphere. In: Krumbein, W. E. (ed.), *Environmental Biogeochemistry and Geomicrobiology*, vol. 1. Ann Arbor Science, Michigan, pp. 245–253.

Joseph, I. M., Manes, A., and Ashbel, D. (1973) Desert aerosols transported by khamsinic depressions and their climate effects. *J. Appl. Meteorol.*, **12**, 792–797.

Jost, D. (1970) Survey of the distribution of trace substances in pure and polluted atmospheres. *Pure Appl. Chem.*, **24**, 643–654.

Jost, D. (1974) Aerological studies on the atmospheric sulphur budget. *Tellus*, **26**, 206–212.

Junge, C. E. (1957) Chemical analysis of aerosol particles and of gas traces on the Island of Hawaii. *Tellus*, **9**, 528–537.

Junge, C. E. (1960) Sulphur in the atmosphere. *J. Geophys. Res.*, **65**, 227–237.

Junge, C. E. (1961) Vertical profiles of condensation nuclei in the stratosphere. *J. Meteorol.*, **18**, 501–509.

Junge, C. E. (1963a) *Air Chemistry and Radioactivity*. Academic Press, New York, 382 pp.

Junge, C. E. (1963b) Sulphur in the atmosphere. *J. Geophys. Res.*, **68**, 3975–3976.

Junge, C. E. (1965) *Chemical Composition and Radioactivity of the Atmosphere*. Mir, Moscow, 424 pp.

Junge, C. E. (1972) The cycle of atmospheric gases, natural and man-made. *Q.J.R. Meteorol. Soc.*, **98**, 711–729.

Junge, C. E., and Manson, J. E. (1961) Stratospheric aerosol studies. *J. Geophys. Res.*, **66**, 2163–2182.

Junge, C. E., Chagnon, C. W., and Manson, J. E. (1961) Stratospheric aerosols. *J. Meteorol.*, **18**, 81–108.

Kabel, R. L., O'Dell, R. A., Taheri, M., and Davis, D. D. (1976) *A Preliminary Model of Gaseous Pollutants Uptake by Vegetation*. The Pennsylvania State University, Center for Air Environment Studies. Publication No. 455–76.

Kadowaki, S. (1976) Size distribution of atmospheric total aerosols, sulphate, ammonium and nitrate particulates in the Nagoya area. *Atmos. Environ.*, **10**, 39–43.

Kalyuzhnyi, D. N. (1961) *Sanitary Protection of the Atmospheric Air from Emissions of Ferrous Industry*. Gos. Med. Izd. Ukr. SSSR, Kiev, 183 pp. [in Russian].

Karo, H. A. (1956) World coastline measurement. *Int. Hydr. Res.*, **33**, 131–140.

Katz, M. (1956) City planning industrial plant location and air pollution. In: Magill, P. L., Holden, F. R., and Ackley, C. J. (eds), *Air Pollution Handbook*. McGraw-Hill, New York, pp. 2-1–2-53.

Kellogg, W. W., Cadle, R. D., Allen, E. R., Lazrus, A. L., and Martell, E. A. (1972) The sulphur cycle. *Science* (Wash. D.C.), **175**, 587–596.

Khasanov, A. S., and Rakhmatullina, R. Sh. (1969) On the chemical composition of the atmospheric precipitations over Uzbekistan. In: *Hydrogeology and Engineering Geology of the Arid Zone of the USSR*. Fan, Tashkent, pp. 68–75 [in Russian].

Khusanov, G. Kh., Petrenchuk, O. P., and Drozdova, V. M. (1974) Chemical composition of the atmospheric aerosols in certain regions of Middle Asia. In: *Problems of Atmospheric Diffusion and Air Pollution*. Hydrometeoizdat, Leningrad, pp. 192–200 [in Russian].

Kirkaite, A. A., Shopauskas, K. K., and Gedraitis, B. I. (1974) Chemical composition of fog water. In: *Protection of the Atmosphere from Pollution*. Tr. Inst. Fiz. Mat. Akad. Nauk SSSR, 1, pp. 64–83 [in Russian].

Kiyoura, R., Kuronuma, H., Uwanishi, G. *et al.* (1970) Some opinions on sulphur dioxide as an atmospheric pollutant. *Bull. Tokyo Inst. Technol.*, **98**, 117–120.

Koide, M., and Goldberg, E. D. (1971) Atmospheric sulphur and fossil fuel combustion. *J. Geophys. Res.*, **76**, 6589–6596.

Kornel, K. (1980) Avilag papiripara 1978-ban. *Papiripar*, **24**, 22–26.

Korolev, S. M., and Ryaboshapko, A. G. (1979) Investigation of parameters of dilution and removal of sulfur dioxide and lead during transport above ocean. In: *WMO Symposium on the Long-Range Transport of Pollutants and its Relation to General Circulation Including Stratospheric/Tropospheric Exchange processes.* Sofia, WMO, No. 538, Geneva, pp. 117–124.

Korzh, V. D. (1971) Calculation of ratios of the chemical components of sea water transferred from the ocean to the atmosphere during evaporation. *Okeanologiya*, **11**, No. 5, 881–888 [in Russian].

Korzh, V. D. (1976) Chemical exchange of the ocean with the atmosphere as factor of salt composition formation in river waters. *Dokl. Akad. Nauk SSSR*, **230**, 432–435 [in Russian].

Korzun, V. I., Sokolov, A. A., Burynev, M. I. *et al.* (1974) *World Water Balance and Water Resources of the Earth.* Hydrometeoizdat, Leningrad, 638 pp. [in Russian].

Kovda, V. A., and Sabolsz, I. (1980) *Modelling of Salinization Processes in Saline Soil Formation.* Nauka, Moscow, 262 pp. [in Russian].

Koyama, T., Nakai, N., and Kamata, E. (1965) Possible discharge rate of hydrogen sulfide from polluted coastal belts in Japan. *J. Earth Sci. Nagoya Univ.*, **13**, 1–11.

Kravchenko, I. V. (1959) Saline dust storms. *Priroda* (Moscow), No. 8, 89–92 [in Russian].

Krouse, H. R. (1974) Sulphur isotope abundance studies of the environment. Paper presented at Canadian Sulphur Symposium *30 May–1 June.* University of Calgary.

Krouse, H. R. (1977) Sulphur isotope abundance elucidate uptake of atmospheric sulphur emissions by vegetation. *Nature (Lond.)*, **265**, 45–46.

Kurylo, M. J. (1978) Flash photolysis resonance fluorescence investigation of the reaction of OH radicals with OCS and CS_2. *Chem. Phys. Lett.*, **58**, 238–242.

Kuznetsov, S. I., and Romanenko, V. I. (1968) Microflora of Sivash and evaporative basins of salt-mining. *Mikrobiologiya*, **37**, No. 6, 1104–1108 [in Russian].

Lavery, T. F., Hidy, G. M., Baskett, R. L., and Mueller, P. K. (1980) The formation and regional accumulation of sulfate concentrations on the northeastern United States. In: Singh, J. J., and Deepak, A. (eds), *Proceedings of the Symposium on Environmental and Climate Impact of Coal Utilization* Academic Press, New York, 655 pp.

Lawson, D. R., and Winchester, J. W. (1978) Sulfur and trace element concentration relationships in aerosols from the South American continent. *Geophys. Res. Lett.*, **5**, 195–198.

Lazrus, A. L., Baynton, H. W., and Lodge, J. P. (1970) Trace constituents in oceanic cloud water and their origin. *Tellus*, **22**, 106–113.

Lazrus, A. L., Gandrud, B. W., and Cadle, R. D. (1971) Chemical composition of air filtration samples of the stratospheric sulfate layer. *J. Geophys. Res.*, **76**, 8083–8088.

Lazrus, A. L., Cadle, R. D., Gandrud, B. W., Greenberg, J. P., Huebert, B. J., and Rose, W. J. (1979) Sulfur and halogen chemistry of the stratosphere and of volcanic eruption plumes. *J. Geophys. Res.*, **84**, 7869–7875.

Levkov, L., Markez, A. G., and Sanfiel, R. P. (1975) Measurement of sulfate aerosols in the surface layer of the atmosphere over Cuba. *Hydrologia i meteorologia*, **24**, 75–77 [in Bulgarian].

Lezberg, E. A., Humenik, F. M., and Otterson, D. A. (1979) Sulfate and nitrate mixing ratios on the vicinity of the tropopause. *Atmos. Environ.*, **13**, 1299–1304.

Likens, G. E., and Bormann, F. H. (1974) Acid rain: a serious regional environmental problem. *Science* (Wash. DC), **184**, 1176–1179.

Livingstone, D. A. (1963) *Chemical Composition of Rivers and Lakes.* US Geol. Surv. Prof. Pap., 440-G, 64 pp.

Lodge, J. P., MacDonald, A. J., and Vihman, E. (1960) A study of the composition of marine atmospheres. *Tellus*, **12**, 184–187.

Lodge, J. P., Machado, P. A., Pate, J. B., Sheesley, D. C., and Wartburg, A. F. (1974) Atmospheric trace chemistry in the American humid tropics. *Tellus*, **26**, 250–253.

Lovelock, J. E., Maggs, R. J., and Rasmussen, R. A. (1972) Atmospheric dimethyl sulphide and natural sulphur cycle. *Nature (Lond.)*, **237**, 452–453.

Ludwig, F. L. (1976) Sulfur isotope ratios and the origins of the aerosols and cloud droplets in California stratus. *Tellus*, **28**, 427–433.

Lysak, A. V., and Ryaboshapko, A. G. (1978) Estimate of the atmospheric admixture fallout from a surface source. In: *Atmospheric Pollution as an Ecological Factor*, vol. 39. Hydrometeoizdat, Moscow, pp. 114–122 [in Russian].

McMahon, T. A., Denison, P. J., and Fleming, R. (1976) A long-distance air pollution transportation model incorporating washout and dry deposition components. *Atmos. Environ.*, **10**, 751–761.

Makhon'ko, K. P. (1967) Simplified theoretical notion of contaminant removal by precipitation from the atmosphere. *Tellus*, **19**, 467–476.

Maroulis, P. J., and Bandy, A. R. (1977) Estimate of the contribution of biologically produced dimethyl sulfide to the global sulfur cycle. *Science* (Wash. D.C.), **196**, 647–648.

Maroulis, P. J., Torres, A. L., and Bandy, A. R. (1977) Atmospheric concentration of carbonyl sulfide in the southwestern and eastern United States. *Geophys. Res. Lett.*, **4**, 510–512.

Maroulis, P. J., Torres, A. L., Goldberg, A. B., and Bandy, A. R. (1980) *Atmospheric SO₂ Measurements on Project GAMETAG.* Report, Chemistry Department, Drexel University, Philadelphia, pp. 1081–1082.

Martin, A., and Barber, F. R. (1978) Some observations of acidity and sulphur in rainwater from rural sites in central England and Wales. *Atmos. Environ.*, **12**, 1481–1487.

Matveev, A. A. (1961) Chemical composition of snow in the Antarctic according to observations on the Mirny–Vostok profile. In: *Hydrochemical Materials*, vol. 34. Hydrometeoizdat, Leningrad, pp. 3–11 [in Russian].

Matveev, A. A. (1964) Chemical composition of snow, ice and precipitation in the glacial area of Elbrus. In: *Hydrochemical Materials*, vol. 37. Hydrometeoizdat, Leningrad, pp. 10–22 [in Russian].

Matveev, A. A., Bashmakova, O. I., Tkacheva, V. I., and Krupenya, L. M. (1976) Assessment of precipitation of substances from the atmosphere with dust and rainfalls. In: *Proceedings of the 4th All-Union Hydrology Congress. Water Quality and Scientific Bases of Water Protection.* Hydrometeoizdat, Leningrad, pp. 261–270.

Maul, P. R. (1978) Preliminary estimates of the washout coefficient for sulfur dioxide using data from an East Midland ground level monitoring network. *Atmos. Environ.*, **12**, 2515–2517.

Meszaros, A. (1978a) On the concentration and size distribution of atmospheric sulfate particles under rural conditions. *Atmos. Environ.*, **12**, 2425–2428.

Meszaros, E. (1978b) Concentration of sulphur compounds in remote continental and oceanic areas. *Atmos. Environ.*, **12**, 699–705.

Meszaros, E., and Varhelyi, G. (1975) On the concentration, size distribution and residence time of sulphate particles in the lower troposphere. *Idojaras*, **79**, 267–273.

Meszaros, E., Varhelyi, G., and Haszpra (1978) On the atmospheric sulfur budget over Europe. *Atmos. Environ.*, **12**, 2273–2277.

Miller, D. F. (1978) Precursor effects on SO_2 oxidation. *Atmos. Environ.*, **12**, 273–280.

Miller, N. H. J. (1905) The amounts of nitrogen as ammonia and as nitric acid and of chlorine in the rain-water collected at Rothamsted. *J. Agric. Sci.*, **1**, 280–299.

Milne, I. W., Roberts, D. B., and Williams, D. I. (1979) The dry deposition of sulphur dioxide–field measurements with a stirred chamber. *Atmos. Environ.*, **13**, 373–379.

Möller, D. (1980) Kinetic model of atmospheric SO_2 oxidation based on published data. *Atmos. Environ.*, **14**, 1067–1076.

Morales, C. (1979) *Saharan Dust. Mobilisation, Transport, Deposition.* SCOPE Report 14. Wiley, Chichester, 320 pp.

Nalivkin, D. V. (1969) *Hurricanes, Storms, Tornadoes.* Nedra, Leningrad, 472 pp. [in Russian].

Natusch, D. F., Klonis, H. B., Axelrod, H. D. *et al.* (1972) Sensitive method for measurement of atmospheric hydrogen sulphide. *Anal. Chem.*, **44**, 2067–2070.

Nguyen, B. C., Bonsang, B., and Lambert, G. (1974) The atmospheric concentration of sulphur dioxide and sulphate aerosols over Antarctic, subantarctic areas and oceans. *Tellus*, **26**, 241–249.

Nguyen, B. C., Gaudry, A., Bonsang, B., and Lambert, G. (1978) Reevaluation of the role of dimethyl sulfide in the sulfur budget. *Nature (Lond.)*, **275**, 637–639.

Nyberg, A. (1977) On air-borne transport of sulfur over North Atlantic. *Q.J.R. Meteorol. Soc.*, **103**, 607–615.

OECD (1977) *The OECD Programme on Long Range Transport of Air Pollutants—Measurements and Findings.* Organization for Economic Cooperation and Development, Paris, 326 pp.

Okita, T. (1972) Calculation of rate of absorption of sulphur dioxide by rain and cloud-droplets. *Bull. Inst. Public Health (Tokyo)*, **21**, 9–13.

Omstedt, G., and Rodhe, H. (1978) Transformation and removal processes for sulfur compounds in the atmosphere as described by a one-dimensional time-dependent diffusion model. *Atmos. Environ.*, **12**, 503–509.

Östlund, G. (1959) Isotopic composition of sulphur in precipitation and sea-water. *Tellus*, **11**, 478–480.

Overton, J. H., Aneja, V. P., and Durham, I. L. (1979) Production of sulfate in rain and raindrops in polluted atmospheres. *Atmos. Environ.*, **13**, 355–367.

Penkett, S. A., Jones, B. M. R., and Eggleton, A. E. I. (1979) A study of SO_2 oxidation in stored rainwater samples. *Atmos. Environ.*, **13**, 133–147.

Perhac, R. M. (1978) Sulphate regional experiment in north-eastern United States: 'The SURE Program'. *Atmos. Environ.*, **12**, 641–647.

Peterson, I. T., and Junge, C. E. (1971) Sources of particulate matter in the atmosphere. In: Matthews, W. H., Kellogg, W. W., and Robinson, G. (eds), *Man's Impact on the Climate*. MIT Press, Cambridge, Mass., pp. 310–320.

Petrenchuk, O. P. (1979) *Experimental Studies of the Atmospheric Aerosols.* Hydrometeoizdat, Leningrad, 264 pp. [in Russian].

PGO (Principal Geophysical Observatory) (1970) *Monthly Data on Chemical Composition of the Atmospheric Fall-out for 1962–1965.* PGO, Leningrad, 68 pp.

Pkhalagava, D. M. (1963) On chemical composition of atmospheric precipitation and ice in the Tsei-Don basin. In: *Chemical Geography and Hydrochemistry.* Perm. N2/3, pp. 37–40 [in Russian].

Poldervart, A. (1957) Chemistry of the earth's crust. In: *Earth's Crust.* Izv. Inostr. Lit., Moscow, pp. 130–157 [in Russian].

Prahm, L. P., Torp, U., and Stern, R. M. (1976) Deposition and transformation of sulphur oxides during atmospheric transport over the Atlantic. *Tellus*, **28**, 355–372.

Prince, R., and Ross, F. (1972) Sulphur in air and soil. *Water, Air Soil Pollut.*, **1**, 286–302.

Rabinovish, A. L. (1971) Sulfur isotope composition in sulfate ions of some surface waters of land and factors of its formation. Doctoral Thesis, Novocherkassk, 20 pp.

Rahn, K. A., and McCaffrey, R. J. (1979) Long-range transport of pollution aerosol to the Arctic: a problem without borders. In: *WMO Symposium on the Long-range Transport of Pollutants and its Relation to General Circulation Including Stratosphere/Troposphere Exchange Process*. Sofia, WMO, No. 538, Geneva, pp. 25–35.

Rasmussen, K. H., Taheri, M., and Kabel, R. L. (1975) Global emissions and natural processes for removal of gaseous pollutants. *Water, Air Soil Pollut.*, **4**, 33–64.

Reay, J. S. S. (1973) Design of environmental information systems, Chapter 5. *Air Pollution Monitoring in the United Kingdom*, London, pp. 109–134.

Reiter, E. R. (1975) Stratospheric–tropospheric exchange processes. *Rev. Geophys. Space Phys.*, **13**, 459–474.

Roberts, D. B., and Williams, D. I. (1979) The kinetics of oxidation of sulfur dioxide within the plume from a sulphide smelter in a remote region. *Atmos. Environ.*, **13**, 1485–1499.

Robinson, E., and Robbins, R. C. (1968) *Sources, Abundance and Fate of Gaseous Atmospheric Pollutants*. Final Report, Project PR-6755, Stanford Research Institute, Menlo Park, California, 110 pp.

Robinson, E., and Robbins, R. C. (1970) Gaseous sulphur pollutants from urban and natural sources. *J. Air. Pollut. Control Assoc.*, **20**, 223–235.

Roderick, H. (1975) Projected emission of sulphur oxides from fuel combustion in the OECD area 1972–1985. Paper presented at IPIECA Symposium, Teheran, pp. 1–35.

Rodhe, H (1972a) A study of the sulfur budget for the atmosphere over North Europe. *Tellus*, **24**, 128–138.

Rodhe, H. (1972b) Measurements of sulfur in the free atmosphere over Sweden, 1969–1970. *J. Geophys. Res.*, **77**, 4494–4499.

Rodhe, H. (1978) Budgets and turn-over times of atmospheric sulfur compounds. *Atmos. Environ.*, **12**, 671–680.

Rodhe, H. (1981) Current problems related to the atmospheric part of the sulphur cycle. In: Likens, G. E. (ed.), *Some Perspectives of the Major Biogeochemical Cycles. Scope Report 17*. Wiley, Chichester, pp. 51—60.

Rodhe, H., and Grandell, J. (1972) On the removal time of aeosol particles from the atmosphere by precipitation scavenging. *Tellus*, **24**, 442–454.

Rodhe, H., and Isaksen, I. (1980) Global distribution of sulfur compounds in the troposphere estimated in a height/latitude transport model. *J. Geophys. Res.*, **85**, 7401–7409.

Rodhe, H., Mukolwe, E., and Söderlund, R. (1981) Chemical composition of precipitation in East Africa. *J. Sci. Technol. Kenya*, **2A**, 3–11.

Romanov, N. N. (1961) *Dust Storms in the Middle Asia*. Tashkent, 198 pp. [in Russian].

Ronneau, C., and Snappe-Jacob, N. (1978) Atmospheric transport and transformation rate of sulfur dioxide. *Atmos. Environ.*, **12**, 1517–1521.

Rosen, J. M. (1964) The vertical distribution of dust to 30 km. *J. Geophys. Res.*, **69**, 4673–4676.

Rovinsky, F. Ya., Koloskov, I. A., Cherkhanov, Yu. P., Vorontsov, A. I., Pastukhov, B. V., and Rusina, E. N. (1980) Materials on complex background monitoring of pollution of natural ecosystems. In: *Complex Global Monitoring of Environmental Pollution*. Hydrometeoizdat, Leningrad, pp. 83–91 [in Russian].

Russell, P. B., Hake, R. D. Jr., and Viezee, W. (1976) *Lidar Measurements of the post-Fuego Stratospheric Aerosol*. Stanford Research Institute, Final Report, Menlo Park, Calif., 48 pp.

Ryaboshapko, A. G., and Erdman, L. K. (1978) Residence time of SO_2 in the polluted atmosphere over the ocean. In: *Atmospheric Pollution as an Ecological Factor*, vol. 19, Tr. Inst. Prikl. Geofiz., pp. 90–107 [in Russian].

Ryaboshapko, A. G., Shopauskene, D. A., and Erdman, L. K. (1978) Background levels of SO_2 and NO_2 and the global sulfur balance. In: *Atmospheric Pollution as an Ecological Factor*, vol. 39. Tr. Inst. Prikl. Geofiz., pp. 20–33 [in Russian].

Sandalls, F. J., and Penkett, S. A. (1977) Measurements of carbonyl sulphide and carbon disulphide in the atmosphere. *Atmos. Environ.*, **11**, 197–199.

Santa-Olalla, M. (1973) The European sulphur dioxide emission problem. *Pet. Petrochem. Int.*, **13**, 48–52, 83.

Scriven, R. A., and Fisher, B. E. A. (1975) The long range transport of air-borne material and its removal by deposition and washout. A general consideration. *Atmos. Environ.* **9**, 49–58.

Selezneva, E. S. (1977) Salt exchange between the atmosphere and ocean from the data on chemical composition of atmospheric precipitation. In: *Atmospheric Diffusion and Air Pollution*. Trudy GGO, 387. Hydrometeoizdat, Leningrad, pp. 110–115 [in Russian].

Shaw, R. (1979) Acid precipitation of Atlantic Canada. *Environ. Sci. and Technol.*, **13**, 406–411.

Shepherd, J. G. (1974) Measurements of the direct deposition of sulfur dioxide onto grass and water by the profile method. *Atmos. Environ.*, **8**, 69–74.

Shopauskas, K. K., Gedraitis, B. I., Nemanis, A. P., and Shopauskene, D. A. (1974) On background concentration of sulphur dioxide and nitrogen oxides in the atmospheric air. In: *International Conference on Physical Aspects of Atmospheric Pollution*, 18–20 June, 1974. Abstracts of Reports. Vilnius, pp. 46–47 [in Russian].

Shopauskas, K. K., Linkaitite, E. Yu., Galvonaite, A. V., Navitskas, V. S., Nemanis, A. P., and Turauskaite, V. V. (1976) On intensity of the fallout of chemical admixtures over the territory of the Lithuanian SSR. In: *Content of Admixtures in Atmospheric Precipitations. Atmospheric Aerosols. Atmosphere Protection Against Pollution*, No. 3. Vilnius, pp. 193–203 [in Russian].

Sievering, H., Dave, M., and McCoy, P. (1979) Deposition of sulfate during stable atmospheric transport over Lake Michigan. *Atmos. Environ.*, **13**, 1717–1718.

Sitting, M. (1975) *Environmental Sources and Emissions Handbook*. Noyes Data Corporation, London, pp. 67–72.

Slatt, B. J., Natusch, D. M. S., Prospero, J. M. and Savoie, D. L. (1978) Hydrogen sulfide in the atmosphere of the Northern Equatorial Atlantic Ocean and its relation to the global sulfur cycle. *Atmos. Environ.*, **12**, 981–991.

Slinn, W. G. M., and Hales, J. M. (1970) Phoretic processes in scavenging. *Proc. Symp. Precipitation Scavenging*, Series 22. USAEC Symp., pp. 411–421.

Smith, A. F., Jenkins, D. G., and Cunningworth, D. E. (1961) Measurement of trace quantities of hydrogen sulphide in industrial atmosphere. *J. Appl. Chem.*, **11**, 317–329.

Smith, F. B., and Jeffrey, G. H. (1975) Airborne transport of sulphur dioxide from the U.K. *Atmos. Environ.*, **9**, 643–659.

Solomatina, I. I. (1977) Analysis and generalization of data on the emission of toxic substances in the atmosphere. In: *Tolerance Rates and Control of Industrial Emissions in the Atmosphere*. Hydrometeoizdat, Leningrad, pp. 27–33 [in Russian].

Stepanov, V. N. (1974) *World Ocean*. Znanie, Moscow, 256 pp. [in Russian].

Sugavara, K. (1964) Migration of elements in the hydrosphere and atmosphere. In: *Chemistry of the Earth's Crust*, vol. 2. Nauka, Moscow, pp. 469–478 [in Russian].

Supatashvili, G. D. (1970) Hydrochemical studies of the Caucasian glaciers. In: *Proceedings of the 4th Caucasian Scientific Conference on Studies of the Snow Cover*,

Avalanches and Glaciers of the Caucasus. Hydrometeoizdat, Leningrad, pp. 188–197 [in Russian].

Thode, H. G., and Monster, J. (1970) Sulphur isotope abundance and genetic relations of oil accumulations in Middle East Basins. *Am. Assoc. Pet. Geol. Bull.*, **54**, 627–637.

Torres, A. L., Maroulis, P. J., Goldberg, A. B., and Bandy, A. R. (1980) Atmospheric OCS measurements on project GAMETAG. *J. Geophys. Res.*, **85**, 7357–7360.

Tsunogai, S. (1971) Oxidation rate of sulphite in water and its bearing on the origin of sulphate in meteoric preicpitation. *Geochem. J.* **5**, 175–185.

Tsunogai, S., Fukuda, K., and Nakaya, S. A. (1975) A chemical study of snow formation in the winter-monsoon season: the contribution of aerosols and water vapor from the continent. *J. Meteorol. Soc. Jap.*, Ser. II., **53**, 203–213.

Tsunogai, S., Saito, O., Yamada, K., and Nakaya, S. (1972) Chemical composition of oceanic aerosol. *J. Geophys. Res.*, **77**, 5283–5292.

Varkonyi, T. (1974) Kendiocid-szennyezödesi terkepek a dunantuli iparvidekröl. *Energ. Atomtech.*, **27**, No. 1, 19–22 [in Hungarian].

Vilenskii, V. D., and Koroleva, N. I. (1973). Sulfate content in the glacious shield of Antarctica. In: *Antarctica. Reports of the Committee*, No. 12. Nauka, Moscow, pp. 94–101 [in Russian].

Visser, S. (1961) Chemical composition of rain water in Kampala, Uganda and its relation to meteorological and topographical conditions. *J. Geophys. Res.*, **66**, 3759–3765.

Vityn, Ya. Ya. (1911) Cited in Drozdova, V. M., Petrenchuk, O. P., Selezneva, E. S., and Svistov, P. F. (1974) *Chemical Composition of the Atmospheric Fall-out over the European Part of the USSR.* Hydrometeoizdat, Leningrad, 242 pp. [in Russian].

Votintsev, K. K. (1954) Cited in Drozdova, V. M., Petrenchuk, O. P., Selezneva, E. S., and Svistov, P. F. (1974) *Chemical Composition of the Atmospheric Fall-out over the European Part of the USSR.* Hydrometeoizdat, Leningrad, 242 pp. [in Russian].

Wesely, M. L., Hicks, B. B., Dannevik, W. P., Frisella, S., and Husar, R. B. (1977) An eddy correlation measurement of particulate deposition from the atmosphere. *Atmos. Environ.*, **11**, 561–563.

Whelpdale, D. M. (1978a) Large-scale atmospheric sulphur studies in Canada. *Atmos. Environ.*, **12**, 661–670.

Whelpdale, D. M. (1978b) Atmospheric pathways of sulphur compounds. Monitoring and Assessment Research Centre of SCOPE. *MARC. Report*, No. 7, pp. 1–39.

Whitby, K. T. (1978) The physical characteristics of sulphur aerosols. *Atmos. Environ.*, **12**, 135–159.

Winkler, P. (1975) Chemical analysis of Aitken particles (<0.2 μm radius) over the Atlantic ocean. *Geophys. Res. Lett.*, **2**, 45–48.

Wys, N. J. de., Hill, A. C., and Robinson, E. (1978) Assessment of the fate of sulphur dioxide from a point source. *Atmos. Environ.*, **12**, 633–639.

Zakharov, P. S. (1965) *Dust Storms.* Hydrometeoizdat, Leningrad, 164 pp. [in Russian].

Zamorsky, A. D. (1952) Deposition of saline dust. *Meteorol. Gidrol.*, No. 9, 25–28 [in Russian].

Zverev, V. P. (1968) Role of precipitation in the cycling of chemical elements between the atmosphere and lithosphere. *Dokl. Akad. Nauk SSR*, **181**, No. 3, 716–719 [in Russian].

The Global Biogeochemical Sulphur Cycle
Edited by M. V. Ivanov and J. R. Freney
© 1983 Scientific Committee on Problems of the Environment (SCOPE)

CHAPTER 5
The Sulphur Cycle in Continental Reservoirs

Part I THE SULPHUR CYCLE IN LAKES AND CONTINENTAL RESERVOIRS

M. V. IVANOV

5.1 INTRODUCTION

A quantitative assessment of biogeochemical processes in continental reservoirs is required to model the global sulphur cycle. The most important link in the sulphur cycle is the discharge of sulphur compounds in rivers—the main transport mechanism supplying the world oceans with sulphur from the zone of subaerial weathering of continents. As sulphur is transported to the ocean (Fig. 5.1), its concentration in river-waters is raised by dry and wet atmospheric deposition, and lowered by biological and geochemical processes in water and bottom sediments of rivers, freshwater lakes, and water reservoirs. Considerable amounts of sulphur are also lost to the soil and uppermost horizons of the lithosphere, and are removed from the global cycle in inland drainage basins (Fig. 5.1).

This schematic picture of the transport of sulphur compounds by natural processes is complicated by anthropogenic influences. It is into continental water bodies, and primarily into the river systems, that the bulk of industrial and urban sewage water, with its variety of sulphur compounds, is discharged. Drainage waters of agricultural regions also supply sulphur compounds from fertilizers, persistent pesticides, and livestock residues to river systems (Fig. 5.1).

In Chapter 4 it was shown that the bulk of gaseous sulphur, resulting from metal smelting, combustion of fossil fuels, and other processes involved in the economic utilization of sulphur-containing minerals, is returned to the continents by atmospheric deposition processes. Quite often the dissolved oxides of sulphur in atmospheric precipitation acidify weakly buffered soils and freshwater lakes.

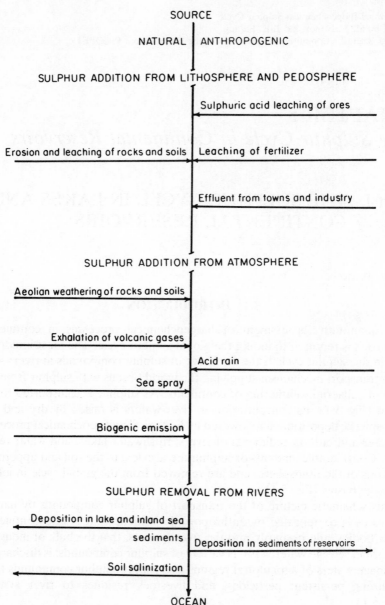

Fig. 5.1   Sulphur cycle processes in continental water-bodies

However, the anthropogenic influence on the sulphur cycle in reservoirs is not confined solely to the increase in sulphur concentration and acidity of the water. In some cases the combined pollution of water with sulphur compounds, organic matter, nitrogen, and phosphorus leads to the activation of a chain of biological processes, including the primary production of organic matter, oxygen consumption during the oxidation of excess organic matter, and production of hydrogen sulphide by putrefaction and sulphate reduction.

In the majority of published reviews on the processes of the global sulphur cycle in water reservoirs, emphasis is placed on one question—the assessment of sulphur flux in river discharge to the ocean. However, some attempts have also been made to divide river sulphate into natural and anthropogenic phases and to assess them separately (Berner, 1971; Friend, 1973; Granat *et al.*, 1976). Among the anthropogenic processes, most attention has been paid to the acidification of freshwater lakes by acid rains.

Our work attempts to estimate quantitatively all the basic processes of the sulphur cycle in continental reservoirs and assesses the natural and anthropogenic contributions to the sulphur cycle.

5.2 THE SULPHUR CYCLE IN LAKES

The most widely occurring form of sulphur in both freshwater and brackish lake-waters and muds is sulphate. Depending on the origin of the lake, the form of the lake basin, climatic conditions, composition of the surrounding rocks, and hydrological regime, the sulphate content of lake-water may vary from several milligrams per litre, in lakes of glacial origin in the humid zone, to a completely saturated solution (with respect to sodium sulphate) in the salt lakes of the arid zone (Hutchinson, 1957; Kuznetsov, 1970; Posokhov, 1972; Nriagu and Hem, 1978).

Freshwater lakes are of primary importance for the economic activities of man and, naturally, receive special emphasis in the works of limnologists.

5.2.1 The Sulphur Cycle in Freshwater Monomictic Lakes

Some information on the sulphur composition of freshwater monomictic[*] lake-waters and muds is given in Table 5.1. Active circulation of water masses is observed in all such lakes in spring and autumn, and thus the water is saturated with oxygen twice yearly. Therefore, anaerobic conditions develop, if at all, only in the hypolimnion of deep-water mesotrophic and eutrophic lakes at the end of the summer stagnation period (Fig. 5.2) or, in some cases, during winter when the lakes are covered with ice.

Despite the low sulphate content of freshwater lakes, and the infrequent, short periods of anaerobiosis, the hydrogen sulphide concentration may reach as much as 13–16 mg litre^{-1} (Table 5.1) during the summer and winter stagnation periods. Large amounts of reduced sulphur are found in the bottom sediments of freshwater lakes, and the numbers of sulphate-reducing bacteria in the muds of such lakes may reach hundreds of thousands of cells per gram of wet silt (Table 5.1).

[*]'Monomictic' refers to a stratified water body that turns over once a year.

Table 5.1 Sulphate, hydrogen sulphide and sulphate-reducing bacteria in the water and sediment of freshwater lakes

Lake	Type of lake	Depth (m)	Water		Sediment		References
			Sulphate sulphur (mg litre^{-1})	Hydrogen sulphide	Hydrogen[a] sulphide (μg g^{-1} wet silt)	Bacteria (10^3 cells g^{-1} wet silt)	
Baikal	Oligotrophic	1300	1.8	0	40–80	2.5–4.0	Goman (1975)
Schahlsee	Mesotrophic	65	8.3	4.0	+	—	Ohle (1954)
Bolshoi Kichier	Mesotrophic	17	14.5	13.2	146–206	200	Chebotarev (1974a)
Beloye	Mesotrophic	4–6	16.3–68.3	0	3–56	+	Sokolova and Sorokin (1957)
Yugdem	Mesotrophic	15	433	0	—	1.2	Kuznetsov (1952)
Plönsee	Eutrophic	40	3.3	4.0	—	—	Ohle (1954)
Rotsee	Eutrophic	16	9–12	16.1	—	—	Bachmann (1931)
Ebergsee	Eutrophic	9.5	8.6	8.0	—	—	Ohle (1934)
Kita-Ura	Eutrophic	10	—	0	150–300	0.2–0.3	Tezuka (1979)
Suwa	Eutrophic	7	—	+	120	+	Tezuka (1979)
Takahokonuma	Eutrophic	7	—	0	20–40	+	Tezuka (1979)

[a]H_2S in sediments was determined either by titration with iodine or by distillation after acidification with HCl.

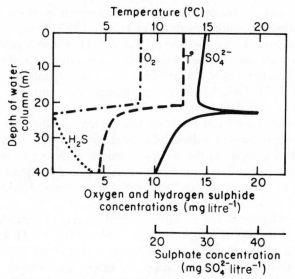

Fig. 5.2 Vertical stratification of oxygen, temperature, hydrogen sulphide, and sulphate in waters of the eutrophic lake Plönsee, 30 September 1952 (Ohle, 1954)

Despite such data, a viewpoint prevails in limnological literature that sulphate-reducing bacteria in the bottom sediments of freshwater lakes play an insignificant role in the sulphur cycle because their activity is limited by a low content of dissolved sulphate (Kuznetsov, 1952, 1970). In recent years, however, data have accumulated which cause us to revise this concept. For example, it has been shown by the use of $^{35}SO_4^{2-}$ that sulphate-reducing bacteria are active in the sediments of freshwater lakes and, in some cases, they produce significant amounts of hydrogen sulphide (Table 5.2).

The subsequent fate of this hydrogen sulphide may vary; part of it may be oxidized at the sediment/water interface, and part may diffuse up the water column. In the latter case, oxidation of hydrogen sulphide occurs at the interface between the aerobic and anaerobic zones of the water column and may result in local increases in sulphate concentration in the lake-water (see Fig. 5.2). However, it appears from the few geochemical analyses available that a considerable part of the hydrogen sulphide formed in the bottom sediments undergoes diagenesis with the production of pyrite and organically bound sulphur, and thus is withdrawn from the cycle.

Recent investigations on Lake Baikal have clarified some aspects of sulphur cycling in freshwater reservoirs. Typical concretions of pyrite and pyrrhotite have been found in the bottom sediments of this oligotrophic freshwater lake (Bondarev, 1974). In addition, quantities of reduced sulphur, both in the form of acid-soluble sulphides and pyrite-sulphur, were present in nearly all of the sediments studied in this lake (Lazo, 1977, 1980) (even in deep water-regions down to 1200–1300 m; Table 5.3). These geochemical findings, together with microbiological data and the results of labelling techniques (Table 5.2), clearly point to the presence of active sulphate reduction in the sediments of Lake Baikal, where the sulphate concentration in water is known to be greater than 2 mgS litre^{-1}.

Several studies have been made on the contents of the various forms of reduced sulphur in bottom sediments of freshwater lakes in Japan and North America. When comparing these data it should be borne in mind that the various authors used different methods of analysis and computations for particular forms of reduced sulphur, and only the contents of acid-soluble sulphide seemed to be estimated by similar methods.

Sediments of freshwater lakes with different degrees of contamination contain appreciable amounts of acid-soluble sulphur (Table 5.4): total reduced sulphur in the muds of the mesotrophic Lake Nakatosuno-Ko amounts to 0.17% in terms of dry silt (Sugawara *et al.*, 1953). The bottom sediments of brackish lakes in Japan are characterized by a markedly higher content of both acid-soluble and pyrite sulphur. The reduced sulphur content of the sediments of continental brackish lakes in Japan is 2%, and is comparable with the amount found in the sediments of the highly productive regions of seas and oceans (See Chapter 6).

Nriagu (1968, 1975), Nriagu and Coker (1976), Nriagu and Harvey (1978), and Nriagu and Hem (1978) determined the different forms of reduced sulphur in lake sediments of North America. In the upper horizons of silt deposits of the eutrophic Lake Mendota (Table 5.5) acid-soluble sulphide is the main form of sulphur. The concentrations of acid-soluble sulphide and total reduced sulphur are much higher in the sediments of the deep part of the lake than in the shallow-water muds. This is explained by the annual generation of anaerobic conditions in the hypolimnion of this lake (Nriagu, 1968).

Figure 5.3 illustrates the distribution of reduced sulphur in a section of the Lake Mendota sediment column. The upper part of the sediment accumulated during the past 150 years and differs markedly in composition and sulphur content from the marl deposits that are lower down the section. Nriagu attributes the changes in the nature of the sediment, and the higher concentrations of sulphur in the upper part of sediments, to the influence of urbanization and economic development of the Mendota basin that began in the first half of the nineteenth century (Nriagu, 1968).

Data on the sediments of two of the Great Lakes—mesotrophic Lake Ontario and eutrophic Lake Erie—are shown in Table 5.6. The reduced sulphur content in the surface horizons of Lake Ontario is 0.16%, while at a depth of 2–6 cm it increases to 0.2%, and then drops to 0.13% and remains practically constant down to a depth of 50 cm. In the uppermost mud horizon

Table 5.2 Sulphate, sulphate-reducing bacteria and sulphate reduction rates in the sediments of freshwater lakes

Lake	Sulphate ($\mu gS\ g^{-1}$ wet sediment)	Bacteria (10^3 cells kg^{-1} wet sediment)	Rate ($\mu gH_2S\ g^{-1}$ wet sediment per day)	References
Beloye	16–68	—	0.0002–0.0011	Sokolova and Sorokin (1957)
Bolshoi Kichier	146	200	4.8–6.2	Chebotarev (1974a)
Nero	11.7–19.3	700	0.01–0.19	
Pleshcheevo	6.3–12.7	20–40	0.016–0.16	Lapteva and Monakova (1976)
Somino	7.3	5000	0.023	
Savelyevskoye	10.7	100	0.005	
Chashnitskoye	7	300	0.014	
Vashutinskoye	10.7	20	0.006	
Ryummikovskoye	5.3	80	0.015	
Baikal, offshore stations	—	2.5–4	0.023–0.213	Goman (1975)
Sister Lake	3.7–4.7	0.6	2.30	Dunnette (1974)

Table 5.3 Reduced sulphur in clay and silt sediments of the oligotrophic Lake Baikal (Lazo, 1977, 1980)

Sampling location	Depth (m)	Thickness of sediment (cm)	Total reduced sulphur (%S, dry weight basis)	Sulphide	Free sulphur	Pyrite	Organic sulphur
				(% of total reduced sulphur)			
135	1200	1020	0.018	16.76	17.15	28.00	38.00
135	1300	800	0.024	7.50	5.10	76.40	11.00
162	1200	400	0.055	17.35	46.10	20.14	15.41
163	1200	320	0.069	25.55	39.25	18.90	16.30
2044	560	850	0.040	3.66	28.00	37.33	31.00
206	240	765	0.051	11.97	27.48	34.28	26.27
213	900	640	0.108	12.25	21.11	49.48	17.16

Table 5.4 Forms of sulphur in overlying-water[a] and sediments of Japanese lakes

Lake type	Overlying water Sulphate (mg litres⁻¹)	Chloride (mg litres⁻¹)	Tested silt horizons (cm)	Sediments Sulphate (%S in dry silt)	Acid-soluble sulphur	Residual sulphur	Total sulphur	References
Freshwater lakes								
Aoki-Ko, oligotrophic	3.7	0.9	0–20	—	0.026–0.044	—	—	Sugarawa et al. (1953)
Kizaki-Ko, mesotrophic	2.1	1.2	0–20	—	0.019–0.023	—	—	
Nakatasuno-Ko, mesotrophic	1.0	1.7	0–21	0.001–0.003	0.021–0.028	0.113–0.147	0.136–0.173	
Brackish lakes								
Nakaumi, polder	—	—	0–20	—	0.002	1.996	2.0	Yoneda and Kouchi (1972)
Togo, polder	—	—	0–20	—	0.002	1.070	1.09	
Lake Aburagafuchi	1150	8330	0–50	0.18–0.23	0.43–1.17	0.05–0.68	1.0–1.7	Sugarawa et al. (1953)
Oe-Gava River	2140	15170	0–60	—	0.43–1.98	—	—	

[a]Water layer lying immediately above the sediment.

Table 5.5 Distribution of sulphur in the upper horizons of Lake Mendota (Nriagu, 1968)

Depth (m)	Total	Acid-soluble	Sulphate	Elemental
		(%S, dry weight basis)		
10.7	0.048	0.028	0.005	0.0021
15.2	0.186	0.102	0.0006	0.0016
15.2	0.187	0.134	0.0012	0.0017
18.3	0.221	0.158	0.0011	0.0013
22.9	0.420	0.350	0.0008	0.0028
23.5	0.397	0.326	0.0009	0.0017
25.3	0.425	0.360	0.0007	0.0036

acid-soluble sulphur has a light sulphur isotopic composition; within the mud column it becomes heavier than pyrite-sulphur by 5–12‰ (Fig. 5.4). This suggests that, in the mud column, acid-soluble sulphur is the product of reduction of heavier sulphate present in pore-water.

From data on the content of reduced sulphur in sediments, and an assumed sedimentation rate of 0.397 mgS cm^{-2} year^{-1} (Kemp *et al.*, 1972; Nriagu and Coker, 1976) the annual accumulation of reduced sulphur in Lake Ontario sediments is estimated to be ~0.082 Tg, more than two-thirds of which was produced by sulphate-reducing bacteria.

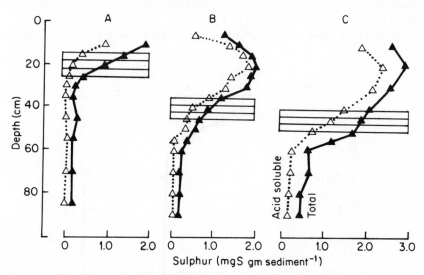

Fig. 5.3 Distribution of acid-soluble (△) and total (▲) sulphur in sediments of Lake Mendota (Nriagu, 1968). The shaded area indicates the column of marl sediments located under the layer of dark reduced sediments. (A, shallow region; B, C, deep regions of lake)

Table 5.6   Pore-water sulphate and reduced sulphur in the sediment of Lake
Ontario (Nriagu and Coker, 1976)

Interval (cm)	Pore water sulphate (mg litre⁻¹)	Acid-soluble	Fixed	Total reduced
		(%S, dry weight basis)		
0–2	14	0.037	0.120	0.157
2–4	5	0.095	0.100	0.195
4–6	1	0.140	0.078	0.218
6–8	<1	0.083	0.056	0.139
8–10	<1	0.078	0.051	0.129
13–15	<1	0.075	0.060	0.135
18–20	<1	0.066	0.078	0.144
28–30	<1	0.078	0.063	0.141
48–50	<1	0.060	0.056	0.116

The sulphur distribution in the bottom sediments of some basins of Lake
Erie is shown in Fig. 5.5. In sediments of this lake, just as in other North
American lakes, a pronounced increase in concentration of reduced sulphur
compounds is observed in the surface (see Fig. 5.5—the central basin) or
subsurface horizons (see Fig. 5.5—eastern and western basins of Lake Erie).
According to Nriagu (1975), the mean concentration of reduced sulphur in
the surface horizons of Lake Erie sediments is 1.4% (dry silt basis). From this
value and the sedimentation rate (Kemp *et al.*, 1972) the amount of sulphur
annually incorporated into the lake sediments in reduced form can be calculated to be ~0.0745 Tg (Nriagu, 1975).

From this review of the sulphur cycle in freshwater lakes it is concluded

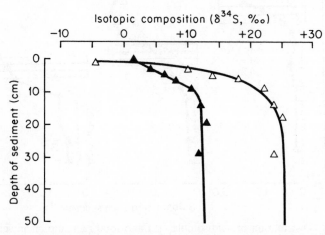

Fig. 5.4   Variation in isotopic composition of acid-soluble
(△) and organic + pyrite (▲) sulphur in a section of the Lake
Ontario sediments. (Nriagu and Coker, 1976)

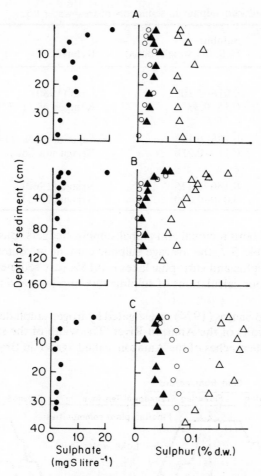

Fig. 5.5 Variation in the amounts of different forms of sulphur in sections of sediments from the western (A), central (B), and eastern (C) basins of Lake Erie. ●, Sulphate in pore-water; ▲, acid-soluble; ○, elemental; △, total. (after Nriagu, 1975)

that, in spite of the low content of dissolved sulphates, the sulphur cycle in these lakes is quite substantial. Usually sulphate reduction takes place only in the sediments, although it can occur in the hypolimnion of deep-water and highly productive lakes.

The geochemical result of sulphate reduction in sediments of freshwater lakes is the accumulation of various insoluble forms of reduced sulphur. However, the amounts accumulated vary widely (Table 5.7). This sulphur is removed from the cycle, and therefore should be considered as a loss item in

Table 5.7   Total reduced sulphur in sediments of freshwater lakes

Lake	Sulphur (%S, dry weight basis)	References
Oligotrophic lakes		
Baikal	0.018–0.108	Lazo (1980)
Huron	0.15–0.38	Kemp *et al.* (1972)
Mesotrophic lakes		
Nakatosuno-Ko	0.136–0.173	Sugawara *et al.* (1953)
Ontario	0.116–0.218	Nriagu and Coker (1976)
Eutrophic lakes		
Mendota	0.186–0.425	Nriagu (1968)
Erie	0.050–0.340	Nriagu (1975)

the estimation of both regional and global sulphur budgets. Based on the few data given in Table 5.7, the average sulphur content in contemporary sediments of mesotrophic and eutrophic lakes is 0.15% (dry sediment basis). This value is used in our calculations of sulphur flux into lake and reservoir sediments.

Brinkmann and Santos (1974) investigated hydrogen sulphide production in high-flood reservoirs of the Amazon River. The width of the stagnant floodplain in the middle reaches of the Amazon, called 'igapo' in Brazil, is 200 km,

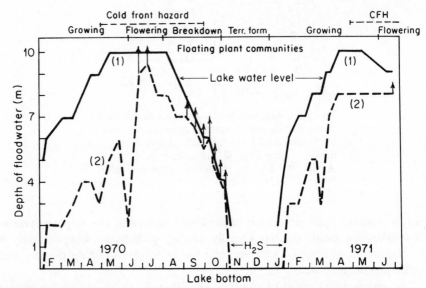

Fig. 5.6   Variations in level of floodwater (1), and the upper limit of dissolved H$_2$S (2) in the floodplain of the Amazon River. Vertical arrows denote the times of H$_2$S emission to the atmosphere (Brinkmann and Santos, 1974)

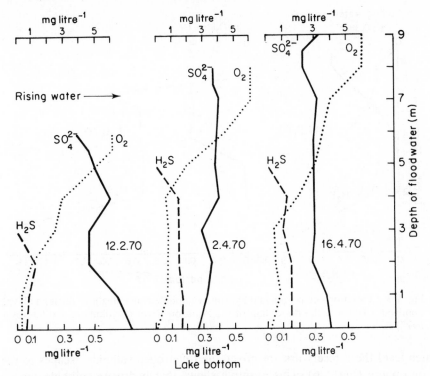

Fig. 5.7 Variations in oxygen, hydrogen sulphide, and sulphate concentrations during rising floodwater levels in the floodplain of the Amazon River (Brinkmann and Santos, 1974)

and this area is flooded with freshet from January to late October (Fig. 5.6). Hydrogen sulphide is produced in the lower horizons of freshet reservoirs in the period of water rise (Fig. 5.7). The hydrogen sulphide concentration gradually increases, and by the end of the high flood period reaches 10–17 mg litre^{-1} (Fig. 5.8).

In the middle of August the level of freshet falls and hydrogen-sulphide-containing waters are exposed to the surface. This results in the degassing of hydrogen sulphide to the atmosphere, usually during September and October (Fig. 5.8). Based on the duration of the degassing process, and the area of floodplain (7500 km^2 according to Lukashova, 1958), the total emission of hydrogen sulphide to the atmosphere was estimated to be 0.2–0.25 Tg (Ivanov, 1979).

During the freshet in floodplain reservoirs the water layer containing hydrogen sulphide is separated from the atmosphere by a layer of water containing oxygen (Fig. 5.6). However, under certain atmospheric conditions, for example during thunderstorms, the upper water horizons of the Amazon freshet reservoirs may cool down; then during the period when freshet is at a

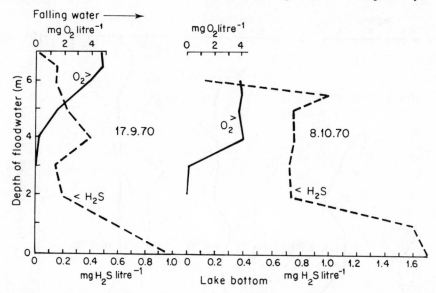

Fig. 5.8 Variations in oxygen and hydrogen sulphide concentrations during falling floodwater levels in the floodplain of the Amazon River (Brinkmann and Santos, 1974)

high level these reservoirs are mixed, and hydrogen sulphide escapes to the atmosphere (Fig. 5.6). This example shows that hydrogen sulphide can be emitted to the atmosphere in significant amounts from freshwater bodies.

5.2.2 The Sulphur Cycle in Meromictic Lakes

The monimolimnions of stratified meromictic* lakes can contain appreciable concentrations of hydrogen sulphide, up to 740–786 mg litre^{-1} (Chebotarev, 1974b; Hutchinson, 1957). In the chemocline in these lakes, intensive chemical and microbiological oxidation of hydrogen sulphide to sulphur and sulphate occurs. The concentration of photoautotrophic bacteria in this zone is so high that the water takes on an intense red or green colour (Kuznetsov, 1952; Sorokin, 1970; Takahashi and Ichimura, 1970; Genovese and Bruni, 1975; Gorlenko *et al.*, 1977).

Two types of meromictic lakes occur most widely: lakes of coastal regions and lakes of karst origin. The first type of meromictic lake is represented by Lake Mogilnoye on Hildin Island (in the Barents Sea) described at the beginning of this century (Isatshenko, 1914), the well-studied Solar Lake, Sinai, Lake Faro in Italy (Genovese and Bruni, 1975; Sorokin and Donato, 1975), and a number of lakes in the coastal region of Antarctica (Matsubaya *et al.*,

---

*A meromictic lake is one that is partly mixed and in which thermal turnover occurs only in the top layer. Bottom layers are stagnant and anaerobic.

Table 5.8 Some characteristics of the sulphur cycle in meromictic lakes

Lake, region	Monolimnion characteristics				Sulphate reduction rate (mgS litre⁻¹ day⁻¹)		Production (mgS m⁻² day⁻²)		References
	Maximal depth	Thickness (m)	Sulphate (mgS litre⁻¹)	Hydrogen sulphide (mgS litre⁻¹)	Water	Sediments	Water	Sediments[a]	
Freshwater lakes									
Sakovo, Vologda region	16.0	11.0	272	11.0	0.016	3.9	176	390	Ivanov (1978)
Konomier, Middle Volga	22.5	12.5	13.7	2.5	0.0008	5.2	10	520	Chebotarev (1975)
Kuznechikha, Middle Volga	20.0	15.0	12.0	16.0	0.020	3.6	300	360	Chebotarev (1975)
Black Kichier, Middle Volga	10.0	5.5	32.7	61.5	0.016	13.3	88	1330	Chebotarev (1974a)
Vae de San Huan, Cuba	20.0	10.0	64	104.0	0.077	3.1	770	310	Romanenko *et al.* (1976)
Brackish lakes									
Repnoye, Ukraine	6.5	1.5	706	112	0.10	15.6	150	1560	Chebotarev *et al.* (1973)
Veisovo, Ukraine	16.5	14.5	997	740	0.043	2.7	625	270	Chebotarev (1975)

[a]Reduced sulphur production in sediments is calculated for 10 cm thickness of mud.

1979; Burton and Barker, 1979). The monimolimnions of these lakes usually contain modified sea-water, and the surface-water layers are markedly fresh. In a number of cases these lakes maintain a constant hydraulic contact with the sea, and the biogeochemical processes occurring in them are similar to processes taking place in gulfs and, especially, in fjords.

Meromictic lakes of karst origin are usually deep-water reservoirs of small area formed in karst gaps in halogenic rocks. Depending on the hydrological regime of the lake, and the salinity of the ground-waters which feed the lake, the sulphate concentration and the level of salinity in the monimolimnions vary widely (Table 5.8). The hydrogen sulphide concentration depends both on the rate of sulphate reduction in lake-water and sediments and on the secondary biogeochemical reactions taking place in the water column and sediments: it is also extremely variable.

In most meromictic lakes, sulphate reduction is most active in the sediment (Table 5.8). However, in some cases, the amount of hydrogen sulphide produced in the water column may be comparable with the amount produced in the sediment or even exceed it.

A further indication that sulphate reduction occurs in the monimolimnions of meromictic lakes is given by the results of microbiological analyses (Fig. 5.9), and by the marked differences in the isotopic composition of hydrogen sulphide obtained from the mud and water columns (Table 5.9). In all

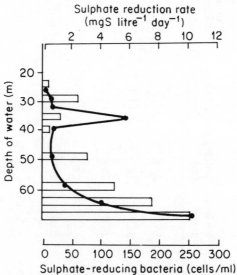

Fig. 5.9 Sulphate reduction rates (●) and distribution of sulphate-reducing bacteria in the water column of Lake Gek-Gel (Sorokin, 1970)

Table 5.9 Concentration and isotopic composition of sulphate and hydrogen sulphide in water and sediments of meromictic lakes (Ivanov, 1978; Matrosov *et al.*, 1978)

| | Water column | | | | Sediments | | | |
| | Sulphate sulphur | Hydrogen sulphide | Sulphate | Hydrogen sulphide | Sulphate | Hydrogen sulphide | Sulphate | Hydrogen sulphide |
Lake	(mg litre^{-1})		(δ^{34}S, ‰)		(mg litre^{-1})		(δ^{34}S, ‰)	
Black Kichier	33	61.5	+24.8	+5.9	132	322	+10.0	+2.0
Great Kichier	17.7	13.1	+19.6	+17.1	42	146	+10.5	+4.9
Sakovo	242	8.2	+15.4	−10.8	750	348	+10.3	−29.7

three karst lakes, hydrogen sulphide from the water column is more enriched in ^{34}S than hydrogen sulphide from the upper sediment horizons. The results of both chemical and isotopic analyses (Tables 5.8 and 5.9) suggest the following sequence of events. Sulphate is added to the deepest part of the karst lake basin from ground-water discharge. Part of this sulphate is reduced to hydrogen sulphide in the sediments and precipitated as pyrite, and the residual sulphate becomes enriched in ^{34}S. This sulphate is transported to the deep water of the lake monimolimnion, where further reduction of sulphate to hydrogen sulphide occurs. Thus, the primary characteristic of the sulphur cycle in meromictic lakes is that sulphate reduction takes place in both the monimolimnion and in the sediment.

The second characteristic of the sulphur cycle in these lakes is that the boundary between oxidized and reduced zones lies not at the sediment/water interface but in the water column in the region of the chemocline. As already mentioned, this region also supports the mass development of micro-organisms capable of oxidizing hydrogen sulphide. Gorlenko *et al.* (1977) have provided a detailed summary of the photo- and chemoautotrophic microflora of meromictic lakes, and have described the results of numerous anal-

Table 5.10   Acid-soluble sulphur in sediments of meromictic lakes

Lake	Depth (m)	Sediment horizon (cm)	Sulphur (mgS litre^{-1} wet sediment)	References
Belovod	11	Surface	506[a]	Kuznetsov (1970)
	15	Surface	678[a]	
	20	Surface	714[a]	
Gek Gel	34	Surface	410	
	39	Surface	960	
	40	Surface	1780	
	50	Surface	4200	Sorokin (1970)
	58	Surface	2700	
	65	Surface	4520	
	85	Surface	3450	
Sakovo	15	0–6	348	Matrosov *et al.* (1978)
		6–12	114	
		10–20	169	
Black Kichier	5	0–2	264	Chebotarev (1974a)
		2–20	422	
		10–20	926	
Repnoye	6.5	0–6	912	Chebotarev *et al.* (1973)
		6–12	788	
		12–18	726	

[a]The content of acid-soluble sulphur in samples (% dry sediment) is 1.035, 0.755, 0.714.

yses of photosynthetic and chemosynthetic activity in the chemocline, using $^{14}CO_2$. The major geochemical result of the activity of these micro-organisms is the complete oxidation of hydrogen sulphide diffusing from the anaerobic zone.

Unlike freshwater lakes, where only a minor part of the sulphur input is buried in sediments in the form of reduced sulphur, meromictic reservoirs seem to be characterized by high concentrations of reduced sulphur in the sediments (Table 5.10). An assessment of the global sulphur flux into the sediments of meromictic lakes cannot as yet be made because the number of these lakes is unknown, and comprehensive analyses of all reduced sulphur compounds in their sediments do not exist. Nevertheless compared with the amount of sulphur in seas and oceans the amount of sulphur in these lakes is probably insignificant, and thus the flux of sulphur to these sediments can be neglected in the global sulphur balance.

5.3 MAN'S CONTRIBUTION TO THE SULPHUR CYCLE IN CONTINENTAL WATER RESERVOIRS

An approximate estimation of the utilization of various sulphur-containing raw materials by man shows that, in the mid 1970s, the amount of sulphur annually extracted in fossil fuels, sulphide minerals, and native sulphur alone, reached 120 Tg. During the processing and utilization of the sulphur-containing raw materials by industry and agriculture, the major part of this sulphur (more than 100 Tg year^{-1}) was added to rivers and continental water reservoirs in the form of gaseous or soluble oxidized compounds, and was then transported to marine reservoirs (see Fig. 5.18).

Three basic aspects of man's contribution to the sulphur cycle in continental water reservoirs have been considered: (1) the increase in the concentration of sulphate and hydrogen sulphide production in freshwater lakes; (2) the increase of reduced sulphur in sediments of these reservoirs; (3) the acidification of lake-water in response to atmospheric acid deposition. We shall supplement the review of these problems with a special treatment of the sulphur cycle in water reservoirs where, according to some authors, large masses of sedimentary material are accumulated, and where sulphate reduction, leading to the conservation of reduced sulphur compounds, occurs.

5.3.1 Sulphate and Hydrogen Sulphide in Lake-Water

This has been studied most extensively in the Great Lakes region of North America (Fig. 5.10). This region, in particular the part belonging to the USA, has been very intensively inhabited and used during the past 100–130 years. The population inhabiting the Great Lakes basin increased from 4.5 million people in 1860 to 16 million in 1900 and to 36.3 million in 1960 (Beeton,

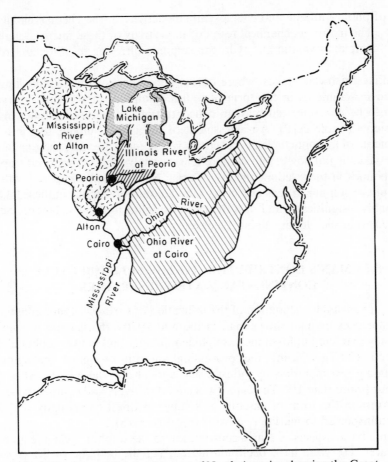

Fig. 5.10   Map of the eastern part of North America showing the Great Lakes and neighbouring basins of the Mississippi, Illinois, and Ohio rivers (after Ackermann *et al.*, 1970)

1965). By 1980, the population exceeded 50 million. According to Ackermann *et al.* (1970), the average sulphate concentration of the Illinois River at Peoria, Ill. and the Ohio River at Cairo, Ill., has risen from 46 to 120 mg litre^{-1} and from 30 to 61 mg litre^{-1}, respectively, during the past 60–70 years. They also report that the sulphate concentration in water from the south-western shore of Lake Michigan was 5 mg litre^{-1} early this century, 14.4 mg litre^{-1} in 1926, and 25.7 mg litre^{-1} in 1967 (Ackermann *et al.*, 1970).

Figure 5.11 illustrates the results of Beeton (1965) who summarized information on sulphate concentration in the waters of the Great Lakes available since 1850. The sulphate concentration in the water of all the lakes, except for Lake Superior whose basin is less populated, has increased

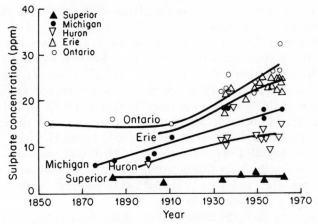

Fig. 5.11 Historical changes in sulphate concentrations in the Great Lakes (after Beeton, 1965)

Table 5.11 Changes in the hydrogen sulphide concentration of the hypolimnion of Lake Belovod from 1928 to 1964 and Lake Black Kichier from 1938 to 1971

Date of analyses	Depth (m)	Hydrogen sulphide (mg litre^{-1})	References
Lake Belovod			
17.03.1928	24	4.47	Dolgov (1955)
20.08.1928	24	4.20	
17.08.1937	20	7.24	Kuznetsov (1942)
09.12.1937	20	8.94	
11.03.1938	20	8.58	
30.07.1938	20	7.68	
02.08.1948	20	6.50	Egorova (1951)
July 1954	20	24.0	Lyalikova (1957)
	24	25.0	
Winter 1955	24	40.0	Ivanov (1956)
July 1958	20	80.0	Kuznetsov (1970)
	24[a]	100.0	
29.07.1964	21	95.0	
	23[a]	152.0	Sorokin (1966)
Lake Black Kichier			
13.09.1938	7	55.8	Kuznetsov (1952)
	9	57.1	
July 1971	8	59.6	Chebotarev (1974a)
	8.75	61.5	

[a]The overlying-water samples may contain an anomalously high concentration due to inclusion of sediment.

throughout the period of observation. Particularly marked is the increase during the past 50–60 years.

According to Shilkrot (1968) and Rossolimo (1977), the sulphate concentration in the hypolimnion of the meromictic Lake Beloye in Kosino, near Moscow, has risen from 2–3 mg litre^{-1} in 1923–27 to 6.6–11.2 mg litre^{-1} in the summer and autumn of 1967. The hydrogen sulphide concentration in the hypolimnion of this lake also increased (Shilkrot, 1968). In 1910–11 hydrogen sulphide was not detected in the water, either during summer stagnation or during winter. However in August 1938, 4.8 mg H_2S litre^{-1} was measured, and in August 1967, 5.6 mg litre^{-1}. Hydrogen sulphide contents of lake-water below the layer frozen from November to April were as follows: 0.02–0.09 mg litre^{-1} in March 1931; 0.03–0.18 mg litre^{-1} in April 1938, and 5.1–7.3 mg litre^{-1} in March 1967.

A greater increase in the concentration of dissolved hydrogen sulphide was observed in the monimolimnion of the meromictic Lake Belovod in the Vladimir region (Table 5.11)—from 4.5 mg litre^{-1} in 1928 to 7.2–9.0 mg litre^{-1} in 1937–38, and 95 mg litre^{-1} in 1964. Lakes Belovod and Beloye are situated in industrial and densely populated regions. By contrast, in the meromictic Lake Black Kichier the hydrogen sulphide concentration has not changed for the past 33 years (see Table 5.11). It may be considered as a control lake since it is situated in the large tracts of forest on the left bank of the Volga River.

5.3.2 Accumulation of Reduced Sulphur in Lacustrine Sediments

Subsurface maxima of reduced sulphur concentrations in the sediments of Lake Mendota, Lake Ontario, and Lake Erie have already been mentioned (Figs 5.4, 5.5 and Tables 5.5, 5.6). Niriagu attributed this phenomenon to anthropogenic contamination of the lakes (Nriagu, 1968, 1975; Nriagu and Coker, 1976). But the most convincing evidence of the influence of anthropogenic processes upon the sulphur cycle in sediments was obtained by Kemp et al. (1972). They studied the distribution of organic and mineral carbon, nitrogen, phosphorus, and reduced sulphur in columns of sediments of three to the five Great Lakes, the oligotrophic Lake Huron, the eutrophic Lake Erie, and the mesotrophic Lake Ontario (Fig. 5.12). Although Kemp et al. (1972) concluded that the reduced sulphur content of the sediments has not changed during the past 100 years, the results in Fig. 5.12 clearly demonstrate the increasing accumulation of practically all of the components studied, from the older to the younger sediments. The total sulphur concentration in sediments from Lake Erie dated 1800–1900 ranged from 0.04 to 0.08%, whereas the concentration in 1965 sediments was 1.1%. Table 5.12 shows that the most marked changes in sulphur occurred in the sediments of the most contaminated lake, Lake Erie, and the smallest changes occurred in those of

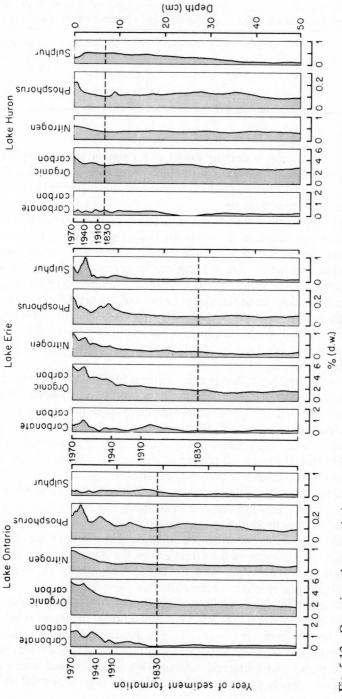

Fig. 5.12 Organic carbon, total nitrogen, phosphorus, and reduced sulphur concentrations in the sediment section of Lakes Erie, Ontario and Huron (after Kemp *et al.*, 1972)

Table 5.12 Ratios of carbon, nitrogen, phosphorus, and sulphur of contemporary 1830 sediments taken from three of the Great Lakes (from **Kemp** *et al.*, 1972)

Lake	Organic carbon	Total nitrogen	Total phosphorus	Total sulphur
Erie	3.2	3.8	2.7	16.8
Ontario	2.7	3.8	2.5	1.6
Huron	1.2	1.3	1.6	1.1

Lake Huron. Thus, there is definite evidence for sulphate reduction in these sediments which has led to sulphide accumulation (see also Nriagu, 1975; Nriagu and Coker, 1976) and which is consistent with large increases in organic carbon in the sediments.

Sorokin (1975) studied the stimulating effect of high concentrations of organic carbon on sulphate reduction in two Alpine lakes—Lugano and Lago-Margory. In the sediments of these lakes, which are grossly contaminated with urban sewage waters and residues from pulp and paper industries, the organic carbon concentrations are 4–5% and 2–3% respectively. Sulphate is reduced in the sediments of both lakes at rates of 2.2 mgH_2S $litre^{-1}$ day^{-1} in Lake Lugano and 14–18 mgH_2S $litre^{-1}$ day^{-1} in Lago-Margory. Appreciable concentrations of acid-soluble sulphur are accumulated—from 0.23 to 0.61% in the sediment of Lake Lugano and from 0.12 to 0.78% (dry silt basis) in Lago-Margory (Sorokin, 1975).

The data presented in the last two subsections indicate that human, industrial, and agricultural activities markedly influence the sulphur cycle in continental reservoirs. First there is a noticeable increase in the concentration of dissolved sulphate in fresh water. When not accompanied by acidification of lake- or river-water, such increases by themselves may not be environmentally detrimental—even a doubling of the natural sulphate levels would result in concentrations of less than 100 mg $litre^{-1}$.

However, an increase in sulphate concentration will often occur simultaneously with an increase in organic matter in the water reservoir, either from exogenous sources or produced in the reservoir itself by photosynthesis. Such conditions are favourable for enhanced sulphate reduction in reservoirs. This is observed in the sediments of the Great Lakes, the Alpine lakes in Italy, and in the waters of Lake Beloye and Lake Belovod, where an appreciable amount of hydrogen sulphide accumulates as an indirect result of anthropogenic influences. As long as lake sediments contain a sufficient amount of reactive iron, hydrogen sulphide will be removed from the ecosystem as iron monosulphides and pyrite. However, if the sediments become iron-deficient, hydrogen sulphide will enter the bottom waters, causing the death of benthic animals and plants and a marked deterioration of the oxygen status of the whole reservoir. Such phenomena have already been observed in Lake

Beloye where, during the last five years, forced aeration of the lake-water with air compressors has been necessary to save the biota from hydrogen sulphide contamination (Rossolimo and Shilkrot, 1971).

5.3.3 The Sulphur Cycle in Water Reservoirs

Intensive construction of artificial water reservoirs over all continents has taken place during the twentieth century and, in particular, during the past 20–30 years. By 1972 their number had exceeded 10 000, and the total area of the water surface of reservoirs had reached 400 000 km^2 (about 20% of the area of the water surface of all lakes and about one-third of the area of freshwater lakes). Water storage in these reservoirs reached 6% of the total reserves of fresh water in lakes (Korzun *et al.*, 1974).

Sorokin (1960, 1961), in a series of investigations carried out on the water reservoirs of the Volga cascade, demonstrated that the sediments, and sometimes the water columns, support extremely active sulphate reduction resulting in the enrichment of reduced sulphur in sediments and its removal from the global biogeochemical cycle. The sediments of three water reservoirs contain large numbers of sulphate-reducing bacteria and high concentrations of reduced sulphur, and exhibit active sulphate reduction. There was a marked increase in the content of hydrogen sulphide and acid-soluble sulphides in the sediments of the Kuibyshevskoye water reservoir between 1957 and 1958 (Table 5.13).

In some parts of the Kuibyshevskoye water reservoir, sulphate reduction was observed in the water column, as well as in the sediments during summer stratification. The rate varied within the range 0.012–0.017 mg litre^{-1} day^{-1} (Sorokin 1960, 1961) which is typical of the level of activity in the waters of mesotrophic meromictic lakes (see Table 5.2); in the bottom-water layer it reached as high as 0.2 mg H$_2$S litre^{-1} day^{-1} (Sorokin 1960). The average content of acid-soluble sulphur in 14 samples of the Kuibyshevskoye water reservoir is 0.14% in terms of dry silt (Sorokin, 1960); i.e. it approximates the total amount of reduced sulphur in lake sediment (Table 5.7). In the sediment samples from the Susakansky and Cheremshansky bays of this water reservoir, the mean concentration of acid-soluble reduced sulphur is 0.35% of dry silt, although in some cases it falls within the range 0.68–1.05%S.

The average concentration of acid-soluble sulphur in sediments of the Rybinskoye reservoir is approximately the same as that in sediments of the Kuibyshevskoye reservoir (viz. 0.14%). This value is markedly lower (0.05%; Sokolova and Sorokin, 1958) in sediments of the Gorkovskoye reservoir during the first year of its operation only. Thus, artificial reservoirs, like the Gorkovskoye and Kuibyshevskoye reservoirs, appear to have supported, from the very beginning, highly active microbiological sulphate reduction which has caused considerable enrichment of reduced sulphur in the

Table 5.13 Some characteristics of the sulphate reduction process in sediments of the Volga water reservoirs in 1958 (Sorokin, 1960, 1961)

Lake	Type of sediment	Station depth (m)	Acid soluble sulphur (mg litre^{-1} wet mud)	Sulphate reducing bacteria (10^3 cells g^{-1})	Sulphate reduction rate[a] (mgH$_2$S litre^{-1} day^{-1})
Rybinskoye, 1955	Grey sediment in the Volga river-bed	14	187	6	0.009
	Grey sediment in the Mologa river-bed	13	181	12	0.049
	Dark-grey sediment in the Sheksna river-bed	12	935	48	0.079
	Peaty brown sediment	11.5	209	42	0.020
Gorkovskoye, 1956	Yellowish sediment of floodlands	8	51.2	363	1.43
	Grey sediment of floodlands	11	115.5	20	1.22
	Floodland soil	3	235.0	910	0.17
	Grey sediment in back water	6	272.0	1	2.94
	Dark-grey sediment in the Sviyaga river-bed	14	68.3	12	0.13
	Grey sediment of Lake Yunetskoye	10	143.5	180	3.94
	Grey sediment of the Arkharovka River	11	421.4	124	0.47
	Black earth	4	27.3	18	0.70
Kuibyshevskoye, 1958	Grey sediment of the Cheremshansky river-bed	18	508.0	120	0.70
	Grey sediment of Cheremshansky Bay	8	835 (288)[b]	41	3.68 (0.19)[b]
	Black sediment of the same region	15	830	16	4.03 (0.71)
	Grey sediment of Susakansky Bay	7	2140 (508)	18	3.95
	Dark-grey sediment	6	365 (143)	31	2.57 (3.90)
	Dark-grey sediment	9	702 (50)	3000	2.87 (0.16)
	Grey sediment of the Volga river-bed	29	266 (34)	25	0.36 (0.15)

[a] Determined with $^{35}SO_4^{2-}$.
[b] The values in brackets were obtained at the same locations in 1959.

sediments. The example of the Rybinskoye water reservoir indicates that this process continues in the sediments of mature water reservoirs which have been operational for decades.

The data of Sorokin (1960, 1961) on the reduced sulphur content of reservoir sediments, support the adopted value for reduced sulphur in the sediments of freshwater lakes (0.15% in terms of dry silt, see above) and suggest that it is also valid for the calculations on the reduced sulphur flux to the bottom sediments of water reservoirs. This value may be taken as the lower limit for the total content of reduced compounds in the sediments of these reservoirs.

5.3.4 Increase in Sulphate Content, and Acidification of Lake-Water due to Atmospheric Deposition

One of the consequences of man-made pollution of the atmosphere is 'acid rain'. Acid rain is formed by precipitation of atmospheric nitrogen and sulphur oxides produced during fossil-fuel combustion and other human activities. The extent of the global emission of oxides of sulphur to the atmosphere is assessed in Chapter 4 of this report. In this section we consider the problems caused by an increase in the sulphate concentration and acidification of lake-water due to acid deposition.

The progressive acidification of atmospheric precipitation during the past 20 years is now established. However, the question as to which component of the anthropogenic pollution is responsible is still hotly debated. Sulphur clearly plays a marked and sometimes a dominant role. Odén (1976) demonstrated a relation between decreasing pH of atmospheric precipitation and increasing amounts of sulphur precipitated in some regions of Sweden, Denmark, and Norway (Fig. 5.13).

A number of reports (e.g. Wright and Gjessing, 1976; Almer *et al.*, 1974; Beamish and Harvey, 1972; Seip and Tollan, 1978) have noted a marked acidification of lake-water in Scandinavia and the Laurentian shield of North America. The average values for lake-water pH have dropped to 5.0, with the annual rate of decrease being 0.05–0.07 of a unit (Table 5.14). The elevated concentrations of sulphate ions in acidic lake-waters (Table 5.15) show that, in most cases, the decrease in pH is due to sulphuric acid.

A relationship between the sulphate content of lake-water and the distance from the local source of atmospheric contamination has been demonstrated by Nriagu and Harvey (1978). They analysed the sulphate content in the water of 120 lakes (Fig. 5.14) in the environs of Copper Cliff (Ontario, Canada), where the metal-smelting plants add ~0.018 TgS as sulphur dioxide to the atmosphere per day. An appreciable increase in sulphate concentration was observed within a radius of at least 60 km from the source of the anthro-

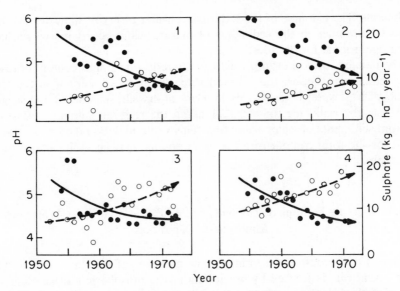

Fig. 5.13   Long-term observations on the pH (●———●) and sulphate
(○– – –○) of atmospheric precipitation at four stations in Scandinavia. 1,
Ås, Norway; 2, Smedbi, Sweden; 3, Plenning, Sweden; 4, Asko, Denmark
(after Oden, 1976). Arrows show the mean trends for pH and sulphur

pogenic pollution. In lakes with buffered water the pH values did not change
significantly, even when the sulphate concentration rose to 40–87 mg litre^{-1}.
In lakes with unbuffered water, distinct acidification was observed which
increased as the distance from the contamination source decreased
(Fig. 5.15). In addition, the isotopic composition of the sulphate–sulphur of
acidic lake-water appeared to be homogeneous and close to the isotopic
composition of sulphur in polluted atmospheric precipitation (Nriagu and
Harvey, 1978).

The data of Nriagu and Harvey (1978) on the Canadian lakes provide a
basis for interpreting the role of water chemistry and local rock composition
on the rate of acidification of lake-water by acid deposition. The waters of
lakes situated in areas of crystalline rocks contain few dissolved salts and a
minimal quantity of bases, and therefore their acidification by acid deposition
occurs rapidly. This has happened in the ultra-freshwater lakes of Norway,
Sweden, and some eastern regions of North America. In regions with a
developed shield of sedimentary rocks, the lake-waters are more highly buffered against the acidifying effect of acid rain (Table 5.14). Therefore, the
first effect of anthropogenic pollution of such lakes via the atmosphere, viz. a
marked increase in sulphate content (Fig. 5.14), is not always associated with
an immediate marked change in pH (Fig. 5.14). It should be borne in mind,
however, that the buffering capacity of lake-water is limited, and that the

Table 5.14 Acidification of the freshwater lakes of Scandinavia and North America (Wright and Gjessing, 1976)

Region	Number of lakes	Early data on pH		Recent data on pH		Mean annual change in pH
Scandinavia						
Central Norway	10	7.3 ± 0.8	(1941)	5.8 ± 0.7	(1975)	−0.05
Western Sweden coast	6	6.6 ± 0.2	(1933–36)	5.4 ± 0.8	(1971)	−0.03
Western Sweden coast	8	6.8 ± 0.4	(1942–49)	5.6 ± 0.9	(1971)	−0.04
West Central Sweden	5	6.3 ± 0.3	(1937–48)	4.7 ± 0.2	(1973)	−0.06
South Central Sweden	5	6.2 ± 0.2	(1933–48)	5.5 ± 0.7	(1973)	−0.03
Southern Sweden	51	6.76 ± 0.14	(1935)	6.23 ± 0.44	(1971)	−0.015
North America						
La Cloche Mts, Ontario, Canada	7	6.3 ± 0.7	(1961)	4.9 ± 0.5	(1972–73)	−0.06
North La Cloche Mts	7	6.6 ± 0.8	(1961)	5.9 ± 0.7	(1971)	−0.07
Adirondack Mts, New York	8	6.5 ± 0.6	(1930–38)	4.8 ± 0.2	(1969–75)	−0.05

Table 5.15   Anion composition of lake-water in regions with acidic and moderately acidic precipitation (Wright and Gjessing, 1976)

Region	Number of lakes	Water pH	Bicarbonate	Chloride	Sulphate (meq litre^{-1})	Nitrate
Lakes in regions with acidic precipitations						
Scandinavia						
Southern Norway	26	4.76	11 ± 26[a]	71 ± 45[a]	100 ± 23[a]	4 ± 2[a]
			11	0	92	4
Western Sweden coast	12	4.37	0	440	200	8
			0	0	155	8
West Central Sweden	4	4.66		170 ± 90	200 ± 70	19 ± 4
				0	180	19
North America						
La Cloche Mts, Ontario	4	4.7	0	22 ± 6	290 ± 40	—
			0	0	290	
Sudbury environs, Ontario	4	4.5	8 ± 2	50 ± 20	800 ± 290	—
			8	0	800	
Lakes in regions with moderately acidic precipitation						
West Central Norway	23	5.2	13 ± 8	46 ± 21	33 ± 8	5 ± 2
			13	0	30	5
Region of experimental lakes, Canada	40	5.6–6.7	60	40	60	1.5
			60	0	55	1.5

[a]On the upper line the average data for the group of lakes are given; on the second line the excess sulphate calculated from the proportion of sulphate to chloride in marine water is shown.

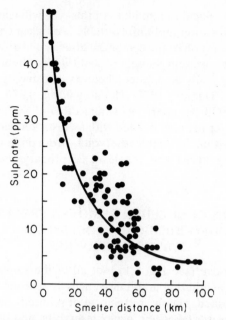

Fig. 5.14 Relationship between sulphate concentration in lake-waters and distance from the local source of atmospheric sulphur – town of Copper Hill near Sudbury, Ontario, Canada. (Nriagu and Harvey, 1978)

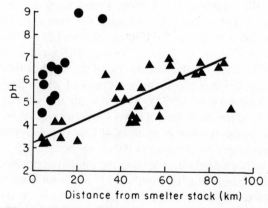

Fig. 5.15 Variation in pH of lake-water with distance from the local source of pollution. ▲, Lakes with poorly buffered water; ●, lakes with well-buffered waters. (Nriagu and Harvey, 1978)

continuous addition of acid rain to lakes of this type will lead, sooner or later, to appreciable acidification and unfavourable ecological consequences.

A detailed consideration of the ecological effects of lake-water acidification is not the aim of the present subsection, and it has already been discussed comprehensively by many ecologists, chemists, and biologists (Almer *et al.*, 1974; Beamish and Harvey, 1972; Hendrey *et al.*, 1976; Schofield, 1976; Seip and Tollan, 1978; Kramer, 1978; and others). Suffice it to say that a decrease in lake-water pH is associated with a drastic reduction in the number of biological species found in water reservoirs—from phytoplankton to fishes. The biomass of organisms and the level of primary production are also depleted.

5.4 ESTIMATES OF SULPHUR REMOVED FROM THE GLOBAL CYCLE BY BIOGEOCHEMICAL PROCESSES IN CONTINENTAL WATER RESERVOIRS

As mentioned above (see Fig. 5.1), not all of the sulphur which finds its way into rivers by weathering and human activity reaches the ocean. Part of the sulphur is deposited, mainly in the form of reduced compounds, with the sediments of freshwater lakes and water reservoirs, and part accumulates in inland drainage basins, for example the Caspian Sea. In both cases this sulphur is withdrawn from the contemporary sulphur cycle by its burial in sediments.

5.4.1 Sulphur Flux to the Sediments of Freshwater Lakes and Water Reservoirs

Uspensky (1970) estimates the average sedimentation rate in lakes to be $0.18 \, g \, cm^{-2} \, year^{-1}$. This value approximates the lowest of the annual sedimentation rates calculated for the Great Lakes—Lake Erie, $0.54 \, g \, cm^{-2}$; Lake Ontario, $0.32 \, g \, cm^{-2}$; Lake Huron, $0.15 \, g \, cm^{-2}$ (Kemp *et al.*, 1972)—and may be used for the assessment of the sulphur flux to the bottom sediments of all freshwater lakes, the total area of which may be as high as $1\,236\,000 \, km^2$ (Korzun *et al.*, 1974). Assuming an area of reduced sediments equivalent to 75% of the total area of lakes (Nriagu and Coker, 1976; Nriagu, 1975), and an average sulphur content of 0.15%, we estimate the total sulphur flux to sediments of freshwater lakes to be $2.5 \, TgS \, year^{-1}$. Similar assumptions have been made in assessing the annual deposition of sulphur in sediments of water reservoirs. Glimph's (1973) data indicate that 1800 Tg of sediments are accumulated annually in water reservoirs in the USA, with a total area of $60\,000 \, km^2$ (Gorshkov, 1980). By extrapolation, the total world water reservoirs (area $400\,000 \, km^2$) will annually accumulate 20 000 Tg of sediments and 13.5 TgS. The sulphur flux to the sediments of all freshwater lakes and reservoirs, therefore, totals $16 \, TgS \, year^{-1}$.

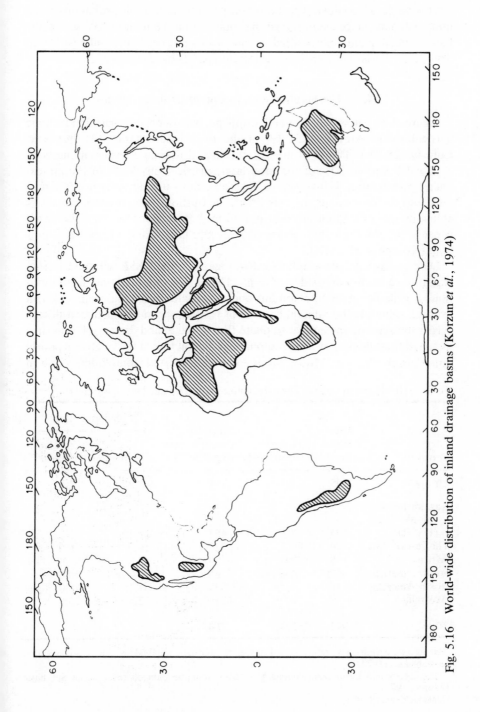

Fig. 5.16 World-wide distribution of inland drainage basins (Korzun *et al.*, 1974)

This value is considered to be tentative only and, in all probability is an underestimate, since even though the total sulphur content in lake sediments has not been adequately studied, in most cases it exceeds the value (0.15%S, dry weight basis) used in the calculations (see Table 5.7)

5.4.2   The Sulphur Flux to Inland Drainage Basins

Figure 5.16 shows that inland drainage basins occupy vast land areas, particularly in Australia (51.3% of the total land), Africa (31.9%), and Eurasia (28.5%). The total area of the inland drainage basins amounts to $30.2 \times 10^6 \, km^2$ or 20% of the total land surface. Only 7.6% of the atmospheric precipitation is involved in the global moisture turnover in these regions, and rivers accounts for only 2% of the outflow. Nevertheless, considerable amounts of sulphur are transported in these regions by river discharge, because the total mineral content of river and lake-waters of arid regions is high (Korzun *et al.* 1974).

Data on the sulphate ion flux for two regions of the USSR without external runoff have been published by Zverev (1971). For other continents information is available on the total ion flux and river-water discharge, from which an average value for the mineral content of the water can be calculated (Table 5.16). In assessing the annual sulphate flux to the inland drainage basins, it is assumed that the proportion of sulphate in the ion flux of the relatively poorly mineralized waters of Africa, South America, and Asia approximates that in

Table 5.16   Estimates of the annual sulphur flux to inland drainage basins

Territory	River[a] flux (km^3)	Ion flux[a] (10^6 tonnes)	Mineral concentration (mg litre^{-1})	Sulphate flux (Tg year^{-1})	Sulphate flux (% of ion flux year^{-1})
Europe[b]	305	89.0	291	18.95	21.3
Kazakhstan and Middle Asia[c] (USSR)	104	47.5	457	11.40	24.0
The rest of Asia	354	80.5	227	16.0	20.0
Africa	158	29.0	183	5.8	20.0
South America	59	4.0	68	0.8	20.0
North America	15	5.0	333	1.25	25.0
Australia	9	10.0	1110	2.5	25.0
Total	1004	265.0	264	56.7	—

[a] All figures, except those referring to Europe, Kazakhstan, and Middle Asia, are taken from Korzun *et al.* (1974).
[b] The flux from the European continent is taken from the data on the Caspian Sea basin (Zverev, 1971).
[c] Data of Zverev (1971).

the river water of the Caspian Sea basin (about 20%), and that the more mineralized river-waters of the North American and Australian territories without runoff to the oceans have a sulphate content close to that of the waters of Kazakhstan and Middle Asia (i.e. 25% of the total ion flux). The annual flux was calculated to be 19 TgS.

The total amount of sulphur removed from rocks by weathering, includes sulphate that is dissolved, and subsequently redeposited in the form of salts. Within the USSR, the salt-affected areas occupy 3.5×10^6 km^2, and the amount of sulphate accumulated over this territory is about 4.7 TgS (Zverev, 1971). Thus, the total amount of sulphur removed annually from the global biogeochemical cycle by processes occurring in continental water reservoirs amounts to 35 TgS, of which 19 TgS are transported to the water reservoirs of the inland drainage regions, and 16 TgS are fixed in the bottom sediments of freshwater reservoirs.

Part II SULPHUR FLUX FROM CONTINENTS TO OCEANS

M. V. IVANOV, V. A. GRINENKO AND A. P. RABINOVICH

As shown in Chapter 4, about 100.5 TgS year^{-1} is transferred from continents to oceans through the atmosphere. In this section we attempt to assess the amount of sulphur transported to oceans in river runoff and to estimate the relative contribution of various natural and anthropogenic processes to this flux.

5.5. SULPHUR FLUX IN RIVER RUNOFF

Estimates of global sulphur transport in river runoff vary between 73 TgS year^{-1} (Robinson and Robbins, 1970 and 136 TgS year^{-1} (Friend 1973). This broad spectrum of estimates is the result of different methodological approaches to the assessment of this flux. Another reason for the variability is that some authors take into account both natural and man-made sulphur emissions, while others consider only the natural sulphur sources.

As the steadily increasing consumption of sulphur-containing fuels and sulphur minerals is followed by an increase in atmospheric and hydrospheric pollution, the assessment of natural sulphur emissions is of paramount importance for a better understanding of the influence of man on the global biogeochemical sulphur cycle, and for predicting future trends in these fluxes from models of industrial and agricultural development.

5.5.1   Natural Sulphur Flux in River Runoff

To estimate the total amount of sulphur transferred to the oceans in rivers, the total water discharge and the average sulphate concentration of river-water must be known. Values for these two parameters have been extensively modified over the years as new data have been acquired. This is especially true for estimates of the total water flow. In 1960 the annual river flow was estimated to be 18 300 km^3 (Eriksson, 1960), while more recent information suggests that the river flux from the land to the ocean is 42 400 km^3 year^{-1} (Korzun *et al.*, 1974).

Livingstone (1963) estimated the average content of various ions in the river-waters of some continents from a large body of data dating back to the first half of this century: the lowest average sulphate concentration was found in Australian rivers (Table 5.17). From more recent data, however, we calculate the average concentration to be 15 mg litre^{-1}. These calculations were based on the mean sulphate concentration for 10 rivers of New South Wales which have a total annual discharge of ~10 km^3 (Garman, 1980) and the mean sulphate concentration in water of the Murray River (Garman, 1980) which has a mean discharge of 10.5 km^3 year^{-1} (Korzun *et al.*, 1974). While this information represents only 7.5% of the total river discharge of Australia, it is nevertheless more representative than Livingstone's estimate which was based on relatively few data for the Rose and Murray rivers.

Data for other continents (Table 5.17) demonstrate that the mean sulphate concentration in river-waters of Europe and North America is substantially higher than those of other continents. This may be due to anthropogenic sulphur pollution of river-waters in the two most industrialized world regions (Berner, 1971).

Livingstone's (1963) data (Table 5.17) suggest that the sulphate : chloride

Table 5.17 Average sulphate and chloride concentrations and sulphate:chloride ratios in river-waters of different continents (Livingstone, 1963)

Continent	Sulphate	Chloride	Sulphate:chloride ratio
	(mg litre^{-1})		
Europe	24.0	6.9	3.48
North America	20.0	8.0	2.50
South America	4.8	4.9	0.98
Asia	8.4	8.7	0.96
Africa	13.5	12.1	1.11
Australia	2.6	10.0	0.26

Note: According to the data of Gibbs (1972) the mean concentrations of sulphate and chloride in the rivers of South America are 4.3 and 5.4 mg litre^{-1} respectively, giving a sulphate:chloride ratio of 0.796.

ratio in river-waters may be used to assess the amount of anthropogenic sulphate in river-waters of North America and Europe. The sulphate : chloride ratio in river-waters from three large but poorly industrialized continents, South America, Asia, and Africa, approximates 1 (0.96–1.11). However, in the waters of North America and Europe the ratios are much higher—2.5 and 3.48 respectively. A similar relationship is observed in river-waters of the USSR (Table 5.18): in the relatively uncontaminated waters of the Siberian rivers (basins of the Kara Sea and East Siberian Sea) the sulphate : chloride ratio approximates 1, while in the waters of the European part of the USSR this ratio varies from 1 (in the Baltic Sea basin) to 3.28 (in the Caspian Sea basin). It is noteworthy that, of all the rivers which discharge into the ocean from the Asian part of the USSR, the highest sulphate : chloride ratio in river-water (2.32) is found in the river discharge into the seas surrounding the Soviet Far East (Table 5.18).The Amur River, which accounts for 40% of the total river runoff from this region, flows through the most densely populated and industrialized area of eastern Siberia. It is suggested that the raised sulphate content in the total river runoff of the Far East is due to a marked anthropogenic impact.

With the exception of the Parana and Amazon rivers in South America, the chemistry of river runoff in other world regions has been investigated to a lesser extent. Systematic analyses of water from the Middle Parana, upstream of the Argentine cities of Santa Fe and Parana, were carried out from March 1970 to December 1972 by Depetris (1976). According to his data, the average annual fluxes of sulphate and chloride in the non-polluted waters of this river are 1.5 Tg and 2.8 Tg respectively (sulphate : chloride ratio = 0.535).

The Amazon River accounts for 53% of the total runoff from the South American continent and for 16% of the world runoff to the oceans. This relatively unpolluted river discharge 17 Tg of sulphate and 22 Tg of chloride

Table 5.18 Average sulphate and chloride concentrations and sulphate:chloride ratios in river-waters flowing into the different seas of the USSR (Alekin and Brazhnikova, 1964)

Sea	Sulphate	Chloride	Sulphate : chloride ratio
	(mg litre^{-1})		
Chukotsk, East Siberian, and Laptev seas	17.5	17.7	0.99
Kara Sea	8.8	7.3	1.20
Barents and White seas	14.8	5.0	2.96
Baltic Sea	7.0	4.0	1.75
Black Sea and Sea of Azov	41.5	16.5	2.51
Caspian Sea	62.1	18.9	3.28
Bering, Okhotsk, and Japan seas	5.8	2.5	2.32

into the ocean annually (average sulphate : chloride ratio = 0.773). Variations in sulphate and chloride concentrations in the Amazon River during one year, based on data of Gibbs (1972), are shown in Fig. 5.17. The monthly discharge of chloride, in general, varies in accordance with changes in water flow. From May to October, however, relatively less sulphate is discharged and the sulphate : chloride ratio drops from 0.94–1.20 (in January–April) to as low as 0.43 in August. By the end of the year the monthly discharges of sulphate and chloride return to the levels near those of January (sulphate : chloride ratio of 0.93).

The observed variation in sulphate and chloride fluxes suggests that, from May to October, a considerable amount of sulphate is withdrawn from the waters of the Amazon and is thus removed from the cycle. The data of Brinkmann and Santos (1974) (Fig. 5.6, 5.7, and 5.8) do indeed show that, from February up to late December, intensive sulphate reduction occurs in the marshy floodlands of the Middle Amazon. The results reported in Fig. 5.17 suggest that this process continues during the whole flood period,

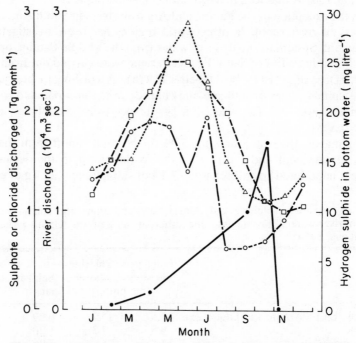

Fig. 5.17   Average monthly water discharge of the Amazon River (□), sulphate (○) and chloride (△) fluxes, and hydrogen sulphide concentration (●) in the bottom water of floodplain reservoirs (some data from Gibbs, 1972 was used in the preparation of this figure)

and show that the hydrogen sulphide concentration reaches 17 mg litre^{-1} of bottom water by the end of October. Some hydrogen sulphide escapes to the atmosphere (Brinkmann and Santos, 1974), and some is fixed in the form of sulphides that are slowly incorporated into the sediments of flood lakes. Thus, in the Amazon floodlands, mechanisms are present for the partial removal of sulphate–sulphur from river-water to the atmosphere and apparently to the soil and lithosphere, which account for the relatively low sulphate concentration in this large river. The mean value for the sulphate : chloride ratio for the six months' period when sulphate reduction is least intense (January–April and November–December) is 0.964, which is close to the sulphate : chloride ratio in river-waters of non-industrialized continents (Table 5.17) and the vast undeveloped regions of Soviet Siberia (Table 5.18).

On the basis of (1) Livingstone's (1963) data for river waters of various continents; (2) those of Alekin and Brazhnikova (1964) for the rivers of various regions of the USSR, and (3) the more recent data on the South American rivers (Gibbs, 1972; Depetris, 1976), an upper estimate can be made of the annual flux of sulphate–sulphur in river-water from continents to oceans, excluding man-made pollution. This equals, by weight, the chloride flux and amounts to 104.1 TgS year^{-1} (Table 5.19).

The major processes contributing to the sulphur content in rivers are shown in Fig. 5.18. Among the natural processes a large contribution is made from atmospheric sulphur in precipitation, or by dry deposition and adsorption. The flux of sulphur from the atmosphere to the continents is estimated in the present work to be 84 TgS year^{-1} (see Chapter 4). This value includes 12 TgS

Table 5.19   Sulphur flux to the oceans as dissolved sulphate in rivers

Continent	Territory with flow to ocean (10^6 km^2)[a]	Annual flow (10^3 km^3)[a]	Average sulphate content (mg litre^{-1})	Flux Sulphate ($TgSO_4^{2-}$ year^{-1})	Flux Sulphur (TgS year^{-1})
North America	19.5	7.84	8.0[b]	62.72	20.9
South America	16.4	11.70	4.3[c]	50.31	16.77
Eurasia	39.1	16.40	8.4[d]	137.76	45.92
Africa	20.5	4.11	13.5[e]	55.48	18.49
Australia and Oceania	4.8	2.37	2.6[e]	6.16	2.05
Total	100.3	42.42	—	312.43	104.13

[a]Korzun *et al.* (1974).
[b]Calculated from chloride concentration (Table 5.17).
[c]Gibbs (1972).
[d]Livingstone's (1963) data for Asia.
[e]Livingstone (1963).

in dry deposition in desert areas without water discharge to the ocean. Consequently, the total atmospheric sulphur flux from continents to the oceans in rivers is 72 TgS year^{-1} (Figs. 4.7, 4.8), 47 TgS of which represent anthropogenic sulphur and 25 TgS from natural sources (Fig. 5.18).

Another important natural process providing the ocean with sulphur is the erosion of soils and rocks. Production from this source is estimated to be 79.1 TgS year^{-1} (see the scheme in Fig. 5.18). This value is the difference

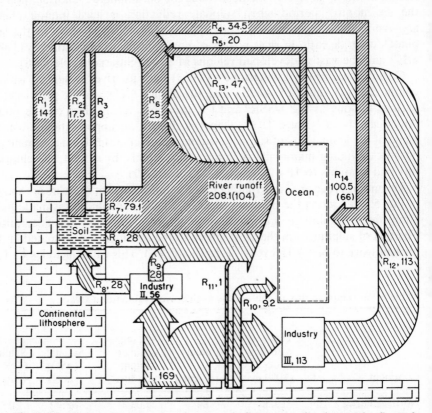

Fig. 5.18 Major natural and anthropogenic fluxes contributing to the flow of sulphate sulphur from the continents to the oceans. Figures by the indices denote the value of the flux in TgS year^{-1}. Symbols: R_1–R_7, natural sulphur fluxes: R_1, volcanic; R_2, biogenic; R_3, dust emission; R_4, flux from the continental atmosphere to the oceans; R_5, flux from the oceans to continents; R_6, total flux from the atmosphere to the continents; R_7, water erosion; R_8, fertilizer sulphur transferred to soil drainage waters; R_9, sewage water; R_{10}, underground sulphur flux; R_{11}, sewage waters from mining and manufacturing; R_{12}, anthropogenic flux to the atmosphere; R_{13}, atmospheric sulphur precipitating on to continents; R_{14}, anthropogenic and natural sulphur transferred from continents to oceans. I, Sulphur output from lithosphere as a result of mining; II, processing of pyrites, native sulphur, and sulphur-containing gases in chemical industry; III, fossil fuel combustion and sulphide ore processing

between the total input of sulphur to rivers due to natural processes other than weathering (Table 5.19) and the amount of non-anthropogenic sulphur transferred from the atmosphere to continents and rivers.

5.5.2 Sulphur Flux to the Oceans in Ground Waters

Dissolved sulphur is transferred from continents to oceans, not only by river runoff but also by ground-waters. Jamalov *et al.* (1977) estimated that the amount of water delivered to the ocean annually in underground flow is about 2400 km^3, while according to Korzun *et al.* (1974), it is 2200 km^3.

In the following calculations an average figure of 2300 km^3 year^{-1} has been chosen. We assume that these waters fall within the calcium bicarbonate type of underground waters in which the average sulphate concentration is 12 mg litre^{-1} (see Table 5.20). Such a mean value may be adopted for all ground-waters except those of the continental salinization zone (Shvartzev, 1978). On the basis of this mean value the sulphur flux to the oceans in underground flow is 9.2 TgS year^{-1} (Fig. 5.18). This calculated value should be added to the non-anthropogenic sulphur flux to the oceans through river runoff (Table 5.19), giving a total flux of dissolved sulphur from the continents of 113.3 TgS year^{-1}.

Table 5.20 Sulphate and chloride in underground waters of various landscape or climatic regions of the globe (Shvartzev, 1978)

Landscape or climatic region	Sulphate		Chloride		Sulphate: chloride ratio
	Number of analyses	(mg litre^{-1})	Number of analyses	(mg litre^{-1})	
Region of permafrost	7 655	5.58	7 655	5.58	1.0
Region of moderately humid climate	7 032	14.70	7 032	10.90	1.35
Tropical and subtropical regions	914	8.87	914	10.60	0.84
Mountain regions	4 778	18.30	4 778	6.13	2.98
Mean value in ground-waters of the zone of leaching	20 379	11.90	20 379	8.30	1.43
Ground-waters of the zone of continental salinization	1 825	328	1 825	202	1.62

5.5.3 Man-made Contribution

Estimates of the anthropogenic sulphur flux to the world oceans in river runoff are also given in Fig. 5.18. These estimates are based on data for the world output of various sulphur-containing minerals that are used as fuel and raw materials.

Sulphur pollution of the atmosphere by the combustion of fossil fuels is discussed in detail in Chapter 4. The total annual emission of anthropogenic sulphur into the atmosphere is estimated to be 113 TgS year^{-1}, and part of this sulphur (47 TgS year^{-1}) is deposited on continents and contributes to the composition of the rivers (Fig. 5.18). Further pollution of the atmosphere and hydrosphere is caused by the products and wastes of chemical industries which are using ever-increasing amounts of sulphur-containing raw materials such as elemental sulphur, pyrite, and hydrogen sulphide.

There has been a rapid growth in world sulphur production during the past two decades and, in 1975, it was close to 52 Tg (Nriagu, 1978). This rapid growth was due mainly to the increasing application of sulphuric acid in fertilizer production. More than 85% of the sulphur produced is used to make sulphuric acid, about 60% of which is used for the production of super-phosphate and ammonium sulphate utilized in agriculture (Nriagu, 1978). Other industries which consume large quantities of sulphuric acid include the chemical industry, steel manufacturing, petroleum refining, and the cellulose industry. In some countries sulphuric acid is also used in the mining industry for the leaching of copper and uranium from ores.

Other sulphur compounds are used in industry in diverse ways: for example native sulphur for the improvement of alkaline soils and in the production of fungicides and insecticides, sulphur dioxide in the pulp and paper industry, and as a preservative in some branches of the food industry. Carbon disulphide and some other sulphur compounds are widely used in the production of synthetic materials. The rapid growth in sulphur consumption, mainly in fertilizer production, began during the 1960s, and over the past decade the average annual increase in sulphur consumption approached 8.5% (Nriagu, 1978). In 1974–75, the total global sulphur consumption was 47.5 TgS, and growth rates in developing countries were noticeably higher than in industrialized countries (Table 5.21). The same tendency is anticipated for the forthcoming decade and, according to some estimates, the total consumption will exceed 65 TgS in 1985 (Table 5.21).

Based on Nriagu's (1978) estimates for sulphur consumption in 1980 (56 TgS) and its utilization in various industries (Fig. 5.19), we estimate that 28 TgS were added to soil in fertilizers in that year. The sulphate component of both phosphate and nitrogen fertilizers is partially used by plants while the remainder may be leached from soil in drainage waters and transferred to rivers. The removal of sulphur in harvested crops is assessed as 6 TgS year^{-1} (Kilmer, 1979). After being digested by man and animals this sulphur is also discharged into sewers and finally into rivers. Thus, the content of anthropogenic sulphates in rivers may increase by a value equivalent to the annual consumption of sulphur fertilizers (28 TgS year^{-1}; Fig. 5.18).

Table 5.21 Sulphur use in different regions of the world (Nriagu, 1978)

		Use	
		1974–75	1985 (estimated)
Region	Annual increase in use 1965–75 (%)	(TgS year^{-1})	
North America	—	13.6	17.7
USSR and eastern Europe	8	12.0	16.5
Western Europe	3	11.7	14.4
Asia and Oceania[a]	12	6.0	8.4
Africa and Middle East	2	2.5	5.0
South America	10	1.7	3.6
World total	8.5	47.5	65.6

[a]Data for India, Pakistan, and Bangladesh.

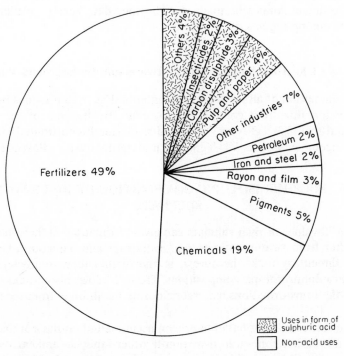

Fig. 5.19 Sulphur consumption in various branches of industry during 1970 (after Cote, 1970)

The remainder of the sulphur used in industry (28 TgS year^{-1} in 1974–80) is transferred to rivers in drainage waters from industrial centres and urban areas (Fig. 5.18). Thus, about 169 TgS year^{-1} are transferred from the lithosphere as a result of the extraction and processing of various raw materials. About 56 Tg of this amount enter the rivers in sewage and drainage waters, and 113 Tg are transferred to the atmosphere and then partially returned to the continents by precipitation.

Sulphate contamination of the hydrosphere also occurs due to weathering of residues in mining regions where deposits of various metal sulphides, sulphur-containing coal, and native sulphur are exploited. Acid drainage waters from various mining operations are generated in many regions, but sulphuric acid input due to acid weathering has been assessed, as far as we know, only for the Appalachian coal basin in the USA. According to Barton (1978), ~0.004 Tg of sulphuric acid (0.0013 TgS), in terms of concentrated ore, are transferred to the rivers of this region daily. Thus, 0.48 TgS produced by acid weathering is carried annually to the rivers in the Appalachian basin. We have arbitrarily used a value of twice this figure for the flux of this type of anthropogenic sulphur from the lithosphere to the hydrosphere (Fig. 5.18). Obviously, the actual value of this flux is far greater than 1 TgS year^{-1}, but a precise estimate cannot be made until further data become available from different mining regions of the world.

5.5.4 The Total Sulphur Flux in Rivers and Underground Waters

The summation of all the fluxes presented in Fig. 5.18 provides the annual sulphate ion flux from continents to oceans. By the end of the 1970s it amounted to 217.3 TgS year^{-1}. Natural river runoff contributed 104.1 TgS, anthropogenic emissions 104 TgS, and underground water flow 9.2 TgS.

5.6 SULPHUR ISOTOPIC COMPOSITION OF RIVER-WATER SULPHATE

The δ^{34}S values of river sulphate can give information on the contributions of sulphur from various sources, e.g. evaporites and oxidation of sulphides from sedimentary rocks, because it is known that these sources have very different sulphur isotopic compositions. The δ^{34}S values also provide information on the transformations undergone during the journey from the source to the sea.

When sufficient information has been accumulated on the isotopic composition and sulphur content in river runoff, other inputs to and emissions from the ocean, it will be possible to prepare a sulphur budget for the ocean. Both mass and isotopic fluxes must balance in any realistic budget for the sulphur cycle.

5.6.1 Variation in Isotopic Composition Along Rivers

Sulphate from almost all of the left-bank tributaries of the Volga (the Vetluga, the Koshaga, the Kama, and the Sok) which drain a territory with widely developed Permian evaporites, has rather high $\delta^{34}S$ values ($+7.7$ to $+10.5\%o$). The $\delta^{34}S$ value for sulphate in the Oka River, which also drains the developed Permian rocks in its lower reaches, falls within the same range.

On the other hand, the right-bank tributaries of the Middle Volga (the Sura and Sviyaga) have a different isotopic sulphur composition ($\delta^{34}S = 0.0$ to $+2.2\%o$). The isotopic composition of sulphate from the Volga water itself does not vary greatly ($+5.5$ to $+6.2\%o$), and it is only below the Volga's confluence with the Oka that its increases to $+7.3\%o$ (Veselovsky *et al.*, 1964).

In rivers with a lower water volume, fluctuations in $\delta^{34}S$ due to variation in the isotopic composition of the rocks in the drainage basin can be more pronounced. The upper basin of the Kuma River, and its tributary the Podkumok, are located at the northern slope of the Caucasian ridge—which is characterized by sedimentary rocks of the Upper Jura and the Cretaceous period. The $\delta^{34}S$ values of sulphates from these rocks and ground-waters vary from $+11.8$ to $+12.5\%o$ (Pankina *et al.*, 1966). Sulphates from the upper reaches of the Kuma, Podkumok, and Eshkanon rivers are also enriched in[34]S (Veselovsky *et al.*, 1969).

In the foothills of the basin, where Oligocene rock with isotopically light sulphides occurs ($\delta^{34}S = -13.5\%o$, Ronov *et al.*, 1974), the sulphates in the Kuma and Podkumok rivers are also markedly enriched in [32]S (Veselovsky *et al.*, 1969).

Hitchon and Krouse (1972) reported on the isotopic composition of sulphate–sulphur of river- and lake-waters, using the reservoirs of the Mackenzie River drainage basin in western Canada as an illustration. The basin of this river includes various rocks: carbonates and evaporites make up 23% of the basin area, argillaceous shale, 40%, and igneous and metamorphic rocks, 29% (Reeder *et al.*, 1972). Figure 5.20 depicts the results of sulphur and isotopic analyses of 52 samples of surface waters from the Mackenzie River basin taken during summer and grouped according to the geology of the respective sub-basins. Samples taken from regions of exposed evaporite deposits (black squares) generally show a wide range of sulphate concentrations and moderately heavy $\delta^{34}S$ values. Isotopically heavier sulphates occur in poorly mineralised waters draining Paleozoic rocks (Fig. 5.20, solid circles). Surface-water-draining deposits of sulphidized argillaceous schists (Fig. 5.20, black triangles) are characterized by intermediate sulphate concentrations (20–110 mg litre^{-1}) and negative $\delta^{34}S$ values.

Thus, the results of Hitchon and Krouse (1972) clearly indicate that, in the absence of appreciable anthropogenic pollution, the isotopic composition of rock sulphur has a pronounced effect upon the isotopic composition of

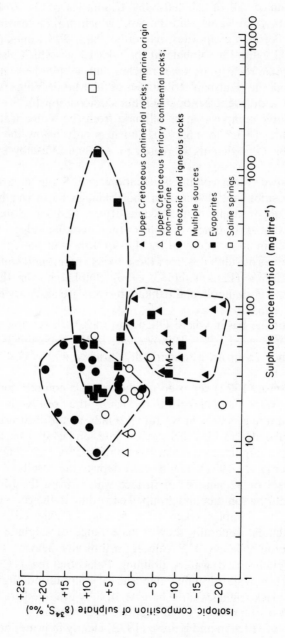

Fig. 5.20 Scatter diagram showing the relationship between $\delta^{34}S$ sulphate and sulphate concentration for 52 surface waters from the Mackenzie River drainage basin, Canada: Point M-44 in this figure indicates the isotopic ratio for sulphate–sulphur from the Mackenzie River delta. (from Hitchon and Krouse, 1972)

surface-water sulphate. However, the major water artery of the basin—the Mackenzie River itself—was represented by only one sample from the delta region. Sulphate in the sample had a δ^{34}S value of $-8.1^0/oo$. If this value is representative, it would suggest that the major part of the sulphur flux in summer originates from soil organic matter, with an isotopic composition in the range -28.0 to $+4.6^0/oo$ (Lowe *et al.*, 1971; Hitchon and Krouse, 1972), or from the oxidation of isotopically light sulphides.

Figure 5.21 presents the frequency distribution of the δ^{34}S values for sulphate in waters of the Mackenzie River basin. Most of the values are between 0 and $+10^0/oo$, a range which is typical of sulphur in rain-water. However, the above-mentioned associations of δ^{34}S values of water sulphate with the types of rocks in the drainage sub-basins, and the high salt (203 mg litre^{-1}) and sulphate (31.4 mg litre^{-1}) concentrations in the waters of the Mackenzie River compared to other Canadian rivers (Reeder *et al.*, 1972), demonstrate conclusively that rain-water sulphate does not play an important part in the sulphur budget of this river system. The extremely wide range of δ^{34}S values ($+19.5$ to $-21.5^0/oo$) of the surface waters of this region reinforces this conclusion.

Some information is available on the variation in isotopic composition of sulphate–sulphur from two small Italian rivers, the Arno and Sergio (Longinelli and Cortecci, 1970), and the river Jordan in Israel. In both Italian rivers, throughout all observation periods, sulphate increased in concentration and ^{34}S content as the waters flowed from source to mouth (Fig. 5.22).

Nissenbaum (1978) attempted to assess the influence of atmospheric sulphur upon the isotopic composition of river sulphate from data for the river Jordan. Figure 5.23 clearly shows that the isotopic composition of the small

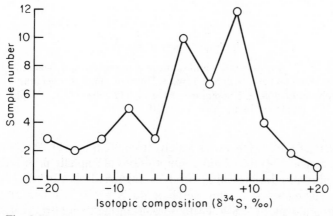

Fig. 5.21 Frequency distribution curve for the isotopic ratio of sulphate in surface waters of the Mackenzie River basin. (Hitchon and Krouse, 1972)

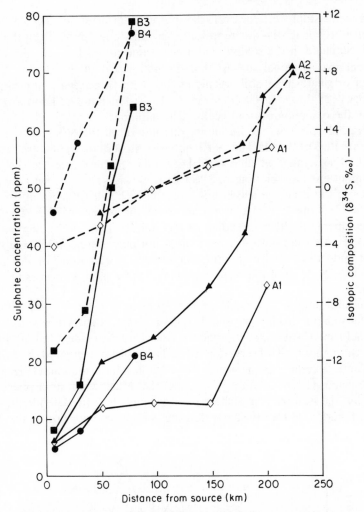

Fig. 5.22 Isotopic ratios and sulphate concentrations downstream in the Arno (A) and Sergio (B) rivers. Dates of samples collection; 1, 7 February, 1969; 2, 5 September 1968; 3, 9 October 1968; 4, 7 March 1969. (Longinelli and Cortecci, 1970)

source rivers of the Jordan varies widely, from +2.6 to +17.8⁰/oo. According to Nissenbaum (1978) the isotopic compositions of sulphate in surface waters from the northern Jordan valley are best explained by a model (Fig. 5.24) involving mixing of sea-water sulphate ($\delta^{34}S \cong +21^0/oo$) with rain-water derived ground-water ($\delta^{34}S = +4$ to $+9^0/oo$). In the area north of Lake Kinneret, sulphate is almost exclusively derived from recharging rain-water, $\delta^{34}S$ values being determined by the relative contributions of Jordan River sul-

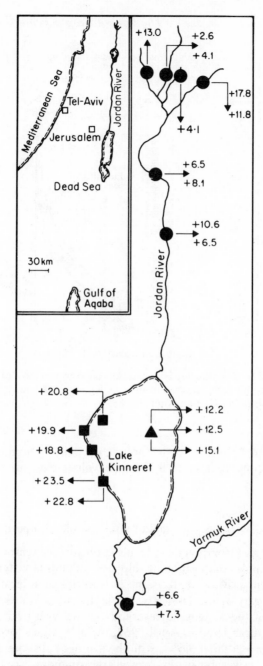

Fig. 5.23 Sampling sites and isotopic composition of sulphate from the Jordan River and Lake Kinneret (Israel). ●, River-water; ▲, lake-water from three horizons; 0, 20, and 42 m above the sediment surface; ■, spring water (after Nissenbaum, 1978)

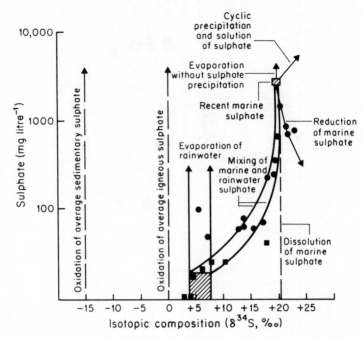

Fig. 5.24   Relationship between isotopic composition and sulphate concentration. ●, Samples from Lake Kinneret; ■ samples north of Lake Kinneret (Nissenbaum, 1978)

phate and sulphate springs debouching at the lake bottom. Saline springs around Lake Kinneret have $\delta^{34}S$ sulphate values close to that of marine sulphate.

5.6.2   Temporal Variation in Amount and Isotopic Composition of Sulphate

Veselovsky *et al.* (1966) made a detailed study of the temporal variation in amount and isotopic composition of sulphate–sulphur in water samples from the Don and Volga rivers. Water samples were taken at least once a month from the deep part of the Don River near the town of Aksai, and at the overflow weir of the Volgograd reservoir on the Volga River. During the period from October 1960 to April 1962, the $\delta^{34}S$ values for sulphate in the Don varied from +1.9 to +6.6‰, with the average value being +4.1‰ (Fig. 5.25); more extensive observations revealed even more variation in $\delta^{34}S$ values (−2.7 to +8.5‰; Rabinovich and Veselovsky, 1974).

A wide range of $\delta^{34}S$ values (+4.0 to +7.3‰) was also obtained for sulphate from the Volga River (Fig. 5.26). The isotopic composition of the

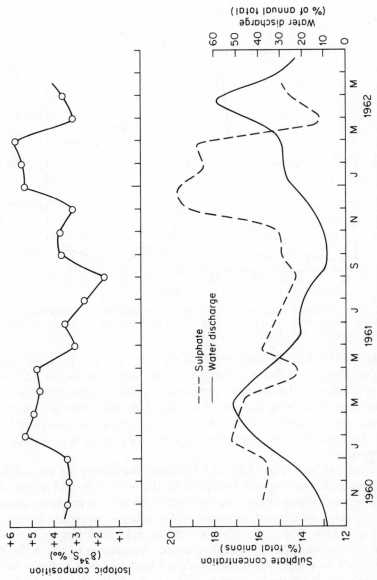

Fig. 5.25 Seasonal variation in water discharge, sulphate concentration, and isotopic composition of the sulphate in water from the Don River (after Rabinovich and Veselovsky, 1974)

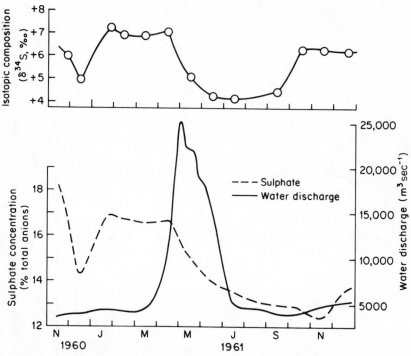

Fig. 5.26   Seasonal variation in water discharge, sulphate concentration, and isotopic composition of the sulphate in water from the Volga River (after Rabinovich and Veselovsky, 1974)

sulphate and the chemical composition of the water phase in the lower flow of the Volga River are usually within the ranges exhibited by the Kuibyshev reservoir. However, the chemical composition of the river-waters is not constant throughout the year because of marked flow from both the Kuibyshev and Volgograd reservoirs. Therefore, the water masses formed in different seasons preserve to some extent their specific features such as isotopic and salt composition.

It is apparent from Figs. 5.25 and 5.26 that the changes in the sulphate content and isotopic composition depend on the water flow and season. The sulphate concentrations of the waters of both rivers are low in late summer. In winter the sulphate content increases and reaches a maximum before the spring flood (Figs. 5.25 and 5.26). Isotopically light sulphates are typical of the summer period, while in winter the sulphur becomes enriched in ^{34}S. This indicates that in winter, when the rivers are iced up, their water is fed by mineralized ground-waters, carrying isotopically heavy sulphates leached from sulphate-bearing evaporites. In summer the amount of evaporite sulphate decreases and the river-water gains isotopically light sulphate, either from the atmosphere or from oxidation of sedimentary sulphides.

5.6.3 Characteristics of the Sulphate Flux from the USSR

In spite of the considerable variability in $\delta^{34}S$ values of river-water sulphate due to hydrological, lithological, and seasonal factors, we may calculate a provisional average $\delta^{34}S$ value for sulphate transported in rivers from continents to oceans. Such a calculation, based on the data of Rabinovich and Grinenko (1979), can be made for two large regions of the USSR, the European and the Asian.

Data obtained during three years of observation on estuaries of the six largest rivers of the Asian part of the USSR are listed in Table 5.22. The mean $\delta^{34}S$ value for sulphate is calculated from the sulphate flux as the weighted mean value for the observation period. Quite different results have been obtained for the rivers of the European part of the USSR (Table 5.23), where the mean $\delta^{34}S$ sulphate is near 5.6‰.

Two important factors influence the isotopic composition of sulphate from the two regions: In the European part of the USSR fairly light evaporites are abundant ($\delta^{34}S = +13.5‰$), while isotopically heavy Cambrian evaporites ($\delta^{34}S = +28‰$) are characteristic of Siberia. In addition, it is believed that anthropogenic sulphur markedly affects the isotopic composition of the sulphate in river runoff in the European part of the USSR. It is probably not due to chance that the lightest $\delta^{34}S$ values are obtained for polluted rivers such as the Kura, Danube, Ural, and Kuban (Table 5.23).

On the other hand, the mean isotopic composition of sulphate–sulphur ($\delta^{34}S = +9.5‰$) from the least polluted northern rivers (the Pechora and Severnaya Dvina) closely approximates that of Permian sulphate ($\delta^{34}S = +11‰$) which is widely distributed in the basins of these rivers.

Table 5.22   The total sulphur flux and $\delta^{34}S$ values for sulphate in the largest rivers of the Asian part of the USSR (1971–74)

River	Sulphate flux (TgS year^{-1})	$\delta^{34}S$ value Range (‰)	$\delta^{34}S$ value Mean (‰)
Aral Sea basin			
Syr Daria	1.3	+6.7 to +10.5	+9.4
Amu Daria	2.5	+6.1 to +12.5	+10.2
Total	3.8	—	+9.92
Pacific and Arctic Ocean basins			
Amur	1.0	+6.0 to +10.6	+6.7
Lena	3.5	+2.5 to +32.0	+14.0
Yenisei	1.8	+5.4 to +15.8	+10.9
Ob	1.3	+5.6 to +11.6	+7.6
Total	7.6	—	+11.16
Total	11.4	—	+10.75

Table 5.23   Total sulphur flux and $\delta^{34}S$ values for sulphate in the largest rivers of the European part of the USSR (1969–72)

River	Sulphate flux (TgS year^{-1})	$\delta^{34}S$ value Range ($^0/oo$)	$\delta^{34}S$ value Mean ($^0/oo$)
Arctic Ocean basins			
Severnaya Dvina	1.2	+6.2 to +13.6	+9.9
Pechora	0.5	−2.4 to +12.2	+8.6
Total	1.7	—	+9.54
Baltic Sea basin			
Neva	0.14	—	+5.4
Zapadnaya Dvina	0.13	+0.1 to +13.1	+11.0
Nieman	0.13	−0.5 to +8.7	+3.6
Total	0.4	—	+6.77
Sea of Azov basin			
Don	0.9	+3.5 to +6.7	+5.8
Kuban	0.2	−0.6 to +9.1	+3.5
Total	1.1	—	+5.39
Black Sea basin			
Danube	1.9	+1.2 to +5.1	+3.9
Dniester	0.1	+2.9 to +8.4	+5.5
Yuzhny Bug	0.02	+5.9 to +13.2	+8.0
Dnieper	0.6	−2.8 to +9.6	+5.5
Rioni	0.1	+1.9 to +13.7	+8.2
Total	2.7	—	+4.55
Caspian Sea basin			
Kura	1.0	−4.8 to +4.5	−0.5
Terek	0.3	+5.0 to +9.2	+5.1
Volga	4.1	+0.9 to +7.4	+6.0
Ural	0.15	+2.9 to +5.3	+3.9
Total	5.6	—	+4.70
Total	11.5	—	+5.55

Table 5.24   Total flux of sulphate–sulphur to the oceans from USSR rivers (Rabinovich and Grinenko, 1979)

Basin	Sulphate flux (TgS year^{-1})	$\delta^{34}S$, mean value ($^0/oo$)
Pacific and Arctic oceans (Asian part)	7.6	+11.16
Baltic Sea	0.4	+6.77
Black Sea	2.8	+4.55
Sea of Azov	1.1	+5.39
Arctic Ocean (European part)	1.7	+9.54
Total for the USSR	13.6	+9.19

The total amount of sulphate–sulphur entering the world oceans from rivers of the USSR, and its mean isotopic composition is given in Table 5.24. The average value obtained for the isotopic composition of sulphate–sulphur, +9.2‰ for the rivers in the USSR with a total flux of 13 TgS may be regarded as characteristic of the isotopic composition of all sulphur transferred from continents to oceans in rivers.

5.7 CONCLUSIONS

Table 5.25 presents the estimates for inputs and outputs of sulphur in river runoff that were schematically portrayed in Fig. 5.1. Justification for these estimates is given in the corresponding sections of this chapter. Sulphur leaching during rock weathering, was obtained by difference.

Finally, in Table 5.26, the absolute and relative sulphur fluxes which characterize three major pathways of sulphur supply to continental reservoirs are compared. They are: (1) weathering and water erosion of rocks (weathered sulphur); (2) various natural processes leading to sulphur emission to the atmosphere and its subsequent deposition (atmospheric sulphur); (3) sulphur entering continental water-bodies as a result of extraction and exploitation by man of various minerals (anthropogenic sulphur). It should be noted that the relative values of the sulphur fluxes for continental water bodies are fairly close to the estimates made by Zverev (1971) in his work devoted to the hydrochemical budget of the USSR, even though the experimental and computational approaches made by him were substantially different from those used in this work.

A major and quite alarming conclusion drawn from the data of Table 5.26 is that the total amount of sulphur involved in the sulphur cycle of continental reservoirs has nearly doubled due to man's activity. For highly industrialized continents such as Europe and North America the amount has increased to a greater extent, by factors of 3.5 and 2.5 respectively (see Table 5.17).

It is also apparent from Table 5.26 that, when compiling budgets of river runoff, geochemists should take into account the large amount of sulphur cycled through the atmosphere.

Table 5.25 Sulphur budget for continental water bodies

Sulphur influx (TgS year^{-1})		Sulphur efflux (TgS year^{-1})	
1. Leaching of fertilizers	28.0	7. Flux to the ocean	208.1
2. Sewage water	28.0	8. Flux into continental water bodies	19.0
3. Acid drainage	1.0	9. Burial of sulphur in bottom	
4. Acid rain	47.0	sediments of lakes and reservoirs	16.0
5. Natural atmospheric	25.0		
6. Water erosion of rocks	114.1		
Totals	243.1		243.1

Table 5.26 Comparative characteristics of three major pathways of sulphur supply to continental water bodies

	Global flux		Flux for the USSR (Zverev, 1971)	
	(TgS year^{-1})	(%)	(TgS year^{-1})	(%)
Sulphur of weathering	114.1	46.9	5.0	35.4
Atmospheric sulphur from natural processes	25.0	10.3	2.2	15.3
Anthropogenic sulphur	104.0	42.8	7.2	49.3
Totals	243.1	100	14.4	100

REFERENCES

Ackermann, W. C., Harmeson, R. H., and Sinclair, R. A. (1970) Some long-term trends in water quality of rivers and lakes. *Trans. Am. Geophys. Union*, **51**, 516–522.

Alekin, O. A., and Brazhnikova, L. V. (1964) *Runoff of Dissolved Compounds from the Territory of the USSR*. Nauka, Moscow, 144 pp. [in Russian].

Almer, B., Dickson, W., Ekström, C., and Hörnström, E. (1974) Effects of acidification on Swedish lakes. *Ambio*, **3**, 30–36.

Bachmann, H. (1931) Hydrobiologische untersuchungen am Rotsee. *Z. Hydrol.*, **5**, 39.

Barton, P. (1978) The acid mine drainage. In: Nriagu, J. O. (ed.), *Sulfur in the Environment*, Pt. 2. Wiley, Chichester, pp. 313–358.

Beamish, R. J., and Harvey, H. H. (1972) Acidification of the La Cloche mountain lakes, Ontario and resulting fish mortalities. *J. Fish. Res. Board Can.*, **29**, 1131–1143.

Beeton, A M. (1965) Eutrophication of the St. Lawrence Great Lakes. *Limnol. Oceanogr.*, **10**, 240–254.

Berner, R. A. (1971) Worldwide sulphur pollution of rivers. *J. Geophys. Res.*, **76**, 6597–6600.

Bondarev, L. G. (1974) *Perpetual Motion: Preliminary Transfer of Matter and Man*. Mysl, Moscow, 158 pp. [in Russian].

Brinkmann, W. L. F., and Santos, U. (1974) The emission of biogenic hydrogen sulfide from Amazonian floodplain lakes. *Tellus*, **26**, 261–267.

Burton, H. R., and Barker, R. J. (1979) Sulfur chemistry and microbiological fractionation of sulfur isotopes in a saline Antarctic Lake. *Geomicrobial. J.*, **1**, 329–340.

Chebotarev, E. N. (1974a) Ecology and geochemical activity of sulfate-reducing bacteria in water bodies. In: *All-Union Conference. Ecology and Geochemical Activity of Microorganisms. Abstracts.* Pushchino [in Russian].

Chebotarev, E. N. (1974b) Microbiological formation of hydrogen sulfide in fresh water karst lakes: Bolshoi Kichiger and Chyornyi Kichiger. *Mikrobiologiya*, **43**, No. 6, 1105–1110 [in Russian].

Chebotarev, E. N. (1975) Geochemical activity of sulfate reducing bacteria. Synopsis of thesis. Moscow, 26 pp. [in Russian].

Chebotarev, E. N., Gorlenko, V. M., and Kachalkin, V. I. (1973) Microbiological process of hydrogen sulfide formation in the Repnoe Lake (Slavyanski Lakes). *Mikrobiologiya*, **42**, No. 3, 537–541 [in Russian].

Cote, P. R. (1970) Sulphur. In: *Canadian Minerals Year Book*, Department of Energy, Mines and Resources, Ottawa, Ontario, pp. 513–526.

Depetris, P. J. (1976) Hydrochemistry of the Parana River. *Limnol. Oceanogr.* **21**, 736–739.

Dolgov, G. I. (1955) Sabinsky lakes. *Tr. Vses. Gidrobiol. Ova.*, **6**, 193–204 [in Russian].

Dunnette, D. A. (1974) Chemical ecology of hydrogen sulfide production in freshwater lake sediment. *Diss. Abstr. Int.*, **B34**, 3834.

Egorova, A. A. (1951) Microbiological studies of Belovod Lake. *Mikrobiologiya*, **20**, No. 2, 103–112 [in Russian].

Eriksson, E. (1960) The yearly circulation of chloride and sulfur in nature; meteorological, geochemical and pedological implications. *Tellus*, **12**, 63–109.

Friend, J. P. (1973) The sulphur cycle in nature. In: Rasool, S. I. (ed.), *Chemistry of the Lower Atmosphere*. Plenum Press, New York, pp. 177–201.

Garman, D. E. J. (1980) Sulfate levels in New South Wales rivers. In: Freney, J. R. and Nicolson, A. J. (eds), *Sulfur in Australia*. Australian Academy of Science, Canberra, pp. 100–104.

Genovese, S., and Bruni, V. (1975) Mineralizing microbial activity in Lake Faro, Italy. *Boll. Pesca Piscic. Idrobiol.*, **30**, 39–56.

Gibbs, R. J. (1972) Water chemistry of the Amazon River. *Geochim. Cosmochim. Acta*, **36**, 1061–1066.

Glimph, L. (1973) Sedimentation of reservoirs. In: Ackermann, W. C., White, G. F. and Worthington, E. B. (eds) *International Symposium on Man-Made Lakes: Their Problems and Environmental Effects*. Geophysical Monograph 17. American Geophysical Union, Washington, D.C., pp. 342–348.

Goman, G. A. (1975) Bacterial reduction of sulfates and oxidation of sulfides in soils of the Baikal Lake. *Gidrobiol. Zh.*, **11**, No. 5, 18–21 [in Russian].

Gorlenko, V. M., Dubinina, G. A., and Kuznetsov, S. I. (1977) *Ecology of Aquatic Microorganisms*. Nauka, Moscow, 288 pp. [in Russian].

Gorshkov, S. P. (1980) The cycle of products of land denudation. In: Ryabchikov, A. M. (ed.), *Cycle of Matter in Nature and its Variation Under the Impact of Man's Economic Activity*. Izv. MGU, Moscow, pp. 34–55. [in Russian].

Granat, L., Rodhe, H., and Hallberg, R. O. (1976) The global sulphur cycle. In: Svensson, B. H. and Söderlund, R. (eds), *Nitrogen, Phosphorus and Sulphur–Global Cycles*. SCOPE Report 7. *Ecol. Bull. (Stockholm)*, **22**, 89–134.

Hendrey, G. R., Baalsrud, K., Traaen, T., Lakke, M., and Raddum, G. (1976) Acid precipitation: some hydrobiological changes. *Ambio*, **5**, 224–227.

Hitchon, B., and Krouse, H. R. (1972) Hydrogeochemistry of the surface waters of the Mackenzie River drainage basin, Canada. III. Stable isotopes of oxygen, carbon and sulphur. *Geochim. Cosmochim. Acta*, **36**, 1337–1357.

Hutchinson, G. E. (1957) *A Treatise on Limnology*, Vol. 1. *Geography, Physics and Chemistry*. Wiley, New York, 1015 pp.

Isatshenko, B. L. (1914) *Investigation of Bacteria of the Arctic Ocean*. GUZIZ, Petrograd, 297 pp. [in Russian].

Ivanov, M. V. (1956) Application of isotopes for investigation of the intensity of sulfate reduction process in Belovod Lake. *Mikrobiologiya*, **25**, No. 3, 305–309 [in Russian].

Ivanov, M. V. (1978) Influence of microorganisms and microenvironment on the global sulfur cycle. In: Krumbein, W. E. (ed.), *Environmental Biogeochemistry and Geomicrobiology*, vol. 1. Ann Arbor Science, Michigan, pp. 47–61.

Ivanov, M. V. (1979) Role of microorganisms in hydrogen sulfide formation. In: *Role of Microorganisms in the Natural Cycles of Gases.* Nauka, Moscow, pp. 114–130.

Jamalov, R. G., Zektser, I. S., and Meskheteli, A. V. (1977) *Underground Runoff to the Seas and the World Ocean.* Nauka, Moscow, 94 pp. [in Russian].

Kemp, L. W., Gray, C. B. J., and Mudrochova, A. (1972) Changes in C, N. P and S in the last 140 years in three cores from lakes Ontario, Erie and Huron. In: Allen, H. E. Kramer, J. R. (eds), *Nutrients in Natural Waters.* Wiley, New York, pp. 251–279.

Kilmer, V. J. (1979) Minerals and agriculture. In: Trudinger, P. A. and Swaine, D. J. (eds), *Biogeochemical Cycling of Mineral-forming Elements.* Elsevier, Amsterdam, pp. 515–558.

Korzun, V. I., Sokolov, A. A., Burynev, M. I. *et al.* (1974) *World Water Balance and Water Resources of the Earth.* Hydrometeoizdat, Leningrad, 638 pp. [in Russian].

Kramer, J. R. (1978) Acid precipitation. In: Nriagu, J. O. (ed.), *Sulfur in the Environment*, Pt. 1. Wiley, Chichester, pp. 325–370.

Kuznetsov, S. I. (1942) Sulfur cycle in lakes. *Mikrobiologiya*, **11**, 5–6 [in Russian].

Kuznetsov, S. I. (1952) *Role of Microorganisms in Cycles in Elements in Lakes.* Izv. Akad. Nauk SSSR, 300 pp. [in Russian].

Kuznetsov, S. I. (1970) *Microflora of Lakes and Its Geochemical Activity.* Nauka, Leningrad, 440 pp. [in Russian].

Lapteva, N. A., and Monakova, S. V. (1976) Microbiological characteristic of lakes of Yaroslavl Region. *Mikrobiologiya*, **45**, No. 4, 717–723 [in Russian].

Lazo, F. I. (1977) Sulfur in benthal sediments of the Baikal Lake. In: *Cycle of Matter and Energy in Water Bodies. Morphology, Lithodynamics, Sedimentation.* Listvenichnoe-on-the-Baikal, pp. 89–92 [in Russian].

Lazo, F. I. (1980) Geochemistry of sulfur in benthal sediments of the Baikal Lake. *Geokhimiya*, No. 1, 109–115 [in Russian].

Livingstone, D. A. (1963) Chemical composition of rivers and lakes. In: *Data of Geochemistry*, 6th edn. Geological survey professional paper 440-G. Washington.

Longinelli, A., and Cortecci, G. (1970) Isotopic abundance of oxygen and sulphur in sulfate ions from river water. *Earth Planet. Sci. Lett.*, **7**, 376–380.

Lowe, L. E., Sasaki, A., and Krouse, H. R. (1971). Variations of sulfur 34 : sulfur 32 ratios in soil fractions in western Canada. *Can. J. Soil Sci.*, **51**, 129–131.

Lukashova, E. N. (1958) *South America.* Uchpedyiz, Moscow, 466 pp. [in Russian].

Lyalikova, N. N. (1957) Investigation of the carbon dioxide assimilation by purple bacteria in Belovod Lake. *Mikrobiologiya*, **26**, No. 1, 92–98 [in Russian].

Matrosov, A. G., Zyakun, A. M., and Ivanov, M. V. (1978) Sulfur cycle in meromictic lakes (microbiological and isotopic data). In: Krumbein, W. E. (ed.), *Environmental Biogeochemistry and Geomicrobiology*, vol. 1. Ann Arbor Science, Michigan, pp. 121–127.

Matsubaya, O., Sakai, H., and Torri, T. (1979) Antarctic saline lakes – stable isotopic ratios, chemical composition and evolution. *Geochim. Cosmochim. Acta*, **43**, 7–25.

Nissenbaum, A. (1978) Sulfur isotope distribution in sulfates from surface waters from the northern Jordan Valley, Israel. *Environ. Sci. Technol.*, **12**, 962–965.

Nriagu, J. O. (1968) Sulfur metabolism and sedimentary environment: Lake Mendota, Wisconsin. *Limnol. Oceanogr.*, **13**, 430–439.

Nriagu, J. O. (1975) Sulphur isotopic variations in relation to sulphur pollution in Lake Erie. In: *Isotope Ratios as Pollutants Source and Behavior Indicators.* Proc. Symp. 1974, Vienna, pp. 77–93.

Nriagu, J. O. (1978) Production and uses of sulfur. In: Nriagu, J. O. (ed.), *Sulfur in the Environment*, Pt. 1. Wiley, Chichester, pp. 1–21.

Nriagu, J. O., and Coker, R. D. (1976) Emission of sulfur from Lake Ontario sediments. *Limnol. Oceanogr.*, **21**, 485–489.

Nriagu, J. O., and Harvey, H. H. (1978) Isotopic variation as an index of sulfur pollution in lakes around Sudbury, Ontario. *Nature (Lond.)*, **273**, 223–224.
Nriagu, J. O., and Hem, J. D. (1978) Chemistry of pollutant sulfur in natural waters. In: Nriagu, J. O. (ed.), *Sulfur in the Environment*, Pt. 2. Wiley, Chichester, pp. 211–270.
Odén, S. (1976) The acidity problem – an outline of concepts. *Water Air Soil Pollut.*, **6**, 137–166.
Ohle, W. (1934) Chemische und physikalische Untersuchungen nord-deutschen Seen. *Arch. Hydrobiol.*, **26**, 4.
Ohle, W. (1954) Sulfat als 'Katalisator' der limnishen Stoffureislaufes. *Vom Wasser*, **21**, 13–32.
Pankina, R. G., Mekhtieva, V. L., Grinenko, V. A., and Churmenteyeva, M. N. (1966) Isotope composition of sulfide and sulfate sulfur in waters of some Sub-Caucasian areas in relation to their genesis. *Geokhimiya*, No. 9, 1087–1094 [in Russian].
Posokhov, E. V. (1972) *Sulfate Waters in Nature*. Hydrometeoizdat, Leningrad, 167 pp. [in Russian].
Rabinovich, A. L., and Grinenko, V. A. (1979) On isotope composition of sulfate sulfur in the river runoff from the territory of the USSR. *Geokhimiya*, No. 3, 441–454 [in Russian].
Rabinovich, A. L., and Veselovsky, N. V. (1974) On isotope composition of sulfur compounds in the Azov Sea. In: *Hydrochemical Materials*, Vol. 60. Leningrad, pp. 41–48 [in Russian].
Reeder, S. W., Hitchon, B., and Levinson, A. A. (1972) Hydrochemistry of the surface waters of the Mackenzie River drainage basin, Canada. I. Factors controlling inorganic composition. *Geochim. Cosmochim. Acta*, **36**, 825–865.
Robinson, E., and Robbins, R. C. (1970) Gaseous sulfur pollutants from urban and natural sources. *J. Air Pollut. Control Assoc.*, **20**, 233–235.
Romanenko, V. I., Peres Eiris, M., Kudryavtsev, M. V., and Aurora Pubienes, M. (1976) Microbiological processes in the meromictic lake Vae de San Huan in Cuba. *Microbiologiya*, **45**, No. 3, 539–546 [in Russian].
Ronov, A. B., Grinenko, V. A., Girin, Yu, P., Savina, L. I., Kazakov, G. A., and Grinenko, L. N. (1974) Influence of tectonic regime on concentration and isotope composition of sulfur in sedimentary rocks. *Geokhimiya*, No. 12, 1772–1798 [in Russian].
Rossolimo, L. L. (1977) *Changes in Limnic Ecosystems under the Impact of Anthropogenic Factor*. Nauka, Moscow, 144 pp. [in Russian].
Rossolimo, L. L., and Shilkrot, G. S. (1971) Effect of forced aeration of a hyperotrophicated lake. *Izv. Akad. Nauk SSSR*, Ser. Geogr., 1971, No. 4, 48–58 [in Russian].
Schofield, C. L. (1976) Acid precipitation: effects on fish. *Ambio*, **5**, pp. 228–230.
Seip, H. M. and Tollan, A (1978) Acid precipitation and other possible sources for acidification of rivers and lakes. *Sci. Total Environ.*, **10**, 253–270.
Shilkrot, G. S. (1968) Hydrochemical regime of a lake at a late stage of anthropogenic eutrophication (example of Beloe Lake). *Gidrobiol. Zh.*, **4**, No. 6, 20–27 [in Russian].
Shvartzev, S. L. (1978) *Hydrochemistry of the Hypergenesis Zone*. Nedra, Moscow, 287 [in Russian].
Sokolova, G. A., and Sorokin, Yu. I. (1957) Bacterial sulfate reduction in muds of Rybinskoe Reservoir. *Mikrobiologiya*, **26**, No. 2, 194–201 [in Russian].
Sokolova, G. A., and Sorokin, Yu. I. (1958) Determination of bacterial sulfate reduction intensity in grounds of Gorkovskoe Reservoir using S^{35}. *Dokl. Akad. Nauk SSSR*, **118**, No. 2, 404–406.

Sorokin, Yu. I. (1960) Bacterial sulfate reduction in Kuibyshevskoe Reservoir. *Tr. Inst. Biol. Vodookhran Akad. Nauk SSSR.*, No. 3, 36–49 [in Russian].

Sorokin, Yu. I. (1961) Process of hydrogen sulfide formation in Volga reservoirs and its influence on oxygen regime. In: *All-Union Conference on Basis of Fish-Breeding in Reservoirs.*1958 Proceedings, pp. 65–70 [in Russian].

Sorokin, Yu. I. (1966) Interrelation of microbiological processes of the sulfur and carbon cycles in the meromictic Belovod Lake. In: *Plankton and Benthos of Inland Water Bodies*, No. 12, 332–335 [in Russian].

Sorokin, Yu. I. (1970) Interrelations between sulphur and carbon turnover in mero-mictic lakes. *Arch. Hydrobiol.*, **66**, 391–446.

Sorokin, Yu. I. (1975) Sulphide formation and chemical composition of bottom sedi-ments of some Italian lakes. *Hydrobiologia*, **47**, 231–240.

Sorokin, Yu, I., and Donato, N. (1975) On the carbon and sulphur metabolism in the meromictic Lake Faro (Sicily). *Hydrobiologia*, **47**, 241–252.

Sugawara, K., Koyama, T., and Kozawa, A. (1953) Distribution of various forms of sulphur in lake-, river- and sea-muds, *J. Earth Sci. Nagoya Univ.*, **1**, 17–23.

Takahashi, M. and Ichimura, S. (1970) Photosynthetic properties and growth of photosynthetic sulfur bacteria in lakes. *Limnol. Oceanogr.*, **15**, 929–944.

Tezuka, Y. (1979) Distribution of sulfate-reducing bacteria and sulfides in aquatic sediments. *Jpn J. Ecol.*, **29**, 95–102.

Uspensky, V. A. (1970) *Introduction to Geochemistry of Petroleum.* Nedra, Lenin-grad, 309 pp. [in Russian].

Veselovsky, N. V., Alekseev, A. P., Goncharova, V. D., Putintseva, V. S., and Polozhentsev, I. F. (1964) Isotope composition of sulfate sulfur in some surface water in the land. In: *Hydrochemical Materials*, vol. 38. Leningrad, pp. 62–76 [in Russian].

Veselovsky, N. V., Bebeshko, M. V., Kozyrev, C. A., Polozhentsev, I. F., Putintseva, V. S., and Rabinovich, A. L. (1966) On the temporal variation of the sulfate sulfur isotope composition in the rivers of Don and Volga. In: *Hydrochemical Materials*, vol. 42. Leningrad, pp. 128–136 [in Russian].

Veselovsky, N. V., Rabinovich, A. L., and Putintseva, V. S. (1969) On the sulfate sulfur isotope composition of the river Kuma water and some of its tributaries. In: *Hydrochemical Materials*, vol. 51. Leningrad, pp. 112–119 [in Russian].

Wright, R. T. and Gjessing, E. T. (1976) Changes in the chemical composition of lakes. *Ambio*, **5**, 219–223.

Yoneda, S., and Kouchi, T. (1972) Characteristics of bottom sediments with special reference to the so-called Hedoro in Japan. 1. *Rep. Lab. Soils Fert. Fac. Agric. Okayama Univ.*, **40**, 45–55.

Zverev, V. P. (1971) On hydrogeochemical balance of the USSR territory. *Dokl. Akad. Nauk SSSR*, **198**, No. 1, 161–163 [in Russian].

The Global Biogeochemical Sulphur Cycle
Edited by M. V. Ivanov and J. R. Freney
©1983 Scientific Committee on Problems of the Environment (SCOPE)

CHAPTER 6
The Sulphur Cycle in Oceans

Part I RESERVOIRS AND FLUXES

I. I. VOLKOV and A. G. ROZANOV

6.1 SULPHUR FLUXES TO THE OCEAN

Most of the sulphur added to seas and oceans is dissolved from land areas and transported in rivers. About 110.8 TgS are discharged to the oceans annually in river-water, mainly in the form of sulphate (see Chapter 5). Part of this sulphur (104.1 TgS) may come from sedimentary rocks and ores exposed to weathering, atmospheric deposition, and, to a lesser degree, volcanism. In addition to the natural sulphur flux, approximately the same amount of sulphur is discharged to the oceans by rivers as a result of anthropogenic processes. This flux is estimated to be 104 TgS year^{-1}. Therefore, the total flux of dissolved sulphate carried from the land to the oceans in rivers amounts to 214.8 TgS year^{-1} (Table 6.1). About 9.3 TgS are also discharged to the oceans each year as suspended matter in rivers, usually in the form of sulphates with a negligible amount as sulphides. About 9.2 TgS year^{-1}, mainly in the form of dissolved sulphate are added to the oceans by ground-water (Chapters 2, 5) and 4.2 TgS year^{-1} are deposited as a result of abrasion and exaration (Chapter 2).

From the difference between the rate of emission of sulphur from the ocean surface to the atmosphere, and the rate of deposition from the atmosphere, the net rate of addition to the oceans from this source is calculated to be 80.5 TgS year^{-1}, 66 Tg of which is anthropogenic sulphur (Chapter 4).

The total flux of sulphur to the oceans estimated from the data given in Table 6.1 amounts of 339.2 TgS year^{-1}. At present almost half of this (170 TgS) is from anthropogenic sources.

One further sulphur flux to the oceans is that resulting from submarine

357

Table 6.1 Sulphur fluxes to the oceans

Source	Form	Flux (TgS year^{-1})	References
River discharge:			
dissolved	SO_4^{2-}	214.8 (104)^a	Chapter 5
suspended	SO_4^{2-}, S^{2-}	9.3	Chapter 5
Underground flow	SO_4^{2-}	9.2	Chapters 2,5
Transport through the atmosphere	SO_4^{2-}, SO_2	94.5 (66)^b	Chapter 4
Underwater volcanoes	S^{2-}, SO_2	6.0	Chapter 2
Ash flow from surface volcanoes on land	S^{2-}, SO_2	1.2	Chapter 2
Abrasion of shores and exaration		4.2	Chapter 2
Total flux to ocean		339.2 (170)	

^{a}Values in brackets represent anthropogenic sulphur
^{b}Including 14 TgS from volcanic sources

hydrothermal activity, which occurs primarily at mid-oceanic rift zones and other areas of tectonic activity. The magnitude of this flux can only be tentatively evaluated (underwater volcanoes appear to contribute ~6.0 TgS year^{-1}).

6.2 SULPHUR IN OCEAN AND SEA-WATERS

6.2.1 Sulphate–Sulphur

Sulphur in ocean and sea-waters is in the form of dissolved sulphates. Sulphate-ion amounts to 7.68% of the total dissolved salts and its concentration varies with changes in total salinity. At an average concentration of 0.904 gS kg^{-1}, the total mass of sulphur in the world ocean waters amounts to $0.904 \times 1.413 \times 10^{21} = 1.277 \times 10^{21}$ gS $= 1.277 \times 10^9$ TgS (Table 6.2).

Sea-water is undersaturated with respect to $CaSO_4$. Dissolved sulphates are partially bound in ion pairs, predominantly $NaSO_4^-$, $MgSO_4^0$, $CaSO_4^0$ and KSO_4^-. Free SO_4^{2-}-ions account for 39–59% of the total sulphate in sea-water; the range results from differences in the dissociation constants used by different workers in their calculations (Garrels and Thompson, 1962; Kester and Pytkowicz, 1969).

The average salinity of the ocean is 35 g kg^{-1}. In marginal zones, and particularly in inland seas which are influenced by the discharge from rivers, the salinity may drop to several grams per kilogram of sea-water. In large gulfs, and inland seas of arid zones it may be as high as 40 g kg^{-1}, and even

Table 6.2 Sulphur in the water of the World ocean

Parameters	Original data	Calculated data	References
Volume of water	137×10^7 km^3	1.37×10^{21} litres	Horn (1969)
Mass of water	1413×10^{18} kg	1.413×10^{12} Tg	Horn (1969)
Sulphate–sulphur concentration	2.712 g SO$_4^{2-}$ kg^{-1}	0.904 gS kg^{-1}	Horn (1969)
Sulphate–sulphur mass	3.832×10^{21} g 0.42×10^{20} mol	1.277×10^9 TgS 1.347×10^9 TgS	Horn (1969) Garrels and Mackenzie (1972)

higher in some area (e.g. the Red Sea deeps, and Persian Gulf). In the Red Sea the increase in salinity is due not only to a high rate of evaporation but also to the discharge of brine into the basins. In some basins separated from the ocean the amount of salts exceeds 200 g kg^{-1} (e.g. the Dead Sea and the Gulf of Kara-Bogaz-Gol in the Caspian Sea). However, the volumes of these basins are far too small to affect substantially the overall balance of sulphates in the world oceans at the present time. If it is assumed that the average salinity of inland sea-water equals that of ocean water, then it can be calculated from the ratio of the volumes of the inland seas and oceans (Stepanov, 1961) that sulphate in inland seas amounts to only 2.7% of that in ocean waters.

6.2.2 Dissolved Sulphide

In stagnant water, some sulphur may be present in the form of dissolved sulphide. The deep water (i.e. below 150 m) of the Black Sea, which is the largest stagnant basin in the world, contains up to 9.6 mg litre^{-1} of sulphide with a mean concentration of 7.5 mg litre^{-1}. Sulphide is also found in the waters of inland seas, lagoons, depressions, fjords, and upwelling areas. Often its appearance is a recurrent phenomenon which is governed by seasonal and climatic conditions. The bulk of dissolved sulphide in sea-water and pore fluids of marine sediments is in the form of HS^{-1} ions (Volkov, 1962; Skopintsev, 1975). The contribution of dissolved sulphide to the sulphur balance is negligible, even in regional cycles. For example, the sulphate and sulphide contents of the Black Sea are 3.07×10^5 TgS and 6.1×10^3 TgS respectively (Skopintsev, 1975).

6.2.3 Biogenic Sulphur Cycle in Waters of the World Oceans

The little information available on the sulphur content of sea animals or plankton suggests that the sulphur content of phyto- and zooplankton approximates 1% dry weight. It it is assumed (Romankevich, 1977) that the annual production of phytoplankton in the world oceans is 20×10^3 Tg of organic carbon or 36×10^3 Tg of organic matter, then 360 TgS are involved the biogenic cycle annually. This figure is some 3.6 times greater than the annual discharge of sulphur in river-water (without anthropogenic sulphur) but amounts to only 0.001% of the sulphate–sulphur in the upper 100 m thick photic layer of the world oceans.

On average, only 10% (2×10^3 Tg year^{-1}) of the organic carbon produced by phytoplankton, reaches the sediments of the oceans; the bulk of organic matter is metabolized in the water column. There is evidence, however, that organic compounds containing sulphur, phosphorus, or nitrogen break down more rapidly than those without these elements (Skopintsev, 1948, 1976; Bogdanov *et al.*, 1971; Almgren *et al.*, 1975; Hartmann *et al.*, 1976). It may be deduced, therefore, that particulate organic matter in bottom waters contains less than 1% sulphur, and that probably not more than 36 Tg of organically bound sulphur reach marine sediments each year. Moreover, between 90% (Gershanovich *et al.*, 1974) and 95% (Romankevich, 1977) of organic matter in reduced sediments is thought to be mineralized at the sediment/water interface. Thus no more than 1.8 Tg of biogenic sulphur per year can be buried with organic matter. Other data (Gershanovich *et al.*, 1974) indicate that about twice this amount of organic matter (3–4 TgS year^{-1}) is buried in Holocene sediments (see Table 6.4).

6.3   CONDITIONS OF SULPHUR FLOW INTO SEDIMENTS

6.3.1   Forms of Sulphur Flow into Sediments

Sulphur in modern ocean sediments exists in various forms. During sedimentation some dissolved sulphate is entrapped in pore-water. In addition, sulphates can occur in the solid phase of sediments as clay mineral sulphate, as sulphate in the carbonate skeletons of organisms, and as anhydrite, gypsum, and barite.

In addition to sulphate, various amounts of soluble sulphide, thiosulphate, and sulphite are found in pore-waters of reduced sediments. Insoluble phases of reduced sulphur in reducing sediments include acid-soluble iron sulphides (e.g. hydrotroilite, mackinawite, and greigite), pyrite, elemental sulphur, and organically bound sulphur. The total amount of reduced sulphur in sediments can be taken to be a measure of the intensity of bacterial sulphate reduction which occurs particularly in sediments with a high content of organic matter. In the majority of cases, the amounts of soluble sulphide, thiosulphate, and sulphite in pore-waters are negligible compared to those of reduced forms of sulphur in the solid phase of sediments and are usually ignored.

At present there are relatively few data on the amounts of the different forms of sulphur in marine sediments. Moreover, data on the absolute mass of sediments being discharged annually to the oceans from rivers and from other sources are unreliable, and only estimates are available for the distribution of sedimentary material over different morphological zones of the world oceans. Therefore, only tentative estimates can be made of the amount of sulphur buried annually in sediments.

6.3.2 Sediment Masses and Distribution Over the Ocean Floor

Most sedimentary material is discharged into the oceans as suspended matter in rivers. However, there are large discrepancies between estimates of suspended matter in annual river discharge. Early measurements of solids discharged in the majority of large rivers of the world yielded a total of 8070 Tg year^{-1} (Lopatin, 1950; Strakhov, 1962). However, based on Lisitsin's (1974) data, the total sediment entering oceanic deposits amounts to about 27 000 Tg year^{-1} (Table 6.3). A similar value (20 000 Tg) was estimated by Gershanovich *et al.* (1974).

The bulk of sedimentary material entering the world oceans from land is deposited at the margins. Maps of absolute rates of sedimentation in the world oceans (Lisitsin, 1971, 1974) show that the highest rates of sedimentation occur near estuaries and deltas, in marginal and inland seas, and on shelves and continental slopes of the open ocean. Lowest rates are found in deep-water regions remote from the shore, for example in the centres of the deep-water basins of the Pacific. As the result of differences between rates of sedimentation there are also great differences between amounts of sediment deposited in the various morphological zones of the oceanic floor (see Table

Table 6.3   Sedimentary material deposited in the ocean (Lisitsin, 1974)

Type of material	Amount (Tg year^{-1})
I. Terrigenous material	
1. Particulate matter discharged in rivers	18 500
2. Glacial discharge	1 500
3. Aeolian material	1 500
4. Erosion of shores	500
II. Biogenic	
1. Carbonates	1 400
2. Silica	500
III. Volcanic	3 000
IV. Material from outer space	10–80
Total	27 000

Table 6.4 Organic carbon accumulation in modern sediments of the world oceans during the Holocene period (Gershanovich *et al.*, 1974)

Morphological zones of ocean floor	Area, (10^6 km^2)	Average thickness of Holocene layer (m)	Sediment mass (10^7 Tg)	Organic carbon	
				Average concentration (%)	Mass (10^3 Tg)
Shelf	26.7	0.8	3.2	0.7	224
Continental slope and rise	76.5	1.5	14.9	1.3	1950
Ocean bed	257.0	0.05	1.9	0.3	54
Total	360.2		20		2228

6.4). Only 9.5% of the sediments are located in the deep oceanic beds which occupy more than 70% of the total floor. The continental slope and rise, occupying about 20% of the area of world ocean sediment, account for more than 70% of the total sediment.

These estimates agree reasonably well with the results of Lisitsin (1974, 1978). However, according to Strakhov (1978), sedimentation rates in pelagic zones are grossly underestimated when determined by isotope geochronology, and therefore the absolute amounts of sediments deposited on the ocean bed are also underestimated. In the absence of more reliable information our calculations on sedimentation of sulphur compounds are based on the data given in Tables 6.3 and 6.4, which suggest that 20 000–27 000 Tg of sediments are deposited annually in the oceans.

6.3.3 Moisture Content of Bottom Sediments

In order to estimate the flux of dissolved sulphate to bottom sediments, information is required on changes in pore-water content during compaction. Such information is provided by data from deep-sea drilling.

The moisture content of oceanic sediments depends on the particle size distribution, the composition, rate of deposition and age of the sediments, depth within the sediment column, and other factors. Common to all types of sediment is an increasing moisture content with increasing content of finer (pelitic) fractions. In sediments with similar textures the moisture content usually decreases from siliceous sediments through terrigenous sediments and carbonates, to red deep-water clays. Shallow-water marine sediments have a higher moisture content than deep oceanic sediments, possibly because of the higher rates of deposition and organic carbon contents of the former. Moisture content decreases sharply only in the upper few metres of oceanic sediments (by 15%) and upper tens of metres (by 20–30%). The lower the moisture content of surface sediments the less it changes with depth.

On the basis of Lisitsin's (1974) data on the surface layer of sediments of various types, and estimates of the moisture content of deep-sea drilling cores, the following average values for moisture contents of oceanic sediments in various morphological zones were used for balance calculations: shelf, 30%; continental slope and rise, 40%; oceanic bed, 50%.

6.3.4 Organic Matter Distribution in Sediments

Analyses of cores from deep-drilling sites show little, if any, change in the pattern of organic matter distribution in oceans since the Miocene period (Geodekyan *et al.*, 1978) and it is possible, therefore, to develop a generalized scheme for organic matter distribution in modern oceanic sediments (Romankevich, 1977).

The organic matter content in modern oceanic sediments varies from trace amounts (hundredths of a per cent in deep-water pelagic sediments of the Pacific Ocean) to substantial quantities (e.g. 22%C in the old Black Sea sediments). Average organic carbon contents for the various morphological zones of the world oceans are shown in Table 6.4. More than two-thirds of the world oceanic beds are covered with sediments containing, on average, 0.3% organic carbon. Studies on Pacific and Atlantic sediments show that reducing processes do not develop in sediments with such low concentrations of organic carbon. High redox potentials (+0.5 to +0.6v) were observed throughout the entire thickness of the sediments (Rozanov *et al.*, 1972, 1976; Volkov *et al.*, 1973, 1975b; Hartmann *et al.*, 1976). The pore-waters contained free oxygen (up to 2.7 mg litre^{-1}) and aerobic nitrifying organisms were active (Pamatmat, 1973; Müller, 1975). In addition, the solid phases of these sediments contain a large reservoir of oxidizing power in the form of tetravalent manganese compounds. Organic matter is mineralized aerobically under these conditions (Equation 1)

$$(CH_2O)_{106}(NH_3)_{16}(H_3PO_4) + 138O_2$$
$$= 106CO_2 + 16HNO_3 + H_3PO_4 + 122H_2O \qquad (1)$$

In sediments with more than 0.3% organic carbon, reducing processes become active, as indicated by the absence of oxygen in pore-water and sediments. Sporadically, small quantities of reduced sulphur compounds appear in some layers of the sediment.

In the absence of oxygen, organic matter is decomposed anaerobically by sulphate-reducing bacteria with sulphate as the electron acceptor (Equation 2)

$$(CH_2O)_{106}(NH_3)_{16}(H_3PO_4) + 53SO_4^{2-}$$
$$= 106HCO_3^- + 16NH_3 + H_3PO_4 + 53H_2S \qquad (2)$$

Well-developed reducing processes, and significant amounts of reduced forms of sulphur, are found in sediments with organic carbon contents of about 1%. The anaerobic decomposition of organic matter follows first-order reaction kinetics.

Organic matter consumption in diagenetic reducing processes (mainly bacterial sulphate reduction) has been estimated by various methods. In one method, the amounts of organic carbon and nitrogen in deeper layers of sediments are compared with the concentration in the upper layer, which is assumed to be the initial value (Emery and Rittenberg, 1952; Starikova, 1956; Lisitsin, 1955). It is also assumed that the rates of sedimentation and the input of organic matter remain constant with time. These assumptions, however, are not always correct and the method thus has severe limitations.

Alternatively, the decomposition of organic matter during reduction of oxidized forms of manganese, iron, and sulphur in sediments may be determined (Strakhov and Zalmanzon, 1955; Strakhov, 1962, 1972, 1976; Bordovskii, 1964; Rozanov et al., 1976). This method, which takes into account variations in organic matter decomposition with time, seems to be the most reliable, but also has limitations. In particular, its application is restricted to reduced sediments on the ocean periphery. Also, the method relies on theoretical stoichiometric relationships between organic matter decomposition and the various oxidants considered (see e.g. Equation 2). Since, in practice, this stoichiometry may not be realized, organic matter consumption can be underestimated.

Nevertheless, both the above methods give basically similar results. Emery and Rittenberg (1952), working on the marginal deposits of California, show that up to 50% of the organic matter is decomposed in the top metre of sediments. Bordovskii (1964), using Strakhov's method, found that 18–46% of organic matter was decomposed in sediments of the Bering Sea. According to Starikova (1956) there is a 20% decrease in organic carbon in the upper metre of Bering Sea sediments and the rate of decomposition declines by 4% over the next 3 m. Organic carbon metabolized during reduction processes in the sediments of the north-west part of the Pacific near Japan varied from 13 to 35% (Strakhov, 1972). Using the same method Rozanov et al. (1976) found than 16–46% of the initial organic carbon was consumed in the deposits of the Gulf of California, and 10–20% in the deposits of the Californian region of the Pacific.

In all these studies it was observed that the decomposition of organic carbon during reduction processes in sediments is related to the initial content of organic matter. For example in Fig. 6.1, which is based on data for the reduced sediments of the Pacific Ocean and the Okhotsk and Bering seas, there is a positive correlation between the initial content of organic matter and its anaerobic consumption during diagenesis. This suggests that a relatively constant proportion of the initial organic matter is readily metabolizable.

According to the data given in Table 6.4, about 2.5% of the total organic

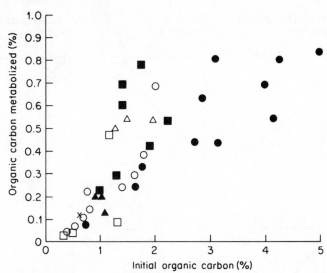

Fig. 6.1 Relationship between organic carbon metabolized during sulphate reduction and the initial organic carbon concentration in sediments. Pacific Ocean—○, Strakhov (1972): ●, Rozanov *et al.* (1976); ▲, Ostroumov and Fomina (1960); ×, Ostroumov and Volkov (1960). Bering Sea— △, Bordovskii (1964). Sea of Okhotsk—■, Ostroumov (1957)

matter in ocean sediments is contained in those of the deep oceans at an average concentration of 0.3%. Other estimates for the relative proportion of total organic carbon in deep ocean sediments are 3.2% for the Pacific Ocean and 13% for the Atlantic Ocean (Romankevich, 1977). Since sediments containing 0.3% organic carbon are aerobic (see above) it can be concluded that the bulk of organic matter in the world oceans (>90% contained in shelf and continental slope and rise sediments) is subject to *anaerobic* diagenesis.

6.4 SULPHATE FLUX FROM OCEAN WATER TO BOTTOM SEDIMENTS

6.4.1 Sulphate in Pore-water

The mean sulphate : chloride ratio in ocean water is 0.1394 (SO_4^{2-}, 2.70 g kg^{-1}; Cl$^-$, 19.35 g kg^{-1}). When sulphate–sulphur is buried in bottom sediments along with interstitial water the sulphate : chloride ratio decreases to 0.133. Several factors, may, however, modify the sulphate content of pore-waters. Chemical changes in the pore-waters may result in sulphate precipitation or dissolution. Sulphate may be reduced to hydrogen sulphide;

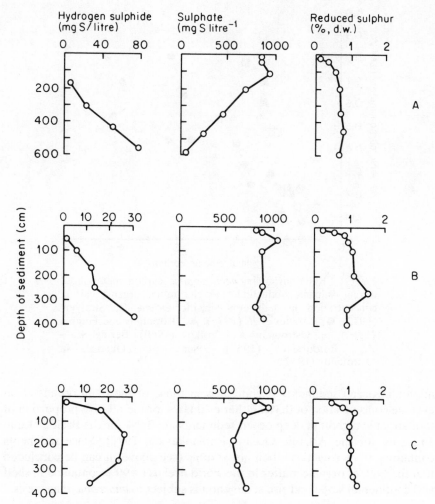

Fig. 6.2   Hydrogen sulphide and sulphate in pore-water and reduced sulphur in reduced sediments of the Pacific Ocean. A, Site 6162, western region; B, site 664, Gulf of California; C, site 668, eastern region

authigenic sulphides can be oxidized to sulphate, and there can be an increase or decrease in sulphate concentration as a result of diffusion, etc.

The sulphate concentration in sediment cores is relatively constant only in pore-waters of pelagic red clays where the total salinity varies within 3⁰/₀₀ throughout most of the sediment (Shishkina, 1972). In a number of cases a slight decrease in sulphate content is observed with depth due to incorporation of pore-water sulphate into the clay phase. On other occasions clay components such as sodium or potassium sulphoaluminates (Na, K) Al_3SO_4

(OH)$_6$ decompose, releasing sulphate into solution. When this occurs the sulphate : chloride ratio in pore-waters of red clays may rise to 0.145.

In pelagic carbonate sediments the sulphate concentration of pore-waters can increase due to the leaching of sulphate from carbonates (section 6.4.3); sulphate : chloride ratios may reach 0.158. The absolute sulphate content in pore-waters is also markedly increased in these sediments. For example, the average sulphate concentration in pore-waters of high-carbonate deposits of the Somali basin in the Indian Ocean is 0.115% compared with 0.09% for the overlying oceanic water of normal salinity (Zhabina *et al.*, 1979).

Microbiological reduction of pore-water sulphates occurs in littoral sediments, and in sediments of other areas which have a high biological productivity and an increased content of organic matter. Figure 6.2 presents a number of characteristic examples of the distribution of sulphate–sulphur, hydrogen sulphide, and its transformation products (reduced sulphur) in anoxic sediments of the Pacific. A feature of anoxic sediments is the sharp increase in reduced sulphur, represented mainly by pyrite, in subsurface horizons.

Sulphate depletion does not occur to the same extent in the pore-waters of all sediments. In some sediments (e.g. Fig. 6.2, sites 664, 668) the sulphate removed by bacterial reduction in the upper layer of the sediments is replenished by diffusion from the overlying water. In others there is almost complete disappearance of sulphate (e.g. Fig. 6.2, site 6162). At site 6162 the sulphate content is extremely low and hydrogen sulphide is the major form of sulphur in pore-water at depth.

In subsequent calculations the following values for pore-water sulphate were used: deep ocean, 0.09% S; continental slope and rise, 0.045% S; shelf sediments, 0.045% S and 0.09% S for regions with high and low sulphate reduction levels respectively.

6.4.2 Sulphates in the Solid Phase of Sediments

Shallow-water and deep ocean marine sediments contain sulphate in the solid phase as well as dissolved sulphate in pore-water.

Figure 6.3 shows the distribution of sulphate–sulphur in two cores of carbonate globigerine sediments from the tropical Pacific (Volkov and Ostroumov, 1960); the sulphate content in the solid phase of these sediments is comparable with that of the pore-water. The sulphate content of pore-waters of carbonate sediments often exceeds that of the overlying water (0.09%) and may reach values as high as 0.125%. This is due to the partial leaching of sulphate from biogenic carbonate to the pore-water (since the pore-water is undersaturated with respect to $CaSO_4$) and possibly to the lower amounts of solid-phase sulphate and the elevated concentrations of pore-water sulphate in the lower horizons of sediment at site 3854. The sulphate

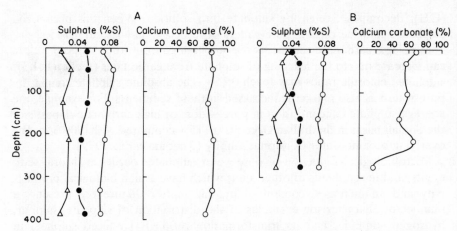

Fig. 6.3   Distribution of sulphate and calcium carbonate with depth in carbonate sediments of the tropical Pacific Ocean. O, Total sulphur; ●, pore-water sulphur; △, solid phase sulphur (A, site 3863; B, site 3854; Volkov and Ostroumov, 1960)

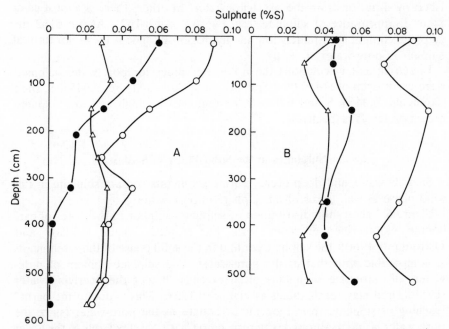

Fig. 6.4   Sulphate distribution in clay sediments of the Pacific Ocean. O, Total sulphur; ●, pore-water sulphur; △ solid phase sulphur (A, site 6163, Japanese Trench; B, site 6175, red clays of the north-western depression)

content of the solid phase of carbonate-free sediments is extremely small
(Fig. 6.4). The vertical distribution of total sulphate and pore-water sulphate
in a core from site 6163 in the Japanese Trench (Volkov *et al.*, 1972) was
influenced by bacterial sulphate reduction. The sulphate content in the solid
phase was not high and did not vary significantly with depth. In the lower
horizons of the sediments virtually all sulphate was in the solid phase. Some
variations of sulphate concentrations in the solid phase and pore-water with
depth is observed in sediments. The total amount of sulphate in a sediment

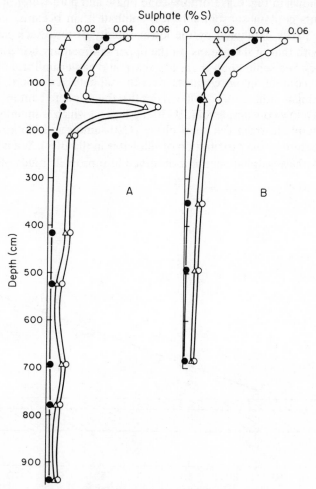

Fig. 6.5 Sulphate distribution in deep sediments of the
hydrogen sulphide zone of the Black Sea. O, Total sulphur; ●, pore-water sulphur; △, solid phase sulphur (A,
site 4751 western region; B, site 4740 central region;
Ostroumov and Volkov, 1964)

also varies depending on changes in these two parameters. Samples from site 6175 in the central part of the north-west basin of the Pacific are representative of deep-water oxidized red clays. The influence of biogenic carbonate on solid phase sulphate is particularly noticeable where carbonate and clay layers alternate in the sediment column. In the clay layers the sulphate content of the solid phase in markedly lower than that of the carbonate layers (see Volkov and Ostroumov, 1960; site 3867).

The sulphate distribution in sulphidic deep-water sediments of the Black Sea is shown in Fig. 6.5. Both the solid phase and pore-water of the Black Sea sediments contain substantially less sulphate than oceanic sediments. This may be partly explained by the lower salinity of the Black Sea water compared with that of the oceans. In the upper sedimentary horizons (recent and old Black Sea sediments) there is a sharp and rather uniform decrease in the sulphur content in pore-waters due to sulphate reduction. Lower in the Neoeuxinian sediments (>200 cm depth), the sulphate content of pore-water reaches a low constant level (20–30 mg litre^{-1}) which is maintained by transfer of sulphate from the solid phase (Ostroumov and Volkov, 1964). The other feature of the distribution of sulphates in the Black Sea is the maximum in solid phase sulphate content observed in sapropelic muds of the old Black

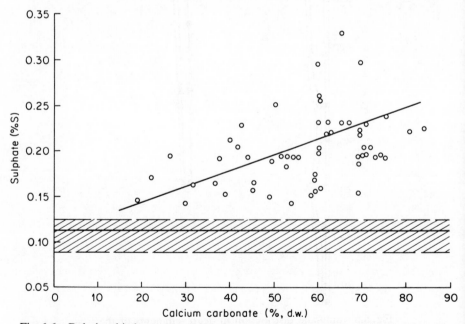

Fig. 6.6 Relationship between sulphate and calcium carbonate concentrations in sediments of the Somali Trench of the Indian Ocean (Zhabina *et al.* 1979). The shaded area represents the range of values obtained for the sulphate concentration in pore-water; mean = 0.115%S

Sea sediments (138–163 cm depth, site 4751). A similar maximum was observed in a core from site 4745 (Ostroumov and Volkov, 1964) suggesting that it was of general occurrence in sapropelic muds of Black Sea sediments.

The form of sulphur in the solid phase of clay marine sediments is still unknown. In mixed clay–calcareous sediments such as those of the Somali basin of the Indian Ocean (Zhabina *et al.*, 1979) some of the sulphate may form part of the carbonate skeletons of organisms, and some may be bound in clay minerals. These sediments are low in organic matter and exhibit low sulphate reduction rates. The sulphate concentration in pore-water of Somali basin sediments remains constant with depth and is equal to that in the overlying water. The total sulphate content of these sediments is always higher than that of oceans and tends to increase with increasing carbonate content (Fig. 6.6). This indicates that, in such mixed clay–calcareous sediments with a high $CaCO_3$ content, most of the solid phase sulphur is present in the carbonate phase.

The sulphate content of biogenic carbonate is discussed in section 6.4.3. Data on solid phase sulphate in clay marine sediments are given in Table 6.5; the average concentration is approximately 0.06%.

6.4.3 Sulphate in Biogenic Marine Carbonate

Data on sulphate sulphur in marine carbonate of biogenic origin are rare. According to Vinogradov (1935, 1937) the $CaSO_4$ concentrations in corals varies from 0.06% to 5.43% ash weight, the greatest amount being in *Gorgonia subfruticose Dana*. The $CaSO_4$ concentration in shells varies over a wide range, from 0.17% to 8.37% ash weight; it varies from 0.36% to 8.37% in *Brachiopoda*, 0.17–1.46% in *Cephalopoda* and 1.45–2.0% in *Ostrea edulis*. In marine plants the average total sulphur concentration is 1.15% (range 0.3–3.3%) while in animals the average is 0.89% and the range 0.45–2.80% (Goldhaber and Kaplan, 1974). Sulphur exists in various forms in living organisms: sulphates and organic sulphur, water- and acid-soluble and tightly bound sulphur, and sulphur of soft tissues and skeletons. The sulphur of skeletons is more stable and better preserved during sedimentation.

Investigations on biogenic carbonate from the Black Sea (Ostroumov and Volkov, 1964), the Pacific Ocean (Kaplan *et al.* 1963; Goldhaber and Kaplan, 1974) and the Azov, Caspian, White and Japanese seas (Mekhtieva, 1974) showed that carbonate skeletons contain $CaSO_4$ in amounts ranging from 0.025% to 0.349% with an average of 0.104% (Table 6.6). The considerable variation in sulphate concentration is due to difference in species of organisms from which the skeletons are derived, and to the dependence of the sulphate content in the skeletons of carbonate organisms on the salinity of sea-water, i.e. on the sulphate content of the growth medium. For example, the shells of fossil molluscs from the Caspian Sea have a sulphate content and sulphur

Table 6.5   Sulphate–sulphur in the solid phase of clay and carbonate–clay sediments

Location of sediments	Type of sediments	Number of samples	Solid phase sulphate		References
			Range (%)	Average (%)	
Pacific Ocean					
North-western part	Reduced	31	0.012–0.165	0.055	Volkov *et al.* (1972)
Gulf of California	Reduced	32	0[a]–0.105	0.041	Volkov *et al.* (1976)
Pelagic red clays	Oxidized	11	0[a]–0.092	0.055	Volkov *et al.* (1976)
Pelagic carbonate–clay	Oxidized	17	0.036–0.089	0.059	Volkov and Ostroumov (1960)
Indian Ocean					Volkov and Ostroumov (1960)
Somali basin, clay–carbonate	Reduced	60	0.031–0.261	0.102	Zhabina *et al.* (1979)
Black Sea					
Recent	Reduced	10	0.019–0.065	0.029	Ostroumov and Volkov (1964)
Old	Reduced	5	0.014–0.172	0.062	Ostroumov and Volkov (1964)
Neoeuxinian	Reduced	14	0[a]–0.026	0.014	Ostroumov and Volkov (1964)
Average of all ocean sediments				~0.06	

[a]Below level of detection.

Table 6.6   Sulphate in biogenic carbonates

Carbonates	Number of samples	Sulphate		References
		Range (%)	Average (%)	
Globigerine sand with admixture of pteropods and corals sand	4	0.053–0.190	0.114	Volkov and Ostroumov (1960)
Live and dead corals (*Madreporaria*), coral sand	5	0.184–0.349	0.233	Volkov and Ostroumov (1960)
Shells of pteropods	3	0.028–0.057	0.041	Volkov and Ostroumov (1960)
Anadara	2	0.088–0.090	0.089	Volkov and Ostroumov (1960)
Venus gallina	2	0.059–0.080	0.070	Ostroumov and Volkov (1964)
Mytilus	5	0.031–0.062	0.047	Ostroumov and Volkov (1964); Goldhaber and Kaplan (1974) Mekhtieva (1974)
Gardium edule	4	0.025–0.051	0.042	Ostroumov and Volkov (1964); Mekhtieva (1974)
Haliotis	1	—	0.046	Goldhaber and Kaplan (1974)
Chactopterus variopedatus	1	—	0.290	Goldhaber and Kaplan (1974)
Chlamys	2	0.080–0.260	0.170	Goldhaber and Kaplan (1974); Mekhtieva (1974)
Didacna terigonoides	1	—	0.092	Mekhtieva (1974)
Dreissenia rostriformis	1	—	0.110	Mekhtieva (1974)
Umbolium suturale	1	—	0.050	Mekhtieva (1974)
Mactra sp.	1	—	0.080	Mekhtieva (1974)
Pecten yescoensis	1	—	0.080	Mekhtieva (1974)
Ostrea giges	1	—	0.127	Mekhtieva (1974)
Glycymeris albolineatus	1	—	0.050	Mekhtieva (1974)
	36	0.025–0.349	0.102	

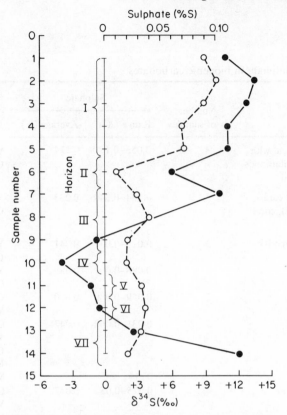

Fig. 6.7 Frequency distribution of sulphate, and its isotopic composition of fossil mollusc shells from ancient Caspian Sea sediments (Mekhtieva, 1974). O--------O, Sulphate; ●———●, δ^{34}S. I, New Caspian horizon; II, Verkhne–Khvolynsky horizon; III, Kazarsky horizon; IV, Pogryansky horizon; V, Verkhne-Bakinsky horizon; VI, Apsheronsky horizon; VII Akchagylsky horizon

isotopic composition (Fig. 6.7) related to the palaeosalinity of the basin (Mekhtieva, 1974; Mekhtieva *et al.*, 1977). The periods when the sulphate concentration is low and the sulphur in shells is relatively enriched in ^{32}S correspond to periods of freshening of the basin.

Data on the sulphate content of biogenic carbonates are collated in Table 6.6. Since the list includes results of sulphate determinations in basins with a low salinity (compared to the ocean), the average sulphate–sulphur concentration of 0.10% may be an underestimate.

6.4.4 Barite Sulphur

Part of the solid phase sulphate in ocean bottom sediments is in the form of barite ($BaSO_4$). Although ocean water is not saturated with respect to barite, local oversaturation may arise in pore-water resulting in the formation of authigenic barite. Barite can also be deposited in sediments as terrigenous or biogenic suspensions. Sea animals, capable of concentrating barium, also transfer barium directly to ocean sediments (Goldberg and Arrhenius, 1958; Chow and Goldberg, 1960; Turekian, 1964; Gurvich *et al.*, 1978).

While most barite formed in ocean sediments is submicroscopic, numerous grains and concretions up to 30 cm in diameter can be found. Barite concretions may be both biogenic (Schulze and Theirfelder, 1905) and hydrothermal (Dean and Schreiber, 1978) in origin.

It is impossible at this time to present a complete picture of barite distribution in oceanic sediments. However, selected phase determinations show that barite is the main form in which barium occurs in terrigenous sediments. Barite accounts for more than two-thirds of the barium in pelagic and metalliferous sediments and in ferro-manganese nodules. In carbonate and siliceous sediments at least 50% of the barium is in the form of barite (Gurvich *et al.*, 1978, 1979). The barium content in marine sediments increases from the margins to the deep oceans. High concentrations of barium are found in sediments in areas with a high biological productivity, in metalliferous sediments, and in ferro-manganese nodules (Boström *et al.*, 1973, 1974; Gurvich *et al.*, 1978, 1979; Skornyakova, 1976).

Data on barium and sulphate distributions in Pacific Ocean sediments (Table 6.7), provide the basis for a preliminary assessment of the average

Table 6.7 Total barium, barite–barium and sulphur, in sediments and ferromanganese nodules of the Pacific Ocean

Sediments	Total barium (%)	Barite Barium (% of total Ba)	Barite Barium (% of sediment)	Sulphur (%S)
Terrigenous	0.129	100	0.129	0.030
Red deep-water clays	0.184	70	0.129	0.030
Carbonate	0.096	50	0.048	0.011
Diatoms	0.451	50	0.226	0.052
Metalliferous				
carbonate-free	0.775	70	0.542	0.163
carbonate-bearing	0.315	70	0.220	0.051
Ferromanganese nodules	0.33	70	0.230	0.053

barite–sulphur content of oceanic sediments. It is assumed that the average barite content of the shelf and continental slope sediments equals that of terrigenous sediments. Using Sverdrup's relation (carbonate sediments : pelagic clays : diatom clays = 48.7 : 38.1 : 14.2) the average barite-S value was derived as follows:

$$\frac{0.011 \times 47.7}{100} + \frac{0.030 \times 38.1}{100} + \frac{0.052 \times 14.2}{100} = 0.024\%S$$

6.4.5 Sulphate Flux to Bottom Sediments

From the data given in section 6.4.1 and in Table 6.8 the annual incorporation of sulphate–sulphur into the pore-water of sediments is estimated to be 6.8–9.9 Tg.

From the mean concentration of sulphate–sulphur (0.06%, Table 6.5) and the total amount of carbonate-free material entering the oceanic deposits (18 600–25 600 Tg year^{-1}, Table 6.3 and 6.4), the sulphate flux from water to the solid phase of non-carbonate sediments is estimated to average 13.3 Tg year^{-1} (range 11.2–15.4).

Since the average sulphate–sulphur content in biogenic carbonate material is 0.104 (Table 6.6), and 1400 Tg of calcium carbonate are deposited in sediments annually (Table 6.3), the annual flux of sulphate–sulphur to sediments in this form is 1.4 Tg. The contribution from barite is assessed at 4.7 TgS year^{-1} (range 4.0–5.4) based on the average barite–sulphur concentration in sediments (0.02%, section 6.4.4) and the amount of sedimentary material deposited (20 000–27 000 Tg year^{-1}).

The total sulphate flux to the sediments is estimated to be 27.8 TgS year^{-1} (range 23.4–32.1, Table 6.9). About half of this sulphur enters the solid phase of carbonate-free sedimentary clays.

The calculated barite flux may require revision. The average content of

Table 6.8 Sulphate flux to pore-water of ocean sediments

Morphological zones of the ocean bed	Sediment flux (Tg dry weight year^{-1})	Calculated moisture (%)	Pore-water (Tg year^{-1})	Pore-water sulphate (%)	Sulphate flux (TgS year^{-1})
Shelf	3 200–4 300	30	1 400–1 800	0.045–0.090	0.6–1.6
Continental slope and rise	14 900–20 100	40	9 900–13 400	0.045	4.5–6.0
Ocean bed	1 900–2 600	50	1 900–2 600	0.090	1.7–2.3
Total	20 000–27 000				6.8–9.9
Average					8.4

Table 6.9 Total sulphate flux from bottom water to sediments of the world oceans

Forms in sediments	Flux (TgS year^{-1})	
	Range	Average
Solid phase of clay sediments	11.2–15.4	13.3
Pore-water	6.8–9.9	8.4
Barite	4.0–5.4	4.7
Biogenic carbonates	1.4	1.4
Total	23.4–32.1	27.8

barite–sulphur in sediments of the world oceans (0.02%) corresponds to 0.08% Ba which would require an annual burial of 16–20 TgBa. However, not all sedimentary Ba exists as barite (6.4.4) and, according to Gordeev and Lisitsin (1978), the annual flux of barium to the ocean in terrigenous river runoff is only 7.75 Tg, of which 7.05 Tg is in the form of suspended matter and 0.70 Tg dissolved. Thus, there must either be an additional flux of barium to the ocean or the data on the average content of barium in oceanic sediments require revaluation.

6.5 FORMATION AND BURIAL OF REDUCED FORMS OF SULPHUR

6.5.1 Distribution and Geochemical Activity of Sulphate-reducing Bacteria

A. Introduction

Microbiological and geochemical aspects of sulphate reduction, and formation of hydrogen sulphide and sulphides in modern sediments of the world oceans, have been discussed in many publications. Most investigators consider bacterial dissimilatory sulphate reduction to be the main mechanism for the formation of reduced sulphur compounds in most modern sediments. Organic sulphur accounts for less than 10% (Goldhaber and Kaplan, 1974) and only small local concentrations of sulphide arise from submarine hydrothermal exhalations. Bacterial sulphate reduction, therefore, is the main compensation for the continuous replenishments of ocean sulphate by river runoff.

Another important consequence of sulphate reduction is the formation of hydrogen sulphide, the accumulation of which results in radical changes in marine geochemistry and sometimes irreversible changes in the ecology of a reservoir. In certain cases intensive production of hydrogen sulphide in shal-

Table 6.10.   The upper horizons of reduced sediments in some deposits of the Pacific Ocean (Volkov *et al.*, 1972, 1976).

Site	6164	6168	6171	6183	673
Depth (m)	5300	6030	6000	4660	3280
Geographic co-ordinate	38°32′N 145°30′W	36°12′N 149°48′W	33°49′N 151°22′W	28°08′N 133°27′W	20°29′N 110°53′W
Upper horizons of reduced sediment: depth (cm) from the water/sediment: interface	10	47	120	415	190
Total reduced sulphur in the upper horizon of reduced sediment (%S)	0.113	0.034	0.035	0.012	0.054

low areas of ocean, or in upwelling zones, may result in its direct emission to the atmosphere.

Typical environments in which sulphate reduction occurs are the sediments of bays, river deltas, fjords, the continental shelf, borderland trenches, the continental rise, inland marginal and mediterranean seas, and deep-sea basins (Goldhaber and Kaplan, 1974). In these sediments reduced sulphur compounds are found even in the topmost layers. In the open ocean reduced sulphur may occur in sediments to a depth of at least 6000 m, but in these sediments sulphate reduction generally begins only at some depth (down to 4 m) from the water/sediment interface (Table 6.10), and reduced sulphur is absent from the upper layers. Thus, only the sediments of the continental shelf, slope, and rise are important for assessing sulphate reduction in the world oceans.

B.   Sulphate reduction in littoral sediments

One of the most extensive studies of sulphate reduction in littoral sediments has been carried out in Limfjorden, a shallow and slightly freshened bay of the North Sea, 1500 km^2 in area, in northern Denmark. The distribution of sulphate-reducing bacteria, and various forms of sulphur, and rates of sulphate reduction, have been monitored for nearly two years at nine sites (Jørgensen, 1977, 1978; Jørgensen *et al.*, 1978).

The vertical profiles of bacteria and sulphate reduction rates in Limfjorden sediments (Fig. 6.8) show that the most intense sulphate reduction takes place in the upper 10 cm. Seasonal variations in reduction rates are largely related to changes in temperature (Fig. 6.9). The average daily sulphate reduction rate in Limfjorden sediments is 254 mgS m^{-2} day^{-1} (Jørgensen,

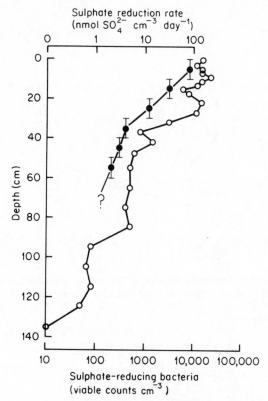

Fig. 6.8 Relation between sulphate reduction rate and number of sulphate-reducing bacteria in a section of the Limfjorden sediments (Jørgensen, 1977). ●———●, Sulphate reduction rate; O———O, number of bacteria

1978). However, only 22 mgS m^{-2} day^{-1} is fixed in the sediments; the remainder diffuses to the oxidized zone where it is subject to chemical and biological oxidation (Fig. 6.10). The annual fixation of reduced sulphur in Limfjorden sediments due to reduction of sulphate is about 8 gS m^{-2}.

Table 6.11 shows rates of sulphate reduction in shallow sediments of different reservoirs; the highest rates are observed in littoral sediments rich in organic matter. It is interesting to note that despite the different climatic zones (ranging from the polar circle to the tropics) the sulphate reduction rates varied only within fairly restricted limits: 0.3–5.2 gS m^{-2} day^{-1} for littoral sediments, and 0.06–0.6 gS m^{-2} day^{-1} for shallow bays, fjords, and lagoons.

According to Woodwell *et al.* (1973) the area of the world's estuaries and shallow bays is 1.75×10^6 km^2 and the average daily sulphate reduction rate

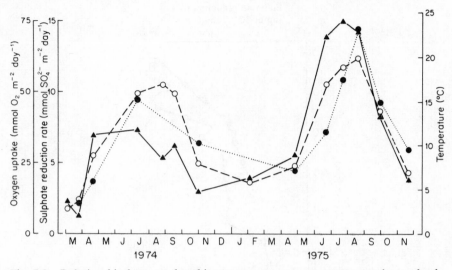

Fig. 6.9   Relationship between benthic water temperature, oxygen uptake, and sulphate reduction rate in upper horizons of the Limfjorden silts during 1974–75 (Jørgensen, 1977). O——O, temperature; ●----●, oxygen uptake; ▲——▲, sulphate reduction rate

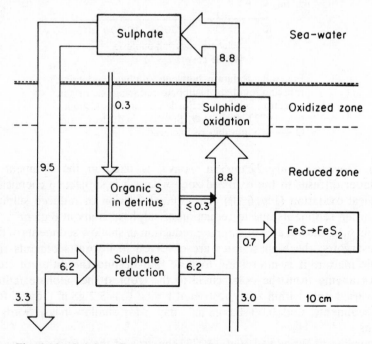

Fig. 6.10   Sulphur fluxes at the sediment/water interface of the Limfjorden sediments (Jørgensen, 1977)

Table 6.11 Sulphate reduction in upper 10 cm of coastal sediments

Region sampled	Depth	Organic carbon (%)	t °C	Rate of sulphate reduction		References
				μgS g^{-1} wet silt day^{-1}	gS m^{-2} day^{-1}	
Barents Sea, sand–argillaceous sediments of the littoral			4.0	9.0–24.3	1.17–3.17[a]	Ivanov and Ryzhova (1960)
North Sea, silt	10–20 m	4.8	13		0.16–0.32	Jørgensen et al. (1978)
North Sea, flooded shore, sands and clays		2–3	20	—	0.64–1.38	Jørgensen et al. (1978)
North Sea, flooded shore, decomposing algae		20–30	16–28	—	1.38–1.60	Jørgensen et al. (1978)
Limfjorden, flooded shore, medium-size sand grains		0.31–1	18–30	—	0.09–0.32	Jørgensen et al. (1978)
Limfjorden, sandy sediments	5–15 m	1–2	10	—	0.16–0.19	Jørgensen et al. (1978)
Limfjorden, silt sediments	5–15 m	10–13	2–4 20		0.06–0.09 0.38–0.64	Jørgensen et al. (1978) Jørgensen et al. (1978)
Sea of Azov, Sivash Bay, shallow sediments rich in organic matter			26–28	15–40	1.95–5.20[a]	Kuznetsov and Romanenko (1968)
Caspian Sea, sand–argillaceous sediments near Krasnovodsk and Kara-Bogaz-Gol Bay			20–22	3.9–8.4	0.5–1.10[a]	Ivanov (1964)
Lizard Island Lagoon, Australia, carbonate sediments			27	0.9–4.8	0.11–0.62[a]	Skyring and Chambers (1976)

[a]Density of sediment at 50% moisture was taken as 1.3.

is 0.11 gS m^{-2}. Therefore, in these ecosystems 70 TgS year^{-1} of sea-water sulphate is reduced and if Jørgensen's observation (Jørgensen, 1978, Jørgensen *et al.*, 1978) that 90% of biogenic sulphide is rapidly recycled is generally applicable, 7 TgS is converted annually to pyrites and other fixed sulphides.

C. Microbiological sulphate reduction in inland sea sediments

The most comprehensive studies on the distribution and geochemical activity of sulphate-reducing bacteria in bottom sediments of mediterranean seas have been undertaken in the Sea of Azov, the Black Sea, and the Baltic Sea. All these reservoirs are situated in the humid zone and are to a certain extent freshened by discharge from the large European rivers. Large cities, industrial and agricultural complexes, and health resorts are located near the coasts and on the rivers, and contribute to intensive contamination of these seas.

The bottom sediments of mediterranean seas are rich in organic matter, and support biological reduction processes including sulphate reduction.

The Sea of Azov

Tolokonnikova (1977) made a detailed study of sulphate reduction in the bottom sediments of the Sea of Azov for three seasons of 1972. She determined numbers of bacteria in the 0–2, 2–5, and 5–10 cm horizons, the sulphate and acid-soluble sulphide concentrations in sediments, and the sulphate reduction rates using ^{35}S-labelled sulphate (Fig. 6.11). Acid-soluble sulphides were detected even in the upper layer of the sediments, where they amount to 588 μg g^{-1} wet weight. During summer temperature stratification, hydrogen sulphide was often present in the bottom waters.

The numbers of sulphate-reducing bacteria and rates of sulphate reduction varied with the season; highest values were observed in summer. The average rates of reduction for the 0–2 cm horizons were 0.61, 4.15 and 0.72 μg g^{-1} for April, July, and October respectively. The lowest rates of sulphate reduction in surface sediments (0.61 μg g^{-1}) were recorded during spring intermixing and aeration of the water. However, in the 2–5 and 5–10 cm horizons, which are better protected and more remote from the oxic zone, the rates were markedly higher; 2.4 and 1.8 μg g^{-1} respectively.

The most intensive sulphate reduction was observed in the eastern and northern parts of the sea, and the least intensive along the Crimean coast and the Arabat spit. Increases in sulphate reduction in the Sea of Azov during the past two decades (Tolokonnikova, 1977), and the fact that organic matter is the limiting factor for sulphate reduction in brackish ecosystems, suggest that the most active sulphate reduction occurs in regions of highest anthropogenic contamination.

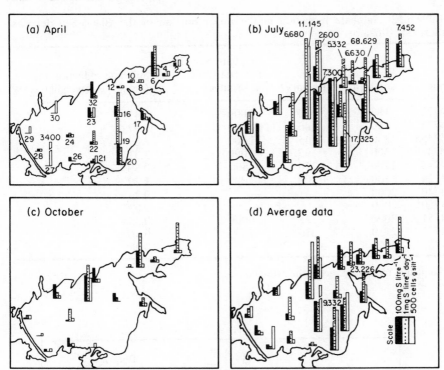

Fig. 6.11 Distribution of sulphide, ▬▬, sulphate reduction rate, (⬚⬚⬚⬚⬚), and sulphate-reducing bacteria (☐), in the uppermost layer of the Sea of Azov sediments in 1972. a, April; b, July; c, October; d, averages for period of observation. 2–32, Sampling sites; numbers at top of columns refer to sulphate reduction rates

The Black Sea

The Black Sea is unique among mediterranean seas, since it is the only reservoir with a large body of water permanently containing hydrogen sulphide. The upper horizons of both near-shore and deep Black Sea sediments contain considerable numbers of sulphate-reducing bacteria (Table 6.12).

Measurements by Sorokin (1962a,b; Table 6.12) of the rates of sulphate reduction in sediments gave an average value of 1.66 μg g^{-1} wet sediment day^{-1} for the shallow part of the sea. Very low values were recorded for sediments of the deep central halistatic areas (sites 4750 and 4753) while the average for the remaining deep-sea sites was 0.65 μg g^{-1} wet weight day^{-1}.

From an analysis of the distribution of bacteria and sulphate reduction, Sorokin (1962b) concluded that most of the viable bacteria in the deep-sea sediments were concentrated within the 0.5 cm horizon. This conclusion appears to conflict with geochemical evidence which shows decreasing pore-

Table 6.12   Sulphate reduction in the upper horizon of the Black Sea deposits (Sorokin, 1962a,b)

Site	Water depth (m)	Eh (mV)	Sulphate-reducing bacteria (cells g^{-1} sludge)	Sulphate reduction rate (μgS g^{-1} wet weight day^{-1})
4772	180	−120	6 000	1.45
4773	250	−145	4 500	0.97
4745a	230	−145	31 000	1.89
4745b	300	−195	50 000	2.31
4745	1700	−210	800	0.89
4754	1700	−200	12 000	1.29
4740	2000	—	—	0.24
4750[a]	2000	—	—	0.027
4751	2000	−225	1 000	0.355
4752	2150	−210	6 000	0.46
4753[a]	2000	—	—	0.085

[a]Located in the central part of the halistatic area.

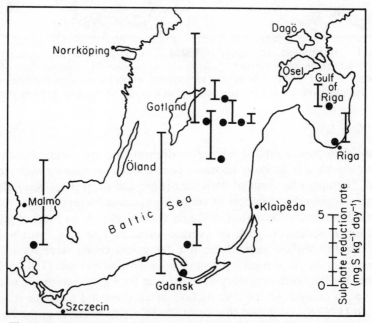

Fig. 6.12   Sulphate reduction rates in upper horizons of the Baltic sea sediments. (Unpublished data from the Biogeochemistry Laboratory of the USSR Academy of Sciences Institute of Biochemistry and Physiology of Microorganisms, from cruise 23a of the research ship *Academician Kurchatov*, 1978)

water sulphate, increasing reduced sulphur, and changes in isotopic composition of sulphides with depth (Shishkina, 1959; Ostroumov 1953a,b; Volkov, 1961a,b, 1964; Vinogradov *et al.*, 1962). Further studies are required to resolve this apparent contradiction.

The Baltic Sea

The distribution of sulphate reduction in the upper horizons of Baltic Sea sediments is shown in Fig. 6.12 (Lein *et al.*, 1982).

In this study all forms of reduced sulphur, including pyrite and elemental sulphur, were analysed for radioactivity after incubation of sediments with $^{35}SO_4^{2-}$. In this respect the analysis differed from those employed in other studies on Mediterranean sea sediments where only free and acid volatile sulphide were analysed. This may explain the high sulphate reduction rates in Baltic Sea sediments (see Table 6.17). It is of interest that the highest rates were found in sediments of the Gulfs of Riga and Gdansk (Fig. 6.12), at the confluence of the rivers Daugava and Vistula, where organic compounds of anthropogenic origin may be discharged.

D. Sulphate reduction in ocean sediments

Sulphate reduction rates in sediments of the western Pacific were determined (using $^{35}SO_4$) during the ninth voyage of the research vessel *Dmitri Mendeleev* (Chebotarev and Ivanov, 1976; Ivanov *et al.*, 1976), and in the Indian Ocean during the 22nd voyage of the *Academician Kurchatov* (Ivanov *et al.*, 1980). These workers analysed acid-soluble sulphide, elemental sulphur, pyrite, and organic sulphur for radioactivity, whereas other research workers analysed acid-soluble sulphide only (Ivanov *et al.*, 1976). In addition, the numbers of sulphate-reducing bacteria at different horizons of the sediment column to a depth of 6 m from the sediment surface were determined. Most of the reduced radioactive sulphur was found in pyrite and organic sulphur, which suggests there is a rapid redistribution of sulphur during diagenesis. It follows, then, that sulphate reduction rates based only on data for acid-soluble sulphide will be underestimates.

The highest rates of sulphate reduction and the largest number of sulphate-reducing bacteria were found in sediments of the Gulf of California and the Mexico shelf, and all sediment layers tested were active (Table 6.13). In general, the rate of reduction decreased with depth of sediment.

Lower rates of sulphate reduction were recorded for the sediments of the continental slope of Mexico (Table 6.14). The rate of sulphate reduction appeared to decrease with increasing depth of ocean water and approached 11 $\mu g\ kg^{-1}\ day^{-1}$ for the top metre of the reduced sedimentary layer. The deep-sea and slightly reduced sediments of the Somali Trench in the Indian Ocean had lower rates of sulphate reduction (Table 6.15).

Table 6.13   Sulphate reduction in Gulf of California and Mexico shelf sediments

Site	Water depth (m)	Sediment horizon (cm)	Eh (mV)	Bacteria, (cells g^{-1})	Sulphate reduction rate (μgS kg^{-1} wet weight day^{-1})
Gulf of California					
663	1760	30	−200	5	19.5
		165	−240	13	4.6
664	1170	70	−190	10	53.9
		150	−250	2	3.4
665	3260	40	−180	1000	102.8
		95	−220	100	62.7
		320	−270	10	10.4
667	2860	10–15	−250	10	53.5
		35–45	−220	100	26.3
		95	−240	2	23.6
Mexico shelf					
666	*120*	34–40	−200	10	33.4
		130–140	−210	10	5.1
		230–240	−260	0	12.8
668	140	15–20	−250	3	55.7
		220	−270	10	10.4
		340	−290	1	2.7

Table 6.14   Sulphate reduction in the continental slope sediments of Mexico

Site	Water depth	Sediment horizon (cm)	Eh (mV)	Sulphate reduction rate (μgS kg^{-1} day^{-1})
669	1000	37	−260	15.7
		90	−250	7.4
670	1450	40	−100	19.7
		70	−310	18.6
		140	−350	20.7
671	2650	30	−40	15.6
		90	−130	6.8
		215	−180	4.7
		305	—	10.0
		401	−150	9.1
672	2900	43	−50	4.1
		75	−50	3.2
		148	−80	1.5
		380	−60	3.8
		500	−60	4.1
		568	−150	5.3

Table 6.15   Sulphate reduction in sediments of the Somali Trench (Ivanov *et al.*, 1980)

Site	Water depth (m)	Sediment horizon (cm)	Sulphate reduction rate (μgS kg^{-1} day^{-1})
1913	5060	90–100	0.48
1915	4656	50–60	3.15
1916	3840	60–80	0.53
1917	2800	200–220	0.11

Table 6.16 gives average values for sulphate reduction rates for four areas of ocean sediments, and an assessment of the amount of sulphur involved in sulphate reduction per square metre of ocean floor. The sediment layer 1 m thick was adopted as the active sulphate reduction zone. The moisture content of this layer was assumed to be 50% and the bulk density 1.3.

The total shelf area of the world oceans is 27.5×10^6 km^2, but only 8×10^6 km^2 (30%) is subject to present-day sediment accumulation (Creager and Stenberg, 1972; Lisitsin, 1978). These are of different grain size and may be coarse-grained, sandy-aleuritic, or shell-carbonate sediments. Coarse-grained sediments probably do not support sulphate reduction because of constant aeration and flushing. In biogenic sediments and carbonate muds of coral reefs and islands, anaerobic conditions may develop, but the low content of iron in these sediments prevents accumulation of significant amounts of reduced sulphur. Hence, sulphide accumulation may be restricted to sandy-aleuritic and pelitic shelf sediments.

Based on the data of Hay (1967) and Emery (1974), cited by Lisitsin (1978), the calculated area of pelitic sediments on the total continental shelf is 1.71×10^6 km^2, and the area of the sandy-aleuritic sediments is 3.35×10^6 km^2. Estimates of annual sulphate reduction in pelitic and sandy-aleuritic sediments of the shelf were made using values of 56 mgS m^{-2} day^{-1} (Mexico shelf; Table. 6.16) and 28 mgS m^{-2} day^{-1}, respectively. Based on these estimates, the total sulphate reduction in shelf sediments amounts to

Table 6.16   Average sulphate reduction rates in the upper 1 m layers of ocean sediments

Location	Sulphate reduction rate	
	(μgS kg^{-1} wet wt day^{-1})	(mgS m^{-2} day^{-1})
Gulf of California	47	61
Mexico shelf	43	56
Mexico continental slope	11	14
Somali Trench	1.4	2

189.5 TgS year^{-1}, including 95.7 TgS year^{-1} in pelitic sediments and 93.8 TgS year^{-1} in sandy-aleuritic sediments.

In sediments of the continental slope, which have an area of 55×10^6 km^2 (Lisitsin, 1974) and a daily sulphide production of 14 mgS m^{-2} (Table 6.16), the annual mass of sulphate sulphur reduced is 280 TgS.

E. Sulphate reduction rates in different sediments of the world oceans

All available values for daily rates of sulphate reduction in sediments of different seas and oceans are listed in Table 6.17: they vary from 0.001 to 40.0 μgS g^{-1} day^{-1}. Typically, shallow, warm, organic-rich sediments of littoral zones and bays show the highest activity. The lowest rates are found in sediments of the open ocean, and especially in slightly reduced deep-sea sediments.

Average values apply to bottom sediments of mediterranean seas, although the coastal areas of these reservoirs are more active.

Table 6.17   Sulphate reduction rates in upper horizons of sediments (μgS g^{-1} day^{-1})

I. Sediments in littorals and shallow bays

Barents Sea littoral	9.0–24.3	Ivanov and Ryzhova (1960)
North Sea littoral	4.92–10.6	Jørgensen *et al.* (1978)
Caspian Sea littoral	3.9–8.4	Ivanov (1964)
North Sea Shallow sediments	1.23–2.5	Jørgensen *et al.* (1978)
Shallow sediments in the Limfjorden	2.9–4.9	
Shallow sediments in Sivash Bay	15.0–40.0	Kuznetsov and Romanenko (1968)
Sediments in the lagoon of Lizard Island, Australia	0.9–4.8	Skyring and Chambers (1976)

II. Sediments in inland seas (average values)

Baltic Sea	1.310[a]	Lein *et al.* (1982)
Sea of Azov	1.800	Tolokonnikova (1977)
Black Sea shelf	1.655	Sorokin (1962a)
The deep-water part of the Black Sea	0.645	Sorokin (1962a)

III. Ocean sediments (average values)

The Gulf of California	0.047[a]	Ivanov *et al.* (1976)
Mexico shelf	0.043	Ivanov *et al.* (1976)
Continental slope	0.011[a]	Ivanov *et al.* (1976)
Somali Trench	0.001[a]	Ivanov *et al.* (1980)
Pacific Ocean, deep sediments	0.001[a]	Lein *et al.* (1976)

[a]Includes all forms of reduced sulphur; other values are for hydrogen sulphide and acid-soluble sulphide only.

6.5.2   Basic Features of Reduced Sulphur Distribution in Sediments

Systematic studies of the geochemistry of sulphur in modern sea sediments began with Ostroumov's (1953a,b) investigations on the Black Sea. Since then the formation, diagenesis, and distribution of reduced sulphur compounds have been studied in the sediments of various basins including the Black, Mediterranean, Baltic, and Okhotsk seas, and the Pacific and Indian oceans.

A. Hydrogen sulphide

In modern reduced sediments hydrogen sulphide ranges from zero to over 100 mg litre^{-1} of pore-water. In sediments with transitional redox conditions hydrogen sulphide is not usually found, but the concentration of reduced compounds of sulphur derived from it may be as high as several parts per thousand. The hydrogen sulphide content of sediments is the result of two opposing processes; generation during bacterial reduction of sulphates, and reaction with reactive iron or its conversion to inorganic and organic derivatives.

In the north-western part of the Pacific Ocean near the coast of Japan, hydrogen sulphide was found at several sites (6158, 6160, 6162, 6163) where the sediments were composed of clays with admixtures of aleurite. At other sites (6159, 6161), dominated by coarse material that is easily flushed and aerated, none was detected despite the occurrence of sulphate reduction. Further towards the open ocean hydrogen sulphide was also absent (Rozanov *et al.*, 1971). The highest concentration found was over 150 mg litre^{-1} (see Fig. 6.15a). By comparison, the highest concentrations in Black Sea sediments are usually less than 100 mg litre^{-1} (Volkov, 1960; Ostroumov *et al.*, 1961). These data indicate that hydrogen sulphide accumulation in sea and oceanic sediments is independent of the redox conditions of the overlying water and is determined solely by specific conditions, for example the organic matter content, in the sediment itself (Strakhov, 1959, 1961).

The concentration of hydrogen sulphide in Pacific Ocean sediments increased with distance from the shore and then decreased. However, the vertical distributions were qualitatively similar at all sites: hydrogen sulphide increased to a maximum with depth and then decreased: the depth at which hydrogen sulphide appeared in the sediments varied from 15 to 170 cm.

In sediments of the Gulf of California, hydrogen sulphide was found at all sites except for one in the littoral zone (depth 120 m, Volkov *et al.*, 1976). It was found in cores taken from the 20–40 cm horizon and varied in concentration from <0.1 to 29 mg litre^{-1}.

The littoral biogenic–terrigenous deposits of the Mexican shore of the Pacific Ocean contained hydrogen sulphide which varied from <0.1 to 26 mg litre^{-1} (Volkov *et al.*, 1976): it appeared in the 20–50 cm layer of sediments, then increased with depth and subsequently decreased. None was detected in hemipelagic sediments.

In the carbonate-rich deposits of the north-western part of the Indian Ocean, hydrogen sulphide was found only at a few sites where it reached

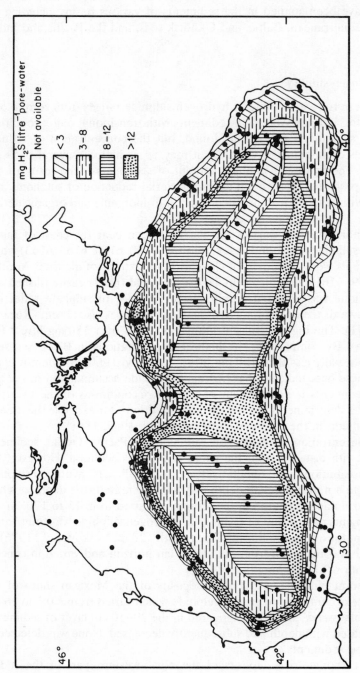

Fig. 6.13   Hydrogen sulphide in the uppermost layer of the Black Sea sediments (Volkov, 1960)

concentrations of no more 0.1 mg litre^{-1} (Zhabina *et al.*, 1979; Ostroumov and Kulumbegashvili, 1977).

Hydrogen sulphide was found in very deep horizons of sediments with high organic matter contents—down to 200–250 m in sediments of the Kurilo-Kamchatska Trench (I. O. Murdmaa, pers. comm.) and to 500 m in those of the north-western littoral of the USA (F. Mangeim, personal communication). Organic-rich sediments of the Gotland basin in the Baltic Sea contained 1.5–24.7 mg litre^{-1} of hydrogen sulphide in the upper layers (0–2, 0–5 cm); it increased to 45.4 mg litre^{-1} at 40 cm depth, but was absent from the basal clays from the later glacial period.

The distribution of hydrogen sulphide in sediments from anaerobic and aerobic zones of the Black Sea has been extensively studied (Volkov, 1960; Volkov and Pilipchuk, 1966). The upper sediment layers (0–2, 0–5 cm) of the hydrogen sulphide zone (water depth over 200 m) contained up to 100 mgH$_2$S litre^{-1} (Fig. 6.13). The highest concentrations were in the deep-sea area, although high concentrations were also observed in sediments at the interface between eastern and western chalistatic areas. Hydrogen sulphide is present throughout the sulphide zones of the Black Sea; in both the modern and old Black Sea and in the upper Neoeuxinian deposits up to the hydrotriol-ite horizon (Volkov, 1964).

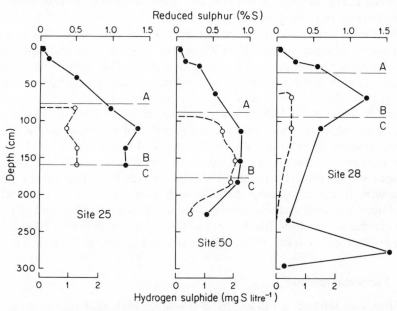

Fig. 6.14 Distribution of hydrogen sulphide and reduced sulphur with depth in sediments of the oxic zone of the Black Sea (Volkov and Pilipchuk, 1966). A, modern sediments; B, old sediments; C, neoeuxinian sediments; O---O, hydrogen sulphide; ●——●, reduced sulphur

While the surface layers (0–5 cm) of sediment in the aerobic zone (water depth less than 200 m) do not usually contain hydrogen sulphide, it may, in some cases, reach 0.1 mg litre^{-1}. However, 24 mgH$_2$S litre^{-1} were found on one occasion near Cape Chanuda in the north-eastern part of the sea at a depth of 178 m in the 0–5 cm layer of the sediments. Hydrogen sulphide does occur at depth in the aerobic zone and reaches 0.5–2 mg litre^{-1} in late Neoeuxinian sediments. It is not found below the hydrotroilite layer of these sediments (Fig. 6.14).

B. Thiosulphate and sulphite

Sulphite and thiosulphate have been detected in the pore-waters of marine sediments. While sulphite is an intermediate in the assimilatory reduction of sulphate and in sulphide oxidation by some thionic bacteria (Trudinger and Loughlin, 1981) its formation in sediments is probably the result of chemical oxidation of hydrogen sulphide (Sorokin, 1964, 1970). Thiosulphate is readily produced by chemical interactions between sulphite and either elemental sulphur or hydrogen sulphide. It is a common product of oxidation of reduced sulphur compounds by thionic bacteria and has been detected in cultures of sulphate-reducing bacteria. Polythionates, which are also produced during sulphur or sulphide oxidation by thionic bacteria, have not been detected in pore-waters.

Sulphite is detectable in sediments only where extremely high concentrations of hydrogen sulphide (30–50 mg litre^{-1}) occur. Only small quantities of thiosulphate were present in the pore-waters of near-shore and hemipelagic sediments of Japan above the horizons containing hydrogen sulphide; but in the sulphide zone itself the concentration of thiosulphate reached several milligrams per litre of interstitial water: it was highest (30–40 mgS litre^{-1}) in sediments containing a great quantity of hydrogen sulphide (site 6160, horizon 515–646 cm; site 6162, horizon 400 cm; Fig. 6.15). Concurrent increases in pore-water thiosulphate and hydrogen sulphide concentrations with depth indicate that thiosulphate was formed during sulphate reduction. Essentially similar patterns have been observed in the Gulf of California, the shelf and continental slope of Mexico, the hydrogen sulphide and aerobic zones of the Black Sea, and the reduced sediments of the Indian Ocean (Ostroumov *et al.*, 1961; Ostroumov and Kulumegashvili, 1977).

C. Elemental sulphur

Elemental sulphur is formed by chemical or biological oxidation of sulphide: the oxidant may be oxygen entering the sediments from the overlying water, or ferric oxide. The most favourable conditions for these processes exist in upper horizons of reduced sediments where oxygen influx is possible and where appreciable quantities of ferric oxide are found.

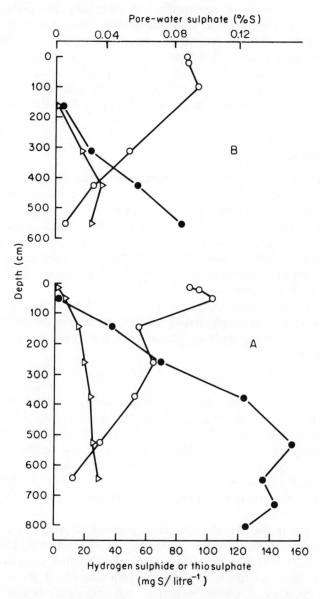

Fig. 6.15 Distribution of hydrogen sulphide, thio-sulphate, and sulphate in pore-waters of continental slope sediments off Japan in the Pacific Ocean (Volkov *et al.*, 1972). A, Site 6160, 1000 m depth; B, site 6162, 4000 m depth. ●——●, hydrogen sulphide; △——△, thio-sulphate; ○———○, sulphate

Elemental sulphur is an intermediate in diagenetic processes and only accumulates in sediments up to a few hundredths of a per cent. It should be noted, however, that if sediments contain greigite (melnikovite) or thiosulphate, these may be decomposed by acid to form elemental sulphur during sediment analysis. Only fresh samples should be used for the extraction of elemental sulphur. Nevertheless, elemental sulphur is an important factor in many of the proposed mechanisms for pyrite formation in sediments (see below).

D. Iron sulphides

Interaction of biogenic hydrogen sulphide with reactive iron is the main mechanism for sulphide fixation in sediments. It is usually assumed that the initial product is colloidal iron monosulphide, i.e. hydrotroilite, the composition of which is $FeS \cdot nH_2S$ or $FeS \cdot nH_2O$ (Sidorenko, 1901; Doelter, 1926; Baas Becking, 1956; Volkov, 1959). This appears to be the precursor of acid-soluble crystalline forms of ferric monosulphide, such as mackinawite, kanzite, and greigite (melnikovite) (Volkov and Ostroumov, 1957b; Volkov, 1964; Berner, 1964; Lein *et al.*, 1978).

In modern sediments pyrite accounts for over 90% of the total reduced sulphur. Ostroumov (1953b) considered that pyrite was formed by a solid phase reaction between hydrotroilite and elemental sulphur (Feld, 1911; equation 3).

$$FeS + S^0 \longrightarrow FeS_2 \tag{3}$$

Others, however, consider that pyrite is formed in a stepwise manner from mackinawite: mackinawite $\rightarrow$ greigite $\rightarrow$ pyrite (Roberts *et al.*, 1969; Berner, 1970; Sweeney and Kaplan, 1973). Nevertheless, the available data suggest that when greigite is formed in sediments in appreciable quantities its pyritization is delayed compared to that in sediments containing little crystalline monosulphide (Volkov, 1961b, 1964; Volkov *et al.*, 1971). At average rates of sulphate reduction most sulphide is immediately converted to pyrite (Ivanov *et al.*, 1976).

Conversion of $FeO(OH)$ to pyrite takes place under weakly acid (pH 6.5) but not weakly alkaline conditions (Berner, 1964; Rickard, 1969, 1975; Roberts *et al.*, 1969; Rozanov, 1973). It is possible that an active form of elemental sulphur for the solid phase reaction mentioned above is formed under acid conditions. Alternatively, acid conditions may favour a solution reaction as in equations (4) and (5).

$$2FeOOH + H_2S \longrightarrow 2Fe^{2+} + S^0 + 4OH^- \tag{4}$$

$$Fe^{2+} + S^0 + H_2S \longrightarrow FeS_2 + 2H^+ \tag{5}$$

While the formation of pyrite from FeS can be formally described by equation (3), there may be intermediate steps. Volkov and Ostroumov (1957a) suggest that thiosulphate is involved (equations 6 and 7).

$$SO_3^{2-} + S^0 \longrightarrow S_2O_3^{2-} \tag{6}$$

$$FeS + S_2O_3^{2-} \longrightarrow FeS_2 + SO_3^{2-} \tag{7}$$

Roberts *et al.* (1969) have proposed a reaction between FeS and disulphane (equations 8 and 9).

$$H_2S + S^0 \longrightarrow H_2S_2 \tag{8}$$

$$FeS + H_2S_2 \longrightarrow FeS_2 + H_2S \tag{9}$$

Rickard (1975) suggested that polysulphide ions could act as sulphur carriers during pyritization. On the basis of Teder's (1971) data on equilibrium constants of polysulphide ions at 25°C and 1 atm, Rickard (1975) concluded that S_4S^{2-} and S_5S^{2-} would be the most stable polysulphide ions under his experimental conditions (pH 7–8, 40°C). Pyritization was envisaged according to equation (10).

$$Fe^{2+} + S_5S^{2-} + HS^- \longrightarrow FeS_2 + S_4S^{2-} + H^+ \tag{10}$$

Hallberg and Wadsten (1980) suggested that pyrite formation from FeS involved iron removal rather than addition of sulphur (equation 11).

$$2FeS \longrightarrow FeS_2 + [Fe] \tag{11}$$

E. Organic sulphur

Ostroumov (1953a) showed that organic derivatives of reduced sulphur were formed in modern sediments during sulphate reduction. The nature of this organic sulphur is still obscure. It appears that the sulphur is not incorporated into microbial protein but into humified organic matter and, together with humic and fulvic acids, it is dissolved by alkaline solutions (Ostroumov, 1953b). Humic acids from Saanich inlet (British Columbia) contain from 3% to 8% of sulphur which appears to be carbon-bonded in the form of C—SH (Brown *et al.*, 1972). In shallow sediments of the inlet, part of the sulphur in humic substances is present as sulphur amino acids which are released on acid hydrolysis.

The rate of formation of organic sulphur in sediments is much slower than that of pyritization (Ivanov *et al.*, 1976). Evidently, organic derivatives of reduced sulphur are formed during diagenesis by interactions between anaerobic decomposition products of organic matter and elemental sulphur or sulphide in the pore-waters.

In sediments of the Pacific Ocean and other basins (Ostroumov, 1953b, 1957; Ostroumov *et al.*, 1961), organic sulphur is the second most important form of reduced sulphur after pyrite and may account for up to 10% of the total reduced sulphur.

6.5.3   Total Reduced Sulphur in Modern Sediments

The various components of the total reduced sulphur (i.e. pyrite, elemental S, sulphide and organic S) in sediments of the Gulf of California and Mexican shore are shown in Fig. 6.16. When the total reduced sulphur is

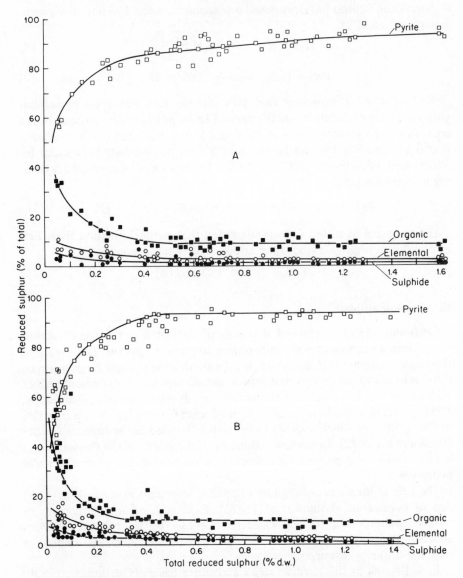

Fig. 6.16   Relationship between sulphide, elemental, organic, and pyrite sulphur and total reduced sulphur in sediments from the Gulf of California (A), and the Mexican coast of the Pacific Ocean (B) (Volkov *et al*., 1976). ●——●, sulphide; ○——○, elemental; ■——■, organic; □——□, pyrite

high, up to 90% is in the form of pyrite and the remainder is mainly organic sulphur (Volkov, 1964; Volkov *et al.*, 1972, 1976). Only at low concentrations of total reduced sulphur (~0.3%) is a substantial part represented by organic sulphur.

As sulphide and elemental sulphur are rapidly transformed, their relative

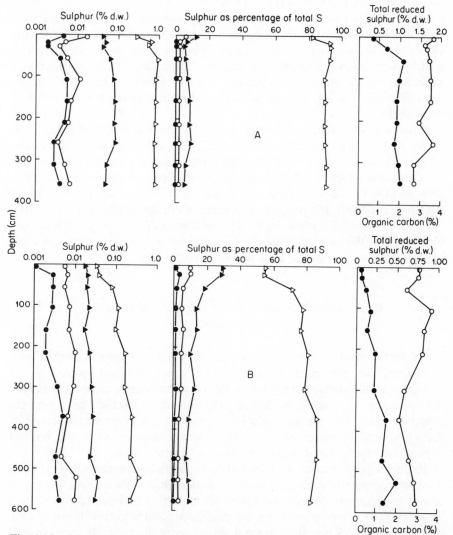

Fig. 6.17 Distribution of organic carbon, sulphide, elemental, organic, pyrite, and total reduced sulphur with depth in sediments from two locations off the Mexican coast of the Pacific Ocean (A, site 668; B, site 672; Volkov *et al.*, 1976). ●——●, sulphide; ○——○, elemental; ▲——▲, organic; △——△, pyrite; ●——●, total reduced; ○——○, organic carbon

concentrations are usually low and they accumulate in sediments only under specific conditions (Volkov, 1961a,b, 1964). The relationships between the reduced forms of sulphur in sediments of the Okhotsk Sea (Ostroumov, 1957) and the north-western part of the Pacific Ocean (Ostroumov and Fomina, 1960; Volkov *et al.*, 1972) resemble those shown in Fig. 6.16. Analogous results were also obtained for sediments of the Somali Trench in the Indian Ocean (Zhabina *et al.*, 1979).

Figure 6.17 shows the distribution of some forms of sulphur and total reduced sulphur with depth in sediments of the western part of the Pacific Ocean. The sediments at site 668, which is nearest to the shore, are characterized by very intensive sulphate reduction. In the first 20–25 cm there is a rapid increase in total reduced sulphur, most of which (up to 90%) is in the form of pyrite. The amounts of sulphide and elemental sulphur are very low throughout.

A more gradual accumulation of reduced sulphur is observed in the sediments at site 672 on the continental slope. These sediments are characterized by a much slower rate of sulphate reduction, and the total reduced sulphur increases progressively down to about 600 cm. Pyrite-sulphur increases from 50% to 60% of the total reduced sulphur at the surface to almost 90% at 400–600 cm depth. The relative proportion of organic sulphur decreases from 30% to 10% over the same depth. The relative proportions of sulphide and elemental sulphur are substantially higher than in core 668, although their absolute concentrations are comparable in both cores.

6.5.4   Relationship between Reduced Sulphur and Organic Matter in Sediments

Sulphate reduction is regulated by the quantity of organic matter in sediments, but the relationship between the two variables is not simple.

Figure 6.18 illustrates the relationship between organic carbon (in surface layers) and total reduced sulphur in sediments of the Pacific Ocean near the Mexican coast. The average reduced sulphur concentration was obtained by taking several measurements in each sediment column below the horizon where the total reduced sulphur reached a constant value.

Reduced sulphur does not accumulate in the pelagic ocean silts (site 673) that are composed of red clays and contain less than 0.5% carbon. Sediments closer to the shore contain more organic matter and there appears to be a general relationship between organic carbon concentration and levels of total reduced sulphur. This relationship is shown more dramatically by sediments of the section from Japan to the north-western depression of the Pacific Ocean (Fig. 6.19).

The levels of total reduced sulphur as a function of organic carbon concen-

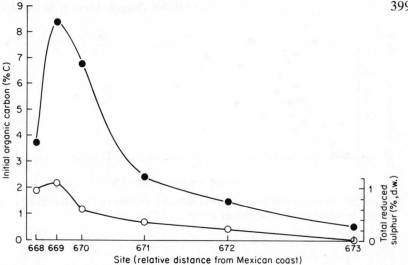

Fig. 6.18 Relationship between initial organic carbon, average reduced sulphur in sediments, and relative distance from the Mexican coast in the eastern depression of the Pacific Ocean (Volkov *et al.*, 1976). ●——●, organic carbon; ○——○, total reduced sulphur

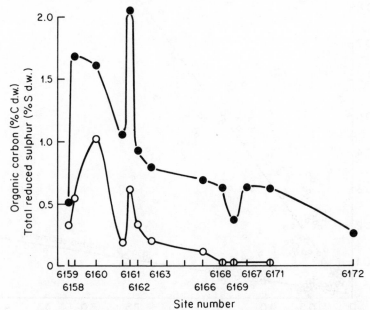

Fig. 6.19 Relationship between initial organic carbon, average reduced sulphur in sediments, and relative distance from Japan to the north-western depression of the Pacific Ocean (Volkov *et al.*, 1972). ●——●, organic carbon; ○——○, total reduced sulphur

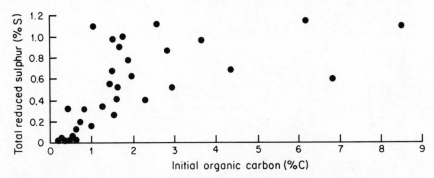

Fig. 6.20 Relationship between total reduced sulphur and initial organic carbon in sediments of the Pacific Ocean and Sea of Okhotsk

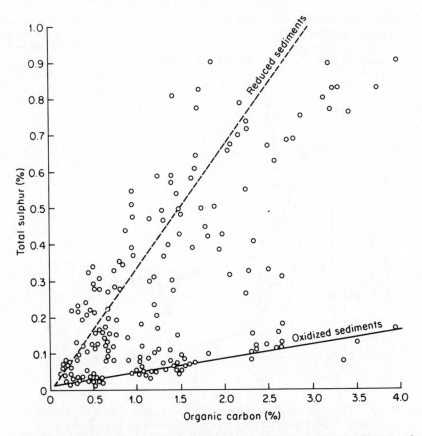

Fig. 6.21 Relation between total sulphur and organic carbon in sediments of the Atlantic Ocean off North-west Africa (Hartmann *et al.*, 1976). ———, oxidized sediments: – – –, reduced sediments

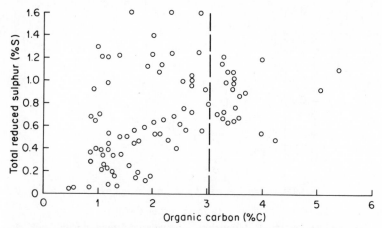

Fig. 6.22 Relationship between total reduced sulphur and organic carbon in sediments of the Pacific Ocean (Volkov *et al.*, 1976). For significance of dividing line, see text; $r = 0.42$

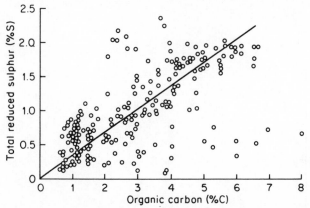

Fig. 6.23 Total reduced sulphur as a function of organic carbon in sediments of the Pacific Ocean near California (Goldhaber and Kaplan, 1974); Reduced S = 0.36 Organic C.

tration for a number of sediments are shown in Figs. 6.20–6.27. The following conclusions can be drawn:

1. Oxidized sediments, low in organic carbon, do not contain significant amounts of reduced sulphur.
2. Sulphate reduction develops in sediments containing 0.5–3% organic matter, and in such sediments there is a reasonable correlation between organic carbon and total reduced sulphur concentrations.
3. Highly reduced sediments containing more than 3% carbon are rich in total reduced sulphur, including hydrogen sulphide, but this reduced sulphur concentration is not correlated with the carbon concentration.

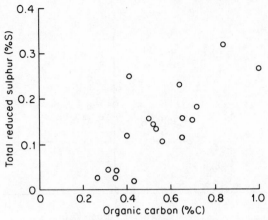

Fig. 6.24 Relationship between mean total reduced sulphur and mean organic carbon in sediment profiles of the Indian Ocean (Zhabina *et al.* 1979; Volkov *et al*, 1981); *r* = 0.76, *n* = 18

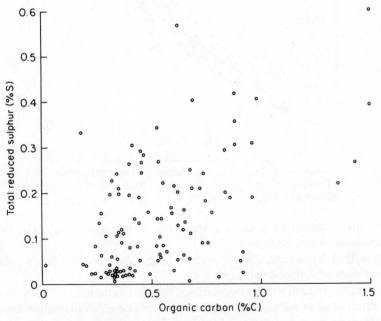

Fig. 6.25 Total reduced sulphur as a function of organic carbon in individual horizon samples of anaerobic sediments of the Indian Ocean (Zhabina *et al.* 1979; Volkov *et al.*, 1981); *r* = 0.55, *n* = 115

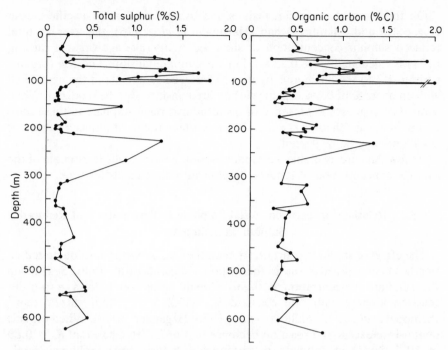

Fig. 6.26 Distribution of total sulphur and organic carbon with depth in sediments from bore-hole 379A in the Black Sea (Calvert and Batchelor, 1978)

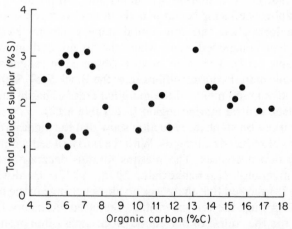

Fig. 6.27 Relationship between total reduced sulphur and organic carbon in sapropelic silts of Old sediments of the Black Sea (Volkov, 1973)

On the basis of these conclusions and using data for the Pacific Ocean (Fig. 6.22) and Atlantic Ocean (Hartmann *et al.*, 1976), the average total reduced sulphur concentration of shelf sediments (average organic carbon, 0.7%) is calculated to be 0.25%S. For the continental slope and rise (organic carbon, 1.3%) the analogous figure is 0.5%. The sediments of the ocean bed with an average of 0.3% organic carbon are considered to be oxidized. While some data suggest the presence of bacterial hydrogen sulphide in these sediments (see, e.g. Zhabina *et al.*, 1979), the total amount of reduced sulphur is small and may be neglected.

As few data are available we cannot present a more realistic estimate of the average concentration of reduced sulphur in sediments at this time.

6.5.5 Relationship between Initial Sulphate in Pore-water and Reduced Sulphur in Sediments

The effect of sulphate concentration on the rate of bacterial sulphate reduction has been studied *in situ* in the surface sediments of the Baltic Sea. When the sulphate concentration is high, sulphate reduction follows zero-order reaction kinetics with rate constant $K_0 = 3.2 \times 10^{-2}$ SO_4^{2-} litre^{-1} day^{-1} (Schippel *et al.*, 1973; Hallberg *et al.*, 1976; Bågander, 1977). When organic matter is in excess, the reaction becomes first-order (rate constant $K_1 = 0.89 \times 10^{-2}$ day^{-1}) at sulphate concentrations below 2 mM (Schippel *et al.*, 1973; Hallberg *et al.*, 1976; Bågander, 1977). These data are in good agreement with the results of laboratory sulphate reduction studies reviewed by Goldhaber and Kaplan (1974).

Microbiological investigations and sulphate reduction rate studies using $^{35}SO_4^{2-}$ show that rates in marine sediments vary within broad limits. The numbers of sulphate-reducing bacteria, and rates of hydrogen sulphide formation, decrease from shelf and continental slope sediments rich in organic matter to the deep pelagic sediments where the organic carbon concentration is low (see Table 6.17). The same trend is observed in sediments of basins with high organic matter concentrations (e.g. the Black Sea). Sulphate reduction is most active in sediments of the marginal areas of basins and decreases towards the deep central marine region (see Table 6.12).

All investigations on sulphate reduction show that the highest rates occur in upper horizons of reduced sediments. With the burial of sediments the rate of sulphate reduction decreases. The reasons for this decrease are not clear. Recent investigations (Romankevich, 1976, 1977; Bordovskii, 1974; Artemyev, 1976) showed that there was no abrupt change in the nature of the organic matter through the vertical profile of reduced sediments. Depletion of organic matter in the course of diagenetic processes is often pronounced, but nevertheless the amount remaining in sediments is quite sufficient to sustain sulphate reduction. In some sediments where sulphate reduction decreases

Table 6.18 Organic carbon and reduced sulphur in sediments of basins with different benthic and pore-water salinities

Basin	Region	Water salinity (%₀)	Organic carbon (%)		Reduced sulphur (%)		Reference
			Surface	At depth	Surface	At depth	
Sea of Azov	Western region of central part of Temruk Bay	11–15	0.33–1.19	0.84–1.28	0.44–1.05	0.59–1.36	Migdisov et al. (1974)
Baltic Sea	Gotland Depression	10–13	1.77–3.84	1.77–7.75	0.46–1.68	1.32–4.86	Present work
Black Sea	Aerobic zone	18	0.8–1.8	1.30–3.66	0[a]–1.38	1.18–1.35	Ostroumov et al. (1961); Volkov and Pilipchuk (1966)
	Hydrogen sulphide zone	20–22	0.92–5.85	0.92–22.65	0.29–2.46	0.72–3.07	Ostroumov (1953a); Volkov (1963, 1973)
Sea of Okhotsk		35	0.32–1.90	0.63–1.90	0[a]–0.20	0.19–1.38	Ostroumov (1957)
Pacific Ocean	(1) Gulf of California	35	1.00–5.40	0.87–5.80	0[a]–0.41	0.77–1.63	Volkov et al. (1976)
	(2) Mexican coast	35	1.37–3.66	0.87–3.49	0[a]–0.36	0.07–1.24	Volkov et al. (1976)
	(3) North-west region	35	0.52–2.10	0.37–1.73	0[a]–0.07	0.07–1.12	Volkov et al. (1972)

[a]Below level of detection.

with depth, the sulphate concentration in pore-water also falls with depth (to $10-20$ mg litre^{-1}, see, e.g. Figs. 6.4, 6.5), whereas in others, for example those of the Gulf of California and the Mexican coast of the Pacific, the sulphate concentration in pore-water does not change (Ivanov *et al.*, 1976; Volkov *et al.*, 1976; see also Fig. 6.2).

Data on reduced sulphur in sediments with different salinity levels are given in Table 6.18 from which it is apparent that accumulation of reduced sulphur in sediments is not related to the initial sulphate concentration of the water. Several studies have shown that the amounts of reduced sulphur in sediments often exceed those of sulphate–sulphur buried with pore-water. For example, Strakhov (1972, 1976) calculated the 'excess' reduced sulphur over sulphate to be $10-15$ times for organic-rich sediments.

The excess sulphur is supplied by diffusion of sulphate from the overlying water. The diffusion is sufficiently rapid to maintain high sulphate concentrations in the upper layers of sediment, even though these are the sites of the most intense sulphate reduction (e.g. see Figs. 6.2, 6.30). These figures illustrate another point, namely, that elevated concentrations of sulphate can develop just above the zone of hydrogen sulphide formation. A similar phenomenon was observed in sediments of the North-west Pacific off Japan and the Gulf of California (Volkov *et al.*, 1972; 1976; Berner, 1964), Newport Bay, and the Santa Catalina basin (Kaplan *et al.*, 1963). The most likely explanation for the increased sulphate levels is that hydrogen sulphide diffusing upwards from the sulphate reduction zone is oxidized, possibly with the help of thionic bacteria, by oxygen diffusing into the sediments from the overlying waters.

Thus, in the most active zones of sulphate reduction in marine sediments sulphate is not rate limiting. In deeper horizons where diffusion is limited, sulphate levels can fall to rate-limiting values. However it may be noted that, because of the high levels of sulphate reduction in the uppermost layers, complete reduction of sulphate at depth would not increase the total reduced sulphur content of sediments by more than 10%.

6.5.6 Accumulation of Reduced Sulphur in Sediments as a Function of Sedimentation Rate

The sedimentation rates in the basins listed in Table 6.18 differ markedly. For example, the average sedimentation rates in the central part of the Azov basin are $100-250$ g cm^{-2} per 1000 years, and in outlying parts >500 g cm^{-2} per 1000 years (Khrustalev and Shcherbakov, 1974). In the Gulf of California, sedimentation rates vary from 78 to 350 g cm^{-2} per 1000 years (Van Andel 1964), and in the shelf and continental slope off the coast of Mexico in the Pacific Ocean, from 2.5 to 15.6 g cm^{-2} per 1000 years (Van Andel, 1964; Lisitsina *et al.*, 1976). Despite such marked differences in rates of sedimentation there are only slight differences in the amounts of reduced sulphur accumulated. It should also be noted that the reduced sulphur contents of the sediments of the Sea of Azov are similar to those of the coast of Mexico in the Pacific Ocean, even though the organic matter content of the latter is markedly higher. Thus, neither absolute sedimentation rates nor the initial sulphate concentration in benthic and pore-waters appears to influence the concentration of reduced sulphur in sediments.

6.5.7 Iron and the Accumulation of Reduced Sulphur in Sediments

An upper limit on the amount of sulphide fixed in a sediment is set by the supply of reactable metal. Figures 6.29–6.32 show distributions of sulphide and reactive iron (acid-reactable plus pyrite) in a number of marine sediments. As expected, the proportion of iron in the form of pyrite (or other iron sulphides) is greater in highly reduced sediments like those of the coast of Japan (Fig. 6.30A), the Gulf of California (Fig. 6.30B), the coast of Mexico (Fig. 6.30C), and the Black Sea (Fig. 6.32), than those of the open ocean (Fig. 6.29).

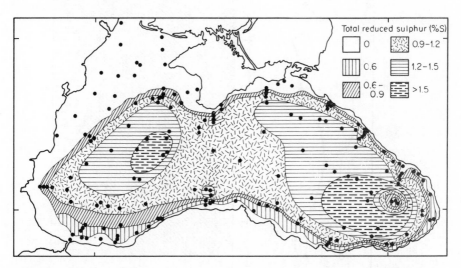

Fig. 6.28   Reduced sulphur in the uppermost sediment layer of the Black Sea. (Ostroumov *et al.*, 1961)

Even in highly reduced sediments, however, not all reactive iron is pyritized even when hydrogen sulphide accumulates in significant amounts. This suggests that not all acid-reactable iron is hydrogen sulphide reactable: some may be in the form of iron silicates.

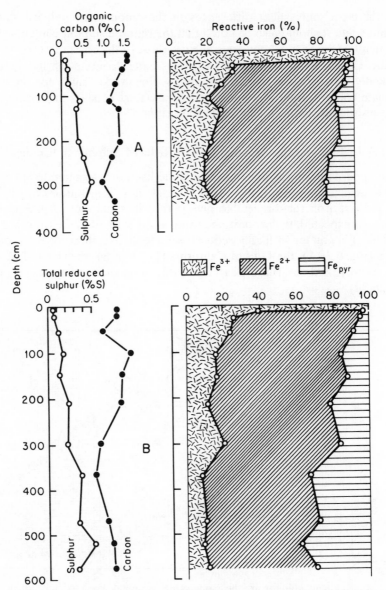

Fig. 6.29 Distribution of reactive iron and reduced sulphur with depth in sediments from transitional zones of the Pacific Ocean (A, site 657; B, site 672). O———O, total reduced sulphur; ●———●, organic carbon

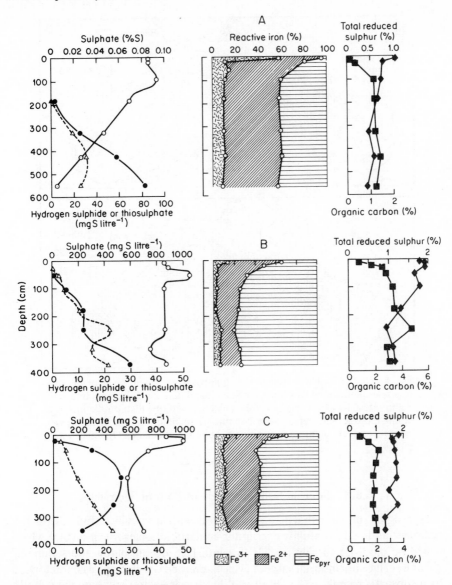

Fig. 6.30 Distribution of sulphur compounds in pore-water, and reactive iron, reduced sulphur, and organic carbon in reduced sediments of the Pacific Ocean (A, site 6162, Japan; B, site 664, California; C, site 668, Mexico). Pore-water: O——O, sulphate; ●——●, hydrogen sulphide; △---△, thiosulphate. Sediments: ▨, ferric iron; ▧, ferrous iron; ☰, pyritic iron; ■——■, total reduced sulphur; ♦——♦, organic carbon

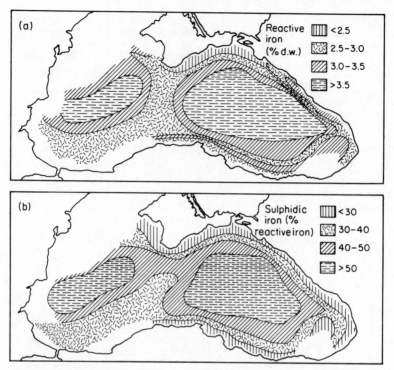

Fig. 6.31   (a) Reactive iron in the uppermost sediment layer of the hydrogen sulphide zone of the Black Sea (calculated for sediments free from
carbonates; Rozanov *et al.*, 1974). (b) Distribution of sulphidic iron (i.e.
monosulphide and pyrite) as a percentage of reactive iron in the uppermost layer of the hydrogen sulphide zone of the Black Sea (Rozanov *et al.*,
1974)

6.5.8   Internal Sulphur Cycle in Bottom Sediments

Hydrogen sulphide that has not reacted with iron will diffuse through a
sediment and will eventually encounter an aerobic/anaerobic interface where
it will be oxidized to sulphate. If this interface occurs below the sediment/
water interface an 'internal sulphur cycle' is created which may be characterized
by a subsurface peak in the depth profile for pore-water sulphate (Fig. 6.30).
An estimate of the intensity of this internal cycle was made from data on
sediments at site 668, situated in the Pacific Ocean on the edge of the shelf of
the Mexico coast (Fig. 6.30).

The sedimentation rate in this region has been variously estimated to be
30 mm per 1000 years (Lisitsina *et al.*, 1976) or 100 mm per 1000 years (Van
Andel, 1964). In the upper part of the sediment, between the 17.5 and 25 cm
horizons, the increase in reduced sulphur is 6 gmS kg^{-1} dry sediment. Accord-

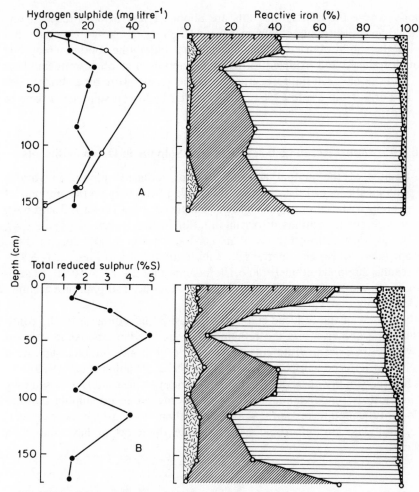

Fig. 6.32 Distribution of hydrogen sulphide, reduced sulphur, and reactive iron with depth in sediments of the Gotland depression of the Baltic Sea (A, site 1972; B, site 1973). ●——●, hydrogen sulphide; ○——○, total reduced sulphur; ▨▨, ferric iron; ▨▨, ferrous iron; ☰, pyritic iron; ▦, monosulphidic iron

ing to Ivanov *et al.* (1976) the rate of hydrogen sulphide formation in the 15–30 cm horizon is 1671 μgS kg^{-1} wet sediment per month or 20 mg kg^{-1} year^{-1}. Assuming a moisture content of 60% this amounts to 50 mgS kg^{-1} dry sediment per year. The times taken to accumulate 7.5 cm (25–17.5) of sediment at rates of 30 and 100 mm per 1000 years are 2500 and 750 years respectively. At a reduced sulphur production rate of 50 mg kg^{-1} year^{-1}, 125 gS kg^{-1} and 37.5 gS kg^{-1} respectively, would accumulate in these periods. These theoretical amounts are about 21 and 6 times respectively, higher than the amount actually measured (6 gS kg^{-1}); and we conclude that most of the sulphur buried in the sediment has undergone continual recycling through reduced and oxidized forms by the action of sulphate-reducing bacteria and thionic bacteria, respectively. Similar results were obtained for sites 667 and 669: from 3% to 11% of the bacterial hydrogen sulphide is fixed in sediments.

6.5.9 Annual Accumulation of Reduced Sulphur in Ocean Sediments

Estimates of the annual accumulation of reduced sulphur in various morphological zones of the oceans were derived from data on the amounts of sedimentary material deposited annually in the ocean (Tables 6.3 and 6.4), the average concentrations of organic matter in the sediments (Table 6.4), and the average concentrations of reduced sulphur corresponding to those concentrations of organic matter (see subsection 6.5.4 and Figs. 6.21–6.25). The results, summarized in Table 6.19, indicate that 20–27 × 10^3 Tg of sediment, containing between 82.5 and 111.3 Tg of reduced sulphur, are accumulated annually.

About 90% of the reduced sulphur is deposited in sediments of the continental slope and rise, and the remainder in shelf sediments. As mentioned elsewhere, open-ocean sediments are largely aerobic and contain insignificant amounts of reduced sulphur (Volkov *et al.*, 1972; Zhabina *et al.*, 1979).

While the validity of the above numbers cannot be accurately assessed they are likely to be underestimates since the average reduced sulphur concentrations of the mediterranean seas are higher than those used in these calculations, and data for anaerobic basins of the open ocean have not been included.

Table 6.19   Annual accumulation of reduced sulphur in ocean sediments

Morphological zones of ocean	Area (10^6 km^2)	Sedimentation rate (10^3 Tg year^{-1})	Average concentration Organic carbon (%)	Average concentration Reduced sulphur (%)	Reduced sulphur flux (TgS year^{-1})
Shelf	26.7	3.2–4.3	0.7	0.25	8.0–10.8
Continental slope and rise	76.5	14.9–20.1	1.3	0.50	74.5–100.5
Ocean bed	257.0	1.9–2.6	0.3	0	0
Total		20.0–27.0			82.5–111.3
Average					96.9

6.6 REDUCED SULPHUR ACCUMULATION IN SEDIMENTS OF MARINE BASINS WITH HIGH RATES OF SULPHATE REDUCTION

In the preceding sections of this chapter it was shown that knowledge of the distribution of sulphur compounds in bottom sediments of the oceans was sufficient to allow only a general description of the quantitative aspects of the sulphur cycle, particularly with respect to the deposition of sulphur in sediments.

A more detailed assessment of sulphur deposition in sediments is possible only for the thoroughly investigated inland seas surrounding the USSR (Black Sea, Sea of Azov, and Baltic Sea) and the Gulf of California.

6.6.1 The Black Sea

Various estimates have been made of the annual hydrogen sulphide production rate in the sediments of the Black Sea. About 5.1 Tg of organic carbon are buried annually in sediments of the Black Sea (Datsko, 1959), roughly 1.2 Tg from the suspended matter in river-water, and the remainder (3.9 Tg) from the water column. Datsko (1959) estimated that about 5.3 Tg of organic carbon per year entered the anaerobic zone of the water layer of the Black Sea in the form of plant and animal remains. It follows then that about 1.4 Tg of organic carbon are mineralized in the anaerobic zone of the water column or at the water/sediment interface. If mineralization were linked solely to sulphate reduction this amount of organic carbon would support the production of about 2 TgH_2S year^{-1}.

Sorokin's (1962b, 1964) data on primary production in the Black Sea, together with $\delta^{13}C$ data, led Deuser (1971) to conclude that 10 gC m^{-2} year^{-1} are metabolized during sulphate reduction, which is equivalent to an annual production of 14 gH_2S m^{-2}. If this value is applied to the total area of sediments of the anaerobic zone down to the 200 m isobath (307 000 km^2), the total annual production of hydrogen sulphide is 4.3 Tg. Detailed descriptions of the computational methods are given in Aizatullin and Skopintsev (1974) and Skopintsev (1975). Based on the water volume discharged annually in the Black Sea through the Bosporus, these authors assessed the annual flux of hydrogen sulphide from sediments to the water column of the Black Sea at 2.4 TgS. However, Sorokin's (1962a) data on rates of sulphate reduction in the water column indicate that the total production may reach 25 gH_2S m^{-2} year^{-1} or 7.67 TgH_2S year^{-1} for the area of 307 000 km^2 (Skopintsev, 1975).

The distribution of reduced sulphur in Black Sea sediments is shown in Fig. 6.28 and Table 6.20. The total amount of material added annually to the bottom sediments of the Black Sea is approximately 220 Tg, of which 149 Tg is particulate matter in river discharge, 60 Tg biogenic carbonate (Shimkus and Trimonis, 1974), and 12 Tg organic matter (Datsko, 1959). If the average concentration of reduced sulphur in sediments is assumed to be 1%, then the annual flux of sulphur to the sediments is 2.2 Tg.

Table 6.20   Average concentration of reduced sulphur (% dry weight)[a] in sediments of different zones of the Black Sea

	Water depth (m)			
	<200	200–1000	1000–2000	>2000
Upper horizons of sediments[b]	0.506 (9)[c]	0.862 (24)	1.194 (18)	1.262 (18)
Deeper horizons of sediments	1.013 (13)	1.280 (5)	1.250 (8)	1.421 (28)

[a] Based on the data of Ostroumov (1953a), Volkov (1961a, 1964), Volkov and Pilipchuk (1966), and Ostroumov *et al.* (1961).
[b] Fifty samples were taken from the 0–5 cm horizon and 19 samples from the 0–10 cm horizon.
[c] The values in brackets are the number of samples analysed.

However, the average concentration of reduced sulphur in sediments of the hydrogen sulphide zone of the Black Sea (>200 m depth) is higher than that of shallow-water sediments (Table 6.20). Using data on sedimentation rates (Ross and Degens, 1974, see Fig. 6.33), and the average sulphur concentrations of the different zones of the Black Sea (Table 6.21), we calculate that the minimum quantities of dry sedimentary material and reduced sulphur deposited annually in sediments of the hydrogen sulphide zone are 75 and 1 Tg respectively, and for shelf sediments, 145 and 1.4 Tg respectively. Thus the total flux of reduced sulphur into the Black Sea sediments is 2.4 TgS year^{-1}.

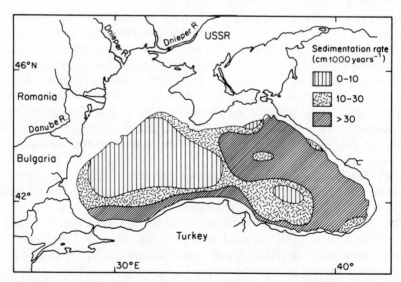

Fig. 6.33   Sedimentation patterns in the deep-water part of the Black Sea (Ross and Degens, 1974)

Table 6.21 Estimated burial of reduced sulphur in sediments of the hydrogen sulphide zone of the Black Sea[a]

Water depth (m)	Area (10³ km)				Sediment flux (Tg year⁻¹)				Average concentration of reduced sulphur (%S)	Reduced sulphur flux (TgS year⁻¹)
	I[b]	II	III	Total	I	II	III	The whole zone		
200–1000	20.0	18.0	3.6	41.6	1.30	4.68	1.40	7.38	1.280	0.094
1000–2000	36.0	23.0	58.2	117.2	2.34	5.98	22.70	31.02	1.250	0.388
2000	46.4	43.5	57.6	147.5	3.02	11.31	22.46	36.79	1.421	0.523
Total	—	—	—	306.3	—	—	—	75.19	—	1.00

[a] Volume/weight of wet sediment is 1.3 in all cases, humidity is 50%.
[b] Areas with sedimentation rates of 5, 20 and 30 cm per 1000 years, respectively.

6.6.2 The Sea of Azov

The Sea of Azov is a shallow bay of the Black Sea with a total area of 38,000 km^2 and a depth of 8 m. According to Tolokonnikova (1977), sediments of this sea actively reduce sulphate (see Table 6.17), and hydrogen sulphide is often found in bottom waters. The average concentration of reduced sulphur in sediments is 1.09% (Migdisov *et al.*, 1974) which is similar to the value found for the sediments of the Black Sea shelf (see Table 6.20).

The annual sedimentation in the Sea of Azov has been variously estimated to be 17.6 Tg (Fedosov, 1961), 21 Tg (Aleksandrov, 1965) and 44.1 Tg (Panov and Spichak, 1961). Using an average rate of sedimentation of 27.5 Tg year^{-1} and a sedimentary sulphur content of 1.1% we estimate the annual flux of reduced sulphur to the sediments to be about 0.3 Tg.

6.6.3 The Baltic Sea

The amounts of reduced sulphur, water, and organic carbon in sediments of the Baltic Sea are summarized in Table 6.22. The average concentration of reduced sulphur is 1.37%. (It is of interest that reduced sulphur in sediments from six of the seven sampling sites exceeded 1% even in the upper layers (0–7 cm).)

The Upper Holocene reduced sediments of the Baltic Sea vary in thickness from 1 to 3.5 m, and were laid down over 8000–10 000 years. From the data

Table 6.22 Reduced sulphur and organic carbon in reduced Holocene sediments of the Baltic Sea[a]

Region	Site	Thickness of Holocene sediment (cm)	Moisture content (%)	Reduced sulphur (%)[b]	Organic carbon (%)[b]
Gotland Basin	2168	200	69.8 (5)[c]	1.83 (5)	4.36 (4)
	2622	250	75.4 (8)	1.59 (5)	3.74 (5)
Central part of the Sea	2631a	233	63.73 (4)	1.186 (4)	3.02 (4)
	2611	175	68.59 (5)	0.996 (5)	3.1 (5)
Arkon Trench	2656	160	66.57 (3)	1.23 (4)	3.93 (4)
Gdansk Trench	2682	341	72.60 (5)	1.60 (5)	5.33 (4)
Bay of Riga	2601	100	73.1 (4)	1.13 (4)	2.09 (3)
Average		208	69.97	1.37	3.65

[a]Unpublished data of Lein, Namsaraev, and Veinstein from the Baltic cruise of the *Academician Kurchatov* (1978).
[b]Dry weight basis.
[c]The value in brackets refers to the number of analyses.

Table 6.23 Sulphur budget in the Baltic Sea sediments[a]

Sedimentation rate (cm per 1000 years)	Bulk density (g cm^{-3})	Water content (%)	Sulphate (%S)[b]	Reduced sulphur (%S)[b]	Area of sediments (10^3 km^2)	Deposition rate (g cm^{-2} per 1000 years)			Accumulation rate (Tg year^{-1})		
						Dry sedimentary material	Reduced sulphur	Sulphate sulphur	Dry sedimentary material	Reduced sulphur	Sulphate sulphur
25	1.45	70	0.05	1.37	187.57[c]	10.87	0.15	0.004	20.4	0.28	0.010
					197.8[d]				32.4	0.44	0.016

[a]Calculated by absolute mass method.
[b]Dry weight basis.
[c]Excluding the Gulf of Finland and the Gulf of Bothnia.
[d]Total area.

on moisture and reduced sulphur concentration (Table 6.22), and assuming an average thickness and sedimentation rate of 2 m and 25 cm per 1000 years, respectively, the calculated amount of reduced sulphur buried annually in Upper Holocene sediments is 0.44 Tg (Table 6.23). The corresponding flux for sulphate is 0.016 TgS year^{-1} (Table 6.23).

An alternative estimation is based on the annual influx of sedimentary material due to shore abrasion, leaching of sea-floor rocks, and river run-off (estimated to be 45.7 Tg year^{-1}). The annual deposition of total and reduced sulphur is calculated to be 0.65 and 0.63 Tg, respectively.

A third method of assessment of the annual deposition of sulphur in sediments is to determine the difference between sulphur influx to the sea from different sources and the efflux from this reservoir. The method of calculation and the individual fluxes are shown in Fig. 6.34. Data on sulphur input from river runoff and from ground-water are taken from Rabinovich and Grinenko (1979) and Gurdelis and Emelyanov (1976), respectively. The atmospheric sulphur flux over the Baltic Sea region is estimated from the data of Ottar (1976) for atmospheric precipitation (Table 6.24), and from the data of Moss (1978) for dry deposition (Table 6.25). Table 6.26 presents data on the amount of sulphate entering and leaving the Baltic Sea in sea-water through the Danish Channels. From the difference between input and output of sulphur (Fig. 6.34) the total sulphur flux from water to sediments is estimated at about 0.7 Tg year^{-1}, with 0.02 Tg buried in the form of sulphate sulphur and 0.68 Tg as reduced sulphur. Thus, three independent methods give rise to annual fluxes of reduced sulphur of, 0.44, 0.63 and 0.68 TgS year^{-1}. In

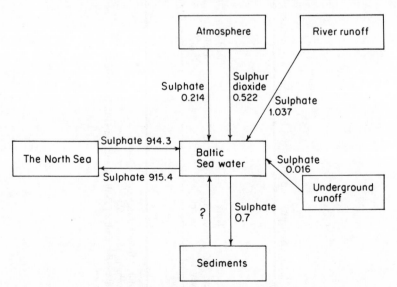

Fig. 6.34   Sulphur fluxes in the Baltic Sea; (TgS year^{-1})

Table 6.24 Sulphate input to the Baltic Sea in atmospheric precipitation (Ottar, 1976)

Region	Sulphate input (gS m^{-2} year^{-1})	Area (10^3 km^2)	Sulphate flux (TgS year^{-1})
Gulf of Bothnia	0.4	117	0.046
Remainder	0.6	280	0.168
Total		397	0.214

Table 6.25 Sulphate input to the Baltic Sea in dry deposition (Moss, 1978)

Region	Sulphate input (gS m^{-2} year^{-1})	Area (10^3 km^2)	Sulphate flux (TgS year^{-1})
Gulfs of Bothnia and Finland	1	147	0.147
Remainder	1.5	250	0.375
Total		397	0.522

Table 6.26 Estimated input and output of sulphate in water flow through the Danish Channels (calculated from the salinity; Chernovskaya *et al.*, 1965)

	Volume (km^3)	Sulphate concentration (mgS litre^{-1})	Sulphate flux (TgS year^{-1})
Output	1659.9	551	915.4
Input	1187.4	770	914.3

subsequent calculations of the flux of reduced sulphur into the sediments of the world oceans the value of 0.6 TgS year^{-1} is used for the Baltic Sea.

6.6.4 The Gulf of California

The average reduced sulphur concentration in sediments of the Gulf of California is 1% (Table 6.27). The sedimentation rates in the Gulf of California are extremely high and vary between 316 cm per 1000 years in the northern part of the Gulf and 12 cm per 1000 years in the southern part (Van Andel, 1964; Emery and Bray, 1962; Lisitsin, 1974).

Table 6.27 Sulphate, reduced sulphur and organic carbon in subsurface sediment horizons of the Gulf of California (Volkov *et al.*, 1976)

Site	Depth (m)	Number of samples	Sulphate (%S)	Reduced sulphur (%)	Organic carbon (%)
662	2400	6	0.118	1.12	2.20
663	1760	5	0.157	1.01	2.28
664	1170	5	0.193	1.18	3.83
665	3260	6	0.049	0.57	2.08
666	120	5	0.051	1.27	1.22
667	2860	5	0.111	0.74	3.50
Average	—		0.113	0.98	2.52

About 75% of the sediment of the Gulf consists of reduced diatomaceous, foraminiferal, and terrigenous deposits (Van Andel, 1964). Assuming that 75% of the material deposited annually in sediments of the Gulf (355 Tg; Van Andel, 1964) enters the reduced zone, then the annual accumulation of reduced sulphur is ~2.66 TgS. The flux of sulphate to this zone is 0.3 TgS.

6.6.5 Sulphur Flux into Sediments of Inland Seas

The data presented in sections 6.6.1–6.6.4 indicate that about 6 Tg of reduced sulphur is deposited annually in sediments of four marine basins: viz. the Black Sea, Sea of Azov, Baltic Sea, and the Gulf of California. This amount is nearly equal to that deposited on the whole continental shelf of the world oceans (Table 6.19). This probably results from the very high reduced sulphur concentrations and sulphate reduction rates for the inland sea sediments (Table 6.28) and the very high rates of sediment formation in the four basins (Lisitsin, 1978).

6.7 TOTAL FLUX OF SULPHUR INTO OCEAN SEDIMENTS AND THE SULPHUR BUDGET OF THE WORLD OCEANS

The total annual flux of sulphur (reduced plus sulphate) into the sediments of oceans is assessed from the data in Tables 6.9 and 6.19, and the rough estimates of biogenic sulphur outlined in section 6.2.3. The results are summarized in Table 6.29.

The total annual accumulation of sulphur in sediments of the world oceans amounts to 106.9–145.4 (average 126.2) TgS. Most of this is derived from the water column in the form of reduced sulphur produced by bacterial sul-

Table 6.28   Total reduced sulphur and sulfate reduction rates in sediments of inland seas and oceanic shelf

Sediment	Concentration (%S)	Sulphate reduction rate[a] (μgS g^{-1} silt day^{-1})	Reference
Black Sea sediments from 2000 m	1.42	0.645	See Table 6.20
Baltic Sea sediments	1.37	1.310	See Table 6.22
Azov Sea sediments	1.09	1.800	Migdisov *et al.* (1974)
35 samples of aleuritic–pelitic shelf sediments from Pacific and Atlantic Oceans	0.87	0.43	Volkov *et al.* (1976), Berner (1970)
29 samples of sand–aleuritic shelf sediments from the Atlantic Ocean	0.13	—	Berner (1970), Hartmann *et al.* (1976), Thode *et al.* (1960)

[a]For references see Table 6.17.

phate reduction. The flux of sulphate–sulphur to sediments is about 22% of the total flux.

In sedimentary rocks, however, sulphate predominates over sulphide (sulphate–S : sulphide–S $\simeq$ 2) due to sulphide oxidation during weathering. From our data, this ratio is 1 : 3.5 for sediments as a whole.

Accumulation of sulphur compounds in modern oceanic sediments is vari-

Table 6.29   Accumulation of sulphur in sediments of the world oceans

Forms	Sulphur flux (TgS year^{-1})	
	Range	Average
Reduced sulphur	82.5–111.3	96.9
Biogenic sulphur	1–2	1.5
Sulphate sulphur:		
in the solid phase of clay sediments	11.2–15.4	13.3
in biogenic carbonates	1.4	1.4
in barite	4.0–5.4	4.7
in pore-water	6.8–9.9	8.4
Total	106.9–145.4	126.2

Table 6.30 Accumulation of carbon and sulphur in sediments of the World ocean

Morphological zones	Area (10^6 km^2)	(%)	Sediment mass (10^3 Tg)	(%)	Organic carbon (TgC year^-1)	(%)	Flux Sulphate (TgS year^-1) Range	(Average)	(%)	Reduced S/sulphate S
Shelf	26.7	7.4	3.2–4.3	16.0	22.4–30.2	10.1	3.5–4.8	4.2	15.1	2.47
Continental slope and rise	76.5	21.2	14.9–20.1	74.5	195.0–263.2	87.5	16.7–22.9	19.8	17.2	4.44
Ocean bed	257.0	71.2	1.9–2.6	9.5	5.4–7.3	2.4	3.2–4.4	3.8	13.7	—
Total	360.2	100	20–27	100	222.8–300.8	100	22.4–32.1	27.8	100	3.54

Flux

	Reduced sulphur (TgS year^-1)			Total sulphur (TgS year^-1)		
	Range	Average	(%)	Range	Average	(%)
	9.0–11.8	10.4	10.5	12.5–16.5	14.6	11.5
	74.5–101.5	88.0	89.4	92.2–124.4	107.8	85.5
	—	—	—	3.2–4.4	3.8	3.0
	83.5–113.3	98.4[a]	100	107.9–145.4	126.2	100

[a]Including 1.5 Tg of organic sulphur.

able. Table 6.30 summarizes data on the annual accumulation of sulphate in the various morphological zones of the ocean floor. Accumulation is most closely related to sediment mass although pelagic sediments are relatively enriched in sulphate. This is probably due to the higher contribution of pore-water sulphate to the total sulphur of oxidized pelagic sediments.

Accumulation of reduced sulphur in oceanic sediments generally correlates with the accumulation of organic carbon except in the case of pelagic sediments where reduced sulphur levels are extremely low. It has already been noted that, while oxidized pelagic sediments occupy about 70% of the area of the world oceans, they contain only about 10% of the total sediment accumulated (see Table 6.30). Thus, in quantitative terms, diagenesis of sulphur is mainly confined to the 90% of reduced oceanic sediments of the continental shelf, slope, and rise.

Conclusions regarding the sulphur balance in the World ocean can be drawn from data given in Tables 6.1, 6.29, and 6.30. Table 6.1 shows that 339.2 TgS is supplied annually to oceans, basically in the form of sulphates. This flux is composed of nearly equal amounts of natural (169.2 TgS) and anthropogenic (about 170 TgS) sulphur. The natural input of sulphur to the oceans is almost balanced by the 126.7 TgS removed each year, mainly in the form of reduced sulphur.

The anthropogenic flux thus represents a net accumulation of sulphur in the oceans. However, as this annual flux is small when compared to the total sulphate in the oceanic reservoir (1.3×10^9 TgS) neither the concentration of sulphate–sulphur, nor the properties of the oceanic water on the whole, will be markedly affected in the foreseeable future.

Part II THE MASS-ISOTOPIC BALANCE OF SULPHUR IN OCEANIC SEDIMENTS

A. Yu Lein, V. A. Grinenko, and A. A. Migdisov

6.8 FRACTIONATION OF SULPHUR ISOTOPES IN SEA SEDIMENTS

In modern sediments the isotopes of sulphur are fractionated during bacterial sulphate reduction with sulphide becoming enriched in ^{32}S. Interpretations of sulphur isotopic patterns are generally based on data from experiments with pure cultures of sulphate-reducing bacteria. These show that there is a general tendency towards increasing fractionation with decreasing specific activity (i.e. rate of reduction per cell) of the bacteria (Chambers and

Trudinger, 1979). Furthermore, the degree of isotopic differentiation between sulphate and sulphide depends on whether a system is open (unlimited sulphate supply) or closed (limited sulphate supply). In the latter case sulphate becomes enriched in ^{34}S during the course of reduction (due to preferential metabolism of $^{32}SO_4^{2-}$) so that sulphides produced early in the reaction will be highly enriched in ^{32}S relative to the sulphate remaining towards the end. Moreover, late-stage sulphides will be enriched in ^{34}S compared with those formed early in the reaction.

These considerations also apply to sulphate reduction in sediments, but here a further variable is introduced by differential diffusion of isotopically different sulphate ions from overlying waters into sulphate-deficient porewaters. Various attempts have been made to express in mathematical form the effects of these factors on isotope fractionation (Jørgensen, 1978; Goldhaber and Kaplan, 1980).

The distribution of sulphur isotopes in sea and ocean sediments has been studied by a number of investigators (e.g Thode *et al.*, 1960; Vinogradov *et al.*, 1962; Hartmann and Nielsen, 1969; Berner, 1964, 1972, 1974; Kaplan *et al.*, 1963; Sweeney and Kaplan, 1973; Migdisov *et al.*, 1974; Goldhaber and Kaplan, 1974; Lein *et al.*, 1976). All of these authors regard bacterial sulphate reduction as important in sulphur isotope geochemistry.

The wide range of $\delta^{34}S$ values in oxidized and reduced forms of sulphur in sediments is the result of complex interacting geological, hydrodynamic, lithological, physical, physicochemical, and biological factors, most of which have not been studied in detail. Moreover, the amount of isotopic information is limited both in the number of analyses and the number of sites studied.

6.8.1   Isotopic Composition of Sulphur Compounds in Reduced Sediments of Inland Seas

Table 6.31 summarizes data on the $\delta^{34}S$ values of pyrite and sulphate sulphur in Upper Holocene sediments of the Black Sea, Baltic Sea, and the Sea of Azov. Only isotopic compositions of sulphate in pore-waters were measured since the changes in isotopic composition during diagenesis mainly apply to the sulphate of pore-water and not that in the solid phase of sediments. The concentrations and isotopic compositions of the other forms of reduced sulphur (organic, elemental, hydrogen sulphide and dissolved sulphide) were not estimated in all cases, but as their contribution to the total amount of reduced sulphur is <10%, they can be neglected in isotopic-mass balance calculations. The distribution of sulphur isotopes in Black Sea sediments is discussed by Vinogradov *et al.*(1962), Grinenko and Grinenko (1974), and Migdisov *et al.* (1974). The latter authors also studied the distribution of sulphur isotopes in the sediments of the Sea of Azov. Similar studies on the Baltic Sea are described by Migdisov *et al.* (1974), Lein *et al.* (1982), and

Table 6.31 Concentration and isotopic composition of sulphate and pyrite sulphur in sediments of the Black, Baltic and Azov seas

Region and sediment type	Locations	Number of isotopic analyses	Pyrite (%S)[a]	Pyrite ($\delta^{34}S$, ‰)	Sulphate (%S)[a]	Sulphate ($\delta^{34}S$, ‰)	Organic carbon (%C)	References
Black Sea								
Hydrogen sulphide zone, 1000–2000 m 0–10 cm horizon	7	11	1.08	−30.8	0.068	+22.5		Vinogradov et al. (1962), Migdisov et al. (1974)
Same region, stabilization zone, 10–200 cm horizon	4	21	1.35	−26.9	0.05	+25.6		Vinogradov et al. (1962)
Baltic Sea								
Western part, Kiel region clay sludges	4	102	0.825	−25.9	0.038	+42.8	2.57	Hartmann and Nielsen (1969)
Eastern and central parts, sandy, aleurite and clay sludges of the upper Holocene	4	49	0.204	−16.2	0.055	+17.6	0.59	Migdisov et al. (1974)
Central and south-western parts, sea sites with depths of 50 m	6	26	1.37	−17.2	0.028	+40.0	3.65	Lein et al. (1982)
Reduced sediments of the upper horizon of the total region, including littoral sediments	8	10	0.728	−18.2	0.037	+19.5	3.43	Lein et al. (1982)
Sea of Azov								
The region near Kerch Bay, sandy, shelly and clay sludges	1	2	1.36	−23.6	0.055	+20.1	0.84	Migdisov et al. (1974)
Remainder	3	14	0.962	+9.0	0.056	+13.6	1.12	

[a]Dry weight basis.

Hartmann and Nielsen (1969). As discussed in section 6.5.1, all of these reservoirs are characterized by high rates of sulphate reduction.

A. The Black Sea

Table 6.31 shows the average isotopic composition of reduced sulphur and sulphate in the sediments of the hydrogen sulphide zone of the Black Sea. Only the 0–10 cm layer of the aerobic zone has been studied (Vinogradov *et al.*, 1962) and the results are not discussed here. Pyrite sulphur in the hydrogen sulphide zone has an average $\delta^{34}S$ value of $-23.9^0/oo$, while the sulphate sulphur is enriched in ^{34}S by $5.6^0/oo$ compared to sea-water.

B. The Baltic Sea

This reservoir is fresher than the Black Sea. The average $\delta^{34}S$ value of sulphate for Kiel Bay is $+42.8^0/oo$, and sometimes reaches $+60^0/oo$ in the upper 40 cm of the sediment (Hartmann and Nielsen, 1969).

Lein *et al.* (1982) examined the isotopic composition of sulphur compounds in 15 m cores of sediments at eight sites in the central and south-western parts of the Baltic Sea at water depths ranging from 46 to 240 m. Some of the data are summarized in Fig. 6.35. The sulphate concentration decreased with depth and was essentially zero at 1.5 m. Accompanying this decrease was an increase in ^{34}S content of both sulphate and pyrite.

C. The Sea of Azov

This is the most shallow and the freshest sea of those under consideration. The salinity and sulphate concentration of the water decrease dramatically from the Gulf of Kerch towards the sea. The isotopic composition of sulphate near the Gulf of Kerch is similar to that of the Black Sea pore-water (Table 6.31). Over the remainder of the sea the $\delta^{34}S$ values for sulphate are more negative and approximate those for sulphate in the water of the Sea of Azov (Rabinovich and Veselovsky, 1974). The $\delta^{34}S$ value for pyrite sulphur is positive (Table 6.31).

D. Sulphur balance in sediments of the inland seas (Table 6.32)

The balance was calculated from data on the rates of accumulation of sulphur (see section 6.6) and the weighted averages of the $\delta^{34}S$ values for pyrite, sulphate, and total sulphur buried in Holocene sediments (Table 6.31).

While there are only minor differences in the pyrite and sulphate concentrations in the sediments of the three seas there are large differences in the

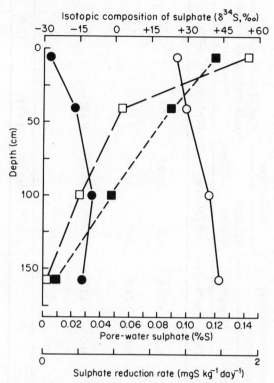

Fig. 6.35 Reduction, concentration, and isotopic composition of sulphate in pore-water, and isotopic composition of pyrite in sediments from the Arcon Trench in the Baltic Sea (site 2656, depth 49 m; Lein, unpublished). □———□, sulphate reduction rate; ■———■, sulphate. δ^{34}S, ○———○, sulphate; ●———●, pyrite

isotopic compositions of these compounds (Table 6.32). The total sulphur buried in the sediments of the Black Sea is isotopically lighter than sea-water sulphate by 41.5‰ while, in the sediments of the Sea of Azov, the difference is only 10‰. This may be due to the different intensities of sulphate reduction in the succession of sediments from the Black Sea through the Baltic Sea to the Sea of Azov. It is in the Black Sea where the lowest rates of reduction and, therefore, the highest fractionation of sulphur isotopes occur (Table 6.17).

Thus, the results obtained in these natural systems are in good agreement with the results of laboratory experiments on the fractionation of sulphur isotopes by bacterial sulphate reduction.

In these inland seas, aleurite–clay and clay sediments with a similar com-

Table 6.32 Sulphur balance in sediments of the inland seas

Sediments	Sedimentation rate (Tg year⁻¹)	Concentration[a] and isotopic composition				Sulphur flux		Flux of total sulphur into sediment and its isotopic composition	
		Reduced sulphur		Sulphate		Reduced sulphur	Sulphate		
		(%S)	(δ^{34}S, ‰)	(%S)	(δ^{34}S, ‰)	(TgS year⁻¹)	(TgS year⁻¹)	(TgS year⁻¹)	(δ^{34}S, ‰)
Black Sea Sediments at depths >200 m	75.2	1.35	−26.9	0.124[b]	+26.6	1.0	0.093	1.093	−22.35
Baltic Sea Reduced sediments over the total sea area	32.4	1.37	−17.2	0.050[b]	+24.3	0.44	0.016	0.456	−15.74
Sea of Azov Sediments of the bulk of freshened water area (depth 11–12 m) (except Kerch region)	27.5	1.0	+9.0	0.056	+13.6	0.27	0.015	0.285	+9.24
Total						1.71	0.124	1.834	−15.80

[a]Dry weight basis.
[b]In pore-water.

position show a gradual decrease in pore-water sulphate and enrichment in ^{34}S with depth. Correspondingly, pyrites become somewhat heavier with depth. This is explained by a limited sulphate supply which is associated with: (1) the reservoirs being freshened; (2) high sedimentation rates which hamper the diffusive exchange between pore-water and overlying water; (3) a high rate of sulphate reduction.

Parallel determinations of the isotopic sulphur composition, sedimentation rate, and sulphate reduction rates in sediments of the Baltic Sea suggest that diagenetic processes in the Holocene sediments of inland seas occur in a system closed to sulphate but nearly open for hydrogen sulphide, because only part of the hydrogen sulphide (15–25%) is fixed in sediments. The total sulphur flux in sediments of these inland seas is characterized by a very small contribution of sulphate sulphur compared with reduced sulphur (Table 6.32).

6.8.2 Isotopic Composition of Sulphur Compounds in Sediments of the Gulf of California

Of all the marginal seas the Gulf of California is the only one for which isotopic data are available (Table 6.33). The concentration of organic matter and sedimentation and sulphate reduction rates in this basin differ markedly from those of other outlying seas. Therefore, data on the distribution of

Table 6.33   Isotopic composition of sulphur in sediments of the Gulf of California

Site	Depth (m)	Horizon (cm)	Sulphate reduction rate (μg kg^{-1} year^{-1})	Pore-water sulphate (%S)	δ^{34}S (⁰/₀₀) Pyrite	δ^{34}S (⁰/₀₀) Sulphate
663	1760	15–30	7.2	0.097	−26.5	+19.4
		45–46	1.7	0.087 ↓	−30.7	—
		245–260	—	0.091	−37.0	+23.8
664	1170	55–70	19.2	—	−35.0	+24.9
		120–135	1.2	0.081 ↓	—	—
		320–340		0.077	−34.0	+30.7
665	3260	35–50	37.2	0.084 ↓	−23.8	+25.7
		120–135	19.2	0.061	−10.8	+35.6
		310	3.6	0.036	−5.5	+43.2
667	2860	30–50	9.6	—	−28.0	+20.4
		95–115	8.4	0.093 ↓	—	+24.4
		190–210	—	0.078	−24.0	+25.3
		240–250	—	0.074	−24.3	+28.0
Average for the zone of stabilization of diagenetic processes indicated by arrows					−24.3	+29.0

sulphur isotopes in sediments of the Gulf of California should not be extra-
polated to all outlying seas.

Several analyses of the isotopic composition of sulphur in sediments of the
Gulf of California were made by Berner (1970), and a detailed study of
sediment sections at four locations has been undertaken by Lein *et al.* (1976).
In many sediments sulphate–sulphur tended to become heavier with
depth, although to a variable degree depending on the sulphate content
(Table 6.33). For example in sediments at site 665, which exhibit a high rate
of sulphate reduction and an appreciable sulphate depletion with depth, $\delta^{34}S$
(sulphate) reaches +43.2% (Fig. 6.36). Accordingly, pyrite in these sediments becomes heavier with depth (Fig. 6.36). A similar, though less pronounced, $\delta^{34}S$ pattern was observed in sediments of site 667.

At two other sites (663, 664: Table 6.33, Fig. 6.37) the concentration of
pore-water sulphate remained more or less constant with depth although its
$\delta^{34}S$ increased by 1–3‰. The sediments at these sites exhibit relatively low

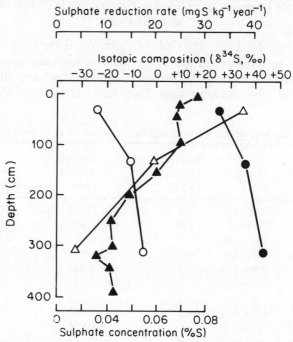

Fig. 6.36 Concentration and isotopic composition of sulphate in pore-water, and isotopic
composition of pyrite in sediments of the Gulf of
California (site 665, depth 3260 m; Lein *et al.*,
1976). △——△, sulphate reduction rate;
▲——▲, sulphate. $\delta^{34}S$, ●——●, sulphate;
○——○, pyrite

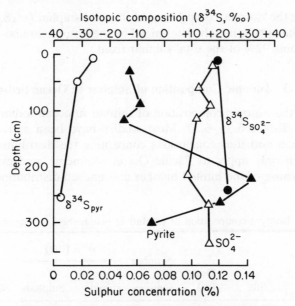

Fig. 6.37   Concentration and isotopic composition of sulphate in pore-water, and pyrite in sediments from the Gulf of California (Lein *et al.*, 1976)

rates of sulphate reduction, but contain on average 1% or more pyrite (Table 6.27). At low sulphate reduction rates there is high isotopic fractionation. As a result there is a general trend towards ^{32}S enrichment in pyrite with depth (Table 6.33, Fig. 6.37).

A. Sulphur balance in sediments of the Gulf of California

The balance (Table 6.34) is based on the data presented in section 6.6.4 and Tables 6.27 and 6.33. From the data presented in Tables 6.33 and 6.34 it

Table 6.34   Sulphur balance in sediments of the Gulf of California

Concentration in sediment (%S)		Sulphur flux into sediments					
		(TgS year^{-1})			($\delta^{34}S$, ‰)		
Reduced sulphur	Sulphate	Reduced sulphur	Sulphate	Total sulphur	Reduced sulphur	Sulphate	Total sulphur
0.98[a]	0.09	2.66	0.30	2.66	−24.3[b]	+29.0[b]	−18.9

[a] From Table 6.27.
[b] From Table 6.33.

follows that the isotopic composition of the total sulphur ($-18.90^0\!/\!00$) buried in these sediments, is largely determined by the isotopic composition of pyrite which contains 92% of the total sulphur fixed.

6.8.3 Isotopic Composition of Sulphur in Ocean Sediments

Data on the isotopic composition of sulphur in ocean sediments are summarized in Tables 6.35–6.37. Most studies have been carried out in the Pacific Ocean and thus conclusions concerning the distribution of sulphur isotopes can only apply to Pacific Ocean sediments, and even then with reserve, because of the limited number and uneven distribution of sampling sites.

Table 6.35   Isotopic composition of sulphur in ocean shelf sediments

Region	Site	Depth (m)	Number of samples	$\delta^{34}S$ ($^0\!/\!00$) Reduced sulphur	Sulphate	References
Pacific Ocean						
Japan coast	6159	264	1	−28.1	—	Ivanov and Lein (1980)
Mexico coast	668	140	3	−24.6	+30.6	Lein *et al.* (1976)
Peru coast	1037	316	2	−13.8	—	Migdisov *et al.* (1974)
	1035–1	260	7	−28.0	+32.3	
Average				−23.6	+31.4	
Atlantic Ocean						
Region of the Orinoco issue (Venezuela)			15	+5.2		Thode *et al.* (1960)

A. Shelf sediments

Four of the five investigated sections of aleuropelitic sediments had a high concentration of reduced sulphur enriched in ^{32}S by an average of $23.6^0\!/\!00$ (Table 6.35). In a relatively homogeneous sediment core from site 668 (Fig. 6.38), $\delta^{34}S$ values of sulphate increased with depth as the sulphate concentration decreased. To a lesser extent the $\delta^{34}S$ values of pyrite also increased with depth. These trends were not so pronounced in other cores, for example those from sediments at site 6159 (Fig. 6.39), that had a more mixed lithological composition with sandy and sandy-aleuritic layers. Shelf sediments are characterized by a lower sulphate content in pore-water than in the ocean and by enrichment in ^{34}S in sulphate by an average $15.0^0\!/\!00$ compared to sea-water sulphate.

Table 6.36   Isotopic composition of sulphur in continental slope sediments

Region	Site	Depth (m)	Number of samples	δ³⁴S (⁰/₀₀) Reduced sulphur	δ³⁴S (⁰/₀₀) Sulphate	References
Pacific Ocean						
Japan coast	6158	490	2	−23.7	+30.9	Ivanov and Lein (1980)
Mexico coast	669	1000	3	−34.7	+22.5	Lein *et al.* (1976)
	672	2900	2	−36.5	+20.2	
Peru coast	1039	1072	2	−29.6	—	
California coast	6248	982	2	−20.8	+21.7	Authors' data
	7131	475	3	−18.6	—	Authors' data
	7184	1000	2	−31.7	+19.4	Authors' data
Average				−28.0	+22.9	Authors' data
Indian Ocean						
Hindustan coast	1960	520		−11.0		Authors' data
	1964	2715		−43.6		Authors' data

Table 6.37   Isotopic composition of sulphur in sediments of the continental rise

Region	Site	Depth (m)	Number of samples	δ³⁴S (⁰/₀₀) Reduced sulphur	δ³⁴S (⁰/₀₀) Sulphate	References
Pacific Ocean						
Japan	6162	4110	4	−20.8	—	Ivanov and Lein (1980)
California	7870	4280	3	−36.4	+19.8	Authors' data
Region near Hawaii	681	4730	2	−33.7	—	Lein *et al.* (1976)
Hess depression		4000		−45.1		Authors' data
Peru and Panama	1047	2410	1	−48.7	—	Authors' data
Trenches	1046	2770	7	−49.3	+20.0	Authors' data
	1043	2800	3	−49.3	—	Authors' data
	1041	3200	1	−49.6	—	Authors' data
Average				−41.6	+20.0	
Indian Ocean						
Somali Trench	1911	5072	9	−41.0	+21.3	Authors' data
	1923	5100	9	−44.5	+20.8	Authors' data

B. Sediments of the continental slope

The distribution of sulphur isotopes in sediments of the Californian border-land was reported in detail by Kaplan *et al.* (1963) and Goldhaber and Kaplan

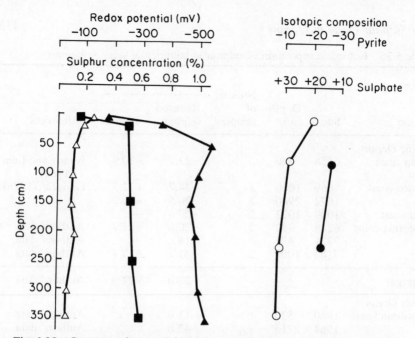

Fig. 6.38 Concentration and isotopic composition of sulphate in pore-water, and pyrite in sediments from the Mexican continental shelf (Lein *et al.*, 1976). Concentrations: △——△, sulphate; ▲——▲, pyrite. δ^{34}S: ○——○, sulphate; ●——●, pyrite. Eh, ■——■

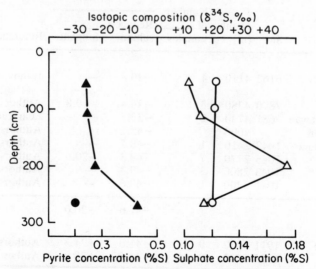

Fig. 6.39 Concentration and isotopic composition of sulphate in pore-water and pyrite in sediments from the Japanese continental shelf (site 6159, depth 264 m, Volkov *et al.*, 1972; Lein *et al.*, unpublished). Concentrations: △——△, sulphate; ▲——▲, pyrite. δ^{34}S: ○——○, sulphate; ●——●, pyrite

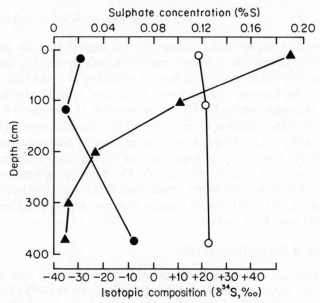

Fig. 6.40 Concentration and isotopic composition of sulphate in pore-water, and isotopic composition of pyrite in sediments off the Californian coast (site 6248, depth 982 m; Kaplan *et al.*, 1963). Concentrations: ▲———▲, sulphate. $\delta^{34}S$: O———O, sulphate; ●———●, pyrite

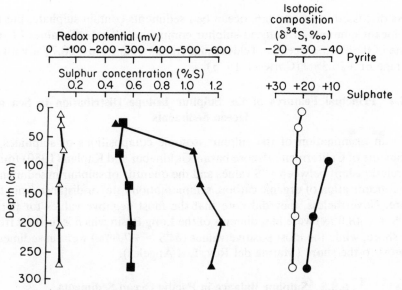

Fig. 6.41 Concentration and isotopic composition of sulphate in pore-water and pyrite in sediments from the Mexican continental slope (site 669, depth 1000 m, Lein *et al.*, 1976). Concentrations: △———△, sulphate; ▲———▲, pyrite. $\delta^{34}S$: O———O, sulphate; ●———●, pyrite. ■——■, Eh

(1980). Those δ^{34}S values which seem reliable suggest that, in sediments where the sulphate content decreases markedly with depth (e.g. site 6248, Fig. 6.40), the δ^{34}S values of sulphate in pore-water and of pyrite increase with depth. Where the isotopic composition of sulphate remains constant with depth, pyrite becomes enriched in ^{32}S (see section 6.8.3). The distribution and isotopic composition of sulphur in sediments of the Masatlan region (Mexico) is illustrated in Fig. 6.41. The sulphate concentration in pore-water remained constant and there was only a slight increase in δ^{34}S (sulphate) with depth; δ^{34}S (pyrite) also increased slightly with depth. On average, sediments of the Pacific Ocean continental slope contain sulphate that is only 3‰ heavier than ocean sulphate (Table 6.36); the ^{34}S content of pyrite sulphur is higher than in shelf sediments (see Table 6.35).

C. Sediments of the continental rise

The concentration and isotopic composition of sulphate in pore-water of sediments of this oceanic zone did not change with depth: the δ^{34}S value of the pore-water sulphate was the same as that for sea-water sulphate ($+20‰$), see site 1046 (Fig. 6.42).

Reduced sulphur, which represents only a small fraction of the total sulphur in these sediments, is highly enriched in ^{32}S (see Fig. 6.42 and Table 6.37).

D. Ocean bed

As discussed in section 6.6, ocean bed sediments contain sulphate, but no significant amounts of reduced sulphur compounds. The δ^{34}S values of sulphate of the Pacific Ocean red clays range from $+19.6‰$ to $+20.7‰$, with an average of $+20.3‰$ (Cortecci, 1975).

6.8.4 Principal Features of the Sulphur Isotope Distribution in Sea and Ocean Sediments

In an examination of the sulphur isotopic compositions of sulphides in sediments of Californian offshore basins Goldhaber and Kaplan (1980) found no relationships between δ^{34}S values and the quantity of sulphur in sediments, the concentration of organic carbon, sedimentation rate, or distance from the shore. Nevertheless, they did note that the most negative values for pyrite (δ^{34}S $= -48.0‰$) were in sediments of the Long Basin which is furthest from the shore, while the most positive values (δ^{34}S $= -7.0‰$) were in sediments nearest to the shore (Marina del Rey, Los Angeles).

6.8.5 Sulphur Balance in Pacific Ocean Sediments

The mass-isotopic balance of sulphur in Pacific Ocean sediments is presented in Table 6.38. Areas of morphological zones are taken from the

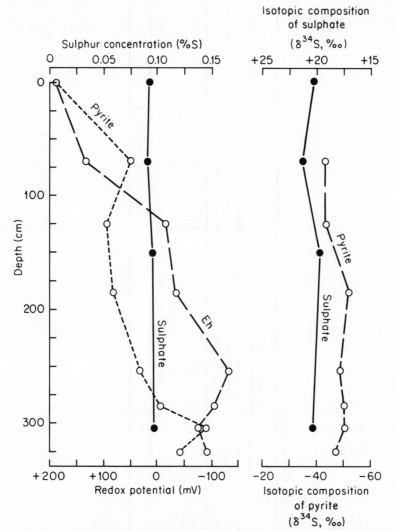

Fig. 6.42 Change in the concentration and isotopic composition of sulphate in pore-water and pyrite with depth in sediments of the Peru Trench (2770 m) (unpublished data of Migdisov *et al.*)

Oceanographic Encyclopedia (1974) and Lisitsin (1978). Calculations of the accumulation rates of dry sediment, were based on Gershanovich *et al.* (1974). The data presented in Table 6.38 show the following: (1) a regular decrease in $\delta^{34}S$ of reduced sulphur with increasing distance from the coast; (2) a maximum ^{34}S-enrichment in pore-water sulphate in the shelf zone; (3) enrichment of ^{32}S in reduced sulphur ($\delta^{34}S = -30.3\%$) and a slight en-

Table 6.38 Sulphur balance in Pacific Ocean sediments

Morphological zones	Depth (m)	Area of zone (10^6 km^2)	Accumulation rate of dry sediments, (Tg year^{-1})	Concentration and isotopic composition				Sulphur flux (TgS year^{-1})		Total sulphur flux and its isotopic composition	
				Reduced sulphur		Sulphate		Reduced	Sulphate	(TgS year^{-1})	(δ^{34}S, $^0/00$)
				(%S)	(δ^{34}S, $^0/00$)	(%S)	(δ^{34}S, $^0/00$)				
Shelf	0–200	10.2	421	0.65	−23.6	0.11	+31.4	3.0	0.5	3.5	−16.4[a]
Continental slope	200–3000	26.4	2059	0.60	−28.0	0.12	+22.9	12.3	2.8	15.1	−18.6[a]
Continental rise	3000–4000	37.3	2909	0.15	−41.6	0.15	+20.0	4.4	4.5	8.9	−10.4[a]
Bed	>4000	104.2	169	no[b]	no[b]	0.14	+20.0	no[b]	0.2	0.2	+20.0[a]
								19.7 (−30.3)[a]	8.0 (+21.6)[a]	27.7	−15.4[a]

[a] Weighted average (δ^{34}S, $^0/00$).
[b] Not analysed.

richment of ^{34}S in sulphate (δ^{34}S = +21.6‰) buried in Pacific Ocean sediments; (4) a dependence of the isotopic composition of total buried sulphur on the sulphate concentration in sediment: the greater the concentration, the heavier the isotopic composition of total sulphur; (5) total sulphur buried in Pacific Ocean sediments is enriched in ^{32}S (δ^{34}S = −15.4‰).

6.8.6   Sulphur Balance in Sediments of the World Oceans

A tentative mass-isotope balance for sulphur in the world oceans has been calculated by extending the results obtained for Pacific Ocean sediments. The Pacific Ocean makes up a half of the area of the world oceans and in it are buried about one-fifth of the reduced sulphur and over one-third of the sulphate sulphur deposited annually in sediments (see Tables 6.29, 6.30, and 6.39). Nevertheless, it should be noted that the sedimentation conditions in the Pacific Ocean have a number of features that are unique to this basin.

Based on the data of Tables 6.39 and 6.40, the weighted average δ^{34}S value for reduced sulphur in sediments of the world oceans was estimated to be −27.3‰. The flux of pore-water sulphate to the world ocean sediments equals 8.4 TgS year^{-1} (Table 6.8), and the flux of solid phase sulphate is

Table 6.39   Isotopic composition of reduced sulphur in sediments with different sulphate reduction rates

	Sulphate reduction rate[a] (μgS g^{-1} day^{-1})	δ^{34}S (‰)
Minor mediterranean seas	0.645–1.80	−17.0[b]
Gulf of California	0.047	−24.3[c]
Pacific Ocean shelf	0.043	−23.6[d]
Pacific Ocean continental slope	0.011	−28.0[e]
Deep-water sediments of Somali Trench and Pacific Ocean	0.001	−41.6[f]

[a]Data from Table 6.17.
[b-f]Data from Tables 6.32, 6.33, 6.35, 6.36, 6.37 respectively.

Table 6.40   Isotopic composition of reduced sulphur in sediments of the world oceans

Morphological zones	Sulphur flux (TgS year^{-1})	δ^{34}S (‰)
Shelf	10.4	−23.6
Continental slope and rise	88.0	−28.0
Total reduced sulphur	98.4	−27.5

Table 6.41 Isotopic composition of sulphate sulphur in sediments of the world oceans

Morphological zones	Sulphate[a] flux (TgS year^{-1})	$\delta^{34}S^b$ (⁰/₀₀)
Pore-water sulphate		
Shelf	1.1	+31.4
Continental slope and rise	5.3	+21.5
Ocean bed	2.0	+20.0
	8.4	+22.4[b]
Solid-phase sulphate		
Total ocean	19.4	+20.0
Total sulphate sulphur	27.8	+20.7[b]

[a] From Tables 6.8 and 6.9.
[b] Weighted average from Table 6.38.

19.4 TgS year^{-1} (Table 6.9). The total flux of sulphate sulphur from bottom water to the world ocean sediments is 27.8 TgS year^{-1} (Table 6.9). From isotopic data on pore-water sulphate in sediments of the different morphological zones of the Pacific Ocean the calculated weighted average for the isotopic composition of sulphate in pore-water of the world ocean sediments is +22.4⁰/₀₀. If the isotopic composition of sulphate sulphur in the solid phase of sediments is equal to that of sea-water sulphate (i.e. +20⁰/₀₀), then the weighted average for the isotopic composition of total buried sulphate is +20.7⁰/₀₀ (Table 6.41). From these data it can be calculated that the isotopic composition of the total sulphur buried in the world ocean sediments is −16.9⁰/₀₀. This value is comparable with the average isotopic composition of the total sulphur in clay rocks of the lithosphere (Chapter 2).

REFERENCES

Aizatullin, T. A., and Skopintsev, B. A. (1974) Studies of the oxidation rates of H₂S in the Black Sea. *Okeanologiya*, **14**, No. 3, 403–420 [in Russian].

Aleksandrov, A. N. (1965) Peculiarities of sediment accumulation in the Azov Sea. Synopsis of Thesis. Rostov-on-Don, 24 pp. [in Russian].

Almgren, T. L., Danielsson, L. G., Dryssen, D., Johannson, T., and Nyquist, G. (1975) Release of inorganic matter from sediments in a stagnant basin. *Thalassia Jugosl.*, **11**, 1–2.

Artemyev, V. E. (1976) Carbon hydrates in benthal sediments of the ocean In: *The Biogeochemistry of Diagenesis of Ocean Sediments*. Nauka, Moscow, pp. 20–58 [in Russian].

Baas Becking, L. (1956) Biological processes in the estuarine environment. VI. The state of the iron in the estuarine mud iron sulphides. *K. Ned. Akad. Wet. Versl. Gewone. Vergad. Afd. Natuurkd.*, **59B**, 181–189.

Bågander, L. E. (1977) Sulfur fluxes at the sediment–water interface – an in-situ study of closed systems, Eh and pH. Ph.D. Thesis, Dept. Geology, Stockholm Univ., Microbial Geochem., Publ. No. 1, 136 pp.

Berner, R. A. (1964) Distribution and diagenesis of sulphur in some sediments from the Gulf of California. *Mar. Geol.*, **1**, 117–140.

Berner, R. A. (1970). Sedimentary pyrite formation. *Am. J. Sci.*, **268**, 1–23.

Berner, R. A. (1972) Sulfate reduction, pyrite formation and the oceanic sulfur budget. In: *The Changing Chemistry of the Oceans. Nobel Symp.*, **20**, 347–361.

Berner, R. A. (1974) Kinetic models for the early diagenesis of nitrogen sulfur, phosphorus and silicon in anoxic marine sediments. In: Goldberg, E. (ed.), *The Sea*, vol. 5. Wiley, New York, pp. 427–450.

Bogdanov, Yu. A., Lisitsin, A. P. and Romankevich, E. A. (1971) Organic matter of marine suspensions and bottom sediments. In: *Organic Matter of Recent and Fossil Deposits*. Nauka, Moscow, pp. 35–103 [in Russian].

Bordovskii, O. K. (1964) *Accumulation and Transformation of Organic Matter in Marine Sediments*. Nedra, Moscow, 128 pp. [in Russian].

Bordovksii, O. K. (1974) *Organic Matter of Marine Sediments in Early Diagenesis*. Nauka, Moscow, 104 pp. [in Russian].

Boström, K., Joensku, O., and Brohm, I. (1974) Plankton: Its chemical composition and its significance as a source of pelagic sediments. *Chem. Geol.*, **14**, 255–271.

Boström, K., Joensku, O., Moore, C., Boström, B., Dalziel, M., and Horowitz, A. (1973) Geochemistry of Ba in pelagic sediments. *Lithos*, **6**, 159–174.

Brown, F S., Beadecker, M. J., Nissenbaum, A., and Kaplan, I. R. (1972) Early diagenesis in a reducing fjord, Saanich Inlet, British Columbia. III. Changes in organic constituents of sediment. *Geochim. Cosmochim. Acta*, **36**, 1185–1203.

Calvert, S. E., and Batchelor, C. H. (1978) Major and minor element geochemistry of sediments from Hole 279A, Leg 42B, Deep Sea Drilling Project. In: *Initial Report on the Deep Sea Drilling Project*, vol. 42, Pt 2. Washington DC, pp. 527–539.

Chambers, L. A., and Trudinger, P. A. (1979) Microbiological fractionation of stable sulfur isotopes: a review and critique. *Geomicrobiol. J.*, **1**, 249–293.

Chebotarev, E. N. and Ivanov, M. V. (1976) Distribution and activity of sulfate-reducing bacteria in benthic sediments of the Pacific Ocean and California Gulf. In: *The Biogeochemistry of Diagenesis of Ocean Sediments*. Nauka, Moscow, pp. 68–74. [in Russian].

Chernovskaya, E. N., Pastukhova, N. M., Buinevich, A. G., Kudryavtseva, M. E., and Auninsh, E. A. (1965) *The Hydrochemical Regime of the Baltic Sea*. Hydrometeoizdat, Leningrad, 168 pp. [in Russian].

Chow, T., and Goldberg, E. (1960) On marine geochemistry of barium. *Geochim. Cosmochim. Acta*, **20**, 192–198.

Cortecci, G. (1975) Isotopic analysis of sulfate in a South Pacific core. *Mar. Geol.*, **19**, M69–M74.

Creager, J. S., and Stenberg, R. W. (1972) Some specific problems in understanding bottom sediment distribution and dispersal on the continental shelf. In: Swift, D. J. P., Drake, D. B., and Pilkey, O. H. (eds), *Shelf Sediment Transport: Process and Pattern*. Dowden, Hutchinson and Ross, Stroudsburg, Pennsylvania, pp. 347–371.

Datsko, V. G. (1959) *Organic Matter in Waters of the USSR South Seas*. Izd. Akad. Nauk SSSR, Moscow, 272 pp. [in Russian].

Dean, E., and Schreiber, B. (1978) Authgenic barite leg 41 Deep Sea Drilling Project. In: *Initial Report of the Deep Sea Drilling Project*, vol. 41. Washington, DC, pp. 915–925.

Deuser, W. G. (1971) Organic carbon budget of the Black Sea. *Deep-Sea Res.*, **18**, 995–1004.

Doelter, C. (1926) Hydrotroilit. In: *Handbuch der Mineralchemie*, Vol. 4. Verlag Von Theodor Steinkopff, Dresden–Leipzig, 526 pp. [in German].

Emery, K. O. (1974) Latitudinal aspects of the law of the sea and of petroleum production. *Ocean Developm. and Intern. Law. J.*, **2**, 137–149.

Emery, K. O. and Bray, E. E. (1962) Radiocarbon dating of California Basin sediments. *Bull. Am. Assoc. Pet. Geol.*, **46**, 1839–1856.

Emery, K. O., and Rittenberg, S. C. (1952) Early diagenesis of California Basin sediments in relation to origin of oil. *Bull. Am. Assoc. Pet. Geol.*, **36**, 735–806.

Fedosov, M. V. (1961) Some characteristic features of sediment formation in the Azov Sea. In: *Recent Marine Sediments*. Izd. Akad. Nauk. SSSR, Moscow, pp. 504–511 [in Russian].

Feld, W. (1911) Über die Bildung von Eisenbisulfid (FeS$_2$) in Losungen und Entstehung der naturlichen Pyritlager. *Z. Angew. Chem.*, **24**, 97–103.

Garrels, R. M, and Mackenzie, F. T. (1972). A quantitative model for the sedimentary rock cycle. *Mar. Chem.*, **1**, 27–41.

Garrels, R. M., and Thompson, M. E. (1962) A chemical model for seawater at 25°C and one atmosphere total pressure. *Am. J. Sci.*, **260**, 57–66.

Geodekyan, A. A., Trotsyuk, V. Ya, and Marina, M. M. (1978). Organic carbon in submarine stratosphere. In: *Accumulation and Transformation of Organic Matter of Recent and Fossil Sediments*. Nauka, Moscow, pp. 18–27 [in Russian].

Gershanovich, D. E., Gorshkova, T. I., and Konyukhov, A. I. (1974) Organic matter in contemporary sediments of submerged marginal parts of continents. In: *Organic Matter of Recent and Fossil Sediments and Methods of its Invesigation*. Nauka, Moscow, pp. 63–80 [in Russian].

Goldberg, E., and Arrhenius, G. (1958) Chemistry of Pacific pelagic sediments. *Geochim. Cosmochim. Acta*, **13**, 153–212.

Goldhaber, M. B., and Kaplan, I. R. (1974) The sulfur cycle. In: Goldberg, E. (ed.), *The Sea*, vol. 5. Wiley, New York, pp. 569–655.

Goldhaber, M. B., and Kaplan, I. R. (1980) Mechanisms of sulfur incorporation and isotope fractionation during early diagenesis in sediments of the Gulf of California. *Mar. Chem.*, **9**, 95–143.

Gordeev, V. V. and Lisitsin, A. P. (1978) Average chemical composition of the world rivers suspensions and the feeding of oceans with sedimentary river material. *Dokl. Akad. Nauk. SSSR*, **238**, No. 1, 225–228 [in Russian].

Grinenko, V. A., and Grinenko, L. N. (1974) *Geochemistry of Sulfur Isotopes*. Nauka, Moscow, 274 pp. [in Russian].

Gudelis, V. K., and Emelyanov, E. M. (1976) *Geology of the Baltic Sea*. Mokslas, Vilnius, 383 pp. [in Russian].

Gurvich, E. G., Bogdanov, Yu. A., and Lisitsin, A. P. (1978) Barium behaviour in contemporary sediment accumulation in the Pacific. *Geokhimiya*, No. 3, 359–374 [in Russian].

Gurvich, E. G., Bogdanov, Yu. A., and Lisitsin, A. P. (1979) Barium behaviour in contemporary sediment accumulation during formation of metalliferous sediments in the Pacific. *Geokhimiya*, No. 1, 108–126 [in Russian].

Hallberg, R. O., Bågander, L. E., and Engval, A. G. (1976) Dynamics of phosphorus, sulfur and nitrogen at the sediment–water interface. In: Nriagu, J. O. (ed.), *Environmental Biogeochemistry*, vol. 1. Ann Arbor Science, Michigan, pp. 295–308.

Hallberg, R. O., and Wadsten, T. (1980) Crystal data of a new phosphate compound from microbial experiments on iron sulfide mineralization. *Am. Mineral.*, **65**, 200–204.

Hartmann, M., Müller, P. J., Suess, E., and Weijden, C. H. van der (1976) Chemistry of late quaternary sediments and their interstitial waters from the NW African continental margin. *'Meteor' Forschungsergeb.*, Reihe, C, No. 24, 1–67.

Hartmann, M., and Nielsen, H. (1969) δ^{34}S–Werte in rezenten Meeressedimenten und ihre Deutung am Beispiel einiger Sedimentprofile aus der westlichen Ostsee. *Geol. Rundsch.*, **58**, 621–655.

Hays, J. D. (1967) Quaternary sedimentation in the Antarctic ocean. *Oceanography*, **4**, 117–131.

Horn, R. A. (1969) *Marine Chemistry. The Structure of Water and Chemistry of Hydrosphere.* Wiley-Interscience, New York, 568 pp.

Ivanov, M. V. (1964) *Role of Microbiological Processes in Genesis of Sulfur Deposits.* Nauka, Moscow, 368 pp. [in Russian].

Ivanov, M. V. (1979) Distribution and geochemical activity of bacteria in oceanic sediments: In: Volkov, I. I. (ed.), *Chemistry of the Ocean. Geochemistry of Benthic Sediments*, vol. 2. Nauka, Moscow, pp. 312–349 [in Russian].

Ivanov, M. V., and Lein, A. Yu. (1980) Distribution of microorganisms and their role in processes of diagenetic mineral formation. In: *Geochemistry of Sediment Diagenesis in the Pacific Ocean (Trans-oceanic Profile).* Nauka, Moscow, pp. 117–137.

Ivanov, M. V., Lein, A Yu., Belyaev, S. S., Nesterov, A. I., Bondar, V. A., and Zhabina, N. N. (1980). Geochemical activity of sulfate-reducing bacteria in benthal sediments in the North-West Indian Ocean. *Geokhimiya*, No. 8, 1238–1249 [in Russian].

Ivanov, M. V., Lein, A. Yu., and Kashparova, E. V. (1976) Intensity of formation and diagenetic transformation of reduced sulfur compounds in sediments of the Pacific Ocean. In: *The Biogeochemistry of Diagenesis of Ocean Sediments*, Nauka, Moscow, pp. 171–178 [in Russian].

Ivanov, M. V., and Ryzhova, V. N. (1960) Intensity of hydrogen sulfide formation in some sediments of the Barents Sea littoral. *Dokl. Akad. Nauk. SSSR*, **130**, No. 1, 187–188 [in Russian].

Jørgensen, B. B. (1977) The sulfur cycle of a coastal marine sediment (Limfjorden, Denmark). *Limnol. Oceanogr.*, **22**, 814–832.

Jørgensen, B. B. (1978) A comparison of methods for the quantification of bacterial sulfate reduction in coastal marine sediments. *Geomicrobiol. J.*, **1**, 11–64.

Jørgensen, B. B., Hansen, M. H., and Ingvorsen, K. (1978) Sulfate reduction in coastal sediments and the release of H_2S to the atmosphere. In: Krumbein, W. E. (ed.), *Environmental Biogeochemistry and Geomicrobiology*, vol. 1. Ann Arbor Science, Michigan, pp. 245–253.

Kaplan, I. R., Emery, K. O., and Rittenberg, S. C. (1963) The distribution and isotopic abundance of sulphur in recent marine sediments of southern California. *Geochim. Cosmochim. Acta*, **27**, 297–331.

Kester, D. R. and Pytkowicz, R. M. (1969) Sodium, magnesium and calcium sulfate ion-pair in seawater at 25°C. *Limnol. Oceanogr.*, **14**, 686–692.

Khrustalev, Yu. P., and Shcherbakov, F. A. (1974) *Late Quaternary Sediments of the Azov Sea and Conditions of their Accumulation.* Izv. RGU, Rostov-on-the-Don. 149 pp. [in Russian].

Kuznetsov, S. I., and Romanenko, V. I. (1968) Microflora of Sivash and evaporative basins of salt-mining. *Mikrobiologiya*, **37**, No. 6, 1104–1108 [in Russian].

Lein, A. Yu., Kudryavtseva, A. I., Matrosov, A. G., and Zyakun, A. M. (1976) Isotope composition of sulfur compounds in sediments of the Pacific Ocean. In: *The Biogeochemistry of Diagenesis of Ocean Sediments.* Nauka, Moscow, pp. 179–185 [in Russian].

Lein, A. Yu., Sidorenko, G. A., Volkov, I. I., and Shevchenko, A. Ya. (1978)

Diagenetic makinawite, melnikovite (greigite) and pyrites in sediments of a profile through the Pacific and sediments of the Californian Gulf. *Dokl. Akad. Nauk. SSSR*, **238**, No. 3, 698–700 [in Russian].

Lein, A. Yu., Vainshtein, M. B., Namsaraev, B. B., Kashparova, E. V., Matrosov, A. G., Bondar, V. A., and Ivanov, M. V. (1982) Biogeochemistry of anaerobic diagenesis of recent sediments of the Baltic Sea. *Geokhimiya*, No. 3, 428–440.

Lisitsin, A. P. (1955) Distribution of organic carbon in sediments of the Bering Sea. *Dokl. Akad. Nauk SSSR*, **103**, No. 2, 299–302 [in Russian].

Lisitsin, A. P. (1971) Rate of contemporary sediment accumulation in oceans. *Okeanologiya*, **11**, No. 6, 957–968 [in Russian].

Lisitsin, A. P. (1974) *Sediment Formation in Oceans*. Nauka, Moscow, 438 pp. [in Russian].

Lisitsin, A. P. (1978) *Processes of Oceanic Sedimentation. Lithology and Geochemistry*. Nauka, Moscow, 392 pp. [in Russian].

Lisitsina, N. A., Butuzova, G. Yu., Dvoretskaya, O. A. (1976). Benthal sediments of the profile through the Pacific Ocean. *Litol. Polezn. Iskop.*, No. 6, 31–46 [in Russian].

Lopatin, G. V. (1950) Erosion and runoff of aluviums. *Priroda (Moscow)*, No. 7, 19–28 [in Russian].

Mekhtieva, V. L. (1974) Use of sulfur isotope composition of fossil mollusk shells for determining the palaeohydrochemical conditions of ancient water basins. *Geokhimiya*, No. 11, 1682–1687 [in Russian].

Mekhtieva, V. L., Pankina, R. G., and Gurieva, S. M. (1977). Sulfur isotope composition of dissolved sulfate as index of hydrochemical features of water bodies. In: *Chemico-Oceanological Investigations*, Nauka, Moscow, pp. 46–51 [in Russian].

Migdisov, A. A., Cherkovsky, S. L., and Grinenko, V. A. (1974). Dependence of sulfur isotope composition in humid sediments on conditions of their formation. *Geokhimiya*, No. 10, 1482–1502 [in Russian].

Moss, M. R. (1978) Source of sulfur in the environment; the global sulfur cycle. In: Nriagu, J. O. (ed.), *Sulfur in the Environment*, Pt. 1. Wiley, New York, pp. 22–50.

Müller, P. J. (1975) Diagenese stickstoffhaltiger organischer substanzen in oxischen und anoxischen marinen sedimenten. '*Meteor*' *Forschungsergeb.*, Reihe C, No. 22, 1–60.

Ostroumov, E. A. (1953a) Method of determination of sulfur forms in sediments of the Black Sea. *Trudy Inst. Okeanol. Akad. Nauk SSSR*, **7**, 57–69 [in Russian].

Ostroumov, E. A. (1953b) Forms of sulfur compounds in sediments of the Black Sea. *Trudy Inst. Okeanol. Akad. Nauk SSSR*, **7**, 70–90 [in Russian].

Ostroumov, E. A. (1957) Sulfur compounds in benthal sediments of the Okhotsk Sea. *Trudy Inst. Okeanol. Akad. Nauk SSSR*, **22**, 139–157 [in Russian].

Ostroumov, E. A., and Fomina, L. S. (1960) Forms of sulfur compounds in benthic sediments of the North-West Pacific. *Trudy Inst. Okeanol. Akad. Nauk SSSR*, **32**, 206–214 [in Russian].

Ostroumov, E. A., and Kulumbegashvili, V. A. (1977) Thiosulfates and sulfites in porewater of sediments in the West Indian Ocean. *Dokl. Akad. Nauk SSSR*, **233**, No. 1, 218–221 [in Russian].

Ostroumov, E. A., and Volkov, I. I. (1960) On forms of sulfur compounds in bottom sediments of the Pacific Ocean near the New Zealand coast. *Trudy Inst. Okeanol. Akad. Nauk SSSR*, **42**, 117–124 [in Russian].

Ostroumov, E. A., and Volkov, I. I. (1964) Sulfates in benthic sediments of the Black Sea. *Trudy Inst. Okeanol. Akad. Nauk SSSR*, **67**, 92–100 [in Russian].

Ostroumov, E. A., Volkov, I. I., and Fomina, L. S. (1961) Distribution of sulfur forms in benthic sediments of the Black Sea. *Trudy Inst. Okeanol. Akad. Nauk. SSSR*, **50**, 93–129 [in Russian].

Ottar, B. (1976) Monitoring long-range transport of air pollutants: the OECD study. *Ambio*, **5**, 203–206.

Pamatmat, M. M. (1973) Benthic community metabolism on the continental terrace and the deep sea in the North Pacific. *Int. Rev. Gesamten Hydrobiol.*, **58**, 345–368.

Panov, D. G., and Spichak, M. K. (1961) Conditions of sediment accumulation in the Azov Sea. In: *Contemporary Marine and Oceanic Sediments.* Izv. Akad. Nauk SSSR, Moscow, pp. 512–520 [in Russian].

Rabinovich, A. L., and Grinenko, V. A. (1979) On isotope composition of sulfate sulfur in the river runoff from the territory of the USSR. *Geokhimiya*, No. 3, 441–454 [in Russian].

Rabinovich, A. L., and Veselovsky, N. V. (1974). On isotope composition of sulfur compounds in the Azov Sea. In: *Hydrochemical Materials*, vol. 60. Hydrometeoizdat, Leningrad, pp. 41–48 [in Russian].

Rickard, D. T. (1969) The chemistry of iron sulfide formation at low temperatures. *Stockholm Contrib. Geol.*, **20**, 68–95.

Rickard, D. T. (1975) Kinetics and mechanism of pyrite formation at low temperatures. *Amer. J. Sci.*, **275**, 636–652.

Roberts, W. M. B., Walker, A. L., and Buchanan, A. S. (1969) The chemistry of pyrite formation in aqueous solution and its relation to the depositional environment. *Miner. Deposita*, **4**, 18–29.

Romankevich, E. A. (1976) Organic matter in benthic sediments to the east of Japan and its impact on the redox processes. In: *The Biogeochemistry of Diagenesis of Ocean Sediments.* Nauka, Moscow, pp. 5–19 [in Russian].

Romankevich, E. A. (1977) *Geochemistry of Organic Matter in Ocean.* Nauka, Moscow, 256 pp. [in Russian].

Ross, D. A., and Degens, E. T. (1974) Recent sediments of Black Sea. In: Degens. E. T., and Ross, D. A. (eds), *The Black Sea Geology, Chemistry and Biology.* AAPG, Tulsa, Oklahoma, pp. 183–199.

Rozanov, A. G. (1973) Experimental study of conditions of low-temperature iron sulfides formation. *Trudy Inst. Okeanol. Akad. Nauk SSSR*, **63**, 172–184.

Rozanov, A. G., Sokolov, V. S., and Volkov, I. I. (1972) Forms of iron and manganese in sediments of the North-west Pacific Ocean. *Lithol. Polezn. Iskop.*, No. 4, 26–39 [in Russian]

Rozanov, A. G., Volkov, I. I., Sokolov, V. S., Pushkina, Z. V., and Pilipchuk. M. F. (1976) Redox processes in sediments of the Californian Gulf and close part of the Pacific Ocean. (Iron and manganese compounds) In: *The Biogeochemistry of Diagenesis of Ocean Sediments.* Nauka, Moscow, pp. 96–135 [in Russian].

Rozanov, A. G., Volkov, I. I., Habina, N. N., and Yagodinskaya, T. A. (1971). Hydrogen sulfide in sediments of coastal slope of N.W. Pacific. *Geokhimiya*, No. 5, 543–550 [in Russian].

Rozanov, A. G., Volkov, I. I., and Yagodinskaya, T. A. (1974) Forms of iron in surface layer of Black Sea sediments. In: Degens, E. T., and Ross, D. A. (eds), *The Black Sea—Geology, Chemistry and Biology*, AAPG, Tulsa, Oklahoma, pp. 532–541.

Schippel, F. A., Bågander, L. E., and Hallberg, R. O. (1973) An apparatus for subaquatic in situ measurements of sediment dynamics. *Stockholm Contrib. Geol.*, **24**, 103–108.

Schulze, F. E., and Theirfelder, H. (1905) Uber Bariumsulfat in Meertieren. *S.B. Ges. Naturf. FR. Berl.*, 2–4.

Shimkus, K. M., and Trimonis, E. S. (1974) Modern sedimentation in Black Sea. In: Degens, E. T. and Ross, D. A. (eds), *The Black Sea—Geology, Chemistry and Biology*. AAPG, Tulsa, Oklahoma, pp. 247–278.

Shishkina, O. V. (1959) Sulfates in porewaters of the Black Sea. *Trudy Inst. Okeanol. Akad. Nauk SSSR*, **33**, 178–193 [in Russian].

Shishkina, O. V. (1972) *Geochemistry of Marine and Oceanic Waters*. Nauka, Moscow, 228 pp. [in Russian].

Sidorenko, M. (1901) Petrographic data on contemporary sediments in the Khadjbey estuary and lithologic surface composition of sediments of Kuyalnitsko-Khadjbey Giresveni. *Zap. Novoross. Obshch. Est.*, **24**, No. 1, 97–119 [in Russian].

Skopintsev, B. A. (1948) Modification of nitrogen and phosphorus contents in suspended terrigenous particles in aqueous medium. *Izv. Akad Nauk SSSR*, Ser, Geogr. Geofiz., **12**, No. 2, 107–118 [in Russian].

Skopintsev, B. A. (1975) *Formation of Contemporary Chemical Composition of the Black Sea Waters*. Hydrometeoizdat, Leningrad, 336 pp. [in Russian].

Skopintsev, B. A. (1976) Mineralization regularities of the organic matter of dead phytoplankton. In: Soviet–Swedish Symposium on the Pollution of the Baltic. *Ambio*, **2**, 45–54.

Skornyakova, N. S. (1976) Chemical composition of ferromanganese concretions of the Pacific Ocean. In: *Ferro-manganese Concretions of the Pacific Ocean*. Nauka, Moscow, pp. 190–240 [in Russian].

Skyring, G. W., and Chambers, L. A. (1976). Biological sulphate reduction in carbonate sediments of a coral reef. *Aust. J. Mar. Freshwater Res.*, **27**, 595–602.

Sorokin, Yu. I. (1962a) Microflora of the Black Sea bottom. *Mikrobiologiya*, **31**, No. 5, 899–903 [in Russian].

Sorokin, Yu. I. (1962b) Experimental study of the bacterial reduction of sulfates in the Black Sea with the help of S^{35}. *Mikrobiologiya*, **31**, No. 3, 402–410 [in Russian].

Sorokin, Yu. I. (1964) On the primary production and bacterial activities in the Black Sea. *J. Cons. Cons. Int. Explor. Mer.*, **29**, 41–60.

Sorokin, Yu, I. (1970) Experimental investigation of the rate and mechanism of oxidation of hydrogen sulfide in the Black Sea using S-35. *Okeanologiya*, **10**, 37–46.

Starikova, N. D. (1956) Organic matter in the depth of sediments of the Bering Sea. *Dokl. Akad. Nauk SSSR*, **106**, No. 3, 519–522 [in Russian].

Stepanov, V. N. (1961) Basin dimensions of the world ocean and its principal parts. *Okeanologia*, **1**, No. 2, 213–219 [in Russian].

Strakhov, N. M. (1959) Iron forms in sediments of the Black Sea and their importance for the theory of diagenesis. In: *Investigation of Sediment Diagenesis. Izd. Akad. Nauk SSSR*, Moscow, pp. 92–119 [in Russian].

Strakhov, N. M. (1962) *Principles of the Lithogenesis Theory*. Izd. Akad. Nauk SSSR, Moscow, vol. 1, 212 pp. vol. 2, 572 pp [in Russian].

Strakhov, N. M. (1961) Problems of sulfur transformation factors in sediments of the Black Sea. In: *Contemporary Marine and Oceanic Sediments*. Izd. Akad. Nauk SSSR, Moscow, pp. 634–644 [in Russian].

Strakhov, N. M. (1972) Balance of reductive processes in sediments of the Pacific Ocean. *Lithol. Polezn. Iskop.*, No. 4, 65–92 [in Russian].

Strakhov, N. M. (1976) *Problems of Geochemistry of Contemporary Oceanic Lithogenesis*. Nauka, Moscow, 299 pp. [in Russian].

Strakhov, N. M. (1978) Investigation of terrigenous sedimentation in the oceans. *Akad. Nauk SSSR*, Ser. Geol., No. 7, 16–38 [in Russian].

Strakhov, N. M., and Zalmanzon, E. S. (1955) Distribution of mineralogic iron forms in sedimentary rocks and its importance for lithology. *Izv. Akad. Nauk SSSR*, Ser. Geol., No. 1, 34–51 [in Russian].

Sweeney, R. E., and Kaplan, I. R. (1973) Pyrite framboid formation: Laboratory synthesis and marine sediments. *Econ. Geol.*, **68**, 618–634.

Teder, A. (1971) The spectra of aqueous polysulfide solutions. Part II: The effect of alkalinity and stoichiometric composition at equilibrium. *Ark. Kemi.*, **31**, 173–198.

Thode, H. G., Harrison, A. G., and Monster, J. (1960) Sulfur isotope fractionation in early diagenesis of recent sediments. *Bull. Am. Assoc. Pet. Geol.*, **44**, 1809–1817.

Tolokonnikova, L. I. (1977) Intensity of sulfate reduction in the Azov Sea. *Mikrobiologiya*, **46**, No. 2, 352–357 [in Russian].

Trudinger, P. A., and Loughlin, R. E. (1981) Metabolism of simple sulphur compounds. In: Neuberger, A. and van Deenen, L. L. M. (eds), *Comprehensive Biochemistry: Amino Acid Metabolism and Sulfur Metabolism*, vol. 19A. pp. 165–256. Elsevier, Amsterdam.

Turekian, K. (1964) The geochemistry of the Atlantic ocean basin *Trans. N.Y. Acad. Sci.* **26**, 312–330.

Van Andel, T. H. (1964) Recent marine sediments of the Gulf of California. In: Van Andel, T. H., and Shor, G. G. (eds), *Marine Geology of the Gulf of California*, vol. 3. Tulsa, Oklahoma, pp. 216–310.

Vinogradov, A. P. (1935) *Chemical Elemental Composition of Marine Organisms*, vol. 1. Izd. Akad. Nauk SSSR, Leningrad, 223 pp. [in Russian].

Vinogradov, A. P. (1937) *Chemical Elemental Composition of Marine Organisms*, vol. 2. Izd. Akad. Nauk SSSR, Leningrad, 225pp. [in Russian].

Vinogradov, A. P., Grinenko, V. A., and Ustinov, V. I. (1962) Isotope composition of sulfur compounds in the Black Sea. *Geokhimiya*, No. 10, 851–873 [in Russian].

Volkov, I. I. (1959) Free hydrogen sulfide and iron sulfide in silt sediments of the Black Sea. *Dokl. Akad. Nauk SSSR*, **126**, No. 1, 163–166 [in Russian].

Volkov, I. I. (1960) Distribution of free hydrogen sulfide in sediments of the Black Sea. *Dokl. Akad. Nauk SSSR*, **134**, No. 3, 676–679 [in Russian].

Volkov, I. I. (1961a) Regularities of formation and transformation of sulfur compounds in sediments of the Black Sea. In: *Recent Marine and Oceanic Sediments*. Izd. Akad. Nauk SSSR, Moscow, pp. 577–596 [in Russian].

Volkov, I. I. (1961b) Free hydrogen sulfide and some products of its transformation in sediments of the Black Sea. *Trudy Inst. Okeanol. Akad. Nauk SSSR*, **50**, 29–67 [in Russian].

Volkov, I. I. (1961c) Iron sulfides, their interrelations and transformation in sediments of the Black Sea. *Trudy Inst. Okeanol. Akad. Nauk. SSSR*, **50**, 68–92 [in Russian].

Volkov, I. I. (1962) Hydrogen sulfide state in waters and sediments of the Black Sea. *Trudy Inst. Okeanol. Akad. Nauk SSSR*, **54**, 39–46 [in Russian].

Volkov, I. I. (1964) Regularity of formation and chemical composition of iron sulfide concretions in sediments of the Black Sea. *Trudy Inst. Okeanol. Akad. Nauk SSSR*, **67**, 111–134 [in Russian].

Volkov, I. I. (1973) Chemical elements in the thickness of deep-water sediments of the Black Sea. *Trudy Inst. Okeanol. Nauka, Moscow*, **63**, 148–171 [in Russian].

Volkov, I. I., and Ostroumov, E. A. (1957a) Forms of sulfur compounds in pore-waters of the Black Sea. *Geokhimiya*, No. 4, 337–345 [in Russian].

Volkov, I. I., and Ostroumov, E. A. (1957b) Iron sulfide concretions in sediments of the Black Sea. *Dokl. Akad. Nauk SSSR*, **116**, No. 4, 645–648 [in Russian].

Volkov, I. I., and Ostroumov, E. A. (1960) Distribution of sulfates in benthic sediments of the Pacific Ocean. In: *Oceanologic Investigations*, No. 2, 61–70. Izd. Akad. Nauk SSSR, Moscow [in Russian].

Volkov, I. I., and Pilipchuk, M. F. (1966) Sulfur compounds in the oxygen zone sediments of the Black Sea. *Litol. Polezn. Iskop.*, No. 1, 66–77 [in Russian].

Volkov, I. I. and Tikhomirova, A. A. (1966) Forms of iron in deposits of the oxic zone of the Black Sea. *Litol. Polezn. Iskop.*, No. 4, 24–37 [in Russian].

Volkov, I. I., Rozanov, A. G., and Sokolov, V. S. (1973) Redox processes of sediments diagenesis in the North-West Pacific Ocean. In: *International Geochem. Congress. 1st Sedimentary Processes Proceedings*, vol. 4, Pt. 2. VINITI, Moscow [in Russian].

Volkov, I. I.,Rozanov, A. G., and Sokolov, V. S. (1975a) Forms of manganese, iron and sulfur in the sediments of the Tyrrhenian Sea. In: *Hydrological and Geological Investigation of the Mediterranean and Black Sea*, Moscow, pp. 257–284 [in Russian].

Volkov, I. I., Rozanov, A. G., and Sokolov, V. S. (1975b) Redox processes in diagenesis of sediments in the Northwest Pacific ocean. *Soil Sci.*, **119**, 28–35 [in Russian].

Volkov, I. I., Rozanov, A. G., and Yagodinskaya, T. A. (1971) Pyrite micro-concretions in the sediments of the Black Sea. *Dokl. Akad. Nauk SSSR*, **197**, No. 1, 195–198 [in Russian].

Volkov, I. I., Rozanov, A. G., Zhabina, N. N., and Fomina, L. S. (1976) Sulfur compounds in sediments of the Californian Gulf and the adjoining part of the Pacific. In: *The Biogeochemistry of Diagenesis of Ocean Sediments*. Nauka, Moscow pp. 136–170.[in Russian].

Volkov, I. I., Rozanov, A. G., Zhabina, N. N., and Yagodinskaya, T. A. (1972) Sulfur in the Pacific sediments to the East of Japan. *Litol. Polenzn. Iskop.*, No. 4, 50–64 [in Russian].

Volkov, I. I., Zhabina, N. N., Demidova, T. P., Sokolov, V. S., Morozov, A. A., and Pushkina, Z. V. (1981) Sulphur compounds in the Arabian part of the Indian Ocean. *Geokhimiya*, No. 10, 1569–1590 [in Russian].

Woodwell, G. M., Rich, P. H., and Entholl, C. H. (1973) Carbon in estuaries. In: Woodwell, G. M. and Pecan, E. D. (eds), *Carbon in the Biosphere*. AEC Symp. Ser., NTIF, IF-Dep. of Commer., Springfield, Virginia, pp. 221–239.

Zhabina, N. N., Demidova, T. P. and Morozov, A. A. (1979) Sulfur compounds in sediments of the Somalian depression of the Indian Ocean. *Geokhimiya*, No. 12, 1868–1883. [in Russian].

The Global Biogeochemical Sulphur Cycle
Edited by M. V. Ivanov and J. R. Freney
© 1983 Scientific Committee on Problems of the Environment (SCOPE)

CHAPTER 7
Major Fluxes of the Global Biogeochemical Cycle of Sulphur

M. V. IVANOV

7.1 NATURE OF THE PROBLEM AND OBJECTIVES

Prior to the commencement of the SCOPE Sulphur Project, several reviews had been published on the global biogeochemical sulphur cycle with included estimates of the amounts of sulphur in the atmosphere, hydrosphere, and lithosphere, as well as assessments of the individual fluxes from sphere to sphere (Eriksson, 1963; Friend, 1973; Kellogg *et al.*, 1972; Robinson and Robbins, 1970; Granat *et al.*, 1976). Most of the discussion in those reviews was devoted to sulphur turnover in the atmosphere and some fluxes were only considered to balance the atmospheric cycle. Consequently, many fluxes were estimated from an analysis of the processes occurring in the atmosphere. Because the lithosphere and hydrosphere were not as well studied as the atmosphere, such an approach was justified. However, this led to significant differences in estimates of some biogeochemical processes, as can be clearly seen from the data presented in Table 7.1.

In the assessment of the atmospheric cycle of sulphur, some fluxes were estimated from limited factual material using a number of *a priori* assumptions, and thus there is an obvious need for independent estimates of the individual fluxes within the global biogeochemical cycle of sulphur. Therefore, the first objective of our work was the collection of primary information on concentrations of various forms of sulphur in separate reservoirs and geospheres, and the assessment of fluxes between geospheres, using a number of methods and approaches wherever possible.

The increased interest shown in the global biogeochemical cycles in recent years has not been prompted solely by the vigorous increase in information in this field or because of the endeavour of scientists to obtain quantitative data on the cycles of the important elements on a global scale. This work has also

449

Table 7.1 Major fluxes of the global biogeochemical sulphur cycle (TgS year^{-1})

Flux	Nature of flux	Eriksson (1960)	Robinson and Robbins (1968)	Kellogg et al. (1972)	Friend (1973)	Granat et al. (1976)	Present report Natural	Anthropogenic	Total
1. Continental part of the cycle									
P_1	Emission to the atmosphere from fuel combustion and metal smelting	39	70	50	65	65	—	113	113
P_2	Volcanic emission	—	—	1.5	2	3	14	—	14
P_3	Aeolian emission	—	—	—	—	0.2	20	—	20
P_4	Biogenic emission	110	68	—	58	5	17.5	—	17.5
P_5	Atmospheric transport of oceanic sulphate	5	4	4	4	17	20	—	20
P_6	Deposition of large particles from the atmosphere	—	—	—	—	—	12	—	12
P_7	Washout from the atmosphere, surface uptake and dry deposition	240	116	111	121	71	25	47	72
P_8	Transport to the oceanic atmosphere	—	26	5	8	18	34.5	66	100.5
P_9	Weathering	15	14	—	42	66	114.1	—	114.1
P_{10}	River runoff to the world oceans	80	73	—	136	122	104.1[a]	104	208.1

P_{11}	Underground runoff to world oceans	—	—	—	—	—	9.2	—	9.2
P_{12}	River runoff to continental water-bodies	—	—	—	—	—	35	—	35
P_{13}	Marine abrasion of shores and exaration	—	—	—	—	—	6.8	—	6.8
P_{14}	Pollution of rivers with fertilizers	10	11	—	26	—	—	28^b	28
P_{15}	Effluents from chemical industry	—	—	—	—	—	—	28	28
P_{16}	Acid mine waters	—	—	—	—	—	—	1	1
	2. Oceanic part of the cycle (see above, P_2)								
P_{17}	Volcanic emission	170	30	—	—	—	14	—	14
P_{18}	Biogenic emission	45	44	—	48	27	23	—	23
P_{19}	Marine sulphate	200	96	47	44	44	140	—	140
P_{20}	Washout, surface uptake, and dry deposition			72	96	73	258	—	258
P_{21}	Burial of reduced sulphur in sediments	—	—	—	—	—	111.4	—	111.4
P_{22}	Burial of sulphate in sediments	—	—	—	—	—	27.8	—	27.8

^{a}To the ionic runoff should be added another 4.8 TgS for particulates in river runoff (see section 6.1).
^{b} About 5.6 TgS are removed from farm lands annually in harvested crops; after use as food this sulphur is discharged in river runoff mainly as domestic sewage and effluents from live stock farms.

been stimulated by fears that the increased cycling of sedimentary sulphur by man's industrial and agricultural activity might severely and unpredictably affect the environment, not only in industrialized regions but on the global scale as well.

Estimates of the total anthropogenic sulphur emissions, or its components in the sulphur cycle, appear relatively simple to obtain on both regional and global scales. To do this it is merely necessary to obtain data on the extraction of various raw materials containing sulphur and to be acquainted with the processing and use of these raw materials by various branches of industry and agriculture. Such information, and especially data on anthropogenic fluxes of sulphur to the atmosphere, is available in most published papers on the sulphur cycle (Table 7.1). The variability of the estimates is due more to the use of data for different years than to differences in the methods used by the various authors.

The quantitative assessment of the different natural emissions of sulphur appears to be much more difficult. Factual material is lacking and often there is no suitable method available for their estimation. When preparing this report, the authors endeavoured in all cases to give separate estimates for anthropogenic and natural sulphur emissions, and paid special attention to the quantitative analysis of the natural sulphur emissions.

Finally, another objective of this report was an attempt to assess, in parallel with the sulphur cycle, fluxes of some other elements involved in sulphur cycling. As can be seen from Fig. 7.1, practically all of the major reactions of the sulphur cycle involve living organisms. The sulphur cycle is, therefore, closely related to the carbon cycle.

The amount of carbon involved in biogenic processes of the sulphur cycle varies with the type of organism. For example, in bacterial chemosynthesis, which is characterized by a low coefficient of energy use for the assimilation of carbon dioxide, relatively little carbon is metabolized to organic forms. However, during the photo-assimilation of carbon dioxide by anaerobic bacteria, when sulphur compounds are used as electron donors, the amounts of oxidized sulphur and assimilated carbon are comparable. In the course of anaerobic reduction 24 g of organic carbon are mineralized for every 32 g of sulphate sulphur reduced. Thus, in ecosystems with photolithotrophic bacteria and sulphate-reducing bacteria significant amounts of carbon are transformed, and consequently these organisms must be regarded as active biogeochemical agents in the carbon cycle as well as participants of the sulphur cycle. Photosynthesis by plants is another important geochemical activity affecting both carbon and sulphur cycles (Fig. 7.1).

The sulphur cycle also interacts with the cycles of various metals through a few key reactions and determines the geochemical fate of a number of metals. Hydrogen sulphide formed during sulphate reduction reacts with many metal ions to form metal sulphides which have low solubility in water; hydrogen

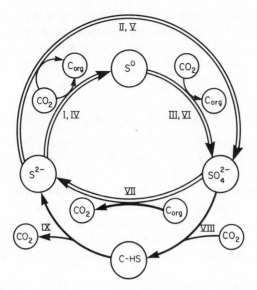

Chemoautotrophic bacteria

I. $2H_2S + O_2 = 2S^0 + 2H_2O$

II. $H_2S + 2O_2 = SO_4^{2-} + 2H^+$

III. $2S^0 + 3O_2 + 2H_2O = 2SO_4^{2-} + 2H^+$

Photoautotrophic bacteria

IV. $2H_2S + CO_2 = 2S^0 + (CH_2O) + H_2O$

V. $H_2S + 2CO_2 + H_2O = SO_4^{2-} + (CH_2O) + 2H^+$

VI. $S^0 + 2CO_2 + 2H_2O = SO_4^{2-} + 2(CH_2O) + 2H^+$

VII. Sulphate-reducing bacteria

$$SO_4^{2-} + 2C_{org} = S^{2-} + 2CO_2$$

VIII. Photosynthesis

IX. Putrefaction

Fig. 7.1   Interactions of biological processes in sulphur and carbon cycles

sulphide is thus responsible for the immobilization of metals. In the zone of weathering, on the other hand, bacterial and chemical oxidation of metal sulphides and elemental sulphur produces sulphuric acid which mobilizes metals. During the intensive acid weathering which has been observed, for example, in volcanogenic sulphide deposits (Lein and Ivanov, 1970; Ivanov, 1971), sulphuric acid decomposes not only ores, but also bedrock. As a result, sulphate solutions are enriched with Al^{3+}, Fe^{2+}, Fe^{3+}, Ti^{2+} and other elements.

Finally, many reactions of the sulphur cycle are of tremendous importance for processes occurring in the oxygen cycle. In numerous water bodies an important part of the dissolved oxygen is used in the oxidation of hydrogen sulphide and elemental sulphur, and thus these reactions significantly influence the oxygen cycle in water bodies (Kuznetsov, 1970).

For these reasons the authors focused their attention on the poorly investigated problems of the global biogeochemical sulphur cycle, in particular the direct determination of fluxes, the separate assessment of natural and anthropogenic components of the individual fluxes, and the interactions of the sulphur cycle with the cycles of a number of other elements.

7.2   FLUXES OF ANTHROPOGENIC ORIGIN

The global biogeochemical cycle of sulphur is summarized in Fig. 7.2. The total amount of sulphur extracted from the lithosphere by all types of mining is 169 Tg year^{-1} and most of this sulphur (about 113 Tg) is contained in fossil fuels and sulphide ores. During combustion of fuel and smelting of ores sulphur oxides are emitted to the continental atmosphere (flux P_1) where they are mixed with sulphur compounds from natural sources (fluxes P_4, P_5). Part of this sulphur is deposited on the surface of continents in precipitation or dry deposition (flux P_7) and the remainder is transported to the oceanic atmosphere (flux P_8).

Two other fluxes, P_{14} and P_{15}, result from the chemical processing of sulphur-containing raw materials in various industries. The sulphur-containing raw materials for these industries are supplied by the mining industry in the form of pyrite, elemental sulphur, and natural gas and amount to 56 TgS year^{-1}.

About half of the sulphur produced in the world is used for production of fertilizers, and consequently about 28 TgS enter the soil annually (flux P_{14}) (Cote, 1970). Some of the fertilizer sulphur is leached from soil in drainage waters and joins the river runoff, while about 5.6 TgS is removed annually from farm lands in harvested crops (Kilmer, 1979). Sulphur in these crops is also mineralized and enters the river runoff when the crops are ingested and excreted. The remainder of the sulphur extracted from pyrite, sulphur, and gas deposits is used in chemical industry and it, too, is finally discharged in sewage water and joins the river runoff (flux P_{15}). As can be seen from the data condensed in Table 7.1, the flux from chemical plants in sewage water was not taken into account by other investigators of the global sulphur cycle.

In addition to the two important anthropogenic fluxes through the soil, another 47 Tg of anthropogenic sulphur is taken up by the oceanic surface, or deposited in the ocean by precipitation or dry deposition (see flux P_7). Thus, 104 Tg of anthropogenic sulphur is supplied to the ocean each year in river runoff.

7.3   NATURAL SULPHUR IN RIVER RUNOFF

Details of the estimation of the natural sulphur flux with river runoff (P_{10}) are given in Chapter 5. Two peculiarities distinguish our assessment from

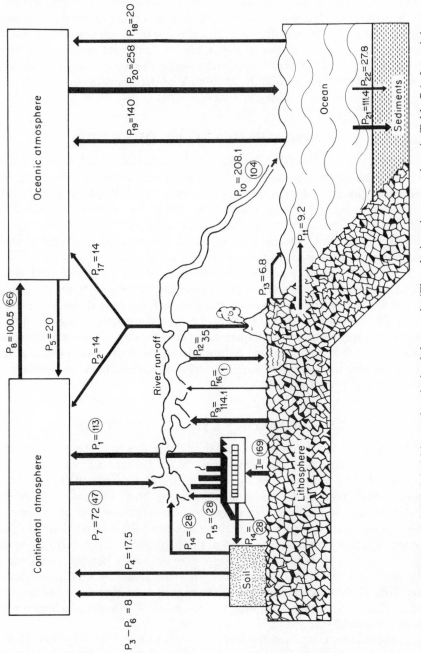

Fig. 7.2 Major fluxes of the global biogeochemical sulphur cycle (Flux designations are given in Table 7.1. I = mining of ores, fossil fuels etc.)

those made earlier. Firstly, we used the new, higher values for the total river runoff, published recently (Korzun *et al.*, 1974). Consequently, the magnitude of the natural sulphur flux is markedly higher than previously published estimates (Table 7.1). Secondly, we assessed the sulphur transported by riverwater to inland water-bodies as well as that to the world oceans (P_{12}). This flux amounts to one-third of the sulphur entering the oceans, and is the reason for the substantial increase in the magnitude of the flux with river runoff due to water erosion and leaching of rocks (flux P_9).

7.4   ATMOSPHERIC CYCLING OF SULPHUR

Our estimates of nearly all fluxes of the atmospheric sulphur cycle are greater than those of previous investigators. The substantial increase in the anthropogenic input figures, up to 113 TgS year^{-1} (flux P_1), is conditioned not only by the increased economic activity but also by the use of better up-to-date information on the consumption of sulphur-containing fuels and metal sulphides (Cullis and Hirschler, 1980), and by including the emission of sulphur-containing ashes to the atmosphere (section 4.2.5).

Analysis of new data on the chemical composition of precipitation over oceans resulted in a substantial increase in the flux P_{20} (sections 4.3.3 and 4.3.4) and necessitated an increase in the value for marine sulphate emission (P_{19}). The considerable increase in magnitude of the volcanic emission (fluxes P_2 and P_{17}) substantiated in Chapter 2, is due to the fact that we not only took into account the sulphur emission during large, but fairly rare volcanic eruptions, but also that given off from fumaroles of volcanic fields and that emitted during quiescent periods.

7.5   THE SULPHUR CYCLE IN THE OCEAN

The sole previous attempt to assess the sulphur flux from the water column to reduced sediments of the hemipelagic 'blue' clay type was undertaken by Berner (1972). Using 0.13% for the content of reduced sulphur in these sediments (Clarke, 1924), and assuming that the 'blue' clays cover 15% of the sediment area in the world oceans with a sedimentation rate of 15 cm per 1000 years, Berner calculated that these sediments accumulate about 6 Tg of reduced sulphur annually. Chapter 6 of this book carries a detailed analysis of the distribution of various forms of sulphur in sections of benthic sediments in different oceanic regions obtained during the past decade by Soviet, American, and West German researchers. Their findings allowed us to assess the amount of sulphate and reduced sulphur buried annually in bottom sediments of oceans (fluxes P_{21} and P_{22}).

Consideration of the fluxes obtained enables us to draw a number of interesting conclusions. The first point worthy of note is the close relation between the amount of natural sulphur transferred annually from continents to oceans (153 TgS, the sum of fluxes P_8, P_{10}, P_{11}, P_{13}, P_{17} minus P_5) with that buried annually in oceanic bottom sediments ($\sim$140 TgS, fluxes P_{21} and P_{22}).

The second conclusion which is of paramount importance for lithology is that the major part of the sulphur removed from the sulphur cycle in the ocean is in the form of reduced compounds. This conclusion is also supported by the results given in Table 7.2 which compares the mean reduced sulphur concentration in the zone of stabilization of diagenetic processes in the sediment layers with the intensity of sulphate reduction and the isotopic composition of reduced sulphur in the sediments. These results show that in all cases the reduced sulphur is enriched with ^{32}S compared to sea-water sulphate, a fact which suggests that the reduced sulphur is of microbial origin. The maximum sulphur concentrations were found in sediments of the small mediterranean seas which are characterized by the highest rates of sulphate reduction; in addition the reduced sulphur from the sediments of these waterbodies is the most highly enriched in ^{34}S.

Farther off shore there is a notable decrease in the sulphate reduction rate and the reduced sulphur concentration in sediments and a progressive accumulation of the light isotope in reduced sulphur (Table 7.2). If the sulphide in oceanic sediments is produced by microbial reduction, then these results are to be expected.

Table 7.2 Sulphate reduction rates, reduced sulphur concentrations and isotopic composition of pyrite in reduced sediments

Region sampled	Depth (m)	Sulphate reduction[a] rate (mgS kg^{-1} wet silt day^{-1})	Reduced sulphur (% dry silt)	Isotopic composition[b] ($\delta^{34}S$, ‰)
Sea of Azov	5–30	1.800	1.092[c]	+9.0
Baltic Sea	50–240	1.310	1.37[d]	−17.2
Black Sea, deep-water region	Over 2000	0.645	1.421[e]	−26.9
Pacific Ocean shelf	Up to 200	0.043	1.108[f]	−23.6
Pacific Ocean continental slope	1000–2900	0.011	0.628[g]	−28.0
Deep-water sediments of the Pacific off the Hawaiian coast	4730	0.001	0.363[g]	−33.7
Deep-water sediments of the Indian Ocean (Somali Trench)	Up to 5100	0.001	0.033[h]	−41.0

[a]See Table 6.17. [b]See Tables 6.31, 6.32, 6.35–6.37. [c]Migdisov *et al.* (1974). [d]See Table 6.22. [e]See Table 6.20. [f]Volkov *et al.* (1976). [g]Lein *et al.* (1976). [h]Zhabina *et al.* (1979).

As the major form of reduced sulphur buried in ocean sediments is pyrite (see Section 6.5), we can calculate that about 100 Tg of iron are also sulphidized annually in these sediments. Most of the reduced sulphur is formed by sulphate-reducing bacteria; thus we can calculate the amount of organic carbon mineralized by these heterotrophic micro-organisms from equation (7) in Fig. 7.1, i.e. 24 g of organic carbon is consumed for each 32 g of sulphide produced. Consequently, the formation of 111 Tg of reduced sulphur in oceanic sediments requires a minimum of 83 Tg organic carbon per year. Chapter 6 gives numerous examples which indicate that only part of the hydrogen sulphide formed by microbial sulphate reduction is bound as pyrite or other insoluble sulphur compounds during sediment diagenesis. The remaining hydrogen sulphide is either oxidized in the internal sulphur cycle of the sediments or diffuses to the water column.

It follows that the real consumption of organic carbon during sulphate reduction in oceanic muds must greatly exceed the minimum estimate of 83 TgC year^{-1}. Hence, the data obtained suggest that a large-scale process of organic matter mineralization is carried out by sulphate-reducing bacteria in reduced bottom sediments. Previous studies of the global cycle evidently underestimated the scope and geochemical consequences of this process. The carbon dioxide formed in this process is enriched in the light isotope of carbon and takes an active part in the geochemical and mineralogical processes occurring in the reduced sediments of oceans. The increase in alkalinity of pore-water is essentially due to the dissolution of this carbon dioxide; part of this migrates in the form of bicarbonate to the overlying water and part is used for the formation of diagenetic carbonate minerals such as calcite, protodolomite, and rhodochrosite (Lein *et al.*, 1975; Lein, 1978; Ivanov and Lein, 1980; Volkov *et al.*, 1979).

7.6 USE OF ISOTOPIC DATA FOR SOURCE IDENTIFICATION

During the review of data on the global biogeochemical cycle of sulphur, numerous analyses on the isotopic composition of sulphur in various natural reservoirs, lithosphere, hydrosphere, and atmosphere, were considered (see Chapters 2 and 4–6). The results obtained characterize the $\delta^{34}S$ values of the total sulphur in: (1) various rocks of the sedimentary envelope of the earth's crust (Table 2.13); (2) river runoff from the Soviet Union (Table 5.22); (3) sediments of the Pacific Ocean (Table 6.38); (4) precipitation from different regions (Table 4.26).

The generalized data indicative of certain reservoirs and fluxes are condensed in Table 7.3. Unfortunately, the information available is inadequate for the estimation of individual fluxes on the global scale. Firstly, the isotopic composition of the volatile compounds of sulphur of biogenic origin emitted to the atmosphere from continents and oceans is not known (fluxes P_4 and

Table 7.3 Isotopic composition of sulphur in rocks, river-water, and oceanic sediments

Material	Isotopic composition $(\delta^{34}S, \text{ }^0\!/\!_{00})$
Rocks from the sedimentary envelope of continents (Table 2.13)	−2.5
Rocks from the sedimentary envelope of oceans (Table 2.13)	+4.8
Sulphate in river runoff from the Asian part of the USSR (Table 5.22)	+10.75
Sulphur buried in sediments of the Pacific (Table 6.38)	−15.4

P_{18}). Secondly, the isotopic composition of sulphur compounds in soil has not been studied in sufficient detail to enable us to determine the emissions from this source, e.g. aeolian emission (flux P_3).

Since the isotopic composition of the sulphur compounds used in the economic activities of man varies within broad limits (Fig. 1.3), it is extremely difficult to estimate the average $\delta^{34}S$ value for one of the most important fluxes — removal of the sulphur-containing raw materials from the lithosphere. Even though examples were given in Chapters 4 and 5 which showed that the isotopic composition of the atmosphere and freshwater basins is dependent upon the isotopic composition of the pollutant sulphur in some industrialized regions, the assessment of the average isotopic composition of the fluxes P_1, P_{14}, and P_{15} is not possible at the present time because information is lacking on the average isotopic composition of the raw materials used in energy production, metallurgy, and chemical industry.

Nevertheless, the results cited in this book demonstrate the efficiency with which isotopic data can be used for the identification of the sources of pollution of atmospheric and freshwater systems with sulphur.

7.7 FUTURE TRENDS

In preparing the present report the authors analysed the literature which attempted to forecast future variations in anthropogenic fluxes of sulphur. The projected use of sulphur-containing raw materials in the chemical industry is plotted in Fig. 2.31. It appears from this projection that, compared with 1975, the output of elemental, pyrite, and gaseous sulphur, as well as the production of sulphuric acid will have increased by at least 30% by the year 1985. Nriagu (1978) points out that the increased demand for food will lead to an increase in the land used for agriculture and a greater requirement for fertilizer sulphur, especially in the developing countries of Africa and South America: possibly only one decade will be needed for a doubling of the

consumption of sulphur. It follows that there will be a corresponding increase
in the sulphur pollution of soil drainage waters and river-water and that new
regions of the world will be affected. Figure 7.3 shows the expected increase
in sulphur dioxide emissions for various projections of energy use. Because of
the decreasing stocks and ever-increasing prices of gaseous and liquid fuels
the most realistic trend is for a considerable increase in the use of coal. With
existing systems for the purification of gaseous fuel such a scenario presents

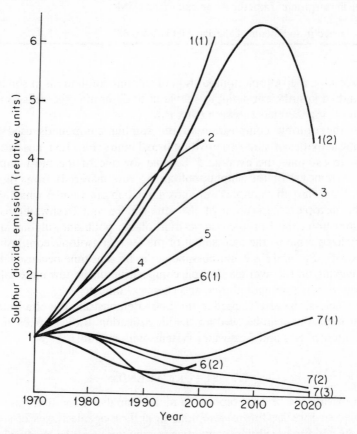

Fig. 7.3   Predicted trends for the global anthropogenic emission of
sulphur dioxide into the atmosphere (after Ryaboshapko *et al.*
1978). 1, Spaite and Harrington (1971): 1(1), fossil-fuel version,
1(2), fossil-fuel–nuclear version; 2, Shvedov (1976) using coal and
oil; 3, Ryaboshapko *et al.* (1978); 4, Roderick (1975); 5, Kellogg *et
al.* (1972); 6, Land (1971); 6(1), fossil fuel version without treatment, 6(2), with treatment; 7, Smil (1975): 7(1), fossil fuel version
with treatment, 7(2), fossil-fuel–nuclear version 7(3), nuclear version

gloomy prospects for further pollution of the environment with sulphur on a global scale.

Comparison of the natural and anthropogenic fluxes of sulphur shows that the total amount of sulphur entering the atmosphere and hydrosphere has doubled due to man's activity. This means that the natural biogeochemical systems for sulphur dioxide removal from the atmosphere and for sulphate and hydrogen sulphide removal from the hydrosphere must function at twice the rate operating in the period preceding the Industrial Revolution.

If we take into account the irregular distribution of the sources of anthropogenic pollution and the short residence time of the sulphur compounds in the atmosphere, then the threat of regional crises caused by the sulphur pollution of the atmosphere, hydrosphere, and pedosphere becomes quite real.

The fall-out of acid precipitation over large territories of Europe, North America, and parts of Asia, acidification of lacustrine water and soil, increasing sulphate contents, and the appearance of hydrogen sulphide in rivers, lakes, and even sea-waters indicate that in many cases the natural processes for sulphur removal cannot cope with the steadily increasing flux of sulphur.

Possible solutions to the existing problems include the development of improved methods of fuel purification, especially for the removal of sulphur from coal, and the design and application of new economic systems for the scrubbing of sulphur oxides from flue gases. An alternative which may be inevitable is to make greater use of solar and other energy alternatives, thus reducing the dependence on coal and oil.

REFERENCES

Berner, R. A. (1972) Sulfate reduction, pyrite formation and the oceanic sulfur budget. In: *The Changing Chemistry of the Oceans. Nobel Symp.*, **20**, 347–361.

Clarke, F. W. (1924) *The Data of Geochemistry*, 4th Edn. Government Printing Office, Washington DC, 841 pp.

Cote, P. R. (1970) Sulphur. In: *Canadian Minerals Year Book*. Department of Energy Mines and Resources, Ottawa, Ontario, pp. 513–526.

Cullis, C. F. and Hirschler, M. M. (1980) Atmospheric sulfur: natural and man-made sources. *Atmos. Environ.*, **14**, 1263–1278.

Eriksson, E. (1960) The yearly circulation of chloride and sulfur in nature; meteorological, geochemical and pedological implications, Pt. 2. *Tellus*, **12**, 63–109.

Eriksson, E. (1963) The yearly circulation of sulphur in nature. *J. Geophys. Res.*, **68**, 4001–4008.

Friend, J. P. (1973) The global sulfur cycle. In: Rasool, S. I. (ed.), *Chemistry of the Lower Atmosphere*. Plenum Press, New York, pp. 177–201.

Granat, L., Rodhe, H., and Hallberg, R. D. (1976) The global sulphur cycle. In: Svensson, B. H. and Söderlund, R. (eds), *Nitrogen, Phosphorus and Sulphur – Global Cycles*. SCOPE Report 7, *Ecol. Bull. (Stockholm)*, **22**, 89–134.

Ivanov, M. V. (1971) Bacterial processes in the oxidation and leaching of sulphide–sulphur ores of volcanic orgin. *Chem. Geol.*, **7**, 185–211.

Ivanov, M. V., and Lein, A. Yu. (1980) Distribution of microorganisms and their role in process of diagenetic mineral formation. In: Ostroumov, E. A. (ed.), *Geochemistry of Sediment Diagenesis in the Pacific Ocean* (Transoceanic Profile), Nauka, Moscow. pp. 117–137 [in Russian].

Kellogg, W. W., Cadle, R. D., Allen, E. R., Lazrus, A. L. and Martell, E. A. (1972) The sulfur cycle. *Science* (Wash. DC), **175**, 587–596.

Kilmer, V. J. (1979) Minerals and agriculture. In: Trudinger, P. A. and Swaine, D. J. (eds), *Biogeochemical Cycling of Mineral-forming Elements*, Elsevier, Amsterdam, pp. 515–558.

Korzun, V. I., Sokolov, A. A., Burynev, M. I., *et al.* (1974) *World Water Balance and Water Resources of the Earth* Hydrometeoizdat, Leningrad, 638 pp. [in Russian].

Kuznetsov, S. I. (1970) *Microflora of Lakes and Its Geochemical Activity*. Nauka, Leningrad, 440 pp. [in Russian].

Land, G. W. (1971) Fossil fuel: national energy supply and air pollution. Paper, ASME, No. WA/Fu-4, 1–15.

Lein, A. Yu. (1978) Formation of carbonate and sulfide minerals during diagenesis of reduced sediments. In: Krumbein, W. E. (ed.), *Environmental Biogeochemistry and Geomicrobiology*, vol. 1, Ann Arbor Science, Michigan. pp. 339–354.

Lein, A. Yu., and Ivanov, M. V. (1970) Processes of natural sulfur oxidation in volcanogenic ores of the Kurilo-Kamchatka region. In: *Geology and Mineralogy of the Weathering Crust*. Nauka, Moscow, pp. 182–213.

Lein, A. Yu., Kudryavtseva, A. I., Matrosov, A. G. and Zyakun, A. M. (1976) Isotope composition of sulfur compounds in sediments of the Pacific Ocean. In: *The Biogeochemistry of the Diagenesis of Ocean Sediment*. Nauka, Moscow, pp. 179–185 [in Russian].

Lein, A. Yu., Logvinenko, N. V., Volkov, I. I., Ivanov, M. V., and Trubin, A. I. (1975) Mineral and isotopic composition of diagenetic carbonate minerals and concretions from the Gulf of California reduced muds. *Dokl Akad. Nauk SSSR*, **224**, 426–429 [in Russian].

Migdisov, A. A., Cherkovsky, S. L., and Grinenko, V. A. (1974) Dependence of sulfur isotope composition in humid sediments on conditions of their formation. *Geokhimiya*, No. 10, 1482–1502 [Russian].

Nriagu, J. O. (1978) Production and uses of sulfur. In: Nriagu, J. O. (ed.), *Sulfur in the Environment*, Pt. 1. Wiley, Chichester, pp. 1–21.

Robinson, E., and Robbins, R. C. (1968) *Sources, Abundance and Fate of Gaseous Atmospheric Pollutants*. Final Report. Project PR-6755 Stanford Res. Inst. Menlo Park, California, 110 pp.

Robinson, E., and Robbins, R. C. (1970) Gaseous sulfur pollutants from urban and natural sources. *J. Air Pollut. Control. Assoc.*, **20**, 233–235.

Roderick, H. (1975) Projected emission of sulfur oxides from fuel combustion in the OECD area 1972–1985. In: *IPIECA, Symposium*. Tehran, pp. 1–35.

Ryaboshapko, A. G., Shopauskene, D. A., and Erdman, L. K. (1978) Background levels of SO_2 and NO_2 and the global sulfur balance. In: *Pollution of the Atmosphere as an Ecological Factor*. Hydrometeoizdat, Moscow, pp. 20–33. [in Russian].

Shvedov, V. P. (1976) Progress of energetics and pollution of the biosphere. In Styro, B. (ed.), *Physical Aspects of the Atmospheric Pollution*. Vilnius, 'Mokslas' pp. 5–9 [in Russian].

Smil, V. (1975) Energy and air pollution: U.S.A. 1970–2020. *J. Air Pollut. Control Assoc.*, **25**, 233–236.

Spaite, P. W. and Harrington, R. E. (1971) Abatement goes global *Power Eng.*, **75**, 42–45.

Volkov, I. I., Logvinenko, N. V., Sokolova, E. G., and Lein, A. Yu. (1979) Rhodochrosite. In: Kholodov, V. N. (ed.), *Lithology and Geochemistry of Sediments of the Pacific (Transoceanic Profile)*. Nauka, Moscow, Proceedings of GIN, Vol. 334, pp. 85–91 [in Russian].

Volkov, I. I., Rozanov, A. G., Zhabina, N. N., and Fomina, L. S. (1976) Sulfur compounds in sediments of the California Gulf and the adjoining part of the Pacific. In: *The Biogeochemistry of the Diagenesis of Ocean Sediments*. Nauka, Moscow, pp. 136–170 [in Russian].

Zhabina, N. N., Demidova, T. P., and Morozov, A. A. (1979) Sulfur compounds in sediments of the Somalian depression of the Indian Ocean. *Geokhimiya*, No. 12, 1868–1883 [in Russian].

Index

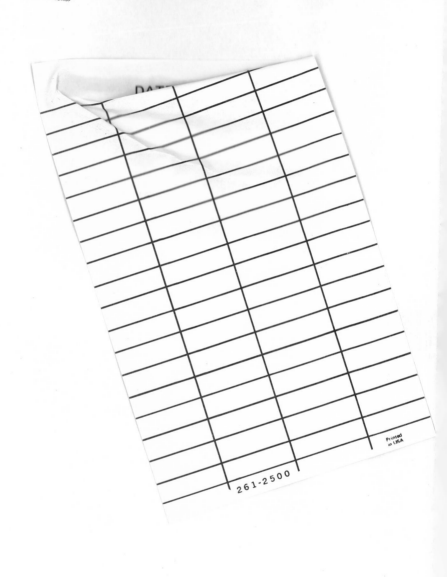

DATE

261-2500

Printed
in USA